Polar Lakes and Rivers

Polar Lakes and Rivers

Limnology of Arctic and Antarctic Aquatic Ecosystems

EDITED BY

Warwick F. Vincent and Johanna Laybourn-Parry

OXFORD

UNIVERSITY PRESS

Great Clarendon Street, Oxford OX2 6DP

Oxford University Press is a department of the University of Oxford.
It furthers the University's objective of excellence in research, scholarship,
and education by publishing worldwide in

Oxford New York

Auckland Cape Town Dar es Salaam Hong Kong Karachi
Kuala Lumpur Madrid Melbourne Mexico City Nairobi
New Delhi Shanghai Taipei Toronto

With offices in

Argentina Austria Brazil Chile Czech Republic France Greece
Guatemala Hungary Italy Japan Poland Portugal Singapore
South Korea Switzerland Thailand Turkey Ukraine Vietnam

Oxford is a registered trade mark of Oxford University Press
in the UK and in certain other countries

Published in the United States
by Oxford University Press Inc., New York

British Library Cataloguing in Publication Data

Data available

Library of Congress Cataloging in Publication Data

Data available

Typeset by Newgen Imaging Systems (P) Ltd., Chennai, India
Printed in Great Britain
on acid-free paper by
Antony Rowe, Chippenham, Wiltshire

ISBN 978–0–19–921388–7 (Hbk)   978–0–19–921389–4 (Pbk)

10 9 8 7 6 5 4 3 2 1

Preface

From the summit of the tumulus I saw the ice ahead of us in the same condition…a long blue lake or a rushing stream in every furrow.

Peary, R.E. (1907). *Nearest the Pole*, p. 220. Hutchinson, London.

We marched down a narrow gap, cut through a great bar of granite, and saw ahead of us a large lake, some three miles long. It was of course frozen, but through the thick ice covering we could see water plants, and below the steep cliffs the water seemed very deep.

Taylor, G. (1913). The western journeys. In Huxley, L. (ed.), *Scott's Last Expedition*, vol. II, p. 193. Smith Elder & Co., London.

From the early explorers onwards, visitors to the Arctic and to Antarctica have commented with great interest on the presence of lakes, wetlands, and flowing waters. These environments encompass a spectacular range of conditions for aquatic life, from dilute surface melt ponds, to deep, highly stratified, hypersaline lakes. Many of these high-latitude ecosystems are now proving to be attractive models to explore fundamental themes in limnology; for example, landscape–lake interactions, the adaptation of plants, animals, and microbes to environmental extremes, and climate effects on ecosystem structure and functioning. Some of these waters also have direct global implications; for example, permafrost thaw lakes as sources of greenhouse gases, subglacial aquatic environments as a planetary storehouse of ancient microbes, and Arctic rivers as major inputs of fresh water and organic carbon to the world ocean.

For more years than we care to admit, the two of us have talked about the need for a text on high-latitude lakes and rivers that compared and contrasted the two polar regions. Whereas the Arctic and Antarctic have much in common, they also have distinct differences. Within the polar research community, scientists typically work exclusively in either the Arctic or in Antarctica. Both of us have conducted research in both polar regions, and this has impressed upon us the remarkable diversity of high-latitude aquatic ecosystems, and their striking commonalities and differences. The Arctic and Antarctica are currently the focus of unprecedented public and political attention, not only for their natural resources and geopolitical significance, but also because they are continuing to provide dramatic evidence of how fast our global environment is changing. Moreover, 2007–2008 marked the fourth International Polar Year (IPY), so the time seemed right to turn the talk into action. However, if it had not been for Ian Sherman of Oxford University Press, and his persuasiveness and encouragement at the American Society for Limnology and Oceanography meeting in Santiago de Compostela in 2005, we would probably not have embarked on this book.

One of the valuable opportunities provided by this book project has been to bring together groups of Arctic and Antarctic scientists for many of the chapters. People who ordinarily would not have found themselves collaborating have shared their knowledge and expertise from the two polar regions. We hope that this collaboration will foster further joint ventures among our colleagues, and that it will encourage a more pole-to-pole approach towards high-latitude ecosystems, in the spirit of IPY.

This book is intended for both the specialist and the more general reader. To assist the latter we have included a glossary of terms. The color plates are also intended to provide a better picture of the habitats and organisms to those unfamiliar with them. We asked the authors to adopt a tutorial approach for nonspecialists, to limit their citations to be illustrative (rather than exhaustive) of key concepts and observations, and, where possible, to consider differences and similarities between the Arctic and Antarctic. We have greatly appreciated their willingness to be involved in this project and the excellence of their contributions.

In addition to thanking the contributing authors to this volume, we express our gratitude to Ian Sherman, Helen Eaton, and other staff at Oxford University Press for their expert help in bringing this volume through to completion; the many reviewers of the manuscripts; Janet Drewery at Keele University and Tanya Adrych at the University of Tasmania for assistance in manuscript preparation; our students, postdoctoral fellows, and other researchers who have worked with us on the limnology of Arctic and Antarctic lakes and rivers; and our research funding and logistics agencies, including the Natural Sciences and Engineering Research Council (Canada), the Canada Research Chair program, the Canadian Network of Centers of Excellence program ArcticNet, Polar Shelf Canada, the Natural Environment Research Council (UK), the Engineering and Physical Sciences Research Council (UK), the Leverhulme Trust, and the Australian, UK, New Zealand, Spanish, and US Antarctic programmes.

Warwick F. Vincent and
Johanna Laybourn-Parry
2008

About International Polar Year

The International Polar Year (IPY) 2007–2008 represents one of the most ambitious coordinated international science programmes ever attempted. Researchers from over sixty countries and a broad range of disciplines are involved in this two-year effort to study the Arctic and Antarctic and explore the strong links these regions have with the rest of the globe.

Researchers in IPY 2007–8 have made a commitment to raising awareness about the polar regions and increasing the accessibility of science. This book is part of an internationally endorsed IPY outreach project.

For more information, please visit www.ipy.org.

Contents

3 **High-latitude paleolimnology**   **43**
Dominic A. Hodgson and John P. Smol

4 **The physical limnology of high-latitude lakes**   **65**
Warwick F. Vincent, Sally MacIntyre, Robert H. Spigel, and Isabelle Laurion

16 Direct human impacts on high-latitude lakes and rivers   291

Martin J. Riddle and Derek C.G. Muir

17 Future directions in polar limnology   307

Johanna Laybourn-Parry and Warwick F. Vincent

Contributors

Ian A.E. Bayly, 501 Killiecrankie Road, Flinders Island, Tasmania 7255, Australia

Brent C. Christner, Department of Biological Sciences, Louisiana State University, Baton Rouge, LA 70803, USA

Kirsten S. Christoffersen, Freshwater Biological Laboratory, University of Copenhagen, Helsingørsgade 5, DK-3400 Hillerød, Denmark

J. Brian Dempson, Fisheries and Oceans Canada, Science Branch, 80 East White Hills Road, St. John's, NL A1C 5X1, Canada

Peter T. Doran, Department of Earth and Environmental Sciences, University of Illinois at Chicago, Chicago, IL 60607, USA

Eduardo Fernández-Valiente, Departamento de Biología, Darwin 2, Universidad Autónoma de Madrid, 28049 Madrid, Spain

Jacques C. Finlay, Department of Ecology, Evolution, and Behavior & National Center for Earth-Surface Dynamics, University of Minnesota, St. Paul, MN 55108, USA

Christine M. Foreman, Department of Land Resources & Environmental Sciences, Montana State University, Bozeman, MT 59717, USA

Andrew G. Fountain, Departments of Geology and Geography, Portland State University, Portland, OR 97201, USA

Pierre E. Galand, Unitat de Limnologia - Departament d'Ecologia Continental, Centre d'Estudis Avançats de Blanes - CSIC, 17300 Blanes, Spain

John A.E. Gibson, Marine Research Laboratories, Tasmanian Aquaculture and Fisheries Institute, University of Tasmania, Hobart, Tasmania 7001, Australia

Michael N. Gooseff, Department of Civil & Environmental Engineering, Pennsylvania State University, University Park, PA 16802, USA

Ian Hawes, World Fish Centre, Gizo, Solomon Islands

John E. Hobbie, The Ecosystems Center, Marine Biological Laboratory, Woods Hole, MA 02543, USA

Dominic A. Hodgson, British Antarctic Survey, High Cross, Madingley Road, Cambridge CB3 0ET, UK

Clive Howard-Williams, National Institute of Water and Atmosphere Ltd, 10 Kyle Street, Riccarton, Christchurch 8011, New Zealand

Erik Jeppesen, National Environmental Research Institute, Aarhus University, Department of Freshwater Ecology, Vejlsøvej 25, DK-8600 Silkeborg, Denmark

Mahlon C. Kennicutt II, Office of the Vice President for Research, Texas A&M University, College Station, TX 77843–1112, USA

Scott F. Lamoureux, Department of Geography, Queen's University, Kingston, ON K7L 3N6, Canada

Isabelle Laurion, Institut national de la recherche scientifique, Centre Eau, Terre et Environnement, 490 rue de la Couronne, Québec City, QC G1K 9A9, Canada

Johanna Laybourn-Parry, Institute for Antarctic and Southern Ocean Studies, University of Tasmania, Hobart, Tasmania 7001, Australia

Michael P. Lizotte, Aquatic Research Laboratory, University of Wisconsin Oshkosh, Oshkosh, WI 54903–2423, USA

W. Berry Lyons, Byrd Polar Research Centre, Ohio State University, 1090 Carmack Road, Columbus, Ohio 43210–1002, USA

Sally MacIntyre, Department of Ecology, Evolution and Marine Biology & Marine Sciences Institute, University of California, Santa Barbara, CA 93106, USA

Diane M. McKnight, Institute of Arctic and Alpine Research, Institute of Arctic and Alpine Research, University of Colorado, Boulder, CO 80309–0450, U.S.A

Daryl L. Moorhead, Department of Environmental Sciences, University of Toledo, 2801 W. Bancroft, Toledo, OH 43606, USA

Derek C.G. Muir, Water Science and Technology Directorate, Environment Canada, Burlington, ON L7R 4A6, Canada

Marjut Nyman, Department of Biological and Environmental Sciences, FIN-00014 University of Helsinki, Finland

David A. Pearce, British Antarctic Survey, High Cross, Madingley Road, Cambridge CB3 0ET, U.K.

Bruce J. Peterson, The Ecosystems Center, Marine Biological Laboratory, Woods Hole, MA 02543, USA

Reinhard Pienitz, Départment de Géographie & Centre d'Études Nordiques, Laval University, Québec City, QC G1V 0A6, Canada

Michael Power, Department of Biology, 200 University Avenue West, University of Waterloo, Waterloo, ON N2L 3G1, Canada

John C. Priscu, Department of Land Resources & Environmental Sciences, Montana State University, Bozeman, MT 59717, USA

Antonio Quesada, Departamento de Biología, Darwin 2, Universidad Autónoma de Madrid, 28049 Madrid, Spain

Milla Rautio, Department of Biological and Environmental Science, P.O. Box 35, FIN-40014 University of Jyväskylä, Finland

James D. Reist, Fisheries and Oceans Canada, 501 University Crescent, Winnipeg, MB R3T 2N6, Canada

Martin J. Riddle, Australian Antarctic Division, Channel Highway, Kingston, Tasmania 7050, Australia

John P. Smol, Department of Biology, Queen's University, Kingston, ON K7L 3N6, Canada

Robert H. Spigel, National Institute of Water and Atmosphere Ltd, 10 Kyle Street, Riccarton, Christchurch 8011, New Zealand

Michael Studinger, Lamont-Doherty Earth Observatory of Columbia University, 61 Route 9W, Palisades, NY 10964–8000, USA

Lars J. Tranvik, Department of Ecology and Evolution, Limnology, BMC, Uppsala University, SE-751 23, Sweden

Slawek Tulaczyk, Department of Earth and Planetary Sciences, University of California, Santa Cruz, CA 95064, USA

Warwick F. Vincent, Département de Biologie & Centre d'Études Nordiques, Laval University, Québec City, QC G1V 0A6, Canada

CHAPTER 1

Introduction to the limnology of high-latitude lake and river ecosystems

Warwick F. Vincent, John E. Hobbie, and Johanna Laybourn-Parry

Outline

Polar lakes and rivers encompass a diverse range of aquatic habitats, and many of these environments have broad global significance. In this introduction to polar aquatic ecosystems, we first present a brief summary of the history of lake research in the Arctic and Antarctica. We provide an overview of the limnological diversity within the polar regions, and descriptions of high-latitude rivers, lakes, and lake districts where there have been ecological studies. The comparative limnology of such regions, as well as detailed long-term investigations on one or more lakes or rivers within them, have yielded new perspectives on the structure, functioning, and environmental responses of aquatic ecosystems at polar latitudes and elsewhere. We then examine the controls on biological production in high-latitude waters, the structure and organization of their food webs including microbial components, and their responses to global climate change, with emphasis on threshold effects.

1.1 Introduction

Lakes, ponds, rivers, and streams are prominent features of the Arctic landscape and are also common in many parts of Antarctica (see Appendix 1.1 for examples). These environments provide diverse aquatic habitats for biological communities, but often with a simplified food-web structure relative to temperate latitudes. The reduced complexity of these living systems, combined with their distinct physical and chemical features, has attracted researchers from many scientific disciplines, and high-latitude aquatic environments and their biota are proving to be excellent models for wider understanding in many fields including ecology, microbiology, paleoclimatology, astrobiology, and biogeochemistry. In northern lands, these waters are important hunting and fishing grounds for indigenous communities. They also provide drinking water supplies to Arctic communities and are a key resource for certain industries such as hydro-electricity, transport, and mining.

In addition to their striking limnological features, high-latitude aquatic environments have broad global significance; for example, as sentinels of climate change, as refugia for unique species and communities, as sources of greenhouse gases and, in the case of the large Arctic rivers, as major inputs of freshwater and organic materials to the World Ocean. There is compelling evidence that high-latitude regions of the world are experiencing more rapid climate change than elsewhere, and this has focused yet greater attention on many aspects of the polar regions, including their remarkable inland waters.

Whereas Antarctica and the Arctic have much in common, their aquatic ecosystems are in many ways dissimilar. Both southern and northern high-latitude regions experience cold temperatures, the pervasive effects of snow and ice, low annual inputs of solar radiation, and extreme seasonality in their light and temperature regimes. However, Antarctica is an isolated continent (Figure 1.1) whereas the Arctic is largely the northern extension of continental land masses (Figure 1.2) and this has major implications for climate, colonization, and biodiversity. Arctic catchments often contain large stocks of terrestrial vegetation, whereas Antarctic catchments are usually devoid of higher plants. This results in a much greater importance of allochthonous (external) sources of organic carbon to lakes in the Arctic relative to Antarctica, where autochthonous (within-lake) processes likely dominate. Given their proximity to the north-temperate zone, Arctic waters tend to have

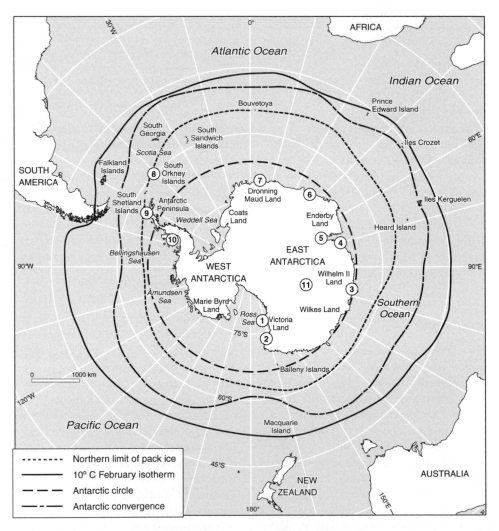

Figure 1.1 The Antarctic, defined as that region south of the Antarctic Convergence, and the location of limnological sites referred to in this volume. 1, Southern Victoria Land (McMurdo Dry Valleys, Ross Island ponds, McMurdo Ice Shelf ecosystem); 2, northern Victoria Land (Terra Nova Bay, Cape Hallett); 3, Bunger Hills; 4, Vestfold Hills and Larsemann Hills; 5, Radok Lake area (Beaver Lake); 6, Syowa Oasis; 7, Schirmacher Oasis; 8, Signy Island; 9, Livingstone Island; 10, George VI Sound (Ablation Lake, Moutonnée Lake); 11, subglacial Lake Vostok (see Plate 1). Base map from Pienitz *et al.* (2004).

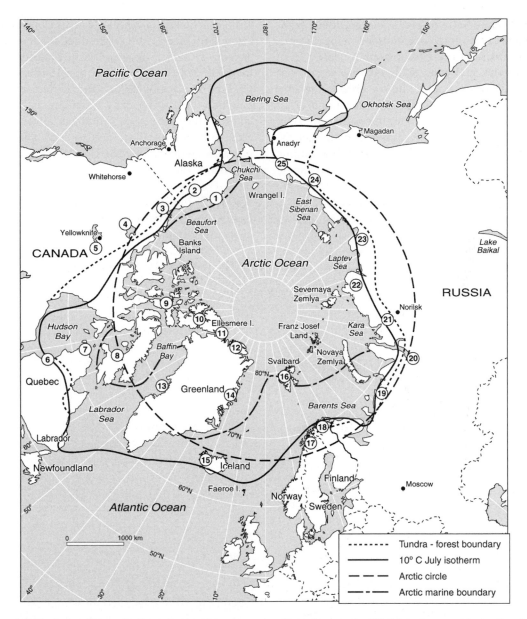

Figure 1.2 The Arctic, which can be demarcated in various ways such as the treeline or by the 10°C July isotherm, and the location of limnological sites referred to in this volume. 1, Barrow Ponds, Alaska; 2, Toolik Lake Long-Term Ecological Research (LTER) site, Alaska; 3, Mackenzie River and floodplain lakes, Canada; 4, Great Bear Lake; 5, Great Slave Lake; 6, Northern Québec thaw lakes and Lac à l'Eau Claire (Clearwater Lake); 7, Pingualuk Crater Lake (see Chapter 2); 8, Amadjuak Lake and Nettilling Lake; 9, Cornwallis Island (Char Lake, Meretta Lake, Amituk Lake); 10, Ellesmere Island (Lake Romulus, Cape Herschel ponds); 11, Ward Hunt Lake and northern Ellesmere Island meromictic lakes; 12, Peary Land, northern Greenland; 13, Disko Island, Greenland; 14, Zackenberg, Greenland; 15, Iceland lakes (e.g. Thingvallavatn, Thorisvatn, Grænalón); 16, Svalbard lakes; 17, Kuokkel lakes, northern Sweden; 18, Lapland lakes, Finland; 19, Pechora River, Russia; 20, Ob River; 21, Yenisei River; 22, Lake Tamyr; 23, Lena River; 24, Kolyma River; 25, Lake El'gygytgyn. Base map from Pienitz *et al.* (2004).

more diverse animal, plant, and microbial compositions, and more complex food webs, than in Antarctica. Fish are absent from Antarctic lakes and streams, and many south polar lakes are even devoid of zooplankton. Insects (especially chironomids) occur right up to the northern limit of Arctic lakes and rivers, but are restricted to only two species in Antarctica, and then only to specific sites in the Antarctic Peninsula region. The benthic environments of waters in both regions have some similarities in that microbial mats dominated by cyanobacteria are common throughout the Arctic and Antarctica. Aquatic mosses also occur in lakes and streams of both regions, but higher plants are absent from Antarctic waters. These similarities and differences make the comparative limnology of the polar regions particularly attractive for addressing general questions such as the factors controlling the global biogeography of aquatic plants, animals, and microbes, the limiting factors for biological production, the causes and consequences of food-web complexity, and the responses of aquatic ecosystems to environmental change.

1.2 History of polar limnology

From the earliest stages of development of limnology as a science, it was realized that high-latitude lakes would have some distinctive properties. The pioneer limnologist, François-Alfonse Forel, surmised that water temperatures in polar lakes would never rise above 4°C as a result of the short summer and low solar angle at high latitudes, and thus the lakes would circulate only once each year (Forel 1895, p.30). In G. Evelyn Hutchinson's classification of polar lakes, he pointed out that these 'cold monomictic' lakes occur at both high latitudes and high altitudes (Hutchinson and Löffler 1956). Some low Arctic lakes are also dimictic (circulating twice) and some polar lakes with salinity gradients never circulate entirely (meromictic; see Chapter 4 in this volume). During the 1950s and 1960s, actual measurements of the thermal regimes of polar lakes began in Alaska, USA (Brewer 1958; Hobbie 1961), Greenland (Barnes 1960), and Antarctica (Shirtcliffe and Benseman 1964).

The earliest work on polar aquatic ecosystems was descriptive and came from short expeditions to specific sites. For example, Juday (1920) described a zooplankton collection from the Canadian Arctic expedition 1913–1918 as well as a cladoceran collected in 1882 at Pt. Barrow, presumably during the First International Polar Year. From the 1950s onwards there were many observations made in lakes in the Arctic and Antarctic; almost all of these were summer-only studies. A notable exception was the work by Ulrik Røen at Disko Island, Greenland, on Arctic freshwater biology (e.g. Røen 1962). Process studies increased during the 1960s and 1970s but the most valuable insights came from intensive studies where many processes were measured simultaneously or successively, and for long periods of time.

The projects of the International Biological Programme (IBP) were funded by individual countries, beginning in 1970, to investigate the biological basis of productivity and human welfare. The many aquatic sites included two Arctic lake sites, ponds and lakes at Barrow, Alaska, and Char Lake, northern Canada (for details see Appendix 1.1). At both sites, all aspects of limnology were investigated from microbes to fish for 3–4 years. It was focused, question-based research at a scale of support and facilities that enabled scientists to go far beyond descriptive limnology and investigate the processes and controls of carbon and nutrient flux in entire aquatic systems. The Barrow project included both terrestrial and aquatic sections (Hobbie 1980) whereas the Char Lake project focused on the lake, with comparative studies on nearby Lake Meretta that had become eutrophic as a result of sewage inputs (Schindler et al. 1974a, 1974b). While the nine principal investigators on the Barrow aquatic project worked on many ponds, they all came together to make integrated measurements on one pond; when 29 scientists began sampling in this pond, the investigator effect was so large that an aerial tramway had be constructed.

The success of the IBP led to a US program of integrated ecological studies at 26 sites, mostly in the USA. This Long-Term Ecological Research (LTER) program includes sites at Toolik Lake, Alaska, and the McMurdo Dry Valleys, Antarctica (see further details in Appendix 1.1). The observations at Toolik began in 1975 and in the McMurdo Dry Valley lakes in the late 1950s. Each LTER project is reviewed every 6 years but is expected to continue for decades; each is expected to publish papers, support graduate students and collect data which are accessible to all on the Internet. The long-term goal of the Arctic LTER is to predict the effects of environmental change on lakes, streams, and tundra. The overall objectives of the McMurdo LTER are to understand the influence of physical and biological constraints on the structure and function of dry valley ecosystems, and to understand the modifying effects of material transport on these ecosystems.

The IBP and LTER projects illustrate the whole-system and synthetic approaches to limnology. The long-term view leads to detailed climate data, data-sets spanning decades, whole-system experiments, and a series of integrated studies of aspects of the physical, chemical, and biological processes important at the particular sites. Whereas there is a need for ongoing studies of this type, there is also a need for extended spatial sampling; that is, repeated sampling of many polar sites, to understand the effects of different geological and climatic settings throughout the polar regions. Other lake districts with important limnological records for Antarctica (Figure 1.1) include Signy Island and Livingston Island (Byers Peninsula; Toro *et al.* 2007) in the maritime Antarctic region, the Vestfold Hills, and the Schirmacher Oasis. Lake studies have now been conducted in many parts of the circumpolar Arctic (Figure 1.2), including Alaska, Canada, northern Finland, several parts of Greenland, Svalbard, Siberia, and the Kuokkel lakes in northern Sweden. Flowing waters have also received increasing attention from polar limnologists; for example, the ephemeral streams of the McMurdo Dry Valleys and the large Arctic rivers and their lake-rich floodplains.

Several special journal issues have been published on polar lake and river themes including high-latitude limnology (Vincent and Ellis-Evans 1989), the paleolimnology of northern Ellesmere Island (Bradley 1996), the limnology of the Vestfold Hills (Ferris *et al.* 1988), and the responses of northern freshwaters to climate change (Wrona *et al.* 2006). Books on regional aspects of polar limnology include volumes on the Schirmacher Oasis (Bormann and Fritsche 1995), McMurdo Dry Valleys

(Green and Friedmann 1993; Priscu 1998), Alaskan freshwaters (Milner and Oswood 1997), Siberian rivers (Zhulidov *et al.* 2008), and Siberian wetlands (Zhulidov *et al.* 1997). The rapidly developing literature on subglacial aquatic environments beneath the Antarctic ice sheet has been reviewed in a volume by the National Academy of Sciences of the USA (National Research Council 2007). Pienitz *et al.* (2004) present multiple facets of Antarctic and Arctic paleolimnology, with emphasis on environmental change, and current changes in Antarctic lake and terrestrial environments are summarized in Bergstrom *et al.* (2006).

1.3 Limnological diversity

The Antarctic, defined as that region south of the Polar Frontal Zone or Antarctic Convergence (which also delimits the Southern Ocean) contains several coastal areas where lakes, ponds, and streams are especially abundant (Figure 1.1), as well as vast networks of subglacial aquatic environments. Lake and river ecosystems are common throughout the Arctic (Figure 1.2), which can be delimited in a variety of ways: by the northern treeline, the 10°C July isotherm, or the southern extent of discontinuous permafrost (for permafrost map definitions, see Heginbottom 2002), which in the eastern Canadian Arctic, for example, currently extends to the southern end of Hudson Bay (http://atlas.nrcan.gc.ca/site/english/maps/environment/land/permafrost). Of course, all of these classifications depend on climate, which is changing rapidly. These northern lands include the forest-tundra ecozone, sometimes referred to as the Sub-Arctic or Low Arctic, which grades into shrub tundra, true tundra, and ultimately high Arctic polar desert. Appendix 1.1 provides a brief limnological introduction to many of the polar rivers, lakes, or lake districts where there have been aquatic ecosystem studies.

Collectively, the polar regions harbour an extraordinary diversity of lake types (Plates 1–9) ranging from freshwater to hypersaline, from highly acidic to alkaline, and from perennially ice-covered waters to concentrated brines that never freeze. The diverse range of these habitats is illustrated by their many different thermal regimes in summer, from fully mixed to thermally stratified over a 40°C span of temperatures (Figure 1.3). This physical diversity is accompanied by large variations in their chemical environments, for example from oxygen supersaturation to anoxia, sometimes within the same lake over time or depth. Permafrost thaw lakes (thermokarst lakes and ponds; Plate 8) are the most abundant aquatic ecosystem type in the Arctic, and often form a mosaic of waterbodies that are hot spots of biological activity in the tundra, with abundant microbes, benthic communities, aquatic plants, plankton, and birds. In the Mackenzie River delta for example, some 45000 waterbodies of this type have been mapped on the floodplain, with varying degrees of connection to the river (Emmerton *et al.* 2007; Figure 1.4), while in the Yukon River delta the total number of thaw lakes and ponds has been estimated at 200000 (Maciolek 1989). Most thaw lakes are shallow, but lake depth in the permafrost increases as a square root of time, and the oldest lakes (>5000 years) can be up to 20m deep (West and Plug 2008). Shallow rock-basin ponds are also common throughout the Arctic (e.g. Rautio and Vincent 2006; Smol and Douglas 2007a) and Antarctica (e.g. McKnight *et al.* 1994; Izaguirre *et al.* 2001).

Certain lake types are found exclusively in the polar regions, for example solar-heated perennially ice-capped lakes (e.g. northern Ellesmere Island lakes in the Arctic, McMurdo Dry Valley lakes in Antarctica; Figure 1.3), and the so-called epishelf

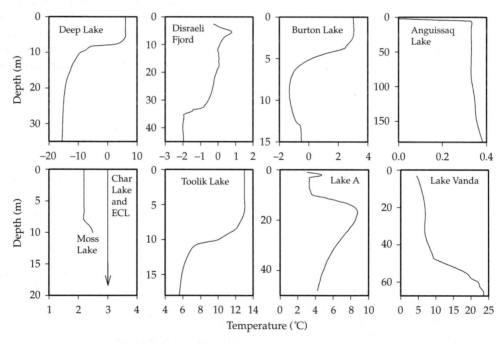

Figure 1.3 From sub-zero cold to solar-heated warmth: the remarkable diversity of summer temperature and mixing regimes in high-latitude lakes. Deep Lake is a hypersaline lake in the Vestfold Hills (15 January 1978; Ferris *et al.* 1988); Disraeli Fiord, northern Ellesmere Island, at the time of study was an epishelf lake with a 30-m layer of freshwater dammed by thick ice floating on sea water (10 June 1999; Van Hove *et al.* 2006); Burton Lake is a coastal saline lake in the Vestfold Hills that receives occasional inputs of sea water (30 January 1983, Ferris *et al.* 1988); Anguissaq Lake lies at the edge of the ice cap in northwest Greenland and convectively mixes beneath its ice cover in summer (19 August 1957; Barnes 1960); Moss Lake on Signy Island (9 February 2000; Pearce 2003), Char Lake in the Canadian Arctic (isothermal at 3°C to the bottom, 27.5m, on 30 August 1970; Schindler *et al.* 1974a), and El'gygytgyn Crater Lake (ECL) in Siberia (isothermal at 3°C to 170m on 1 August 2002; Nolan and Brigham-Grette 2007) are examples of cold monomictic lakes that mix fully during open water in summer; Toolik Lake, northern Alaska, is dimictic, with strong summer stratification (8 August 2005; see Figure 4.6 in this volume); Lake A is a perennially ice-covered, meromictic lake on northern Ellesmere Island (1 August 2001, note the lens of warmer sub-ice water; Van Hove *et al.* 2006); and Lake Vanda is an analogous ice-capped, meromictic system in the McMurdo Dry Valleys with more transparent ice and water, and extreme solar heating in its turbid, hypersaline bottom waters (27 December 1980; Vincent *et al.* 1981).

lakes, tidal freshwater lakes that sit on top of colder denser seawater at the landward edges of ice shelves; for example Beaver Lake, Antarctica, and Milne Fjord in the Arctic. Networks of subglacial aquatic environments occur beneath the thick ice of the Antarctic ice cap (Plate 1), and include the vast, deep, enigmatic waters of Lake Vostok. Ephemeral rivers and streams are found around the margins of Antarctica with biota that are active for only a brief period each year (Plate 6). Flowing surface waters play a much greater role in the Arctic where there are extensive catchments and some of the world's largest rivers that discharge into Arctic seas (Plate 7). Many polar lakes are classed as ultra-oligotrophic or extremely unproductive, whereas some are highly enriched by animal or human activities.

1.4 Controlling variables for biological production

What factor or combination of factors limits biological production in high-latitude freshwater

ecosystems? This question is not only of interest to polar limnologists, but it may also provide insights into the controlling variables for aquatic productivity at other latitudes. Such insights are especially needed to predict how inland water ecosystems will respond to the large physical, chemical, and biological perturbations that are likely to accompany future climate change (see Section 1.6 below).

1.4.1 Water supply

The availability of water in its liquid state is a fundamental prerequisite for aquatic life, and in the polar regions the supply of water is severely regulated by the seasonal freeze–thaw cycle. For a few ecosystem types, this limits biological activity to only a brief period of days to weeks each year, for example the ephemeral streams of Antarctica and meltwater lakes on polar ice shelves. For many high-latitude lakes, however, liquid water persists throughout the year under thick snow and ice cover, and even some shallow ponds can retain a

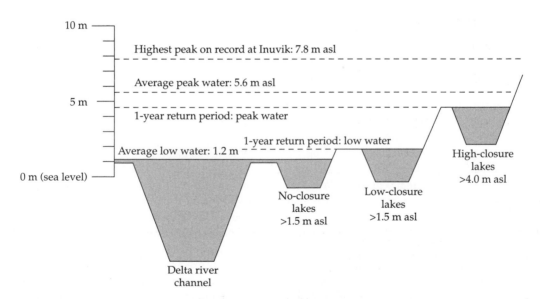

Figure 1.4 Lake classification in the Mackenzie River delta according to the extent of isolation from the river. Large Arctic rivers carry 3300 km³ of freshwater to the Arctic Ocean each year and during their spring period of peak flow they recharge vast areas of flood-plain lakes. The Mackenzie River delta in the Canadian western Arctic has some 45 000 lakes of more than 0.0014 km², and its total open-water surface area (including the multiple channels of the river) in late summer is about 5250 km². Climate change is modifying the amplitude and duration of seasonal flooding and therefore the connectivity of these lakes with the river (Lesack and Marsh 2007). Redrawn from Emmerton *et al.* 2007, by permission of the American Geophysical Union. asl, above sea level.

thin layer of water over their benthic communities in winter (e.g. Schmidt *et al.* 1991). The larger rivers of the Arctic are fed from their source waters at lower latitudes and they continue to flow under the ice during autumn and winter, albeit at much reduced discharge rates. Thus many polar aquatic ecosystems are likely to be microbiologically active throughout the year, but with strong seasonal variations that are dictated by factors other than or in addition to water supply. During summer, the availability of meltwater for aquatic habitats is favored by the continuous, 24-h-a-day exposure of snow and ice to solar radiation, combined with the slow rates of evaporative loss at low temperatures. Polar streams and rivers are fed by melting glaciers and snow pack, and for the large Arctic rivers the peak snowmelt in spring gives rise to extensive flooding of their abundant floodplain lakes and generates a vast interconnected freshwater habitat (Plate 8).

1.4.2 Irradiance

The polar regions receive reduced amounts of incident solar radiation relative to lower latitudes (annual solar irradiance drops by about 50% over the 50° of latitude from 30° to 80°), and this effect is compounded by the attenuating effects of snow and ice on underwater irradiance. This limits the total annual production in Arctic and Antarctic aquatic ecosystems, and it has a strong influence on the seasonality of photosynthesis, which ceases during the onset of winter darkness and resumes with the first return of sunlight. Underwater irradiance does not, however, appear to be the primary variable controlling the large variation among lakes in daily primary production by the phytoplankton during summer. Polar lakes may show an early spring maximum in phytoplankton biomass and photosynthesis immediately beneath the ice, with pronounced decreases over summer despite increased irradiance conditions in the upper water column (Tanabe *et al.* 2008, and references therein), but likely decreased nutrient availability (Vincent 1981). Nearby lakes such as Meretta and Char in the Canadian High Arctic, and Fryxell and Vanda in the McMurdo Dry Valleys, show contrasting phytoplankton biomass concentrations and photosynthetic rates despite similar incident irradiances but large differences in nutrient status (see Chapter 9).

1.4.3 Low temperature

Contrary to expectation, some polar aquatic habitats have warm temperatures in summer, and some even remain warm during winter. Shallow thaw lakes can heat to 10°C or above, and the surface waters of northern lakes with high concentrations of light-absorbing dissolved organic matter and particles may undergo diurnal heating, with temperatures rising to >15°C. The large Arctic rivers begin more than 1000 km further south of their discharge point to the Arctic Ocean, and these waters can warm during summer over their long transit to the sea; for example, the Mackenzie River can be up to 17°C at its mouth, despite its far northerly latitude of 69°N. Stratified, perennially ice-covered lakes can heat up over decades to millennia via the solar radiation that penetrates the ice (Vincent *et al.* 2008, and references therein; see also Chapter 4). Examples of these solar-heated, meromictic lakes are known from both polar regions, and in these waters the deep temperature maxima lie well above summer air temperatures and up to 70°C above winter temperatures. 'Warm' of course is a relative term, and even liquid water temperatures of 0°C beneath the ice of most polar lakes, or −2°C in the subglacial Antarctic lakes capped by ice many kilometers thick, provide hospitable thermal conditions for biological processes relative to the extreme cold of the overlying atmosphere (down to −89°C at Vostok station in winter).

Most polar aquatic habitats experience water temperatures close to 0°C for much of the year. Many of the organisms found in these environments appear to be cold-tolerant rather than cold-adapted, and the cool ambient conditions likely slow their rates of metabolism and growth. Although cold temperatures may exert an influence on photosynthesis and other physiological processes, it does not preclude the development of large standing stocks of aquatic biota in some high-latitude waters. Conversely, lakes with warmer temperatures do not necessarily have higher phytoplankton and production rates (e.g. compare Lake Vanda with Lake Fryxell in the McMurdo Dry Valleys; Vincent 1981).

1.4.4 Nutrient supply

Several features of polar lakes and their surrounding catchments result in low rates of nutrient delivery for biological production, especially by their plankton communities. The combination of low temperature, low moisture, and freezing constrains the activity of soil microbes and slows all geochemical processes including soil-weathering reactions. This reduces the release of nutrients into the groundwater and surface runoff, which themselves are limited in flow under conditions of extreme cold. The severe polar climate also limits the development of vegetation, which in turn reduces the amount of root biomass, associated microbes (the rhizosphere community), and organic matter that are known to stimulate weathering processes (Schwartzman 1999). Nutrient recycling rates are also slowed by low temperatures within Arctic and Antarctic waters. Additionally, the presence of ice cover inhibits wind-induced mixing of polar waters throughout most of the year. This severely limits the vertical transport of nutrients from bottom waters to the zone immediately beneath the ice where solar energy is in greatest supply for primary production. It also results in quiescent, stratified conditions where inflowing streams can be short-circuited directly to the outflow without their nutrients mixing with the main body of the lake (see Chapter 4), and where nutrient loss by particle sedimentation is favored.

Several lines of evidence indicate that nutrient supply exerts a strong control on phytoplankton production in polar lakes, in combination with light and temperature. First, large variations in primary production occur among waters in the same region, despite similar irradiance and temperature regimes, but differences in nutrient status (see above). Second, in stratified waters, highest-standing stocks of phytoplankton and primary production rates are often observed deep within the ice-covered water column where light availability is reduced, but nutrient supply rates are greater (Plate 14; details in Chapter 9). Third, waterbodies in both polar regions that have received nutrient enrichment from natural or human sources show strikingly higher algal biomass stocks and production rates; for example, penguin-influenced ponds on Ross Island (Vincent and Vincent 1982) and Cierva Point, Antarctica (Izaguirre et al. 2001), and high Arctic Meretta Lake, which was enriched by human sewage (Schindler et al. 1974b). Finally, nutrient bioassays show that high-latitude plankton assemblages respond strongly to nutrient addition; for example Toolik Lake, Alaska (O'Brien et al. 1997); lakes in northern Sweden (Holmgren 1984); Ward Hunt Lake in the Canadian Arctic (Bonilla et al. 2005); and McMurdo Dry Valley lakes (Priscu 1995). However, although nutrients may impose a primary limitation on productivity through Liebig-type effects on final yield and biomass standing stocks, there may be secondary Blackman-type effects on production rates per unit biomass, and thus specific growth rates (see Cullen 1991). Low temperatures reduce maximum, light-saturated photosynthetic rates, and to a lesser extent light-limited rates, and low irradiances beneath the ice may also reduce primary production. These effects may be further compounded by nutrient stress, which limits the synthesis of cellular components such as light-harvesting proteins and photosynthetic enzymes to compensate for low light or low temperature (Markager et al. 1999).

1.4.5 Benthic communities

In many high-latitude aquatic ecosystems, total ecosystem biomass and productivity are dominated by photosynthetic communities living on the bottom where the physical environment is more stable, and where nutrient supply is enhanced by sedimentation of nitrogen- and phosphorus-containing particles from above, nutrient release from the sediments below and more active bacterial decomposition and nutrient recycling processes than in the overlying water column. Some of these communities achieve spectacular standing stocks in polar lakes, even under thick perennial ice cover (Plate 10); for example, cyanobacterial mats more than 10 cm in thickness, and algal-coated moss pillars up to 60 cm high in some Antarctic waterbodies (Imura et al. 1999). These communities fuel benthic food webs that lead to higher trophic levels, including fish and birds in Arctic lakes. The phytobenthos may be limited more by habitat and light availability than by nutrients. For example, bioassays of microbial mats in an Arctic lake showed no effect of nutrient enrichment over 10 days, while the phytoplankton showed a strong growth response (Bonilla *et al.* 2005). On longer timescales, however, even the benthic communities may respond to nutrients, as shown by the shift of Arctic river phytobenthos to luxuriant moss communities after several years of continuous nutrient addition (Plate 7; Bowden *et al.* 1994; for details see Chapter 5), and longer-term shifts in benthic diatom communities in a sewage-enriched Arctic lake (Michelutti *et al.* 2007).

1.5 Food webs in polar lakes

There is no typical food web for polar lakes (see Chapters 11–15). Instead there is a continuum of types of food web ranging from low Arctic lakes

with well-developed zooplankton and fish communities to high Arctic and Antarctic lakes with flagellates, ciliates, and rotifers at the top of the food web. The structure and diversity of the various food webs depend primarily on the trophic state of the lake and secondarily upon biogeography. Thus, some Antarctic lakes could likely support several types of crustacean zooplankton but few species have reached the continent (see Chapter 13), and many lakes are devoid of crustaceans.

Toolik Lake, Alaska (68°N, see Appendix 1.1), is an example of an oligotrophic low Arctic lake (Figure 1.5). The planktonic food web is based on small photosynthetic flagellates and on the bacteria that consume mainly dissolved organic matter from the watershed. Small flagellates (e.g. *Katablepharis*, Plate 11) consume bacteria and some of these (e.g. the colonial flagellate *Dinobryon*, Plate 11) are also

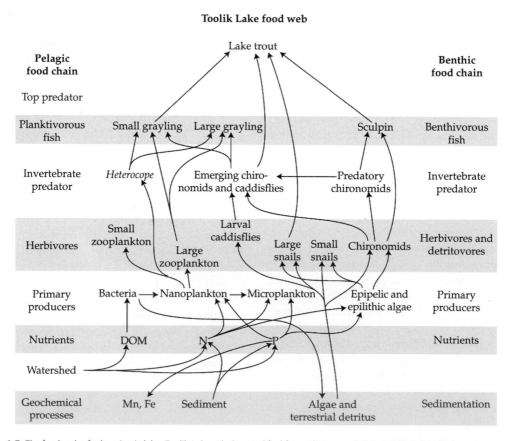

Figure 1.5 The food web of a low Arctic lake: Toolik Lake, Alaska. Modified from O'Brien *et al.* (1997). DOM, dissolved organic matter.

photosynthetic (mixotrophic). These forms are consumed by seven species of crustacean zooplankton and eight species of rotifer. Zooplankton, in turn, are consumed by some of the five species of fish. Although this appears to be a conventional lake food web, it differs from the usual in several ways. First, almost all of the primary productivity is by nanoplankton. Second, the low primary productivity supports only a few zooplankton, not enough to control the algal abundance by grazing. Third, the sparse zooplankters are not abundant enough to support the fish growth. Therefore, the food web based on phytoplankton ends with zooplankton.

Stable-isotope analysis of the Toolik Lake food web reveals that the fish rely on the benthos as their main source of energy. In this benthic food web the energy passes from diatoms on rocks and sediments into snails and from detritus into chironomid larvae. Small fish eat the chironomids and are consumed in turn by the lake trout that also consume snails. A similar benthic-based food web supporting fish was also described by Rigler (1978) from Char Lake (75°N, see Appendix 1.1), where there was but one species of copepod and Arctic char (often spelled charr; *Salvelinus alpinus*; Plate 13) as the only fish species. Char occur in some of the northernmost lakes in North America, for example Lake A at 83°N, where their diet may also depend on benthic invertebrates.

Antarctic lakes are species poor and possess simplified and truncated planktonic food webs dominated by small algae, bacteria, and colorless flagellates (Figure 1.6 and Chapter 11). There are few metazoans and no fish. The phytoplankton are often both photosynthetic and mixotrophic. The latter are species that can ingest bacteria as well as photosynthesize. One important pathway in both Antarctic and Arctic lakes is the microbial loop which may be defined as carbon and energy cycling via the nanoplankton (protists less than 20 μm in size). It includes primary production plus production of dissolved organic carbon (DOC) by planktonic organisms as well as uptake of DOC by bacteria, with transfer to higher trophic levels via nanoflagellates, ciliates, and rotifers (see Vincent and Hobbie 2000). In many Arctic and Antarctic lakes, picoplanktonic (<2 μm in diameter) and filamentous species of cyanobacteria are also components of the phytoplankton and microbial loop. Examples of cyanobacteria include *Synechococcus* sp. in Ace Lake in the Vestfold Hills (Powell *et al.* 2005) and Char Lake, Lake A and Lake Romulus in

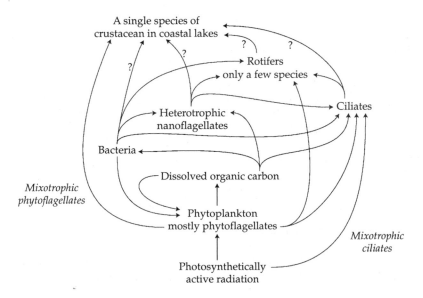

Figure 1.6 The typical planktonic food web in continental Antarctic lakes.

the high Arctic (Van Hove *et al.* 2008), and thin oscillatorians (filamentous cyanobacteria) in the deep chlorophyll maximum of Lake Fryxell, McMurdo Dry Valleys (Spaulding *et al.* 1995).

Lakes on coastal ice-free areas, like the Vestfold Hills (see Appendix 1.1), usually have a single planktonic crustacean. In freshwater and slightly brackish lakes sparse populations of the endemic Antarctic cladoceran *Daphniopsis studeri* occur (Plate 12) whereas in marine-derived saline lakes the marine copepod *Paralabidocera antarctica* is found. The McMurdo Dry Valleys lie inland and much further south (see Appendix 1.1). Here planktonic crustaceans are lacking, although a few copepod nauplii and two species of rotifers have been found in the benthos.

The great similarity between the food webs of lakes at both poles, in terms of structure and diversity, is closely related to the trophic state of the lakes (Hobbie *et al.* 1999). In this scheme, Type I lakes are ultra-oligotrophic; that is, they have very low primary productivity, and support only algae, bacteria, nanoflagellates and ciliates (e.g. ice-shelf lakes; Plates 2 and 3). Type II lakes are more productive and contain microzooplankton such as rotifers. Far northerly Arctic lakes such as Ward Hunt Lake and some McMurdo Dry Valley lakes fall into this category (Plate 4). With Type III lakes such as Char Lake, the increase in productivity allows copepods to survive. The most productive type of lake, Type IV, includes both copepods and Cladocera, much like a temperate lake. Ponds at Barrow fall into this category although they freeze completely and so have no fish. However, how can we explain the occurrence of fish (Arctic char) in the most northerly lakes that are Type I and Type II in the level of productivity? The answer is that the char are consuming the chironomid larvae of the more productive benthic food web. In contrast, continental Antarctic lakes contain neither insect larvae nor fish.

1.6 Polar lakes and global change

The polar regions are now experiencing the multiple stressors of contaminant influxes, increased exposure to ultraviolet radiation, and climate change (Schindler and Smol 2006; Wrona *et al.*

2006; Plates 13–16) and these impacts are likely to become more severe in the future. Global circulation models predict that the fastest and most pronounced increases in temperature over the course of this century will be at the highest latitudes (Plate 16; Meehl *et al.* 2007) because of a variety of feedback processes that amplify warming in these regions. These include the capacity for warm air to store more water vapour, itself a powerful greenhouse gas, and the reduced albedo (reflection of sunlight) as a result of the melting of snow and ice, leaving more solar energy to be available for heating. Major changes are also predicted in the regional distribution of precipitation, with increased inputs to many parts of the Arctic and Antarctica (Plate 16; Meehl *et al.* 2007), and an increased frequency of precipitation as rainfall, even at the highest latitudes where such events are unusual.

High-latitude lakes have already begun to show striking impacts of climate change. These include loss of perennial ice cover, increasing duration of open water conditions, increasing water temperatures, stronger water-column stratification, and shifts in water balance, in some cases leading to the complete drainage or drying-up of lakes and wetlands. For many polar aquatic ecosystems, small changes in their physical, chemical, or biological environment induced by climate can be amplified into major shifts in their limnological properties. Rather than slow, deterministic changes through time accompanying the gradual shift in air temperature, these threshold effects can result in abrupt step-changes in ecosystem structure and functioning.

1.6.1 Physical thresholds

The most critical threshold is that affecting the integrity of lake basins and the presence or absence of standing water. The degradation of permafrost soils in Siberia has resulted in collapse and drainage of many lake basins, and the complete disappearance of many waterbodies (Smith *et al.* 2005). A shift in the precipitation/evaporation balance in parts of the High Arctic has resulted in the complete drying up of ponds, perhaps for the first time in millennia (Smol and Douglas 2007a). In other regions, the accelerated melting of permafrost over the last 50 years has created new basins for lakes and ponds, and increased development of shallow water ecosystems (Payette *et al.* 2004; Walter *et al.* 2006). An analysis of long-term changes of Mackenzie River floodplain lakes indicates that climate change is having disparate effects on their connectivity with the river. On average, the lower-elevation lakes (low-closure lakes; Figure 1.4) are being flooded for longer periods of time, whereas the highest-elevation lakes (high-closure lakes; Figure 1.4) are less flooded and may eventually dry up because of reduced ice dams and associated reductions in peak water levels in the river (Lesack and Marsh 2007).

For some polar lakes, ice dams from glaciers and ice shelves can be the primary structures retaining freshwater. Gradual warming can eventually cross the threshold of stability of these structures, resulting in catastrophic drainage. For example, the break-up of the Ward Hunt Ice Shelf in 2002 resulted in complete drainage and loss of an epishelf lake had probably been in place for several thousand years (Mueller *et al.* 2003).

The surface ice cover of polar lakes is also a feature subject to threshold effects. Many lakes in Antarctica and some Arctic lakes retain their ice covers for several years, decades, or longer. The loss of such ice results in changes in mixing regime and a complete disruption of their limnological gradients (Vincent *et al.* 2008). It also results in a massive increase in solar radiation; for example, order-of-magnitude increases in ultraviolet exposure that far exceed stratospheric ozone effects (Vincent *et al.* 2007), but also increased light supply for photosynthesis and more favorable conditions for the growth of benthic communities.

The persistent low temperatures in high-latitude lakes and ponds limit their water-column stability during open water conditions. In the coldest locations, there is insufficient heating of the water column to exceed the maximum density of water at 3.98°C, and the lakes remain free-mixing throughout summer (cold monomictic). Increased warming will result in the crossing of that threshold and a complete change in summer structure with the development of thermal stratification (dimictic conditions). These changes have far-reaching implications, including for light supply

to the phytoplankton, gas exchange, and biogeochemical processes. Recent changes in mixing and stratification patterns have been inferred from fossil diatom records in some lakes from Finnish Lapland, with evidence of increased productivity and the development of cladoceran communities (Sorvari *et al.* 2002).

1.6.2 Biogeochemical thresholds

The arrival of shrubs and trees in a catchment can result in a major step-increase in terrestrial plant biomass and hence the quantity of fulvic and humic materials in the soil, in turn resulting in a substantial increase in the concentration of particulate and colored dissolved organic material (CDOM) in lake waters (e.g. Pienitz and Vincent 2000). At CDOM concentrations less than $2 \, mg \, L^{-1}$, small changes in vegetation and hence CDOM can cause disproportionately large changes in the underwater light penetration and spectral regime, especially ultraviolet exposure. This vegetation change may also cause other changes that accelerate shifts in biogeochemistry, for example the increased development of root biomass, rhizosphere microbial activity, and soil weathering, and decreased albedo accompanied by increased soil heating and deepening of the permafrost active layer. Another biogeochemical threshold is that associated with water column anoxia. Once a lake fully depletes the oxygen in its bottom waters during stratification, large quantities of inorganic phosphorus, as well as iron and manganese, may be released from the lake's sediments. This increased internal loading can result in a sudden acceleration of eutrophication.

1.6.3 Biological thresholds

The extirpation of certain high Arctic taxa may occur if critical thresholds of tolerance are exceeded. Conversely new species may arrive in the catchments (e.g. the arrival of shrub and tree species as noted above) or in the rivers and lakes. For example, analyses of range distributions and climate change scenarios have shown that warmwater fish species, such as the smallmouth bass, *Micropterus dolomieu*, will shift northwards into the Arctic, with negative impacts on native fish communities (Sharma *et al.* 2007). More subtle effects may take place through changes in animal behavior. For example, changes in the migration patterns of Arctic char may accompany increased water temperatures, with negative effects on fish productivity and size distribution, and on native fisheries (see Chapter 14).

Paleolimnological analyses of sediment cores from Cape Herschel, Ellesmere Island, showed that there was abrupt biotic change in the mid-nineteenth century at this site, as indicated by the quantity and composition of fossil diatoms (Douglas *et al.* 1994). This effect has been detected subsequently in sediment cores from many parts of the circumpolar Arctic (e.g. Antoniades *et al.* 2007), with differences in the exact timing and magnitude of change among lakes and locations, as expected (Smol *et al.* 2005; see Plate 15). These striking effects on community structure have been attributed primarily to changes in the duration and extent of lake and pond ice cover, as well as related limnological changes (Smol *et al.* 2005; Smol and Douglas 2007b; see also Chapter 3).

1.7 Conclusions

High-latitude lakes, rivers, and wetlands are a major focus for limnological research not only because of their remarkable diversity and intrinsic importance in polar biomes, but also because of their value as compelling models for understanding aquatic ecosystem processes in general. These environments provide habitats for plants, animals, and microbes that are adapted to or at least tolerate persistent low temperatures, freeze–thaw cycles, and marked seasonal and interannual variations in energy and nutrient supplies. However, the resilience of these biological communities is now being tested severely as they face the multiple stressors associated with local and global human impacts, notably contaminant effects, increased ultraviolet exposure, and climate change. The physical characteristics of polar lakes and rivers depend strongly on ice and the freezing and melt cycle, and small variations in air temperature can radically alter their structure and functioning. These ecosystems are therefore a sensitive guide to the magnitude and pace of global climate change, as well as key sites for environmental research, monitoring, and stewardship.

Acknowledgements

Our limnological research on high-latitude lakes has been supported by the National Science Foundation (USA); the Natural Sciences and Engineering Research Council (Canada); the Canada Research Chair program; the Network of Centres of Excellence program ArcticNet; Polar Shelf Canada; and the British, Australian, USA, New Zealand, and Spanish Antarctic programs. We thank Kirsten Christoffersen, John P. Smol, and Dale T. Andersen for their valuable comments on an earlier draft, and Marie-Josée Martineau and Serge Duchesneau for assistance in manuscript preparation.

References

Antoniades, D. *et al.* (2007). Abrupt environmental change in Canada's northernmost lake inferred from diatom and fossil pigment stratigraphy. *Geophysical Research Letters* **34**, L18708, doi:10.1029/2007GL030947.

Barnes, D.F. (1960). An investigation of a perennially frozen lake. *Air Force Surveys in Geophysics* **129**.

Bergstrom, D., Convey, P., and Huiskes, A. (eds) (2006). *Trends in Antarctic Terrestrial and Limnetic Ecosystems.* Springer, Dordrecht.

Bonilla, S., Villeneuve, V., and Vincent, W.F. (2005). Benthic and planktonic algal communities in a high arctic lake: pigment structure and contrasting responses to nutrient enrichment. *Journal of Phycology* **41**, 1120–1130.

Bormann, P. and Fritsche, D. (eds) (1995). *The Schirmacher Oasis.* Justus Perthes, Gotha.

Bowden, W.B., Finlay, J.C., and Maloney, P.E. (1994). Long-term effects of PO_4 fertilization on the distribution of bryophytes in an Arctic river. *Freshwater Biology* **32**, 445–454.

Bradley, R.S. (ed.) (1996). Taconite Inlets lakes project. *Journal of Paleolimnology* **16**, 97–255.

Brewer, M.C. (1958). The thermal regime of an Arctic lake. *Transactions of the American Geophysical Union* **39**, 278–284.

Christoffersen, K.S., Amsinck, S.L., Landkildehus, F., Lauridsen, T.L., and Jeppesen, E. (2008). Lake flora and fauna in relation to ice-melt, water temperature and chemistry. In Meltofte, H., Christensen, T.R, Elberling, B., Forchhammer, M.C., and Rasch, M. (eds),

High-Arctic Ecosystem Dynamics in a Changing Climate. Ten years of monitoring and research at Zackenberg Research Station, Northeast Greenland, pp. 371–390. Advances in Ecological Research, vol. 40, Academic Press, San Diego, CA.

Cornwell, J.C. (1987). Phosphorus cycling in Arctic lake sediment: adsorption and authigenic minerals. *Archives für Hydrobiologie* **109**, 161–170.

Cullen, J.J. (1991). Hypotheses to explain high nutrient conditions in the open sea. *Limnology and Oceanography* **36**, 1578–1599.

Doran, P.T., Wharton, R.A., Lyons, W.B., DesMarais, D.J., and Andersen, D.T. 2000. Sedimentology and isotopic geochemistry of a perennially ice-covered epishelf lake in Bunger Hills Oasis, East Antarctica. *Antarctic Science*, **12**, 131–140.

Douglas, M.S.V., Smol, J.P., and Blake, Jr, W. (1994). Marked post-18th century environmental change in high Arctic ecosystems. *Science* **266**, 416–419.

Emmerton, C.A., Lesack, L.F.W., and Marsh, P. (2007). Lake abundance, potential water storage, and habitat distribution in the Mackenzie River Delta, western Canadian Arctic. *Water Resources Research* **43**, W05419.

Ferris, J.M., Burton, H.R., Johnstone, G.W., and Bayly, I.A.E. (eds) (1988). Biology of the Vestfold Hills. *Hydrobiologia* **165**.

Forel, A.-F. (1895). *Le Léman: Monographie Limnologique* Tome 2. F. Rouge, Lausanne.

Franzmann, P.D., Roberts, N.J., Mancuso, C.A., Burton, H.R., and McMeekin, T.A. (1991). Methane production in meromictic Ace Lake, Antarctica. *Hydrobiologia* **210**, 191–201.

Gibson, J.A.E. (1999). The meromictic lakes and stratified marine basins of the Vestfold Hills, East Antarctica. *Antarctic Science* **11**, 175–192.

Gibson, J.A.E. and Andersen, D.T. (2002). Physical structure of epishelf lakes of the southern Bunger Hills, East Antarctica. *Antarctic Science* **14**, 253–262.

Goldman, C.R., Mason, D.T., and Hobbie, J.E. (1967). Two Antarctic desert lakes. *Limnology and Oceanography* **12**, 295–310

Green, W.J. and Friedmann, E.I. (eds) (1993). *Physical and biogeochemical processes in Antarctic lakes.* Antarctic Research Series vol. 59. American Geophysical Union, Washington DC.

Heginbottom, J.A. (2002). Permafrost mapping: a review. *Progress in Physical Geography* **26**, 623–642.

Hobbie, J.E. (1961). Summer temperatures in Lake Schrader, Alaska. *Limnology and Oceanography* **6**, 326–329.

Hobbie, J.E. (ed.) (1980). *Limnology of Tundra Ponds, Barrow, Alaska.* US/IBP Synthesis Series vol. 13. Dowden,

Hutchinson and Ross, Stroudsburg, PA. www.biodiversitylibrary.org/ia/limnologyoftundr00hobb.

Hobbie, J.E., Bahr, M., and Rublee, P.A. (1999). Controls on microbial food webs in oligotrophic Arctic lakes. *Archive für Hydrobiologie, Advances in Limnology* **54**, 61–76.

Holmgren, S.K. (1984). Experimental lake fertilization in the Kuokkel area, northern Sweden: phytoplankton biomass and algal composition in natural and fertilized subarctic lakes. *Internationale Revue der gesamten Hydrobiologie und Hydrographie* **69**, 781–817.

Hutchinson, G.E. and Löffler, H. (1956). The thermal classification of lakes. *Proceedings of the National Academy Sciences USA*, **42**, 84–86.

Imura, S., Bando, T., Saito, S., Seto, K., and Kanda, H. (1999). Benthic moss pillars in Antarctic lakes. *Polar Biology* **22**, 137–140.

Izaguirre, I., Mataloni, G., Allende, L., and Vinocur, A. (2001). Summer fluctuations of microbial planktonic communities in a eutrophic lake–Cierva Point, Antarctica. *Journal of Plankton Research* **23**, 1095–1109.

Jensen, D.B. (ed.) (2003). *The Biodiversity of Greenland—a Country Study.* Technical Report no. 55. Pinngortitaleriffik, Grønlands Naturinstitut, Nuuk, Greenland.

Juday, C. (1920). The Cladocera of the Canadian Arctic Expedition, 1913–1918. In *Report of the Canadian Arctic Expedition 1913–1918.* Vol VII. Crustacea, Part H, Cladocera. 1–8. T. Mulvey, Ottawa.

Kaup, E. (2005). Development of anthropogenic eutrophication in Antarctic lakes of the Schirmacher Oasis. *Verhandlungen Internationale Vereingung der Limnologie* **29**, 678–682.

Lesack, L.F.W. and Marsh, P. (2007). Lengthening plus shortening of river-to-lake connection times in the Mackenzie River Delta respectively via two global change mechanisms along the arctic coast. *Geophysical Research Letters* **34**, L23404, doi:10.1029/2007GL031656.

Maciolek, J.A. (1989). Tundra ponds of the Yukon Delta, Alaska, and their macroinvertebrate communities. *Hydrobiologia* **172**, 193–206.

Markager, S., Vincent, W.F., and Tang, E.P.Y. (1999). Carbon fixation in high Arctic lakes: Implications of low temperature for photosynthesis. *Limnology and Oceanography* **44**, 597–607.

McKnight, D.M., Andrews, E.D., Aiken, G.R., and Spaulding, S.A. (1994). Dissolved humic substances in eutrophic coastal ponds at Cape Royds and Cape Bird, Antarctica. *Limnology and Oceanography* **39**, 1972–1979.

Meehl, G.A. *et al.* (2007). Global climate projections. In Solomon, S. *et al.* (eds.), *Climate Change 2007: The Physical*

Science Basis. Contribution of Working Group I to the Fourth Assessment Report of the Intergovernmental Panel on Climate Change, pp. 747–846. Cambridge University Press, Cambridge.

Michelutti, N., Hermanson, M.H., Smol, J.P., Dillon, P.J., and Douglas, M.S.V. (2007). Delayed response of diatom assemblages to sewage inputs in an Arctic lake. *Aquatic Sciences* doi:10.1007/s00027-007-0928-8.

Milner, A.M. and Oswood, M.W., (eds) (1997). *Freshwaters of Alaska: Ecological Synthesis*. Ecological Studies, Vol 119. Springer, New York.

Mueller, D.R., Vincent, W.F., and Jeffries, M.O. (2003). Break-up of the largest Arctic ice shelf and associated loss of an epishelf lake. *Geophysical Research Letters* **30**, 2031, doi:10.1029/2003GL017931.

National Research Council. (2007). *Exploration of Antarctic Subglacial Aquatic Environments: Environmental and Scientific Stewardship*. The National Academies Press, Washington DC.

Nolan, M. and Brigham-Grette, J. (2007). Basic hydrology, limnology, and meteorology of modern Lake El'gygytgyn, Siberia. *Journal of Paleolimnology* **37**, 17–35.

O'Brien, W.J. *et al.* (1997). The limnology of Toolik Lake. In Milner, A.M. and Oswood, M.W. (eds.), *Freshwaters of Alaska: Ecological Synthesis*, pp. 61–106. Springer, New York.

Payette, S., Delwaide, A., Caccianiga, M., and Beauchemin, M. (2004). Accelerated thawing of subarctic peatland permafrost over the last 50 years. *Geophysical Research Letters* **31**, L18208.

Pearce, D.A. (2003). Bacterioplankton community structure in a maritime Antarctic oligotrophic lake during a period of holomixis, as determined by denaturing gradient gel electrophoresis (DGGE) and fluorescence in situ hybridization (FISH). *Microbial Ecology* **46**, 92–105.

Peterson, B.J. *et al.* (2006).Trajectory shifts in the arctic and subarctic freshwater cycle. *Science* **313**, 1061–1066.

Pienitz, R. and Vincent, W.F. (2000). Effect of climate change relative to ozone depletion on UV exposure in subarctic lakes. *Nature* **404**, 484–487.

Pienitz, R., Douglas, M.S.V., and Smol, J.P. (eds) (2004). *Long-Term Environmental Change in Arctic and Antarctic Lakes*. Springer, Dordrecht.

Poulsen, E.M. (1940). Freshwater Entomostraca. *Zoology of East Greenland, Meddelser om Grønland* **121**, no. 4.

Powell, L.M., Bowman, J.P., Skerratt, J.H., Franzmann, P.D., and Burton, H.R. (2005). Ecology of a novel *Synechococcus* clade occurring in dense populations in saline Antarctic lakes. *Marine Ecology Progress Series* **291**, 65–80.

Priscu, J.C. (1995). Phytoplankton nutrient deficiency in lakes of the McMurdo Dry Valleys, Antarctica. *Freshwater Biology* **34**, 215–227.

Priscu, J.C. (ed.) (1998). *Ecosystem dynamics in a polar desert: the McMurdo Dry Valleys Antarctica*. Antarctic Research Series vol. 72. American Geophysical Union, Washington DC.

Quayle, W.C., Peck, L.S., Peat, H., Ellis-Evans, J.C., and Harrigan, P.R. (2002). Extreme responses to climate change in Antarctic lakes. *Science* **295**, 645.

Rautio, M. and Vincent, W.F. (2006). Benthic and pelagic food resources for zooplankton in shallow high-latitude lakes and ponds. *Freshwater Biology* **51**, 1038–1052.

Riget, F. *et al.* (2000). Landlocked Arctic charr (*Salvelinus alpinus*) population structure and lake morphometry in Greenland – is there a connection? *Polar Biology* **23**, 550–558.

Rigler, F.H. (1978). Limnology in the High Arctic: a case study of Char Lake. *Verhandlungen der Internationalen Vereinigung für Theoretische und Angewandte Limnologie* **20**, 127–140.

Røen, U.I. (1962). Studies on freshwater entomostracan in Greenland II. Locations, ecology and geographical distribution of species. *Meddelser om Grønland* **170**, no 2.

Schindler, D.W. and Smol, J.P. (2006). Cumulative effects of climate warming and other human activities on freshwaters of Arctic and SubArctic North America. *Ambio* **35**, 160–168.

Schindler, D.W., Welch, H.E., Kalff, J., Brunskill, G.J., and Kritsch, N. (1974a). Physical and chemical limnology of Char Lake, Cornwallis Island (75°N lat). *Journal of the Fisheries Research Board of Canada* **31**, 585–607.

Schindler, D.W. *et al.* (1974b). Eutrophication in the high arctic- Meretta Lake, Cornwallis Island (75°N lat). *Journal of the Fisheries Research Board of Canada* **31**, 647–662.

Schmidt, S., Moskall, W., De Mora, S.J.D., Howard-Williams, C., and Vincent WF. (1991) Limnological properties of Antarctic ponds during winter freezing. *Antarctic Science* **3**, 379–388.

Schwartzman, D.W. (1999). *Life, Temperature, and the Earth: The Self-organizing Biosphere*. Columbia University Press, New York.

Sharma, S., Jackson, D.A., Minns, C.K., and Shuter, B.J. (2007). Will northern fish populations be in hot water because of climate change? *Global Change Biology* **13**, 2052–2064.

Shirtcliffe, T.G.L. and Benseman, R.F. (1964). A sun-heated Antarctic lake, *Journal of Geophysical Research* **69**, 3355–3359.

Smith, L.C., Sheng, Y., MacDonald, G.M., and Hinzman, L.D. (2005). Disappearing Arctic lakes. *Science* **308**, 1429.

Smol, J.P. and Douglas, M.S.V. (2007a). Crossing the final ecological threshold in high Arctic ponds. *Proceedings of the National Academy of Sciences of the USA* **104**, 12395–12397.

Smol, J.P. and Douglas, M.S.V. (2007b). From controversy to consensus: making the case for recent climatic change in the Arctic using lake sediments. *Frontiers in Ecology and the Environment* **5**, 466–474.

Smol, J.P. *et al.* (2005). Climate-driven regime shifts in the biological communities of Arctic lakes. *Proceedings of the National Academy of Sciences USA* **102**, 4397–4402.

Sorvari, S., Korhola, A., and Thompson R. (2002). Lake diatom response to recent Arctic warming in Finish Lapland. *Global Change Biology* **8**, 153–163.

Spaulding, S.A., McKnight, D.M., Smith, R.L., and Dufford, R. (1994). Phytoplankton population dynamics in perennially ice-covered Lake Fryxell, Antarctica. *Journal of Plankton Research* **16**, 527–541.

Stanley, D.W. and Daley, R.J. (1976). Environmental control of primary productivity in Alaskan tundra ponds. *Ecology* **57**, 1025–1033.

Stocker, Z.S.J. and Hynes, H.B.N. (1976). Studies on tributaries of Char Lake, Cornwallis Island, Canada. *Hydrobiologia* **49**, 97–102.

Tanabe, Y., Kudoh, S., Imura, S., and Fukuchi, M. (2008). Phytoplankton blooms under dim and cold conditions in freshwater lakes of East Antarctica. *Polar Biology* **31**, 199–208.

Tominaga, H. and Fukii, F. (1981). Saline lakes at Syowa Oasis, Antarctica. *Hydrobiología* **81–82**, 375–389.

Toro, M., Camacho A., Rochera C., *et al.* (2007). Limnological characteristics of the freshwater ecosystems of Byers Peninsula, Livingston Island, in maritime Antarctica. *Polar Biology* **30**, 635–649.

Vadeboncoeur, Y. *et al.* (2003). From Greenland to green lakes: cultural eutrophication and the loss of benthic pathways in lakes. *Limnology and Oceanography* **14**, 1408–1418.

Van Hove, P., Belzile, C., Gibson, J.A.E., and Vincent, W.F. (2006). Coupled landscape-lake evolution in the Canadian High Arctic. *Canadian Journal of Earth Sciences* **43**, 533–546.

Van Hove, P., Vincent, W.F., Galand, P.E., and Wilmotte, A. (2008). Abundance and diversity of picocyanobacteria in high arctic lakes and fjords. *Algological Studies* **126**, 209–227.

Vincent, A., Mueller, D.R., and Vincent, W.F. (2008). Simulated heat storage in a perennially ice-covered high Arctic lake: sensitivity to climate change. *Journal of Geophysical Research* **113**, C04036, doi:10.1029/2007JC004360.

Vincent, W.F. (1981). Production strategies in Antarctic inland waters: phytoplankton eco- physiology in a permanently ice-covered lake. *Ecology* **62**, 1215–1224.

Vincent, W.F. (1988). *Microbial Ecosystems of Antarctica.* Cambridge University Press, Cambridge.

Vincent, W.F., Downes, M.T., and Vincent, C.L. (1981). Nitrous oxide cycling in Lake Vanda, Antarctica. *Nature* **292**, 618–620.

Vincent, W.F. and Vincent, C.L. (1982). Nutritional state of the plankton in Antarctic coastal lakes and the inshore Ross Sea. *Polar Biology* **1**, 159–165.

Vincent, W.F. and Ellis-Evans, J.C. (eds) (1989). High latitude limnology. *Hydrobiologia* **172**.

Vincent, W.F. and Hobbie, J.E. (2000). Ecology of Arctic lakes and rivers. In Nuttall, M. and Callaghan, T.V. (eds), *The Arctic: Environment, People, Policies*, pp. 197–231. Harwood Academic Publishers, London.

Vincent, W.F., Rautio, M., and Pienitz, R. (2007). Climate control of underwater UV exposure in polar and alpine aquatic ecosystems. In Orbaek, J.B., Kallenborn, R., Tombre, I., *et al.* (eds), *Arctic Alpine Ecosystems and People in a Changing Environment*, pp. 227–249. Springer, Berlin.

Walter, K.M., Zimov, S.A., Chanton, J.P., Verbyla, D., and Chapin, F.S. (2006). Methane bubbling from Siberian thaw lakes as a positive feedback to climate warming. *Nature* **443**, 71–75.

Wand, U., Samarkin, V.A., Nitzsche, H.M., and Hubberten, H.W. (2006). Biogeochemistry of methane in the permanently ice-covered Lake Untersee, central Dronning Maud Land, East Antarctica. *Limnology and Oceanography*, **51**, 1180–1194.

West, J.J. and Plug, L.J. (2008). Time-dependent morphology of thaw lakes and taliks in deep and shallow ground ice. *Journal of Geophysical Research* **113**, F01009, doi:10.1029/2006JF000696.

Woo, M.K. (2000). McMaster River and Arctic hydrology. *Physical Geography* **21**, 466–484.

Woo, M.K. and Young, K.L. (2006). High Arctic wetlands: their occurrence, hydrological characteristics and sustainability. *Journal of Hydrology* **320**, 432–450.

Wrona, F.J., Prowse, T.D., and Reist, J.D. (eds) (2006). Climate change impacts on Arctic freshwater ecosystems and fisheries. *Ambio* **35**(7), 325–415.

Zhulidov, A.V., Headley, J.V., Robarts, R.D., Nikanorov, A.M., and Ischenko, A.A. (1997). *Atlas of Russian Wetlands: Biogeography and Heavy Metal Concentrations.* National Hydrology Research Institute, Environment Canada, Saskatoon.

Zhulidov, A.V., Robarts, R.D., Ischenko, A.A., and Pavlov, D.F. (2008). *The Great Siberian Rivers.* Springer, New York (in press).

Appendix 1.1

This section presents illustrative examples of research sites in the Arctic and in Antarctica (see Figures 1.1 and 1.2 for location maps).

A1.1 The Toolik Lake LTER

Toolik Lake and surrounding rivers (Plate 6) and lakes lie in tussock tundra at 68°N, 149°W in the northern foothills of the North Slope of Alaska. Studies began in 1975 and continue under the Long Term Ecological Research (LTER) program (data and bibliography at http://ecosystems.mbl.edu/ARC/). Investigations of Toolik Lake (25 m maximum depth) have included physics, chemistry, and numbers, productivity, and controls of phytoplankton, zooplankton, benthos, and fish, and cycling of carbon, phosphorus, and nitrogen. Whole lakes, large mesocosms, and streams have been treated by fertilization as well as by the introduction and exclusion of predators (O'Brien *et al.* 1997).

Toolik Lake is ice-bound for 9 months (early October until mid-June) and is ultra-oligotrophic, with an annual planktonic productivity of approximately 10 g C m^{-2}. The low rates of organic-matter sedimentation and unusually high concentrations of iron and manganese combine with high amounts of oxygen in the water column to cause strong adsorption of soluble nitrogen and phosphorus by the metal-rich sediments (Cornwell 1987). The algae, nearly all nanoflagellates, are fed upon by seven species of crustacean zooplankton (Hobbie *et al.* 1999). Allochthonous organic matter from land produces a lake DOC level of 6 mg C L^{-1} and Secchi disk depths of 6–7 m. The added DOC means that the microbial food web resembles that of temperate lakes (i.e. $1–2 \times 10^6$ bacterial cells ml^{-1}). Both the microbial food web and the food web beginning with phytoplanktonic algae are truncated as the zooplankton are so rare that their grazing does not control algae or flagellates. Fish in Toolik Lake are dependent upon benthic productivity; isotopes indicate that even the top predators, lake trout, obtain most of their carbon and energy from benthic invertebrates, such as snails and chironomid larvae.

A1.2 Thaw lakes and ponds at Barrow, Alaska

Permafrost thaw lakes (also called thermokarst lakes and ponds) are found throughout the circumpolar Arctic. A three-year integrated study of this ecosystem type took place during the IBP at Barrow, Alaska (71°N, 157°W), where the coastal plain is covered either by large lakes (2–3 m deep) and shallow ponds (≈50 cm deep) or wet sedge tundra. The IBP investigations (Hobbie 1980) included studies on the cycling of nitrogen and phosphorus and the changes in standing stock and productivity of phytoplankton, sediment algae, zooplankton, micro- and macrobenthic invertebrates, bacteria, and emergent sedges which cover one-third of each pond. Results included annual cycles of carbon, nitrogen, and phosphorus and a stochastic model of a whole pond ecosystem.

The Barrow ponds were solidly frozen from late September until mid-June. Mean summer temperatures were 7–8°C (maximum 16°C). Primary productivity was dominated by emergent sedges and grasses (96 g C m^{-2} year^{-1}), while benthic algae (8.4 g C m^{-2} year^{-1}), and phytoplankton (1.1 g C m^{-2} year^{-1}) contributed lesser amounts. The grazing food webs were unimportant (annual production of zooplankton was 0.2 g C m^{-2}) relative to the detritus food web of bacteria, chironomid larvae, and protozoans (annual production of bacteria was approximately 10–20 g C m^{-2}, of macrobenthos 1.6 g C m^{-2}, and of protozoans 0.3 g C m^{-2}).

The phytoplankton productivity of Barrow ponds was phosphorus-limited (Stanley and Daley 1976). Concentrations in the water were extremely low, 1–2 µg P L^{-1}, despite high amounts (25 g P m^{-2}) in the top 10 cm of sediments. The concentrations in the water were controlled by sorption on to a hydrous iron complex. Nitrogen was supplied mostly from the sediments as ammonium at a turnover rate of 1–2 days. The smallest life forms of the ponds did not seem to have any special adaptations to the Arctic. Concentrations and even species of bacteria and protists resembled those of temperate ponds. Metazoans were different in that many forms were excluded (e.g. fish, amphibia, sponges, many types of insects).

Permafrost thaw lakes are widely distributed throughout the tundra. They are formed by the thawing of permafrost and subsequent contraction and slumping of soils. In this way, they are particularly sensitive to present and future climate change. At some northern sites (e.g. Siberia; Smith *et al.* 2005) they are appear to be draining, drying up, and disappearing, whereas in some discontinuous permafrost areas they are expanding (Payette *et al.* 2004). Recent attention has focused especially on their abundant zooplankton populations, value as wildlife habitats, striking optical characteristics (see Plates 8 and 9), and biogeochemical properties, especially production of greenhouse gases (Walter *et al.* 2006).

A1.3 Canadian Arctic Archipelago

A diverse range of lake ecosystem types occur throughout the Canadian high Arctic, ranging from rock basin ponds at Cape Herschel (Smol and Douglas 2007a), to large deep lakes such as Lake Hazen (542 km^2; Plate 5) on Ellesmere Island, and Nettilling Lake (5542 km^2) and Amadjuak Lake (3115 km^2), both on Baffin Island. The most northerly lakes in this region resemble those in Antarctica, with perennial ice cover, simplified foods webs and polar desert catchments (e.g. Ward Hunt Lake, 83°05′N, 74°10′W; Plate 4). Meromictic lakes (saline, permanently stratified waters) are found at several sites, including Cornwallis Island, Little Cornwallis Island, and Ellesmere Island (Van Hove *et al.* 2006), and several of these have unusual thermal profiles that result from solar heating, as in some Antarctic lakes (Vincent *et al.* 2008; Figure 1.3).

Around 1970, Char Lake, a 27-m-deep (74°43′N, 94°59′W) lake on Cornwallis Island, was the site of a 3-year IBP comprehensive study (Schindler *et al.* 1974a; Rigler 1978), with comparative studies on nearby Meretta Lake that had become eutrophic as a result of sewage discharge into it (Schindler *et al.* 1974b). The average yearly air temperature was −16.4°C and summer temperatures averaged 2°C. Although Char Lake is usually ice-free for 2–3 months, the weather is often cloudy so the water temperature rarely exceeds 4°C. Because of the extreme conditions, the lake lies in a polar desert catchment with sparse vegetation, resulting in an unusually low loading of phosphorus. This, plus cold water temperatures and low light levels beneath the ice, results in planktonic production of approximately 4 g C m^{-2} year^{-1}, one of the lowest ever measured. However, benthic primary production of mosses and benthic algae is four-fold higher. One species of copepod dominates the planktonic community; the seven species of benthic Chironomidae account for half of the energy through the zoobenthos. Most animal biomass is found in the Arctic char that feed mainly on the chironomids. Rigler (1978, p. 139) concluded that 'There is little sign of Arctic adaptation in the classical sense. The species that live in Arctic lakes merely develop and respire more slowly than they would at higher temperatures'. However, this study was undertaken before the advent of molecular and other advanced microbiological techniques, and little is known about the microbial food web of Char Lake.

A1.4 Greenland lakes and ponds

The tremendous latitudinal extent of Greenland, from 60° to 83°N, includes a great variety of lakes, ponds, rivers, and streams (Poulsen 1940; Røen 1962; Jensen 2003). Very special features exist, such as saline lakes with old seawater in the bottom and hot and/or radioactive springs. In North Greenland, some lakes are permanently covered

with ice while the lakes in South Greenland have open water for 6 months of the year. There is a gradient of biodiversity and, as expected, the fauna and flora are much reduced in the north. For example, 21 species of vascular aquatic plants are found in southern Greenland and three in northern Greenland. Only 672 species of insects are present in Greenland while over 20 000 are found in Denmark. There is, furthermore, an east–west gradient as exemplified by the zooplankton diversity that decreases from south to north and from west to east. The largest number of freshwater entomostracans is found around Disko Island and at the southern west coast (around 45 species according to Røen 1962). A gradient study of primary productivity in Greenland, Danish, and US lakes showed that the Greenland lakes were all highly oligotrophic and that more than 80% of their total primary productivity took place on benthic surfaces (Vadeboncoeur *et al.* 2003). Arctic char are found throughout Greenland, sometimes with dwarf forms (3–8 cm), medium forms, and large forms (>30 cm) in the same lake (Riget *et al.* 2000). The large forms are often piscivorous on sticklebacks and young char. Medium-sized fish fed mainly on zooplankton while dwarf forms fed mainly on chironomid and trichopteran larvae.

Several areas have been sites of detailed limnological studies during the last few decades because there are field stations and associated infrastructure: Kangerlussuaq (Sønder Strømfjord) and Disko Island in West Greenland, and Pituffik (Thule Airbase) and Peary Land in North Greenland (Jensen 2003). The Danish BioBasis 50-year monitoring program in the Zackenberg Valley in northeastern Greenland (74°N) includes two shallow lakes (<6 m in depth), one with Arctic char. These lakes have been monitored for 10 years and it is evident that phytoplankton and zooplankton biomass is greatest in warm summers when there is deep thawing of the active layer of the soil and more nutrients enter the lakes (Christoffersen *et al.* 2008).

A1.5 Maritime Antarctic lakes

Islands to the north and along the western side of the Antarctic Peninsula experience a climate regime that is wet and relatively warm by comparison with continental Antarctica, and their limnology reflects these less severe conditions. Byers Peninsula (62.5°S, 61°W) on Livingston Island is an Antarctic Specially Protected Area under the Antarctic Treaty and is one of the limnologically richest areas of maritime Antarctica. This seasonally ice-free region contains lakes, ponds, streams, and wetlands. The lakes contain three crustacean species: *Boeckella poppei*, *Branchinecta gaini*, and the benthic cladoceran *Macrothrix ciliate.* The chironomids *Belgica antarctica* and *Parochlus steinenii,* and the oligochaete *Lumbricillus* sp., occur in the stream and lake zoobenthos. Cyanobacterial mats occur extensively, and epilithic diatoms and the aquatic moss *Drepanocladus longifolius* are also important phytobenthic components. The Antarctic Peninsula region is currently experiencing the most rapid warming trend in the Southern Hemisphere, and Byers Peninsula has been identified as a valuable long-term limnological reference site for monitoring environmental change (Toro *et al.* 2007).

Signy Island (60°43'S , 45°38'W) is part of the South Orkney Islands and also experiences the relatively warm, wet maritime Antarctic climate. It has a number of lakes that have been studied for many years by the British Antarctic Survey. The largest is Heywood Lake (area 4.3 ha, maximum depth 15 m), which has undergone eutrophication because its shores provide a wallow for an expanding population of fur seals. The lakes are cold monomictic (Figure 1.3) and contain the planktonic zooplankton species *Bo. poppei* and benthic crustacean species such as *Alona rectangular.* These waters are proving to be excellent study sites for molecular microbiology (e.g. Pearce 2003), and there is limnological evidence that they are responding to recent climate change (Quayle *et al.* 2002). Phycological and limnological studies have also been made at many other sites in the maritime Antarctic region, including King George Island and Cierva Point, an Antarctic Specially Protected Area on the Antarctic Peninsula (e.g. Izaguirre *et al.* 2001).

A1.6 McMurdo Dry Valleys LTER

First discovered by Captain Robert Falcon Scott in 1903, this is the largest ice-free region (about 4800 km²) of continental Antarctica (77°30'S, 162°E). It is best known for its deep lakes that are capped by thick perennial ice (Goldman *et al.* 1967; Green and Friedmann 1993; Priscu 1998). In the most extreme of these, Lake Vida (Victoria Valley), the ice extends almost entirely to the sediments; its 19 m of ice cover overlies a brine layer that is seven times the salinity of seawater with a liquid water temperature of –10°C. Most of the lakes are capped by 4–7 m of ice and are meromictic, with a surface layer of freshwater overlying saline deeper waters. These include Lake Fryxell, Lake Hoare, and Lake Bonney (Plate 4) in the Taylor Valley, Lake Miers in the Miers Valley, and Lake Vanda in the Wright Valley. The latter has a complex water column with thermohaline circulation cells and a deep thermal maximum above 20°C (Figure 1.3). The lakes contain highly stratified microbial communities, often with a deep population maximum of phytoplankton,

and a benthic community of thick microbial mats, but no crustacean zooplankton. The lakes also contain striking biogeochemical gradients and extreme concentrations at specific depths of certain gases and other intermediates in elemental nutrient cycles (see Chapter 8). Ephemeral streams are also common through the valleys (Plate 6), and are fed by alpine or piedmont glaciers. The largest of these, the Onyx River, flows 30 km inland, ultimately discharging into Lake Vanda (see Chapter 5 in this volume). Most of the streams contain pigmented microbial mats dominated by cyanobacteria, typically orange mats largely composed of oscillatorian taxa and black mats composed of *Nostoc commune* (details in Vincent 1988). The valleys are polar deserts that are largely devoid of vegetation, with dry, frozen soils that are several million years old. New Zealand and the USA have conducted research in the region from the 1957/1958 International Geophysical Year onwards, and in 1993 the Taylor Valley was selected as an NSF-funded long term ecological research site (LTER; data and bibliography are given at: www.mcmlter.org). In recognition of the environmental sensitivity of this region, the McMurdo Dry Valleys have been declared an Antarctic Specially Managed Area under the terms of the Antarctic Treaty System.

A1.7 Vestfold Hills

This lake-rich area lies in east Antarctica, at 68°30'S, 78°10'E (Ferris *et al.* 1988). The proximity of Davis Station permits year-round investigations, and consequently the lakes of the Vestfold Hills are among the few polar water bodies for which there are annual data-sets. Unlike most of the lakes of the McMurdo Dry Valleys, the lakes of the Vestfold Hills usually lose all or most of their ice cover for a short period in late summer. Saline lakes carry a thinner ice cover and the most saline, such as Deep Lake (about eight times the salinity of seawater), cool to extreme low temperatures in winter (Figure 1.3) but never develop an ice cover. The meromictic lakes (Gibson 1999) have well-oxygenated mixolimnia (upper waters), whereas the monimolimnia (lower waters that never mix) are anoxic. In contrast, the larger freshwater lakes are fully saturated with oxygen throughout their water columns at all times in the year. Compared with the lakes of the McMurdo Dry Valleys, the lakes of the Vestfold Hills are relatively young; the saline lakes are derived from relic seawater by evaporation, or where they are brackish by dilution. Sulphate reduction occurs in the meromictic lakes, as it does in those of the Dry Valleys. The high reducing capacity of sulphide serves to maintain anoxic conditions in the monimolimnia of these lakes. Large populations of photosynthetic sulphur bacteria and chemolithotrophic thiosulphate-oxidizing bacteria occur in the anoxic waters, and use sulphide or its oxidation product, thiosulphate, as an electron donor (Ferris *et al.* 1988). Methanogenesis occurs in these sulphate-depleted waters (Franzmann *et al.* 1991), with rates up to $2.5\,\mu mol\,kg^{-1}\,day^{-1}$. Ace Lake (Plate 5) is the most studied lake in the Vestfold Hills, largely because it is easily accessed both in summer and winter from Davis Station. It contains stratified microbial communities including flagellates, ciliates and high concentrations of picocyanobacteria (Powell *et al.* 2005). Unlike the McMurdo Dry Valley lakes, it also has crustacean zooplankton.

A1.8 Larsemann Hills

This oasis of ice-free land is an Antarctic Specially Managed Area that lies at 69°25'S and 76°10'E between the Sørsdal Glacier and the Amery Ice Shelf, about 80 km south of the Vestfold Hills. It has about 150 lakes and ponds, most of which are freshwater. They vary in size from Progress Lake (10 ha in area and 38 m deep) to small ponds of a few square metres in area and a depth around 1 m. Geomorphologically the lakes can be classified as supraglacial ponds, lakes and ponds in large glaciated rock basins, and ponds dammed by colluvium (loose sediment that accumulates at the bottom of a slope). Most of them are fed by meltwater from snow banks and a number of them have distinct inflow and outflow streams which flow for around 12 weeks each year in summer. Like the freshwater lakes of the Vestfold Hills, the Larsemann Hills lakes contain sparse phytoplankton populations that are likely to be phosphorus-limited. Most of these clear waters contain luxuriant benthic mats dominated by cyanobacteria.

A1.9 Bunger Hills

This site lies at 66°S, 100°E in Wilkes Land, adjacent to the Shackleton Ice Shelf to the north. It has an area of $950\,km^2$, making it one of the largest oases in Antarctica. It contains hundreds of lakes, both freshwater and saline, in valleys and rock depressions. The freshwater lakes (the largest being Figurnoye, area $14.3\,km^2$) are concentrated in the southern part of the oasis and at its periphery, while the saline lakes are located in the north and on the islands (Gibson and Andersen 2002). Most lakes in the centre of the Bunger Hills lose their ice-cover in summer, whereas those at its margins in contact with glaciers (epiglacial lakes) retain perennial ice caps. Geochemical and sedimentological studies have been conducted on White Smoke Lake, an epishelf lake in the region that is capped by 1.8–2.8 m of perennial ice (Doran *et al.* 2000).

A1.10 Schirmacher and Untersee Oases

The Schirmacher and Untersee Oases lie in Dronning Maud Land at 71°S, 11–13°E (Bormann and Fritsche 1995). The Russian Antarctic Station *Novolazarevskaya* and the Indian Station *Maitri* are located in this region. The largest lake, Lake Untersee, has an area of $11.4\,km^2$ and a maximum depth of 167 m. The lakes are fed by underwater melting of glaciers, and lose water by sublimation from their perennial ice surfaces. Within the Schirmacher Oasis there are over 150 lakes ranging in size from $2.2\,km^2$ (Lake Ozhidaniya), to small unnamed water bodies of less than $0.02\,km^2$ in area. These lakes are small and shallow compared with some of the lakes which occur in the McMurdo Dry Valleys and the Vestfold Hills. The geomorphological diversity of lakes in the Schirmacher Oasis is considerable. There is a supraglacial lake (Taloye, $0.24\,km^2$), which is around 5 m in depth. A number of relatively small epishelf lakes have formed on the northern edge of the oasis (Lake Prival'noye, $0.12\,km^2$; Lake Zigzag, $0.68\,km^2$; Lake Ozhidaniya, $2.2\,km^2$; and Lake Predgornoye, $0.18\,km^2$). Several lakes have formed in tectonically developed glaciated basins, for example Lake Sbrosovoye ($0.18\,km^2$) and Lake Dlinnoye ($0.14\,km^2$). Glacier-dammed and ice-wall-dammed lakes such as Lakes Iskristoye and Podprudnoye also occur as does one morainic lake (Lake 87). All of the lakes are freshwater and carry ice cover for most of the year. The majority become ice-free for a period in summer. The lakes have high transparency with several allowing light penetration to considerable depth; for example, Lakes Verkheneye and Untersee. Apart from several lakes that are subject to anthropogenic influences, the surface waters of the lakes of both oases are nutrient-poor. Total phosphorus levels are low, ranging between 4 and $6\,\mu g\,l^{-1}$ in surface lakewaters of the Schirmacher Oasis and less than $1\,\mu g\,l^{-1}$ in the upper water column of Lake Untersee. However, the bottom waters (>80 m depth) of Lake Untersee are anoxic, with extraordinarily high methane concentrations (around $22\,mmol\,l^{-1}$) that are among the highest observed in natural aquatic ecosystems (Wand *et al.* 2006). Lake Glubokoye receives waste water from the Soviet Station and now has elevated dissolved reactive phosphorus levels of $300\,\mu g\,l^{-1}$ in its deepest water (Kaup 2005).

A1.11 Syowa Oasis

This site lies on the Sôya Coast at 69°S, 39°30'E. Like the Vestfold Hills and Bunger Hills it carries freshwater, saline, and hypersaline lakes. A number of the saline lakes have been studied in some detail (Tominaga and Fukii 1981). Lake Nurume is a meromictic lake (maximum depth 16.6 m) with two distinct haloclines, which makes it similar to well-studied Ace Lake in the Vestfold Hills. The upper waters are brackish and overly waters with a salinity similar to seawater, which in turn overly more saline waters with a salinity equivalent to 1.5 times that of seawater. Summer water temperatures are close to 10°C at the surface, decreasing in the more saline lower waters to 8°C. A temperature maximum occurs at around 3–4 m (15°C). There are several hypersaline lakes (Lake Suribati at six times seawater and Lake Hunazoko at 5.5 times seawater), which are monomictic. Like the hypersaline lakes of the Vestfold Hills they are mixed in winter and develop a stratified water column during the summer. The freshwater lakes of this region are renowned for their luxuriant moss communities. These rise from the cyanobacterial mat-coated benthos at depths from 3 to 5 m and form 'moss pillars' about 40 cm in diameter and up to 60 cm high (Imura *et al.* 1999; see Plate 10). Continuous data-logging in two of the freshwater lakes has shown that phytoplankton biomass increases under the ice with the arrival of first light in spring, drops to low values during summer open water conditions, and may then show a second peak during the fall ice-up period (Tanabe *et al.* 2008).

A1.12 Subglacial aquatic ecosystems

Many liquid-water lakes are now known to occur beneath the Antarctic Ice Sheet (Plate 1; see Chapter 7 in this volume). The first of these to be discovered was Lake Vostok, a deep waterbody nearly the area of Lake Ontario (but more than three times deeper) lying beneath 3700 m of ice. A total of 145 lakes have now been identified, and there are likely to be more. There is evidence that many may be connected either by films of flowing water between the basement rock and ice sheet or by larger subglacial rivers and streams. Lake Vostok, Lake Ellsworth, and Lake Concordia are the main sites of current research activities. There is intense interest in how these environments may act as a lubricant for movement of the overlying ice sheet, as well as their outstanding value for climate records, as potential habitats for microbial growth and evolution, and as astrobiological models for the search for life on icy moons and planets. There is much concern about how they may be explored without contamination or other impacts, and a first set of guidelines has been prepared for their long-term stewardship (National Research Council 2007).

A1.13 Supraglacial aquatic ecosystems

The Arctic and Antarctica contain numerous regions where water lies on glaciers, ice sheets, and ice shelves.

These range from small cryoconite holes, first discovered on the Greenland ice cap and now compelling models for biogeochemical studies, to the extensive systems of meltponds on the Arctic and Antarctic ice shelves (thick landfast ice that floats on the sea) that are a habitat for rich microbial mat communities that are of great interest to astrobiologists (see Chapter 6 in this volume). The major sites of limnological study of ice shelves have been the Ward Hunt and Markham ice shelves in the Canadian High Arctic (Plate 2) and the McMurdo Ice Shelf, Antarctica (Plate 3).

A1.14 Arctic rivers and floodplains

The Arctic contains six large rivers (Pechora, Ob, Yenisei, Lena, and Kolyma rivers in Siberia; the Mackenzie River in Canada, Plate 7) that flow into the Arctic Ocean, with a total discharge that amounts to 11% of the total freshwater drainage into the World Ocean. These rivers flow across extensive fluvial deltas that are underlain by permafrost. The uneven thawing of these regions results in tens of thousands of lake basins that become filled with snow melt and with ice-jammed flood waters from the river (Figure 1.4; Plate 8). These floodplain lakes provide a rich habitat for aquatic plants, animals, and microbes, and play a key role in the biogeochemical processing of river water before its discharge to the Arctic Ocean (see Chapter 5 in this volume). Climate change is affecting the flow regime of many Arctic rivers (Petersen *et al.* 2006) and the connectivity between the flood-plain lakes and their river source waters (Lesack and Marsh 2007).

A1.15 High Arctic wetlands and flowing waters

The Arctic also contains many wetlands and lower-order flowing-water ecosystems. The inflows to Char Lake were examined during the IBP program on Cornwallis Island in the Canadian high Arctic, and were found to contain a low-diversity benthos dominated by cold-water chironomids, notably *Diamesa*; insect groups that are commonly found in temperate-latitude streams such as *Simuliidae* and *Ephemeroptera*, were conspicuously absent from the fauna (Stocker and Hynes 1976). Hydrological studies on McMaster River, also on Cornwallis, showed that infiltration and storage of meltwater into the frozen ground was limited compared with total melt. Much of the meltwater ran down the slopes, and streamflow was high during the melt season, followed by low summer flows influenced by active layer processes (Woo 2000). This has implications for impact of climate warming on high Arctic wetlands, which are dependent upon sustained water input throughout summer. Most of the snow accumulated over 9–10 months is released as melt-water over 2–3 weeks and floods the wetlands while they are still frozen. Patchy wetlands occur where local water sources, especially late-lying snow banks, river flooding, or groundwater inflow into topographic depressions, permit a high water table. The development of vegetation and accumulation of peat at these locations helps preserve the wetland by providing insulation over the permafrost and a porous medium for water infiltration and storage (Woo and Young 2006).

CHAPTER 2

Origin and geomorphology of lakes in the polar regions

Reinhard Pienitz, Peter T. Doran, and Scott F. Lamoureux

Outline

A characteristic and often dominant feature of many polar landscapes is the great diversity and abundance of their standing surface waters. The focus in this chapter is on lakes and ponds. Antarctica has little surface water by comparison with the Arctic, but many saline lakes exist in the ice-free oasis areas, and freshwater ponds and lakes are abundant in the maritime and peripheral Antarctic regions. The aim of this chapter is to provide a brief introduction to the different origins, distinguishing features, and landscape controls that result in the extraordinary diversity of lakes and ponds that exist in both polar regions. The main emphasis is on the description of the geological and geomorphological processes involved in the formation and modification (natural change) of these high-latitude lake ecosystems. Throughout this review, we have drawn on examples from both the Arctic and Antarctic to emphasize the differences and similarities that exist between north and south polar lake ecosystems.

2.1 Introduction

In both polar regions, permanently frozen soils (permafrost) exert a strong influence on catchment properties such as hydrological processes and geochemical interactions (Lamoureux and Gilbert 2004). Similarly, in both regions snow and ice cover are major controls on the structure and functioning of aquatic ecosystems. In this review, we delimit the northern polar zone, or Arctic, using the southern limit of continuous permafrost, whereas the extent of the southern polar zone, or Antarctica, is defined here by the northern limit of pack ice.

High levels of solar radiation reaching high latitudes in spring result in rapid snowmelt. Runoff in late spring typically comprises 80–90% of the yearly total in the Arctic and lasts only a few (2–3) weeks or even less (Marsh 1990). This pulse is not as prevalent in the Antarctic, as snow cover is less, and much of the snow sublimates or recharges soil moisture rather than running off into streams.

Infiltration of the Arctic pulse of water is limited initially by frozen soil and later by permafrost, perennially frozen ground that stays at or below 0°C for at least two consecutive summers (Woo and Gregor 1992). Permafrost may reach depths of 600–1000 m in the coldest areas of the Arctic, becoming discontinuous and patchy (sporadic) in Sub-Arctic regions (Stonehouse 1989). Permafrost is also prevalent in the Antarctic, including the Dry Valleys of the Trans-Antarctic Mountains in south Victoria Land. Cartwright *et al.* (1974) and Decker and Bucher (1977) inferred regional thicknesses of permafrost approaching 1000 m based on local temperatures. However, permafrost is absent below much of the East Antarctic Ice Sheet due to the pressure of the overlying ice which initiates basal melting (Bockheim and Tarnocai 1998). In the McMurdo Dry Valleys of Antarctica, most of the water feeding lakes comes from melting glaciers and follows similar stream paths year after year. Permafrost in the Dry Valleys is relatively thin, presumably because of the presence of the lakes themselves. McGinnis *et al.* (1973) and Cartwright *et al.* (1974) found areas of completely unfrozen sediment beneath thermally stratified saline lakes such as Lake Vanda or Don Juan Pond because of the freezing-point depression due to abundant salts. As permafrost seals the subsoil, spring melt-water may flow over land and enter rivers, or accumulate into the many wetlands, ponds, and lakes characteristic of low-relief tundra environments. Summer sources of water include late or perennial snow patches, glaciers, rain, melting of the upper (active) layer of permafrost, as well as some cases of groundwater discharge (Rydén 1981; Van Everdingen 1990). Groundwater levels and distribution within polar regions are, in general, greatly influenced by bedrock geology, soil thickness, and permafrost layers. Permafrost can control the amount of physical space in the ground available to groundwater as well as its movement within drainage systems.

In the northern hemisphere, ice-sheet advance during the Last Glacial Maximum (LGM) produced a lake-rich postglacial landscape. Using a geographic information system (GIS)-based statistical comparison of the locations of about 200 000 lakes in the northern hemisphere (sized 0.1–50 km², northwards of latitude 45°N) with data from global databases on topography, permafrost, peatlands, and LGM glaciation, Smith *et al.* (2007) revealed the apparent importance of LGM glaciation history and the presence of permafrost as determinants of lake abundance and distribution at high latitudes. According to their study, lake density (i.e. the number of lakes per 1000 km²) in formerly glaciated terrain is more than four times that of non-glaciated terrain, with lake densities and area fractions being on average approximately 300–350% greater in glaciated (compared with unglaciated) terrain, and approximately 100–170% greater in permafrost-influenced (compared with permafrost-free) terrain. The presence of peatlands is associated with an additional approximately 40–80% increase in lake density, and generally large Arctic lakes are most abundant in formerly glaciated, permafrost peatlands (≈14.4 lakes/1000 km²) and least abundant in unglaciated, permafrost-free terrain (≈1.2 lakes/1000 km²).

Lithology and geological setting also exert a strong influence on polar lakes. In many cases, faults and fractures control the location of lakes and the characteristics of basins and drainage networks. Many lakes are located in glacially over-deepened, tectonically controlled sites, whereas others are located in craters or in tectonic troughs. Lakes most strongly influenced by fractures are generally of more elongate shape and usually aligned parallel to each other, for example on Byers Peninsula, Livingston Island, Antarctica (López-Martinez *et al.* 1996). Igneous rocks often show abundant '*roches moutonnées*' (elongate rock hills) and the development of 'rock bars' and 'riegels' that may act as dams for lakes. Differences in geological substrates can also affect the extent of rock weathering and the chemical composition of soil water that ultimately discharges into lakes. There are many significant differences in lake and pond characteristics in various parts of the Arctic, which partly reflect gradients of geology, climate, and vegetation, as well as local impacts and other important factors.

2.2 Lake origins

2.2.1 Wetlands

Wetlands and saturated soils are characteristic features of the Arctic because moisture received from rain and snowmelt is retained in the active layer above the permafrost barrier. Additionally, recent glaciated landscapes often have poorly developed drainage patterns that result in frequent ponded water. Due to the generally higher levels of precipitation at lower latitudes, wetlands are more common in the Low Arctic than in the High Arctic. The size of Arctic wetlands ranges from small strips to very extensive plains such as the West Siberian Lowland and the Hudson Bay Lowlands. Wetlands may co-exist with tundra ponds or lakes of various sizes, since both wetlands and lakes occupy areas with abundant storage. Wetlands tend to occur locally in high concentrations in the form of lowland polygon bogs and fens, peat-mounted bogs, snowpatch fens, tundra pools, or as flood-plain marshes and swamps. More detailed descriptions of these and other Arctic wetland types are provided in Woo

(2000). In polar desert regions there is less snow for meltwater to accumulate; however, ponds and lakes are still present due to the surface ponding of water by the permafrost. In some Arctic regions, winter snow is plentiful and ponds and shallow lakes occupy large areas, often forming networks of static waterways. On ice-scoured uplands and coastal flats the landscape may actually be dominated by ponds and lakes. Coastal plains often feature long, shallow lakes that are separated and aligned by raised beaches and occupy more than 90% of the terrain (Figure 2.1). Although a variety of definitions exist for distinguishing lakes from ponds, we define a pond here as a freshwater basin that is sufficiently shallow so that it freezes totally to the bottom in winter, whereas a lake always has liquid water in its basin.

2.2.2 Ice-dependent lakes

Lakes form also on poorly drained ice, next to glaciers and ice caps, on ice sheets, and between ice sheets and ice-filled moraines (Figure 2.2). Some are long-lived, fed by annual influxes of meltwater, while others are temporary systems that are flooded during the summer melt and liable to sudden drainage through rifts in the ice and other causes.

Figure 2.1 Landsat 7 image of thermokarst thaw lakes located on the northern coastal plain near Barrow, Alaska. The image was acquired on 30 August 2000. Source: Kenneth Hinkel and Benjamin Jones.

Ice-dammed lakes are more common in the Arctic than elsewhere and in Greenland all the larger lakes are of this type. Glaciers or ice dams have formed some of the lakes in Iceland (e.g. Grænalón), and such lakes are occasionally emptied beneath the damming ice, sometimes resulting in *jökulhlaups* (Icelandic for 'glacier-burst' or catastrophic drainage of lakes). Ice-dammed systems develop next to glacier fronts in mountainous terrain or besides ice shelves (epiglacial) and in depressions on glaciers, ice sheets, and ice shelves (supraglacial; e.g. Ward Hunt Ice Shelf in the Canadian High Arctic, McMurdo Ice Shelf in Antarctica; see Vincent 1988 and Plates 2 and 3). Epiglacial lakes are common in Antarctica given that, in most cases, the source of water that sustains the lakes is from glaciers. They can persist for many years with frequent changes in water level and morphology resulting from glacial movements and changing meltwater inputs (Hodgson *et al.* 2004), and are also prone to periodic draining (Mackay and Løken 1974; Smith *et al.* 2005). Supraglacial lakes are often ephemeral, forming during the summer melt. These systems have been proposed as potential refugia for microbiota in the controversial Snowball Earth hypothesis when the Earth underwent extreme freeze–thaw events between approximately 750 million and 580 million years ago (Vincent *et al.* 2000).

Another class of ice-dependent water bodies is 'epishelf lakes'. These are a form of proglacial lakes in which floating ice shelves dam freshwater runoff. Epishelf lakes are hydrologically connected to the ocean and therefore are tidally forced. Some epishelf lakes have saline bottom waters while some (e.g. Doran *et al.* 2000) are fresh throughout. In stratified epishelf lakes, inflows of meltwater are impounded by glacial ice, and the density difference between fresh and salt water prevents the mixing of the runoff with the marine water below. Many examples of epishelf lakes are known from Antarctica, including Moutonnée and Ablation lakes on the coast of Alexander Island, in the Schirmacher Oasis and Bunger Hills (reviewed in Hodgson *et al.* 2004). Beaver Lake located in the Radok Lake area and associated with the Amery Ice Shelf (Prince Charles Mountains, Antarctica) is especially interesting, because it is probably the world's largest epishelf lake (McKelvey and

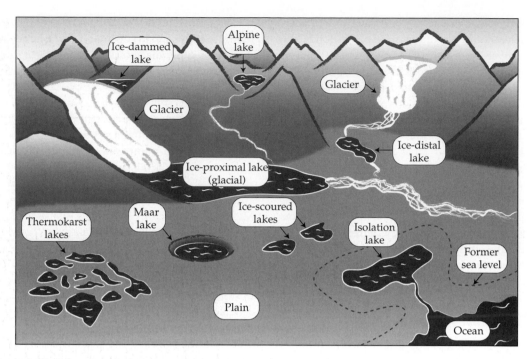

Figure 2.2 Some of the diverse lake types that can be found in the polar regions.

Stephenson 1990; Laybourn-Parry *et al.* 2001). Epishelf lakes are becoming increasingly rare in the Arctic, as one of the few existing systems in Disraeli Fiord was recently destroyed by the break-up of the Ward Hunt Ice Shelf (Mueller *et al.* 2003), and another in Ayles Fiord by the collapse of the Ayles Ice Shelf. In these cases, the freshwater layers escaped and drained into the Arctic Ocean. Only one epishelf lake has been confirmed to still exist in the Arctic, in Milne Fiord. Epishelf lakes can be thought of as windows in a proglacial estuary, where the position of the window along the estuary gradient and the depth of the lake dictate lake chemistry. A review of the formation and dynamics of Antarctic lake ecosystems (including epishelf, epiglacial, subglacial, and supraglacial lakes) and their paleolimnology is provided by Hodgson *et al.* (2004).

2.2.3 Postglacial lakes

Glacial erosion and deposition by former continental and local ice masses have left behind a vast number of lakes in the polar and temperate regions of the world. The irregularity of residual landscapes caused by glacial processes has resulted in a diverse range of lakes, many of which still exist due to the relatively young age of these landscapes (Benn and Evans 1998). In polar regions that have not been subjected to late Pleistocene glaciation, most lakes are of non-glacial origin and are described in other sections of this chapter.

Differential substrate erosion by ice is caused by large-scale patterns of ice flow, as well as by localized flow controlled by topographic relief and the resistance of substrates to erosion. Hence, in areas that have been subject to pronounced ice flow, scouring and erosion of bedrock and surficial sediments show broad-scale linear alignment or radial patterns. This is particularly true where basal ice is at melting temperature and meltwater is available to further increase the entrainment and erosion of subglacial sediments (Benn and Evans 1998). Subject to pre-existing topography, persistent ice flow will differentially erode surfaces and generate quasi-linear troughs, as well as more irregular scour

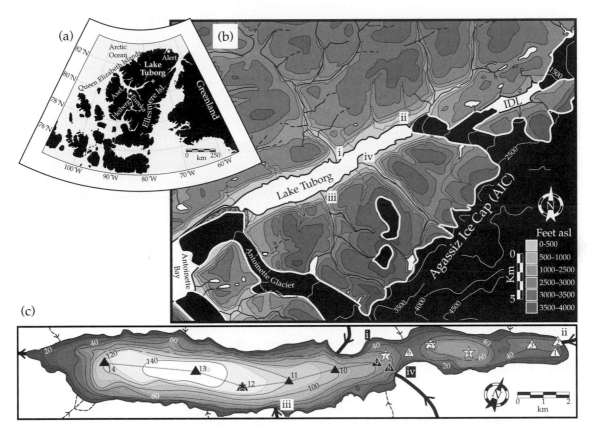

Figure 2.3 Bathymetry of glacially eroded, fiord-like Lake Tuborg on Ellesmere Island, Nunavut, Canada. Solid bathymetric contours are shown at every 20 m. asl, above sea level. From Lewis *et al.* (2007).

depressions. In mountainous terrain, glacial erosion is further constrained by pre-existing topography and focuses erosion in the valley bottom and walls, producing the U-shaped cross-section typical of glacial valleys. As erosion by ice is not constrained longitudinally to a fixed downstream base level, as is the case with fluvial processes, localized scour and erosion generate closed depressions that subsequently form lakes, as exemplified by Lake Tuborg on Ellesmere Island in the Canadian High Arctic (80°59′N, 75°33′W; Figure 2.3). Reduced erosion along glacier margins, together with the frequent deposition of morainal material, further generate sedimentary dams and form lakes behind former ice margins. In this regard, lakes formed by glacial scour that fill valleys are the terrestrial equivalents of fiords that are common features in many formerly glaciated coastal areas.

In regions with resistant bedrock, the differential mechanical strength of the available lithology may result in localized and selective erosion of less-resistant units. Although this may further reinforce pre-existing topography, the result can also be a complex array of subglacial depressions that form lake basins. The resultant landscape can be covered with a large number of lakes that have diverse shapes and morphometries. Excellent examples of these types of lakes are found on the large Precambrian shields of northern Canada and Scandinavia (Figure 2.4).

Glacial processes and the resultant glaciogenic sediments provide a wide range of settings for the ultimate formation of depressions and lakes, mainly through the emplacement of low-permeability tills and kettling. Thick glaciofluvial deposits along the margin of ice can generate sedimentary dams

Figure 2.5 Kame lakes at the edge of Ahklun Mountains, southwestern Alaska, USA. Photograph: S. Lamoureux.

Figure 2.4 Canadian Shield lakes on the Boothia Peninsula (Nunavut) near the northern coast of Canada. Photograph: S. Lamoureux.

on tributary valleys or cross pre-existing drainage patterns on low-relief landscapes. The results are frequent settings for the ultimate formation of lake basins. Similarly, sediments can accumulate around stagnant ice margins and blocks. Referred to as kame deposits, the subsequent ice melt generates depressions that can become lakes with irregular morphology (Figure 2.5).

The proglacial environment is highly conducive to the formation of lakes through high rates of sediment transport and deposition. Proglacial river systems may rapidly form during ice advance, maximum stand, or retreat. Sediments deposited at the mouths of valleys may block pre-existing drainage and form lakes, in much the same way that a landslide or other mass movement may block drainage. In some instances, sediment transported from a tributary valley may form a fan or delta into another lake, effectively separating the former lake into two. In the case of coastal environments, progradation of sediments may isolate marine embayments and result in the ultimate formation of a freshwater or meromictic lake (see Section 2.2.5 below).

Finally, the impact of glaciation on landscapes typically lasts long after ice melts. After ice ablates, the barren landscape is especially susceptible to erosion. The wide variety of glacial sediment deposits are typically reworked extensively during the postglacial period, resulting in increased erosion, sediment yield, and drainage changes. Rivers exhibit increased sediment loads that can fill shallow lake basins and form new fans and deltas into larger water bodies and damming of downstream valleys. The result is relatively rapid alteration of initial lakes and the formation of new lakes. These processes in the deglaciated environment were defined as paraglacial by Church and Ryder (1972) and have been recognized subsequently as important in postglacial landscape evolution in many formerly glaciated regions. In some cases, evidence for continued paraglacial influences has been documented in the Canadian Cordillera more than 10 000 years after deglaciation (Church and Slaymaker 1989).

2.2.4 Thermokarst lakes and ponds

Thermokarst ponds and lakes (also called thaw lakes) are common features of Arctic landscapes, developing in depressions that result from the thawing of permafrost (Figure 2.6; Plates 8 and 9). Most of these water bodies are shallow (<1 m depth), and they are widespread in lowland areas of western and northern Alaska, Canada, and Siberia. For example, thaw lakes comprise approximately 90% of the lakes in the Russian permafrost zone (Walter *et al.* 2006) and are widespread across much of the northern slope of Alaska (Hinkel *et al.* 2005) and the coastal lowlands of Canada (Côté and Burn 2002).

Figure 2.6 Forest-tundra landscape with abundant thermokarst lakes and ponds near treeline in northern Québec. Photograph: I. Laurion (see also Plates 8 and 9).

The process of thermokarst formation is triggered by the degradation of ice wedges, followed by the subsidence of the surface and the presence of ice beneath the sunken area, which leads to the formation of ponds. The pond then accelerates thermokarst formation because the heat released from its water body during winter promotes the formation of a talik (an unfrozen zone in permafrost that is located under a lake or deep pond that results from heat release from the overlying water column, thereby preventing formation of ground frost during winter) below its basin. Eventually, the reduction of permafrost can lead to complete drainage of the pond and subsequent dry-up. Depending on atmospheric conditions, palsas may form after refreezing of the surface and permafrost may form again. Residual depressions and cross-cutting relationships of thaw lakes suggest that they are relatively short-lived features and are sensitive to underlying bedrock, surficial sediment, and climate conditions (Hinkel *et al.* 2005), including predominant wind directions (Côté and Burn 2002).

Studies into the dynamics and evolution of these widespread thaw lakes in the Northern Hemisphere receive great attention as they contribute to a better understanding of their role in the global atmospheric methane budget, a potent greenhouse gas with highest concentrations between latitudes 65°

and 70°N (Hinkel *et al.* 2003; IPCC 2007). Walter *et al.* (2006) linked a 58% increase in lake methane emissions during recent decades to the expansion of thaw lakes in northern Siberia, demonstrating the importance of this feedback to climate warming (Smith *et al.* 2005).

2.2.5 Coastal uplift systems

Glacioisostatic rebound or uplift is the process by which the Earth's crust, once depressed by the weight of overlying ice, begins to rise following the retreat of continental ice sheets. In many polar regions, this rebound is still occurring today, although at a much slower rate than immediately following glacial retreat. In areas such as eastern Hudson Bay (Canada), for example, the landscape continues to rise at a rate of about 1 m per century (Allard and Tremblay 1983). New lakes are thus still being formed as depressed, scoured land emerges from the sea (Fulford-Smith and Sikes 1996; Lamoureux 1999; Saulnier-Talbot *et al.* 2003). Hence, lakes closer to present-day sea level are often younger than those located at higher elevations further inland. For example, coastal Nicolay Lake in the Canadian High Arctic has effectively been a lake for only approximately 500 years. Glacioisostatic submergence of the land to over 100 m or more below current sea level effectively moved coastlines inland by tens of kilometers for much of the Holocene. As a result, the Nicolay Lake basin remained marine until the late Holocene, and freshening and fluvial processes that deliver sediment and related terrestrial materials were captured by upstream lake basins until approximately 500 years before present (Lamoureux 1999). Recent research has demonstrated that this type of lake evolution may follow a number of divergent paths, and results in a range of lake types ranging from fully fresh to hypersaline (Van Hove *et al.* 2006). In many instances, glacioisostatic rebound results in the formation of meromictic saline lakes that become stratified for part of the year (monomictic) or are permanently stratified (meromictic). Sea water that is trapped in the lake during the isolation process is frequently preserved in a dense saline hypolimnion and the resultant stable water column may last for thousands of years, or indefinitely. Numerous examples of coastal meromictic lakes have been documented in the polar regions and are of particular interest due to their unique limnology and sedimentary records (reviewed in Pienitz *et al.* 2004). Long-term preservation of meromixis appears to be dependent on a number of factors, including lake morphometry, ice cover, and the relative loading of freshwater to the lake from the surrounding catchment. Meromictic lakes have a surface freshwater layer that mixes with the wind (mixolimnion), but permanently ice-covered Antarctic lakes remain stratified throughout their water column due to a combination of salt and the protection offered by the ice cover, and therefore the term meromictic may not properly apply.

2.2.6 Meteoritic impact crater lakes

Approximately 174 meteoritic impact craters have been identified so far on the Earth's surface (Earth Impact Database, Planetary and Space Science Centre: www.unb.ca/passc/ImpactDatabase/index.html), and include the Arctic sites El'Gygytgyn (67°50′N, 172°E) and Popigai (71°30′N, 111°E), both in northern Russia, and Lac à l'Eau Claire (Clearwater Lake) in northern Québec (Plate 9). A striking example of a meteoritic impact crater lake in North America is Pingualuk Crater Lake. This crater is located at 61°N, 74°W in the northernmost part of the Ungava Peninsula in northern Québec (Nunavik, Canada). The crater was formed from the impact of a meteorite that entered Earth's atmosphere approximately 1.4 million years ago, as determined by Ar/Ar dating of the impactites collected at the site (Grieve *et al.* 1989). Pingualuk is situated close to the area where the inland ice masses reached maximum thickness during the last (Wisconsinan) glaciation. The crater is believed to have formed a subglacial lake basin under a dome of the Laurentide Ice Sheet with its waters remaining liquid below the ice due to the pressure of the overlying ice cap and to geothermal heating from below, thereby representing conditions similar to those found in modern subglacial systems in Antarctica (e.g. Lake Vostok; see Section 2.3.3). The crater is a circular depression about 400 m deep and 3.4 km in diameter, hosting a lake presently 267 m

deep with no surface outflow (Figure 2.7). The deep sediment infill of the Pingualuk Crater Lake, which escaped scouring and erosion from ice flow (Figure 2.8), promises to yield an uninterrupted, approximately 1.4-million-year Arctic paleoclimate record of several interglacial–glacial cycles. These unique long sediment records may offer unique insights into long-term climatic and environmental change in high-latitude regions, thereby having the potential to significantly enhance our understanding of past climate dynamics in the Arctic (Briner *et al.* 2007; see also Chapter 3 in this volume).

Similarly, Lake El'Gygytgyn in northeastern Siberia, an impact crater created approximately 3.6 million years ago, is the only site with a continuous terrestrial Arctic climate record that covers several interglacial periods (Nowaczyk *et al.* 2002; Brigham-Grette *et al.* 2007) as its lake basin was created in the center of what was to become Beringia, the largest contiguous landscape in the Arctic to have escaped northern-hemisphere glaciation. An overview of papers detailing the remarkable paleolimnological record of the last 250 000 years from El'Gygytgyn Crater Lake is available in Brigham-Grette *et al.* (2007).

2.2.7 Volcanic lakes

One important type of volcanic lake are maar lakes (Figure 2.2), which are shallow, broad, low-rimmed craters formed during powerful explosive eruptions involving magma–water (phreatomagmatic) interaction. The accumulation of material ejected from craters contributes to the formation of their surrounding rims. A few maar lake systems exist in explosion craters on Iceland (Sigurdsson and Sparks 1978), whereas they are quite widespread in Alaska. Zagoskin Lake and all other major lakes on St. Michael Island (Alaska) are maar crater lakes (Ager 2003; Muhs *et al.* 2003). Ranging from 4 to 8 km in diameter, the four Espenberg Maars (Whitefish Maar, Devil Mountain Maar, and North and South Killeak Maars) on the northern Seward Peninsula just south of the Arctic Circle in northwestern Alaska have been described as the largest known maar craters on Earth (Beget *et al.* 1996). They were formed by a series of Pleistocene basaltic eruptions through thick (approximately 100 m) permafrost,

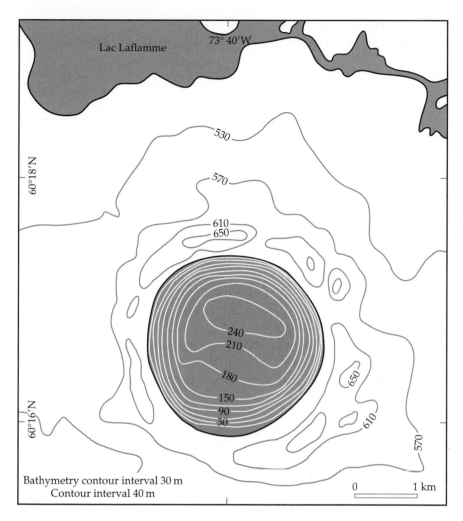

Figure 2.7 Location of Pingualuk Crater Lake within Pingualuit impact structure, Parc National des Pingualuit, Nunavik (northern Québec, Canada). The topographic and bathymetric contour intervals shown are 40 and 30 m, respectively. Pingualuk is a closed lake basin, without any inlet or outlet.

excavated as much as 300 m into older lithologies. There are also volcanic crater lakes (not maars) in the Ingakslugwat Hills in the south-central Yukon Delta (Ager 1982). Other crater lakes (some of which may be maars) are located on Nunivak and St. Lawrence Islands in southwestern Alaska, whereas some sizable lakes in calderas can be found in southwestern Alaska (e.g. Aniakchak crater on the Alaska Peninsula, approximately 3600 years before present). Craters differ from calderas both in size and origin. Craters are much smaller features than calderas and are typically defined as being less

than 1 km in diameter. Although both craters and calderas are most often associated with explosive eruptions, craters are typically formed by the explosive ejection of material in and surrounding the upper part of the conduit, rather than by collapse. Steep-walled pit craters, in contrast, often found on shield volcanoes, are more passive features formed when magma drains from a fissure, leaving overlying lava flows unsupported. Multiple explosive eruptions can form overlapping or nested craters, and adjacent craters may reflect localized areas of eruption along fissures, as seen in rows of small

(a)

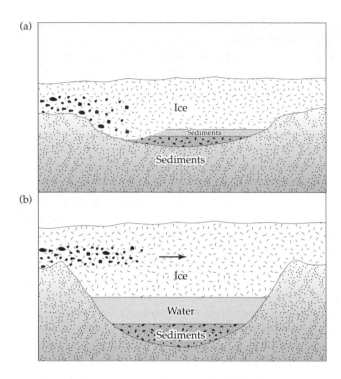

(b)

Figure 2.8 (a) Glacial erosion of bottom sediment deposits in shallow lake basins within formerly glaciated Arctic regions. (b) The situation at Lake Pingualuk (northern Québec, Canada), where lake sediments have escaped glacial erosion due to the extremely deep and steep-walled crater basin, perhaps analogous to Antarctic subglacial lakes (see Chapter 7).

crater (maar) lakes in Iceland that extend for tens of kilometers.

2.2.8 Karst systems

Lakes of karstic origin are relatively rare in high latitudes due to the scarcity of limestone and other rocks suitable for karst development in many locations, in addition to the pervasive permafrost conditions that restrict the infiltration of water into the vadose (unsaturated soil) zone and the absence of effective groundwater systems. The discovery of mineral springs in polar settings (Andersen *et al.* 2002) and groundwater drainage in permafrost locations like the Ungava Peninsula, northern Québec (Lauriol and Gray 1990) suggest that the role of groundwater in permafrost regions may be more extensive than commonly thought and that these regions hold the potential for karst systems. This potential has been realized to some extent by studies carried out on Svalbard (Norway) where karst systems in Paleozoic carbonate and evaporate bedrock maintain open taliks and several karst lakes have been produced (Salvigsen and

Elgersma 1985). Further evidence of karst systems that occurred during hypsithermal (early to mid-Holocene) conditions in the northern Yukon (Clark *et al.* 2004) suggest that karst processes may have contributed to lake formation in the past, although no systems have been specifically recognized to date.

2.2.9 Tectonic lakes

Generally rare in the polar regions, tectonic lake basins tend to be old systems compared to the relatively young age of most high-latitude lakes. Although the origins of Icelandic lakes are diverse, a large number are of tectonic origin, including the two largest Icelandic lakes, Thingvallavatn (84 km²) and Thorisvatn (70 km²). Subglacial systems such as Lake Vostok in Antarctica may also be considered tectonic lakes (Studinger *et al.* 2003; see below).

2.2.10 Lakes of other origins

The flood-plains bordering large Arctic rivers are characterized by meandering or anastomotic streams, oxbow lakes and bayous, marshes, or stagnant pools (Plate 8). Flood-plain lakes that have been cut off from river channels are seasonally flooded with river water (usually in the spring when the river overflows its banks) and can be found in great numbers within the deltas of the Mackenzie and Lena rivers in northwestern Canada and Siberia, respectively. Small, crescent-shaped lakes called oxbow lakes can form in river valleys as the result of meandering. The slow-moving river forms a sinuous shape as the outer side of bends are eroded away more rapidly than the inner side. Eventually a horseshoe bend is formed and the river cuts through the narrow neck. This new passage then forms the main passage for the river and the ends of the bend become silted up, thus forming a bow-shaped lake.

The well-known Icelandic Lake Mývatn (37 km²) is a basin formed in a collapsed lava flow. Other Icelandic lakes have been formed by rock slides or glacial deposits. Lagoon lakes are common on the sandy shores, Hóp (45 km²) being the largest of this type and the fourth largest lake in Iceland.

There are also artificial or human-made lakes or reservoirs within the Arctic that have been created along Arctic rivers for the purpose of hydroelectric power generation. In Canada and Russia, some of these reservoirs are quite large, for example the La Grande complex in northern Québec (9900 km² of flooded surface area) and the Krasnoyarskoye More ('Krasnoyarsk Sea') reservoir on the Yenisey River in Siberia (2000 km²).

2.3 Geographical regions

2.3.1 The circumpolar Arctic

Repeated glaciations during the **Pleistocene** epoch resulted in glaciers and vast inland ice sheets covering much of the eastern and central North American and western Eurasian Arctic regions. These large ice masses carved out the land as they advanced over it, gouging out topsoil and broken rock. The many structural depressions left behind filled with water once climate warming caused glaciers to retreat, forming lakes of glacial origin. A characteristic feature of the Arctic, therefore, is the large numbers of lakes and ponds. In the

Province of Murmansk in Russia, there are more than 100 000 lakes, the largest of which is Lake Imandra with an area of 812 km² and a maximum depth of 67 m (NEFCO 1995). Iceland, Sweden, and Finland have approximately 2.7, 5.2, and 5.8% of their territory occupied by lakes, respectively, whereas only about 0.5% of Alaska is covered by fresh water (CAFF 1994). In the Canadian Arctic north of latitude 60°N, approximately 18–20% of the landscape is covered by surface waters (Prowse and Ommanney 1990). The Canadian Arctic contains more than 28 large lakes in excess of 600 km² in surface area, including Great Bear (31 326 km²) and Great Slave (28 568 km²) lakes located on the mainland (Mackay and Løken 1974). In particular, there are innumerable small lakes in the Arctic (also called ponds). Although a variety of definitions exist for distinguishing lakes from ponds, the latter are generally defined as sites that are sufficiently shallow for their water column to freeze entirely to the bottom in winter, whereas lakes always preserve a lens of liquid water within their basin (see above). Due to the extreme harshness of High-Arctic winter climates, ice thicknesses of between 4 and 5 m are possible, and thus by this definition sites shallower than about 5 m could be classified as ponds. However, most Arctic ponds are much shallower, and are often less than 2 m in depth. Despite their shallow depths, these ponds often dominate Arctic landscapes and are important habitats for waterfowl and other animals.

2.3.2 Coastal Antarctic lakes

The Antarctic ice cap is a dome of ice exceeding 4 km thickness in places, with thinning towards the continental edges. In these ice-marginal areas, some ice-free areas occur that have been variably named oases, dry areas, and dry valleys. Ice-free areas in Antarctica only comprise 0.35% of the continent (Fox and Cooper 1994). Much of this dry land is in the form of nunataks formed by the Trans-Antarctic Mountains. In this review, only the coastal dry areas are of relevance, as most nunataks are at elevations that are too high for the occurrence of ice melt and lake formation.

The majority of Antarctica's lakes are found in coastal oases such as the McMurdo Dry Valleys,

Vestfold Hills, Larseman Hills, Bunger Hills, Schirmacher Oasis, and Syowa Oasis (Plates 4 and 5; see Chapter 1 of this volume for descriptions). The Antarctic Peninsula is also home to many ice-free areas, some containing lakes (Hodgson *et al.* 2004). Many coastal Antarctic lakes have been formed as a direct result of isostatic rebound after postglacial retreat of ice (e.g. Zwartz *et al.* 1998; Hodgson *et al.* 2001; Verleyen *et al.* 2004). Coastal lakes can evolve into either closed (no outflow) or open basins (with outflow), and both types are common in Antarctic oases, as well as on the Antarctic Peninsula and the maritime islands (Hodgson *et al.* 2004). Trapped sea water can be flushed out with fresh water (e.g. Pickard *et al.* 1986), but closed basin lakes typically become saline due to the evaporative concentration of salts over time. Antarctic saline lakes are among the saltiest water bodies on Earth, with lakes in the Vestfold Hills and the McMurdo Dry Valleys having salinities ranging from 4 to 235 g l^{-1} (Spigel and Priscu 1998; Gibson 1999). Lake Vida, which today occupies the center of Victoria Valley, is the largest modern lake in the Dry Valleys and was previously thought to be frozen to its bed (Calkin and Bull 1967; Chinn 1993). However, Doran *et al.* (2003) provided evidence for an extensive brine body beneath its 19-m-thick ice cover sealed from the atmosphere for at least 2800 ^{14}C years. The NaCl brine remains below −10°C year-round and its salinity exceeds seven times that of sea water, thereby representing a unique new type of lake: an ice-sealed lake.

Lake basins formed in scoured rock by glacial erosion and ice-cap retreat are common in the maritime Antarctic islands such as Signy Island, King George Island, and Livingston Island. Another type of Antarctic lake forms where former marine embayments or fiords are dammed by advancing ice shelves and isolated from the sea. In these lakes, the marine water has been replaced over a period of time by glacial meltwaters. Where the ice shelf forms a complete seal, these lakes will be similar to the saline lakes discussed above. Where an incomplete seal is formed and a hydrological connection to the sea persists under the ice shelf, an epishelf lake is formed. Lake Untersee in East Antarctica is one of the largest epiglacial lakes on the continent (Wand and Perlt 1999). White Smoke

Lake in the Bunger Hills, almost 50% of which is bordered by glacier ice and also an epishelf lake, has maintained its rough dimensions for at least 3000 years (Doran *et al.* 2000).

Coastal lakes of lagoon origin are very common in the lower beaches of maritime Antarctica. They have gently sloping edges and are either subcircular or elongate in plan (Jones *et al.* 1993; Cuchí *et al.* 2004), and their salinities tend to be high because of intermittent exchange with sea waters.

2.3.3 Antarctic and Arctic subglacial lakes

Under certain circumstances, liquid water masses can be maintained beneath glaciers and ice caps, such as the Antarctic ice sheet. Subglacial water forms under thick ice sheets because geothermal heat is trapped by the insulating effect of the ice. The pressure of the overlying ice also depresses the freezing point by a few degrees. Water then pools in subglacial depressions to form lakes. Lakes are detectable mainly through aerial radar surveys which have now detected at least 145 individual lakes beneath the Antarctic ice sheet (Siegert *et al.* 2005). New evidence (Wingham *et al.* 2006) in Antarctica suggests that these lakes are interconnected, creating vast subglacial drainage systems. Recent studies show that several subglacial lake systems may have existed beneath portions of the Laurentide Ice Sheet in North America near the last glacial (Wisconsinan) maximum, such as subglacial Lake McGregor in south-central Alberta (Munro-Stasiuk 2003) and possibly Lake Pingualuk in northern Québec (see Section 2.2.6).

The largest and most studied subglacial lake is Lake Vostok, which lies beneath the Russian Vostok Station, in Antarctica (Plate 1). Lake Vostok has been suggested to be in a basin created by tectonic thrusts (Studinger *et al.* 2003), but this interpretation remains controversial because it and the majority of Antarctic subglacial lakes are developed on basement rocks that became part of the stable craton in the Precambrian and there have been few opportunities for more recent faulting to create basins in which subglacial lakes can form. Whether there was a preglacial Lake Vostok is also controversial. For a lake to survive from preglacial times into the present, a fairly specialized set of circumstances would have to occur, but models for this have been put forth (Pattyn *et al.* 2004) and arguments made against the concept (Siegert 2004). Without direct evidence, all we can say about the history of formation of subglacial lakes is that they began forming when ice achieved sufficient thickness to cause the glacial bed to be above the pressure melting point. A full discussion of subglacial lakes can be found in Chapter 7.

2.4 Effects of landscape evolution and climate change on polar lakes

Landscapes continue to evolve, and aquatic ecosystems including lakes and ponds are transient features of the polar regions that can experience rapid changes, from initial formation until the eventual basin filling by abiotic and biotic sediments. Climate change has the potential to affect lake evolution through a variety of processes, especially in the polar regions where even small changes in temperature can have profound impacts on landscape properties such as snowpack, permafrost, glacial melt, and hydrological inputs, as well as soil and vegetation stability. These complex controls are exemplified by the Holocene development of small lakes near Toolik Lake on the north slope of the Brooks Range, Alaska (Hobbie 1980, 1984; O'Brien *et al.* 1997). Glacial deposits with different composition and permeability resulted in the establishment of different tundra vegetation communities. Gradual evolution of the postglacial landscape, coupled with vegetation succession and Holocene climate changes, all combined to generate substantially different lake-watershed systems (Oswald *et al.* 2003).

The climate-induced dynamics of alpine glaciers and ice sheets exert a strong control on landscape evolution and geomorphological processes that determine the abundance, distribution, and form of lake basins. Climate change can modify erosional patterns and landscape morphology in the catchments of lakes and ponds, for example through changes in evaporation/precipitation patterns, water-induced erosion and the thawing of permafrost. In turn, these changes will affect the transport of sediments from the surrounding terrestrial landscape to lakes and therefore the extent and rate of infilling. The glacioisostatic readjustment of formerly glaciated land masses also has the potential to radically alter the influence of marine waters on coastal lakes and lagoons at high latitudes (see above).

Ice-bound lakes are especially sensitive to small variations in climate. In the Canadian High Arctic, for example, the extensive ice shelves that once dammed the northern fiords of Ellesmere Island (e.g. Ayles, Markham, M'Clintock, Ward Hunt) experienced considerable contraction, fracturing, and ultimately break-up during the twentieth century (Vincent *et al.* 2001). This has resulted in the erosion and loss of many ice-dammed epishelf lakes (see above), and most recently the freshwater layer of Disraeli Fiord was completely drained away as a result of the break-up of the Ward Hunt Ice Shelf in 2002 (Mueller *et al.* 2003). Likewise, the catastrophic drainage events or *jökulhlaups* of ice-dammed lakes are likely to become more extensive and frequent in the polar regions (Lewis *et al.* 2007).

Recent so-called change-detection studies show that climate warming is already having profound effects on thermokarst landscapes in the northern hemisphere, with lake expansion occurring, especially in continuous permafrost regions (Osterkamp *et al.* 2000; Christensen *et al.* 2004; Payette *et al.* 2004; Smith *et al.* 2005), but shrinkage or disappearance in regions with discontinuous, sporadic, and isolated permafrost (Yoshikawa and Hinzman 2003; Riordan *et al.* 2006). In the former case, thawing permafrost triggers thermokarsting and associated lake growth, whereas in the latter scenario it enhances water infiltration to the subsurface and underlying groundwater systems through taliks. Thermokarst lakes are thus very fragile systems (sentinels) that are affected by climate change and associated permafrost dynamics in high-latitude regions. Because of this, research into the response of these freshwater bodies to accelerated permafrost thaw and their contribution to greenhouse gas emissions (e.g. release of methane and carbon dioxide to the atmosphere via ebullition; enhanced methanogenesis in sediments through higher lake productivity) is at the forefront of recent and ongoing scientific efforts in the circumpolar Arctic (e.g. Zimov *et al.* 2001; Christensen *et al.* 2004; Walter *et al.* 2006).

2.5 Conclusions

A full understanding of lake formation and distribution throughout the polar regions requires consideration of many factors including geology, topographic relief, glacial history, climate and periglacial processes, substrate permeability, permafrost properties, peatland distribution, groundwater movement, and talik depths, to name just a few. Formerly glaciated lowland environments possess the greatest abundance of lakes in the northern hemisphere, whereas widespread permafrost can result in the persistence of lakes by presenting a barrier to water infiltration to the subsurface. In the Arctic, the spatial and temporal distribution of Arctic lake systems is changing rapidly due to amplified climate warming and permafrost thaw, with anticipated profound impacts on the hydro-ecological processes and the quality and availability of fresh waters in these landscapes (ACIA 2005; IPCC 2007; Plate 16). A key issue in the context of these changes is whether lakes and ponds that often dominate Arctic landscapes will act as sources or sinks of greenhouse gases. With the exception of the Peninsula region, the Antarctic continent has been less impacted by global warming as yet (Doran *et al.* 2002; Thompson and Solomon 2002), but models show (Shindell and Schmidt 2004) that this should change in the near future and the Antarctic lakes and surrounding regions will undergo the same change that the Arctic lakes are experiencing.

Acknowledgments

Our research has been facilitated through financial support provided by grants from the Natural Sciences and Engineering Research Council (NSERC, Canada), the National Science Foundation (USA) grants OPP9211773, OPP9813061, OPP9810219, and OPP0096250, as well as logistic support from the Polar Continental Shelf Project (PCSP, Canada) and Centre d'Études Nordiques (CEN). We thank Serge Duchesneau and Karine Tessier for their assistance with drawing Figures 2.7 and 2.8, and Dr J.P. Smol, Dr D. Antoniades, Dr J. Laybourn-Parry and Dr W.F. Vincent for their comments on the manuscript.

References

ACIA (2005). *Arctic Climate Impact Assessment.* Cambridge University Press, Cambridge. www.acia.uaf.edu.

Ager, T.A. (1982). Vegetational history of western Alaska during the Wisconsin glacial interval and the Holocene. In Hopkins, D.M., Matthews, Jr, J.V., Schweger, C.E., and Young, S.B. (eds), *Paleoecology of Beringia*, pp. 75–93. Academic Press, New York.

Ager, T.A. (2003). Late Quaternary vegetation and climate history of the central Bering land bridge from St. Michael Island, western Alaska. *Quaternary Research* **60**, 19–32.

Allard, M. and Tremblay, G. (1983). La dynamique littorale des îles Manitounuk durant l'Holocène. *Zeitschrift für Geomorphologie* (suppl.) **47**, 61–95.

Andersen, D.T., Pollard, W.H., McKay, C.P., and Heldmann, J. (2002). Cold springs in permafrost on Earth and Mars. *Journal of Geophysical Research* **107E**, doi:10.1029/2000JE001436.

Beget, J.E., Hopkins, D.M., and Charron, S.D. (1996). The largest known maars on Earth, Seward Peninsula, northwest Alaska. *Arctic* **49**, 62–69.

Benn, D.I. and Evans DJA (1998). *Glaciers and Glaciation.* Arnold Publishers, London.

Bockheim, J.G., and Tarnocai, C. (1998). Nature, occurrence and origin of dry permafrost. In *Proceedings 7th International Conference on Permafrost*, 23–27 June 1998, Yellowknife, NWT, Canada, pp. 57–63.

Brigham-Grette, J., Melles, M., and Minyuk, P. (2007). Overview and significance of a 250 ka paleoclimate record from El'gygytgyn Crater Lake, NE Russia. *Journal of Paleolimnology* **37**, 1–16.

Briner, J.P., Axford, Y., Forman, S.L., Miller, G.H., and Wolfe, A.P. (2007). Multiple generations of interglacial lake sediment preserved beneath the Laurentide Ice Sheet. *Geology* **35**, 887–890.

CAFF (1994). *The State of Protected Areas in the Circumpolar Arctic 1994.* Habitat Conservation Report No. 1. Conservation of Arctic Flora and Fauna, Directorate for Nature Management, Trondheim.

Calkin, P.E. and Bull, C. (1967). Lake Vida, Victoria Valley, Antarctica. *Journal of Glaciology* **6**, 833–836.

Cartwright, K., Harris, H., and Heidari, M. (1974). Hydrogeological studies in the dry valleys. *Antarctic Journal of the United States* **9**, 131–133.

Chinn, T.J. (1993). Physical hydrology of the dry valley lakes. In Green, W.J. and Friedmann, E.I. (eds), *Physical and Biogeochemical Processes in Antarctic Lakes.* American Geophysical Union, Washington DC, pp. 1–51.

Christensen, T.R. *et al.* (2004). Thawing sub-arctic permafrost: effects on vegetation and methane emissions. *Geophysical Research Letters* **31**, L04501, doi: 10.1029/2003GL018680.

Church, M. and Ryder, J.M. (1972). Paraglacial sedimentation, a consideration of fluvial processes conditioned by glaciation. *Geological Society of America Bulletin* **83**, 3059–3072.

Church, M. and Slaymaker, O. (1989). Disequilibrium of Holocene sediment yield in glaciated British Columbia. *Nature* **337**, 452–454.

Clark, I.D. *et al.* (2004). Endostromatolites from permafrost karst, Yukon, Canada: paleoclimatic proxies for the Holocene hypsithermal. *Canadian Journal of Earth Sciences* **41**, 387–399.

Côté, M.M. and Burn, C.R. (2002). The oriented lakes of Tuktoyaktuk Peninsula, Western Arctic Coast, Canada: a GIS-based analysis. *Permafrost and Periglacial Processes* **13**, 61–70.

Cuchí, J.A., Durán, J.J., Alfaro, P., Serrano, E., and López-Martínez, J. (2004). Discriminación mediante parámetros fisicoquímicos in situ de diferentes tipos de agua presentes en un área con permafrost (Península Byers, Isla Livingston, Antártida Occidental). *Boletín de la Real Sociedad Española de Historia Natural (seccion. Geología.)* **99**, 75–82.

Decker, E.R. and Bucher, G.J. (1977). Geothermal studies in Antarctica. *Antarctic Journal of the United States* **11**, 102–104.

Doran, P.T., Wharton, R.A., Lyons, W.B., Des Marais, D.J., and Andersen, D.T. (2000). Sedimentology and geochemistry of a perennially ice-covered epishelf lake in Bunger Hills Oasis, East Antarctica. *Antarctic Science* **12**, 131–140.

Doran, P.T. *et al.* (2002). Valley floor climate observations from the McMurdo Dry Valleys, Antarctica, 1986–2000. *Journal of Geophysical Research* **107 (D24)**, 4772, doi:doi:10.1029/2001JD002045.

Doran, P.T., Fritsen, C.H., McKay, C.P., Priscu, J.C., and Adams, E.E. (2003). Formation and character of an ancient 19-m ice cover and underlying trapped brine in an "ice-sealed" east Antarctic lake. *Proceedings of the National Academy of Sciences of the USA* **100**, 26–31.

Fox, A.J. and Cooper, A.P.R. (1994). Measured properties of the Antarctic ice sheet derived from the Antarctic SCAR digital database. *Polar Record* **30**, 201–204.

Fulford-Smith, S.P. and Sikes, E.L. (1996). The evolution of Ace lake, Antarctica, determined from sedimentary diatom assemblages. *Palaeogeography, Palaeoclimatology, Palaeoecology* **124**, 73–86.

Gibson, J.A.E. (1999). The meromictic lakes and stratified marine basins of the Vestfold Hills, East Antarctica. *Antarctic Science* **11**, 175–192.

Grieve, R.A.F., Robertson, P.B., Bouchard, M.A., and Alexopoulos, J.S. (1989). Origin and age of the Cratère du Nouveau-Québec. In Bouchard, M.A. (ed.), *L'Histoire Naturelle du Cratère du Nouveau-Québec.* Collection Environnement et Géologie vol. 7, pp. 59–71. Université de Montréal, Montréal.

Hinkel, K.M., Eisner, W.R., Bockheim, J.G., *et al.* (2003). Spatial extent, age, and carbon stocks in drained thaw lake basins on the Barrow Peninsula, Alaska. *Arctic Antarctic and Alpine Research* **35**, 291–300.

Hinkel, K.M., Frohn, R.C., Nelson, F.E., Eisner, W.R., and Beck, R.A. (2005). Morphometric and spatial analysis of thaw lakes and drained thaw lake basins in the Western Arctic Coastal Plain, Alaska. *Permafrost and Periglacial Processes* **16**, 327–341.

Hobbie, J.E. (1980). *Limnology of Tundra Ponds, Barrow, Alaska.* Dowden, Hutchinson and Ross, Stroudsburg, Pennsylvania.

Hobbie, J.E. (1984). Polar limnology. In F.B. Taub (ed.) *Lakes and Rivers. Ecosystems of the World*, pp.63–104. Elsevier, Amsterdam.

Hodgson, D.A., Vyverman, W., and Sabbe, K. (2001). Limnology and biology of saline lakes in the Rauer Islands, eastern Antarctica. *Antarctic Science* **13**, 255–270.

Hodgson, D., Gibson, J., and Doran, P.T. (2004). Antarctic Paleolimnology. In Pienitz, R., Douglas, M.S.V., and Smol, J.P. (eds), *Long-Term Environmental Change in Arctic and Antarctic Lakes*, pp. 419–474. Springer, Dordrecht.

IPCC (2007). *Intergovernmental Panel on Climate Change, Fourth Assessment Report.* Cambridge University Press, Cambridge.

Jones, V.J., Juggins, S., and Ellis-Evans, J.C. (1993). The relationship between water chemistry and surface sediment diatom assemblages in maritime Antarctic lakes, *Antarctic Science* **5**, 229–348.

Lamoureux, S.F. (1999). Catchment and lake controls over the formation of varves in monomictic Nicolay Lake, Cornwall Island, Nunavut, Canadian Arctic. *Journal of Earth Sciences* **36**, 1533–1546.

Lamoureux, S.F. and Gilbert, R. (2004). Physical and chemical properties and proxies of high latitude lake sediments. In Pienitz, R., Douglas, M.S.V., and Smol, J.P. (eds), *Long-Term Environmental Change in Arctic and Antarctic Lakes*, pp. 53–87. Springer, Dordrecht.

Lauriol, B. and Gray, J.T. (1990). Karst drainage in permafrost environments: the case of Akpatok Island, NWT Canada. *Permafrost and Periglacial Processes* **1**, 129–144.

Laybourn Parry, J., Quayle, W.C., Henshaw, T., Ruddell, A., and Marchant, H.J. (2001). Life on the edge: the plankton and chemistry of Beaver Lake, an ultra-oligotrophic epishelf lake, Antarctica. *Freshwater Biology* **46**, 1205–1217.

Lewis, T., Francus, P., and Bradley, R.S. (2007). Limnology, sedimentology, and hydrology of a jökulhlaup into a meromictic High Arctic lake. *Canadian Journal of Earth Sciences* **44**, 1–15.

López-Martínez, J., Serrano, E., and Martínez de Pisón, E. (1996). Geomorphological features of the drainage system. In *Supplementary text of the Geomorphological Map of Byers Peninsula, BAS Geomap Series*, 5-A, pp. 15–19. British Antarctic Survey, Cambridge.

Mackay, D.K. and Løken, O.H. (1974). Arctic hydrology. In Ives, J.D. and Barry, R.G. (eds), *Arctic and Alpine Environments*, pp. 111–132. Methuen and Co., London.

Marsh, P. (1990). Snow hydrology. In Prowse, T.D. and Ommanney, C.S.L. (eds), *Northern Hydrology: Canadian Perspectives*. National Hydrology Research Institute, Science Report No. 1, pp. 37–61. Environment Canada, Saskatoon.

McGinnis, L.D., Nakao, K., and Clark, C.C. (1973). *Geophysical Identification of Frozen and Unfrozen Ground, Antarctica*. In North American Contribution to the Second International Permafrost Conference, 13–28 July 1973, pp. 136–146. National Academy of Sciences, Washington DC.

McKelvey, B.C. and Stephenson NCN (1990). A geological reconnaissance of the Radok Lake area, Amery Oasis, Prince Charles Mountains. *Antarctic Science* **2**, 53–66.

Mueller, D.R., Vincent, W.F., and Jeffries, M.O. (2003). Break-up of the largest Arctic ice shelf and associated loss of an epishelf lake. *Geophysical Research Letters* **30**, 2031, doi:10.1029/2003GL017931.

Muhs, D.R., Ager, T.A., Been, J., Bradbury, J.P., and Dean, W.E. (2003). A late Quaternary record of eolian silt deposition in a maar lake, St. Michael Island, western Alaska. *Quaternary Research* **60**, 110–122.

Munro-Stasiuk, M.J. (2003). Subglacial Lake McGregor, south-central Alberta, Canada. *Sedimentary Geology* **160**, 325–350.

NEFCO (1995). *Proposals for Environmentally Sound Investment Projects in the Russian Part of the Barents Region, Vol. 1 Non-radioactive Contamination*. AMAP Report. AMAP Secretariat, Oslo.

Nowaczyk, N.R. *et al.* (2002). Magnetostratigraphic results from impact crater Lake El'gygytgyn, north-eastern Siberia: a 300 kyr long high-resolution terrestrial palaeoclimatic record from the Arctic. *Geophysical Journal International* **150**, 109–126.

O'Brien, W.J. *et al.* (1997). The limnology of Toolik Lake. In Milner, A.M. and Oswood, M.W. (eds.), *Freshwaters of Alaska: Ecological Synthesis*, pp. 61–106. Springer, New York.

Osterkamp, T.E. *et al.* (2000). Observations of thermokarst and its impact on boreal forests in Alaska, USA. *Arctic, Antarctic and Alpine Research* **32**, 303–315.

Oswald, W.W., Brubaker, L.B., Hu, F.S., and Kling, G.W. (2003). Holocene pollen records from the central Arctic Foothills, northern Alaska: testing the role of substrate in the response of tundra to climate change. *Journal of Ecology* **91**, 1034–1048.

Pattyn, F., De Smedt, B., and Souchez, R. (2004). Influence of subglacial Vostok lake on the regional ice dynamics of the Antarctic ice sheet: a model study. *Journal of Glaciology* **50**, 583–589.

Payette, S., Delwaide, A., Caccianiga, M., and Beauchemin, M. (2004). Accelerated thawing of subarctic peatland permafrost over the last 50 years. *Geophysical Research Letters* **31**, L18208.

Pickard, J., Adamson, D.A., and Heath, C.W. (1986). The evolution of Watts Lake, Vestfold Hills, East Antarctica, from marine inlet to freshwater lake. *Palaeogeography, Palaeoclimatology, Palaeoecology* **53**, 271–288.

Pienitz, R., Douglas, M.S.V., and Smol, J.P. (eds) (2004). *Long-term Environmental Change in Arctic and Antarctic Lakes*. DPER vol. 8. Springer, Berlin.

Prowse, T.D. and Ommanney, C.S.L. (eds) (1990). *Northern Hydrology: Canadian Perspectives*. National Hydrology Research Institute, Science Report No. 1. Environment Canada, Saskatoon.

Riordan, B., Verbyla, D., and McGuire, A.D. (2006). Shrinking ponds in subarctic Alaska based on 1950–2002 remotely sensed images. *Journal of Geophysical Research – Biogeosciences* **111**, G04002, doi: 10.1029/2005JG000150.

Rydén, B.E. (1981). Hydrology of northern tundra. In Bliss, L.C., Heal, O.W., and Moore, J.J. (eds), *Tundra Ecosystems: a Comparative Analysis*, pp. 115–137. Cambridge University Press, Cambridge.

Salvigsen, O. and Elgersma, A. (1985). Large-scale karst features and open taliks at Vardeborgsletta, outer Isfjorden, Svalbard. *Polar Research* **3**, 145–153.

Saulnier-Talbot, É., Pienitz, R., and Vincent, W.F. (2003). Holocene lake succession and palaeo-optics of a subarctic lake, northern Québec, Canada. *The Holocene* **13**, 517–526.

Shindell, D.T. and Schmidt, G.A. (2004). Southern Hemisphere climate response to ozone changes and greenhouse gas increases. *Geophysical Research Letters*, **31**, L18209, doi:10.1029/2004GL020724.

Siegert, M.J. (2004). Comment on "A numerical model for an alternative origin of Lake Vostok and its exobiological implications for Mars" by N.S. Duxbury, I.A. Zotikov, K.H. Nealson, V.E. Romanovsky, and F. D. Carsey. *Journal of Geophysical Research-Planets* **109**, E02007, doi:10.1029/2003JE002176.

Siegert, M.J., Carter, S., Tabacco, I., Popov, S., and Blankenship, D.D. (2005). A revised inventory of Antarctic subglacial lakes. *Antarctic Science* **17**, 453–460.

Sigurdsson, H. and Sparks, S. (1978). Rifting episode in North Iceland in 1874–1875 and the eruptions of Askja and Sveinagja. *Bulletin of Volcanology* **41**, 149–167.

Smith, L.C., Sheng, Y., MacDonald, G.M., and Hinzman, L.D. (2005). Disappearing Arctic lakes. *Science* **308**, 1429–1429.

Smith, L.C., Sheng, Y., and MacDonald, G.M. (2007). A first pan-Arctic assessment of the influence of glaciation, permafrost, topography and peatlands on northern hemisphere lake distribution. *Permafrost and Periglacial Processes* **18**, 201–208.

Spigel, R.H. and Priscu, J.C. (1998). Physical limnology of the McMurdo Dry Valley lakes. In Priscu, J.C. (ed.), *Ecosystem Dynamics in a Polar Desert: The McMurdo Dry Valleys, Antarctica*, pp. 153–187. American Geophysical Union, Washington DC.

Stonehouse, B. (ed.) (1989). *Polar Ecology*. Chapman & Hall, New York.

Studinger, M. *et al.* (2003). Geophysical models for tectonic framework of the Lake Vostok region, East Antarctica. *Earth and Planetary Science Letters* **216**, 663–677.

Thompson, D.W.J. and Solomon, S. (2002). Interpretation of recent Southern Hemisphere climate change. *Science* **296**, 895–899.

Van Everdingen, R.O. (1990). Groundwater hydrology. In Prowse, T.D. and Ommanney, C.S.L. (eds), *Northern Hydrology: Canadian Perspectives*. National Hydrology Research Institute, Science Report No. 1, pp. 77–101. Environment Canada, Saskatoon.

Van Hove, P., Belzile, C., Gibson, J.A.E., and Vincent, W.F. (2006). Coupled landscape-lake evolution in the Canadian High Arctic. *Canadian Journal of Earth Sciences* **43**, 533–546.

Verleyen, E., Hodgson, D.A., Sabbe, K., and Vyverman, W. (2004). Late Quaternary deglaciation and climate history of the Larsemann Hills (East Antarctica). *Journal of Quaternary Science* **19**, 361–375.

Vincent, W.F. (ed.) (1988). *Microbial Ecosystems of Antarctica*. Cambridge University Press, Cambridge.

Vincent, W.F. *et al.* (2000). Ice shelf microbial ecosystems in the high Arctic and implications for life on snowball earth. *Naturwissenschaften* **87**, 137–141.

Vincent, W.F., Gibson, J.A.E., and Jeffries, M.O. (2001). Ice shelf collapse, climate change and habitat loss in the Canadian High Arctic. *Polar Record* **37**, 133–142.

Walter, K.M., Zimov, S.A., Chanton, J.P., Verbyla, D., and Chapin, III, F.S. (2006). Methane bubbling from Siberian thaw lakes as a positive feedback to climate warming. *Nature* **443**, 71–75.

Wand, U. and Perlt, J. (1999). Glacial boulders 'floating' on the ice cover of Lake Untersee, East Antarctica. *Antarctic Science* **11**, 256–260.

Wingham, D.J., Siegert, M.J., Shepherd, A., and Muir, A.S. (2006). Rapid discharge connects Antarctic subglacial lakes. *Nature* **440**, 1033–1036.

Woo, M.-K. (2000). Permafrost and hydrology. In Nuttall, M. and Callaghan, T.V. (eds), *The Arctic. Environment, People, Policy*, pp. 57–96. Harwood Academic Publishers, Amsterdam.

Woo, M.-K. and Gregor, D.J. (1992). *Arctic Environment: Past, Present and Future*. Department of Geography, McMaster University, Hamilton.

Yoshikawa, K. and Hinzman, L.D. (2003). Shrinking thermokarst ponds and groundwater dynamics in discontinuous permafrost near Council, Alaska. *Permafrost and Periglacial Processes* **14**, 151–160.

Zimov, S.A. *et al.* (2001). Flux of methane from north Siberian aquatic systems: influence on atmospheric methane. In Paepe, R., Melnikov, V., Van Overloop, E., and Gorokhov, V.D. (eds), *Permafrost Response on Economic Development, Environmental Security and Natural Resources*, pp. 511–524. Kluwer Academic Publishers, Dordrecht.

Zwartz, D., Bird, M., Stone, J., and Lambeck, K. (1998). Holocene sea-level change and ice-sheet history in the Vestfold Hills, East Antarctica. *Earth and Planetary Science Letters* **155**, 131–145.

CHAPTER 3

High-latitude paleolimnology

Dominic A. Hodgson and John P. Smol

Outline

This chapter provides an overview of high-latitude paleolimnology. First, we describe the role that paleolimnological studies have played in reconstructing the geomorphological origin and development of high-latitude lakes and the establishment and succession of their biota. Second, we describe how both organic and inorganic components incorporated into lake sediments record changes both within lakes and in the surrounding environment. We illustrate this using examples of studies that have tracked past changes in climate, hydrology, vegetation, sea level, human impacts on fish and wildlife populations, ultraviolet radiation, and atmospheric and terrestrial pollutants. Third, we describe some synthesis studies that have combined paleolimnological data from multiple lakes across the Arctic and Antarctic to identify the magnitude and direction of environmental changes at regional to continental scales. Finally, we discuss some future prospects in high-latitude paleolimnology. The geographic scope of the chapter includes the Arctic north of the tree line (tundra or polar desert catchments), the Antarctic continent, and the Antarctic Peninsula region, with occasional reference to the warmer sub-Arctic and sub-Antarctic regions.

3.1 Introduction

Paleolimnology is concerned with the origin and development of lakes (Cohen 2003; Smol 2008). A critical first step in any paleolimnological study is to determine a lake's origin from geomorphological evidence in the landscape; for example, is it of glacial, marine, volcanic, or tectonic origin? The next steps are to examine the limnology of the lake, and to interpret the information archived in its sediments. These sediments incorporate the products of glacial erosion together with biological and chemical remains from the catchment, water column, and benthos. Laid down year on year, the composition of these sediments typically responds rapidly to changes in the environment such as temperature, precipitation, evaporation, light environment, ice cover, glacial extent, and human impacts (Figure 3.1; Plate 14). Thus lake sediments can archive both the nature and direction of changes within the lake, and also changes in

the surrounding environment. Paleolimnological studies are therefore providing invaluable data on the history of high-latitude lakes with which it is possible to address important questions in limnology, ecology, biogeography, Earth system science, climate, and global environmental change (Pienitz *et al.* 2004).

3.1.1 Why study paleolimnology at high latitudes?

Chapter 1 has described how the high latitudes contain some of the fastest-warming regions on Earth, and will be strongly affected by warming trends in the next decades and centuries (see Plate 16). This is a result of positive feedbacks between albedo loss, caused by reductions in snow and ice extent, and the consequent warming of the atmosphere and ocean. In the northwest High Arctic, temperatures are rising at rates of up to 0.7°C per decade (Weller 1998), sea-ice extent has reduced by 15–20%

Figure 3.1 This 1600-year-old lake sediment core from the Larsemann Hills, East Antarctica, has intact layers of sediments that have accumulated year on year preserving a remarkably detailed record of environmental change. This core includes the remains of filamentous cyanobacterial mats and green algae that are strongly pigmented. These pigments help scientists to reconstruct which phototrophs were present and the growing conditions at the time of deposition. Photograph: D. Hodgson. See Plate 14 for the colour version.

(Johannessen *et al.* 2005), lake- and river-ice duration has decreased by 12 days, and the Greenland ice sheet is losing mass (Schellnhuber *et al.* 2006). In the Antarctic Peninsula temperatures are rising at 0.55°C per decade: six times the global mean. Eighty-seven percent of Antarctic Peninsula glaciers have retreated in the last 60 years (Cook *et al.* 2005), 14 000 km² of ice shelves have collapsed (Hodgson *et al.* 2006a), and the West Antarctic Ice Sheet is losing mass.

With these remarkable facts, there is little doubt (among scientists) that we are in the midst of a rapid climate change. In order to better understand the causes and consequences of this, a considerable volume of research is being carried out in the polar regions, most recently during the International Polar Year (2007–2009). This is not only because climate changes affect the regions themselves, for example through changes in ecosystem function and economic impacts, but also because any directional changes in the cryosphere also have the potential to exert marked changes on the rest of the planet. For example, the polar regions contain ice volumes equivalent to sea-level rises of 7.2 m (Greenland Ice Sheet) and 70 m (Antarctic ice sheets). Melting of these ice masses and other ice sheets and glaciers is therefore an important contributor to rising sea levels, together with the thermal expansion of the ocean. Another example is the impact that changes in the spatial extent of the polar sea ice has on atmospheric temperature, ocean thermohaline circulation, and the atmospheric thermal gradients that drive the Earth's climate and weather systems. This can result in positive feedbacks where, as the ice retreats, the ocean transports more heat towards the pole, and the open water absorbs more sunlight, further accelerating the rate of warming and leading to the loss of more ice. This positive feedback is important as the Earth is losing a major reflective surface (i.e. its albedo is decreasing) and so absorbs more solar energy, enhancing the warming trend. Many researchers believe that there are 'tipping points' or 'critical thresholds' in the Earth's system, beyond which such positive feedbacks would become difficult to stop, potentially accelerating climatic change across the world. An example of a tipping point is the widespread melting of permafrost in the Arctic (already underway), which is releasing methane into the atmosphere; methane is a greenhouse gas that will further accelerate climate warming.

Despite the importance of the polar regions in environmental and climate change research, on account of their remoteness, relatively little long-term instrumental and other monitoring data exist. However, this is changing as new technologies are developing (see Chapter 17 in this volume). It is here that paleolimnology plays a key role by reconstructing past environmental changes in regions that have never been monitored, or extending records in those that have (Schindler and Smol 2006). Lakes are common features of many polar landscapes and their sediments, together with marine sediments and ice cores, enable scientists to piece together how different parts of the Earth system have interacted on timescales longer than the instrumental record. Specifically, these records enable scientists to quantify natural variability, determine whether recent changes are the result of human activities, and permit informed estimates of the most-probable nature and direction of future change.

A key technological development that has unlocked the potential of lake sediment records is radiocarbon dating. In particular, the direct measurement of radiocarbon (^{14}C) by accelerated mass spectrometry (AMS) has enabled small autochthonous macrofossils such as aquatic moss fragments, chironomids, and foraminifera, and extracted organic components such as humic acids, to be analysed (Wolfe *et al.* 2004). These typically assimilate their carbon from the water column and this carbon is often in near-equilibrium with atmospheric carbon dioxide, except in lakes with reservoir or hard-water effects. Measurements of short-lived isotopes such as ^{210}Pb and ^{137}Cs, have provided similar advances in the dating of the most recent sediments (last ~100 years) in some polar settings (Wolfe *et al.* 2004). These methods can be validated through direct comparison with varve chronologies and event horizons in the sediments such as tephra from volcanic eruptions and the presence of persistent organic pollutants. Other dating methods such as Optically Stimulated Luminescence, Uranium-Thorium dating of lacustrine carbonates and paleomagnetic intensity dating provide further chronological control.

Paleolimnologists also have a wide range of methods at their disposal with which to analyze various physical, chemical, and biological markers in lake sediments. These are described in the *Developments in Paleoenvironmental Research* book series, including volume 8, which is dedicated to long-term environmental change in Arctic and Antarctic lakes (Pienitz *et al.* 2004).

In this chapter we divide high-latitude paleolimnological studies into three groups. The first group comprises studies that focus on the geomorphological origin of lakes and their long-term development or ontogeny. The second 'applied' group uses lake sediments to address specific research questions, for example relating to climate and ecosystem changes and human impacts. The third 'synthesis' group employs various meta-analysis and statistical methods to combine multiple studies and reveal wider-scale patterns and trends, and to direct new research and policy priorities. Below we use this classification to provide an overview of high-latitude paleolimnology together with case studies. We do not consider methodological aspects, which are described elsewhere (Pienitz *et al.* 2004).

3.2 Lake geomorphology and ontogeny

3.2.1 Geomorphology

Paleolimnological studies can provide important data on the geomorphological origin and development of high-latitude lakes and their catchments (see Chapter 2). In lakes of glacial origin, deglaciation is often seen as a transition from siliciclastic glacial to organic non-glacial sediments. These organic sediments are deposited as soon as liquid water and sunlight become available, and together with radiocarbon chronologies, provide evidence of when the landscape became ice-free. In combination with cosmogenic isotope dating of moraines and trim lines, these lake-based deglaciation chronologies have been used to document the postglacial retreat of glaciers across both northern- and southern-hemisphere landmasses. For example, in Eurasia a substantial meltwater input to the Fram Strait and the accumulation of organic material in lakes at 15 000 years before present (BP) shows that the Eurasian Arctic Ice Sheet was one of the first northern-hemisphere ice sheets to retreat (Siegert 2001). In North America, paleolimnological mapping and dating of former strandlines and glaciolacustrine sediments shows that the two major ice masses, the Cordilleran Ice Sheet (originating from the Rocky Mountains) and the Laurentide Ice Sheet (originating from northern Canada and coalesced with the Greenland Ice Sheet) began to form large proglacial lakes at their margins about 13 000 years ago and commenced their major retreats just before 8000 years ago. At paleo Lake Agassiz on the Laurentide ice margin, which at its greatest extent may have covered as much as 440 000 km², glaciolacustrine deposits and paleo outlet channels provide evidence of multiple stages in the lake's evolution, including the large discharges of fresh water into the ocean that many scientists link to cooling of the North Atlantic and the onset of the Younger Dryas period at 12 800–11 500 cal years (i.e. converted from radiocarbon dates after calibration) BP (Cohen 2003). Radiocarbon dates of deposits in the Clearwater-Lower Athabasca spillway (Figure 3.2) and Late Pleistocene Athabasca braid delta show one such discharge via the Mackenzie River into the Arctic Ocean at 11 300 cal years BP linked to the Preboreal oscillation, a brief (150–250-year) cooling event in the North Atlantic (Fisher *et al.* 2002).

Reconstructing deglaciation from lake sediments has been relatively straightforward across the major continental landmasses of North America and Eurasia as they possess extensive lake districts that stretch northward from approximately 45°N. The lakes occupy eroded hollows left by the retreating ice, or folds and depressions in the underlying geology. Any limnologist who has flown from Toronto to London on a clear day will have spent several privileged hours captivated by their airborne transect of just one of these vast lake districts spanning from the Great Lakes into the Canadian Shield and across Hudson Bay to Baffin Island.

In contrast, in Antarctica and Greenland the grounded ice extended towards the edge of the continental shelf, and the onset of deglaciation is first recorded in the marine geological record and generally later in the lake sediment record, after the ice streams had retreated across the shelf and up the major fiords. Thus both Antarctica and Greenland still have relatively small areas of ice-free land (0.32% and 20% respectively) occupied by lakes. In the Antarctic, dates for the onset of postglacial lake sedimentation in continental ice-free areas range from 13 500 cal years BP in the Larsemann Hills (Verleyen *et al.* 2004), to 12 400 cal years BP in the Amery Oasis (Wagner *et al.* 2004), 10 000 years BP in the Schirmacher Oasis (Schwab 1998), and 9070 cal years BP in the Windmill Islands (Roberts *et al.* 2004), whereas on the Antarctic Peninsula dates range from before 9500 cal years BP in Marguerite Bay (D.A. Hodgson, unpublished results) to 5000 years BP at Byers Peninsula, South Shetland Islands (Björck *et al.* 1993) and on James Ross Island (Björck *et al.* 1996). In Greenland, deglaciation dates range from 12 100 cal years BP for lakes on Nuussuaq (Bennike 2000), to 10 000 cal years BP at

Figure 3.2 Digital elevation model showing the Clearwater-Lower Athabasca Spillway; one of the outlet channels of paleo Lake Agassiz. Radiocarbon dates of sediment units in cores from this spillway have enabled a reconstruction of the history, timing, and magnitude of discharge events from this and other major outlets of Lake Agassiz. One such discharge took place via the Mackenzie River into the Arctic Ocean at 11 300 cal yr BP and is considered a likely cause of the Preboreal oscillation (Fisher *et al.* 2002).

Sønde Strømfjord and approximately 7400 cal years BP in Lake SFL4, 5 km west of the present-day ice sheet (Anderson *et al.* 2004).

3.2.2 Lake ontogeny

Lake sediments record many of the geological, sedimentological, and biological (flora, fauna, and ecosystem function) developmental stages of lakes and their catchments, a subject called lake ontogeny. In Greenland, together with an increase in organic content, deglaciation is often marked by a period of carbonate-rich sediments resulting from

calcite deposition (Fredskild 1983; Anderson *et al.* 2004 and references therein). Where terrestrial vegetation is present, Arctic lakes can become more dilute and acidic with time, lose ions and nutrients, and accumulate dissolved organic carbon, all as a result of primary succession trapping nutrients in surrounding vegetation and soils. Dissolution models also attribute at least part of postglacial lake acidification, or reduction in alkalinity, to calcite depletion in run-off (Boyle, 2007). These and other developmental stages have been observed in the fossil diatom stratigraphy of lake sediment cores from Glacier Bay, Alaska (Figure 3.3) (Engstrom

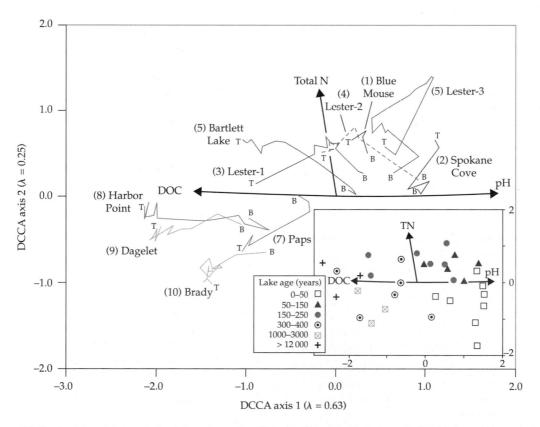

Figure 3.3 The developmental stages in the diatom communities of lakes in Glacier Bay, Alaska, is revealed by this detrended canonical correspondence analysis (DCCA) of modern and fossil diatom assemblages with time trajectories from 10 sediment cores (B, bottom; T, top). The core assemblages were added to the biplot as passive samples; numbers on lake trajectories correspond to each of the 10 study lakes. The inset shows an ordination of modern diatom assemblages (from surface sediments) and associated water-chemistry data chosen by forward selection (a stepwise addition based on explained variance). The arrows depict the direction of increase and maximum variation of the measured environmental variables; other significant variables (not shown) include Na^+, Cl^-, SO_4^{2-}, and total Al. The core data illustrate the main trend of lake development as a progressive loss of pH, alkalinity, and base cations, and a corresponding increase in **DOC**. From Engstrom *et al.* (2000), by permission of Macmillan Publishers Ltd (Nature).

et al. 2000) and the pigment stratigraphy of lakes in Greenland (McGowan *et al.* 2007). Catchment soil development and plant cover are therefore key drivers of nutrient inputs into lakes and also of the supply of dissolved organic matter (DOM). DOM can be particularly important as it absorbs ultraviolet (UV) radiation. In lakes where DOM inputs have declined due to a retreat of the Arctic treeline, algal biomass has declined 10–25-fold (Leavitt *et al.* 2003).

Another common driver of high-latitude lake ontogeny is the changing influence of lake ice and snow cover, which exert strong controls on the abundance and diversity of the plankton and periphyton (Smol 1988; Douglas and Smol 1999; see also Chapters 1 and 9). During relatively cold periods, winter ice cover may persist well into the short polar summer and, in some lakes, a central raft of snow and ice may persist all year round. In these lakes, most of the primary production is concentrated to the shallow-water moat of open water in the littoral zone. With warming, more of the lake is made available and production increases. Once the central raft of ice melts completely, the plankton may flourish and diversity at all levels of the ecosystem increases. Paleolimnologists can follow the development of this ecosystem using a range of morphological and biogeochemical fossils. Morphological fossils include diatom frustules, non-siliceous remains of green algae (e.g. *Pediastrum*), cyanobacteria, and chrysophyte scales and cysts, together with the remains of the fauna such as cladocerans, chironomids and other insect fossils, rotifer and tintinnid loricae, folliculinid tests, rotifers, mites, Anostrachan (fairy shrimp) eggs, copepod eggs, spermotophores and exoskeletal fragments, and foraminifera (Hodgson *et al.* 2004a). Biogeochemical fossils include sedimentary pigments and other molecular components (DNA, lipids) and biogenic silica. Pigments are particularly useful in tracking groups where no morphological remains are preserved (Leavitt and Hodgson 2001) including selected algae and bacteria (Squier *et al.* 2005; McGowan *et al.* 2007). DNA has been used to trace past populations of copepods (Bissett *et al.* 2005), haptophytes, and diatoms (Coolen *et al.* 2004). However, in recently deglaciated environments where do all these species come from?

3.2.3 Origin of the lake biota

Due to the regular succession of Quaternary glaciations, most high-latitude lakes are relatively short-lived compared with lakes in the mid and low latitudes. The interglacial periods in which they form comprise just 10% of the last 400000 years and 50% of the period between 400000 and 800000 years ago. Thus, high-latitude lake districts have experienced multiple phases of biological colonization, succession, and extinction. A major question is whether the lake biotas consist primarily of postglacial colonists arriving by dispersal or whether they survived the glaciations in polar refugia. In the Arctic the more continuous landmasses provide an easy route for both colonizers with a dispersal stage in their life history and/or those regularly vectored in on other organisms such as birds (Bennike 2000); but in the Antarctic the situation is very different. Here the Southern Ocean, steep oceanic and atmospheric thermal gradients, and circumpolar currents and winds have provided formidable barriers to the dispersal of the biota. Thus many species are believed to have become extinct during glaciations and then (re) colonized the Antarctic during interglacials from the maritime and sub-Antarctic islands and from the higher-latitude southern-hemisphere continents (South America, Australasia) (Clarke *et al.* 2005; Barnes *et al.* 2006). However, recent field data from the Antarctic has provided evidence that some zooplankton and phytobenthos have survived on the continent over a single glacial cycle (Cromer *et al.* 2005; Hodgson *et al.* 2005a), so it is likely that non-glaciated lakes, epiglacial lakes, and cryoconite holes, have provided refuges for many of the species now found in Antarctic lakes and account for the presence of numerous endemic species. Endemism in cyanobacteria, diatoms, and other biota on the continent supports this view (Sabbe *et al.* 2003; Taton *et al.* 2006; Vyverman *et al.*, 2007; see Chapter 12) and allows for the possibility that Antarctic lakes may contain species that are relicts of Gondwana (Convey and Stevens 2007; see Chapter 13). The groups that comprise the refuge species and the colonizers are usually differentiated by their life history, in particular the presence or absence of a life-history stage capable of dispersal across the formidable barriers of the polar frontal zone and the circumpolar winds of the Southern Ocean.

3.3 Applied studies: tracking environmental change

As well as providing archives of glaciation, deglaciation, and ontogeny, lake sediments can document changes in local and regional ecosystems and, by inference, changes in the Earth system that drive them. Such studies provide long-term records against which the magnitude and direction of recent climate and environmental changes can be compared. Below we discuss some of the wide range of applied studies in high-latitude paleolimnology.

3.3.1 Climate change

Studies of lake sediments often enable inferences to be made about the nature and direction of past climate change. For example, sediment composition can respond to temperature-related variables, which are most often linked to changes in the duration and extent of lake ice and snow cover. Sediment composition also responds to hydrologically related variables, for example changes in the moisture balance which affect lake salinity, meltwater inputs, and nutrient and dissolved organic carbon (DOC) release from soils (Schindler et al. 1996; Douglas and Smol 1999; Schindler and Smol 2006).

Much of the work on temperature and its interaction with changing ice and snow cover was stimulated by an early study of Douglas et al. (1994) on the sediments of High Arctic shallow ponds at Cape Herschel (Ellesmere Island, Nunavut), which recorded striking changes in diatom assemblages, indicating a longer growing season and the development of moss substrates for periphytic growth over the last approximately 100–150 years, which they attributed to warmer summers and longer ice-free conditions. Reduced ice cover, a longer growing season, and/or stronger thermal stratification patterns can also result in marked changes in the planktonic diatom flux to lake sediments. This has been observed in, for example, Finnish Lapland

where lake ecosystems have experienced not only a shift from **benthic** to planktonic diatom species, but also changes in crysophyte and zooplankton assemblages since the onset of twentieth century Arctic warming (Sorvari et al. 2002). The influence of twentieth-century warming has also been recorded at Slipper Lake in the Canadian Central Arctic treeline region where a marked change in the top 5.0 cm (c. mid-1800s) of cores shows a clear shift to a planktonic diatom assemblage dominated by *Cyclotella stelligera* (Rühland et al. 2003; Rühland and Smol 2005; Smol and Douglas 2007). In five lakes on Baffin Island, rising chlorophyll *a* concentrations also reveal whole-lake production increases in the twentieth century, synchronized with the meteorological record of recent temperature change (Michelutti et al. 2005). Other variables such as sediment accumulation rates and varve thickness can also provide a semi-quantitative estimate of past temperature (Hambley and Lamoureux 2006), as can changes in lake productivity and the abundance and diversity of organisms (Quinlan et al. 2005; Smol et al. 2005). For example, increasing sediment-accumulation rates in maritime Antarctic lakes at Signy Island since the 1950s correspond with a measured increase in atmospheric temperature (Appleby et al. 1995).

Production and temperature are not always tightly coupled, however, particularly in lakes that have clear 'black ice' and no snow cover, where 50% of annual primary production may occur before ice out (Schindler and Smol 2006), and in lakes where the very high light irradiances experienced during the summer can substantially inhibit algal blooms during the ice-free period (Tanabe et al. 2008) (see Chapter 9). To overcome this, catchment-based indicators can be used. For example, at Signy Island, long-term temperature changes have been inferred from the macrofossil remains of orabatid mites in lake sediments, which were more abundant during a warm period in the mid to late Holocene (3800–1400 cal years B.P., Figure 3.4) when conditions favored the development of terrestrial mosses in which the mites thrive (Hodgson and Convey, 2005). In other lakes the effects of warming can be recorded even where no local atmospheric warming has been detected. For example in the Windmill Islands, East Antarctica, a rapid increase

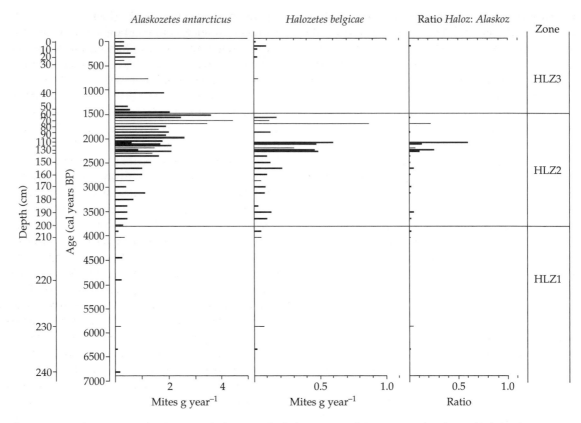

Figure 3.4 Fossil mite stratigraphies in Heywood Lake, Signy Island, showing a population expansion by a factor of 7 during the Antarctic mid-Holocene hypsithermal. This is thought to have increased habitat size, temperature, and moisture availability favoring the expansion of mite populations. From Hodgson and Convey (2005), by permission of the Institute of Arctic and Alpine Research, Boulder, CO, USA.

in the salinity of three coastal lakes (Hodgson *et al.* 2006b) has been attributed to warming of the Southern Ocean and changes in the atmospheric circulation patterns that have brought about a decline in effective precipitation (D.A. Hodgson, unpublished results).

Temperature increases have also brought about thawing of the Arctic permafrost, detected in lakes by an increasing number of shoreline permafrost slumps (Kokelj *et al.* 2005). This is part of a wider predicted positive climate feedback whereby the melting permafrost releases methane from clathrates into the atmosphere, which is predicted to further accelerate the warming.

Sediment composition also responds to hydrologically related variables, such as the atmospheric

moisture budget, which can influence meltwater inputs. These can be inferred from changes in the deposition of stream-derived diatoms (Antoniades and Douglas 2002), or measured as a so-called lotic index (Ludlam *et al.* 1996), an index based on counts of stream-derived compared with total pennate diatoms. Changes in the atmospheric moisture budget can also bring about changes in lake salinity (Pienitz *et al.* 2000; Ryves *et al.* 2002). In East Antarctica, salinity (and lake depth) diatom-based transfer functions now incorporate more than 100 lakes and have been applied to the long-term reconstruction of the past water balance of lakes (Verleyen *et al.* 2004; Hodgson *et al.* 2006b). Water isotopes are also widely used as tracers of present and past hydrology (Edwards *et al.* 2004) as they record changes in water input and evaporation, especially in closed lake systems. For example, stable isotope evidence ($\delta^{18}O$ and $\delta^{13}C$) from authigenic calcite in laminated sediments from West Greenland lakes indicate a strongly negative moisture balance accompanied by lowered lake levels between 7000 and 5600 cal years BP, attributed to decreased precipitation, coupled with higher insolation (Anderson and Leng 2004). At South Georgia (sub-Antarctic) changes in oxygen isotopes from diatom silica ($\delta^{18}O_{Si}$) are attributed to changes in both water temperature and hydrology (Rosqvist *et al.* 1999).

Another method of tracking changes in the hydrological cycle is to study paleoshorelines, which record past periods of lake level change. Some of the most impressive paleoshorelines are found in the McMurdo Dry Valleys region of Antarctica, which are up to 450 m above the present lake level (Hendy 2000). These were created when expansion of the Ross Ice Shelf blocked the valleys during the Last Glacial Maximum. Biogeochemical studies of one of the lakes, Lake Fryxell, have shown that it evaporated to a low-stand before the Ross Ice Shelf retreated after the Last Glacial Maximum and has experienced a series of high and low stands including evaporating to a small pond between approximately 2500 and 1000 cal years BP (Wagner *et al.* 2006). Instrumental studies show hydrological asymmetry in these lakes whereby water levels respond to occasional warm summers by a rapid rise in lake levels of approximately 1 m, which then takes many decades to be drawn down again (Doran *et al.* 2006).

Detailed regional syntheses of high-latitude climate change based on paleolimnological studies are given in Pienitz *et al.* (2004).

3.3.2 Regional vegetation change

Changes in regional vegetation invariably accompany climate change and can be tracked using resistant pollen and spores incorporated into lake sediments. There are many palynological studies in the Arctic that document the postglacial vegetation history of the treeline zone as well as past vegetation changes in tundra ecosystems and

aquatic and wetland plants (Pienitz *et al.* 2004). Workers on such studies often select a range of lake-coring sites spanning key vegetation or latitudinal gradients. The close coupling of catchment soil, vegetation development, and lake-water chemistry means that these studies are critical in understanding the history of lake catchments and the changing geochemistry of high-latitude lakes, particularly the supply of chromophoric DOC, which can be tracked using paleolimnological techniques (Pienitz and Smol 1993), and has a marked influence on the transparency of water columns and algal biomass, as described earlier (Leavitt *et al.* 2003). In contrast, Antarctic pollen studies are relatively rare, primarily because there are only two species of flowering plants in the Antarctic, *Deschampsia antarctica* (Poaceae) and *Colobanthus quitensis* (Caryophyllaceae). Even in the maritime Antarctic at Signy Island, pollen grains in lake sediments are sparse and dominated by long-range components from southern-hemisphere continents (Jones *et al.* 2000). However, further north, in the sub-Antarctic, pollen analyses have been used successfully to track past vegetation trends.

3.3.3 Reconstructing sea-level change

With the continued rise in emissions of greenhouse gases, future sea-level rise has become an important social and political issue. In 2001, the Intergovernmental Panel on Climate Change (IPCC) Third Assessment Report predicted that, by 2100, global warming and consequent melting of the high-latitude ice sheets will lead to a sea-level rise of 9–88 cm (Houghton 2001). Reconstructing the postglacial interaction between these ice sheets and past sea-level rise therefore provides a valuable series of constraints on future ice-sheet behavior.

Using the paleolimnological record, it is possible to construct the history of marine transgressions and hence the relationship between relative sea-level change, ice thickness, global eustatic sea-level change, and isostatic rebound (Long *et al.* 2003; Bentley *et al.* 2005a; Verleyen *et al.* 2005a). High-latitude lakes situated below the Holocene marine limit, which have been isolated from the sea, contain one or more stratigraphic horizons that precisely record the isolation event. These can be identified by, for example, transitions from marine to freshwater diatoms, pigments, and carbon isotope signatures, and dated using radiocarbon technologies. A good example is the relative sea level curves that have been constructed for Disko Bugt, West Greenland (Figure 3.5; Long *et al.* 2003). Such constraining data enable ice-sheet modelers

to accurately reconstruct the past, and better predict the future contribution of the polar ice sheets to global sea-level change. This is a pressing task as predicted increases in global temperature over the coming decades threaten the stability of parts of the Greenland Ice Sheet and possibly the West Antarctic Ice Sheet. Evidence from the last interglacial, which was only very slightly warmer than our own (1–2°C), shows that global sea levels were 5 m higher than they are today (Lambeck and Chappell

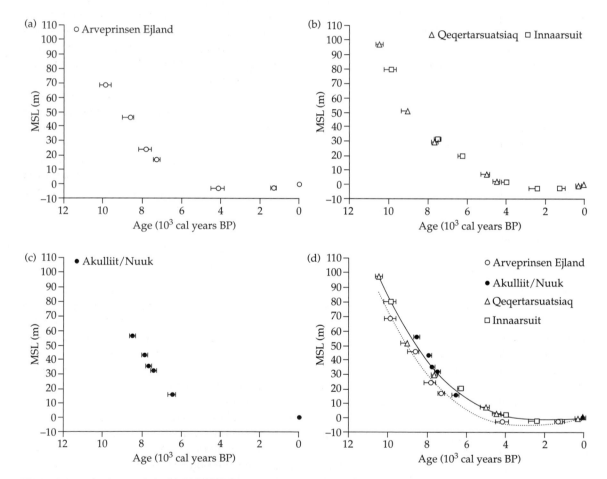

Figure 3.5 Sea-level curves derived from isolation basins at Innaarsuit, a site on the south shores of the large marine embayment of Disko Bugt, west Greenland. The *x* axis shows the date in calibrated radiocarbon years before present, and the *y* axis shows mean sea level (MSL). These relative sea level (RSL) curves show the interaction between ice thickness, global eustatic sea-level change, and **isostatic rebound**. These RSL data record rapid RSL fall from the marine limit (≈108 m) at 10 300–9900 cal years BP to reach the present sea level at 3500 cal years BP. Since 2000 cal years BP, RSL rose by approximately 3 m to the present. These data show that the northern part of Disko Bugt experienced less rebound (≈10 m at 6000 cal years BP) compared with areas to the south. Submergence during the late Holocene supports a model of crustal down-warping as a result of renewed ice-sheet growth during the neoglacial. From Long *et al.* (2003), by permission of Elsevier.

2001), and it is likely that these high-latitude ice reservoirs provided at least some of this water.

Epishelf lakes, which form where fresh water is impounded in ice-free depressions or former marine embayments by ice shelves or glaciers, are also yielding valuable paleolimnological information on the first signs of decay of the ice masses that impound them. For example, fracturing of the largest Arctic ice shelf (Ward Hunt Ice Shelf) on the north coast of Ellesmere Island, Nunavut, Canada, in 2002, caused the drainage of the ice-dammed epishelf lake (Disraeli Fiord), which it impounded. Attributed to rapid regional warming since 1967, this drainage resulted in an influx of marine water and a 96% loss of the fresh and brackish water habitats that were known to support a unique biological community (Mueller *et al.* 2003). Such changes are recorded in epishelf lake sediments and have been exploited in the Antarctic to examine the history of past ice-shelf collapse. Sediments from Moutonnée Lake on Alexander Island have revealed that the currently extant George VI Ice Shelf, the largest ice shelf on the west of the Antarctic Peninsula, broke up in the early Holocene (Figure 3.6), following a period of atmospheric warmth at the same time as an intrusion of warm ocean water under the shelf (Bentley *et al.* 2005b; Hodgson *et al.* 2006a; Smith *et al.* 2006; Roberts *et al.* 2008). Establishing the oceanic and atmospheric conditions that have caused previous collapse events provides scientists with datasets that can be used to help predict when future collapses might occur. The Antarctic ice shelves are important as they help restrain the inland glaciers of the Antarctic ice sheet on account of being pinned to submerged rises in the ocean floor. Without the restraining ice shelves, measurements have shown that the inland ice of Antarctica flows at an accelerated rate into the ocean. Therefore loss of the ice shelves provides an early warning of an increased Antarctic contribution to global sea-level rise.

3.3.4 Tracking past fish and wildlife populations

Arctic salmon, seabirds, and caribou, as well as Antarctic seal and penguin populations have all been tracked using paleolimnological methods. In the Arctic, (responsible) salmon fisheries managers

seek to maintain catches at a sustainable level, and so there is considerable value in tracking and understanding long-term changes in fish populations. One way of doing this is to track the delivery of marine-derived nutrients into lakes using diatom, pigment, and stable-isotope analyses ($\delta^{15}N$) as has been done for sockeye salmon in Alaska (Finney *et al.* 2002; Gregory-Eaves *et al.* 2003). These marine-derived nutrients are incorporated rapidly into lake-wide nutrient pools, providing an index of whole-lake salmon densities (Brock *et al.* 2006). Changing nutrient inputs have also been used to track the impact of increasing populations of pink-footed resting geese in Svalbard Lakes (Christoffersen 2006), and changing concentrations of fecally derived elements to track historical penguin populations in the Antarctic (Zale 1994; Sun *et al.* 2000) and seabird populations in Greenland (Wagner and Melles 2001). Other studies of past fish and wildlife populations have used morphological fossils. For example, seal hairs in lake sediments have been used to track the changing populations of the Antarctic fur seal, and to evaluate the impact of the nineteenth- and twentieth-century sealing industry in the Antarctic (Hodgson and Johnston 1997; Hodgson *et al.* 1998).

3.3.5 Changes in UV radiation

Depletion of stratospheric ozone has occurred over both high-latitude regions and resulted in an increase in the receipt of erythemal UV radiation at the Earth's surface. UV radiation receipt has increased by 22% in the Arctic since the 1970s and in the Antarctic by 180% (Palmer Station), 310% (McMurdo Dry Valleys), and 480% (South Pole Station) since the early 1980s (Bernhard *et al.* 2000). Antarctic lakes are most susceptible to this change as they generally lack the terrestrial sources of DOC and chromophoric DOM that continue to limit UV radiation penetration in many Arctic systems, even after the disappearance of long-term snow and ice cover (Pienitz and Vincent 2000; Leavitt *et al.* 2003). In the Antarctic, the low to undetectable levels of DOC measured in the water columns reveal a close coupling between the pigment composition of benthic microbial mats and lake depth, with shallower lake microbial mats incorporating

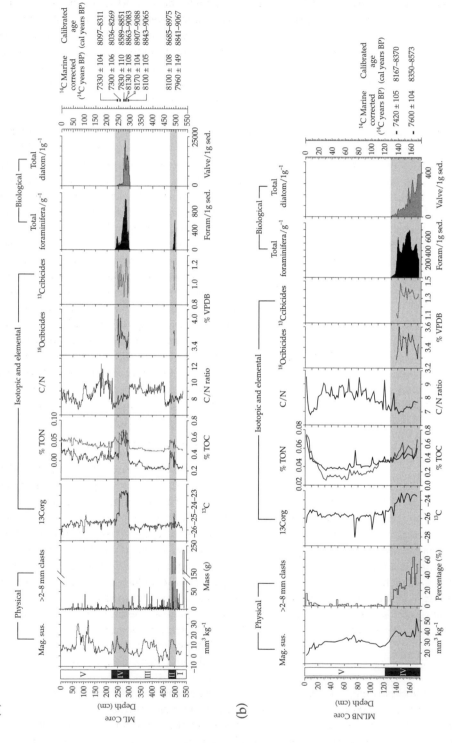

Figure 3.6 Summary diagram of the stratigraphic record from epishelf Moutonnée Lake, Alexander Island, Antarctic Peninsula, showing key multiproxy paleolimnological data from (a) Moutonnée Lake main basin (ML) and (b) Moutonnée Lake North Basin (MLNB). Proxies include: magnetic susceptibility (Mag. sus.), grain size, $\delta^{13}C$ on organic matter in bulk sediment ($^{13}Corg$); total organic carbon (**TOC**), total organic nitrogen (**TON**), C/N ratio performed on bulk sediment, foraminiferal isotopes ($\delta^{18}O$ and $\delta^{13}C$), and diatom and foraminiferal analyses (shown per gram of sediment). Zones 4 and 5 represent the same stratigraphic zones in both cores. These data provide robust evidence for one period of past ice-shelf absence during the early Holocene (shaded grey). The timing of this period has been constrained by 10 accelerated mass spectrometry (AMS) ^{14}C dates performed on mono-specific foraminifera samples. These dates suggest that the George VI Ice Shelf was absent between c.9600 and c.7730 cal years BP. This early **Holocene** collapse immediately followed a period of maximum **Holocene** warmth that is recorded in some Antarctic ice cores and coincides with an influx of warmer ocean water on to the western Antarctic Peninsula (AP) shelf at c.9000 cal years BP. From Smith *et al.* (2007) and Roberts *et al.* (2008), by permission of Elsevier.

higher concentrations of UV-screening pigments such as scytonemin (Hodgson *et al.* 2004b). Paleolimnological analyses of these same pigments have enabled both Holocene (Verleyen *et al.* 2005b) and Pleistocene reconstructions of UV radiation receipt (Hodgson *et al.* 2005b). Perhaps the most intriguing study is that of Lake Reid in the Larsemann Hills where long-term reconstructions of UV radiation receipt showed that the highest concentrations were experienced in the lake during the last glacial period and not, as might be anticipated, during the current period of ozone depletion (Hodgson *et al.* 2005b). Ultimately UV radiation is probably most important in high-latitude lakes for its impact on ecosystem structure and biomass. This is most evident in the benthic cyanobacteria where the synthesis of carotenoids and screening pigments is one of multiple attributes that favor their growth and dominance in many high-latitude clear-water lakes (Vinebrooke and Leavitt 1996; Hodgson *et al.* 2004b).

3.3.6 Atmospheric and terrestrial pollutants

Mid-latitude water-quality problems such as lake acidification and cultural eutrophication are not yet recorded in most high-latitude lake sediments, except for example where they are immediately downwind of a point source of acid (Weckström *et al.* 2003) or downstream of a sewage outfall (Douglas and Smol 2000). Instead, being remote from point sources of pollution, high-latitude lakes are ideally suited to tracking the long-range atmospheric transport of industrial pollutants (Muir and Rose 2004). For example, studies of a 3000-year lake sediment record on King George Island showed that lead concentrations (derived from fossil-fuel combustion) have significantly increased during the last 200 years, and especially in the last 50 years (Sun and Xie 2001). A similar pattern is found in Alaska where lead deposition has increased approximately 3-fold over the same period (Fitzgerald *et al.* 2005).

Some pollutants are found at higher than background concentrations as a result of focusing by biological vectors. For example, marine-derived nutrients are present at high concentrations in salmon nursery lakes (Krümmel *et al.* 2003), and in

ponds draining seabird colonies. Many pollutants that have never been used in Arctic regions (e.g. insecticides and other persistent organic pollutants, mercury, and other metals) have also been found in elevated concentrations due to bioaccumulation along long food chains. In the Arctic, migratory seabirds play an important role in the creation of these pollution hot-spots by transporting marine contaminants to their nesting sites. In Devon Island (Nunavut, Arctic Canada), studies of pond surface sediments near nesting sites found up to 10 times the amount of hexachlorobenzene (HCB), 25 times the amount of mercury, and 60 times the amount of dichlorodiphenyltrichlorethane (DDT) compared with the ponds with little or no influence from seabird populations (Blais *et al.* 2005).

Human cultural changes have also caused pollution of specific high-latitude lakes at a local scale. Examples include the marine-derived ^{15}N and changes in diatom assemblages in ponds at Somerset Island, Nunavut, attributed to thirteenth-century Inuit whaling activities (Douglas *et al.*, 2004), and in Antarctica, elevated silt inputs to littoral sediments and increases in nutrients and heterotrophic microbial activity in a lake have been attributed to the establishment of a nearby scientific base (Ellis-Evans *et al.* 1997; see Chapter 16).

3.4 Synthesis studies

Although remote from major centres of population, the high latitudes are providing a clear indication that climate warming and other human activities are having major global impacts. For much of the year, the subpolar regions' ecosystems are within a few degrees of the freezing point, so a small shift in temperature regimes can have widespread ecosystem impacts. These include changes in the amount and duration of ice and snow cover, changes in river hydrology from accelerated glacier and permafrost melting, and declining percentages of precipitation falling as snow. The cumulative effects of these stresses will be far more serious than those caused by changing climate alone.

Large-scale syntheses of high-latitude paleolimnological data are an important component of measuring these changes and address the limitations of long-term monitoring and survey measurements and satellite data, which often only span the last few decades. For example, fossil diatoms and other biotic indicators enumerated from approximately the last 200 years of sediment accumulation in cores from 42 lakes and ponds (representing 55 different profiles) across the circumpolar Arctic have shown increases in diatom assemblages characteristic of extended open-water periods, with more habitat differentiation and complexity in littoral zones and an increase in plankton in deep lakes (Figure 3.7), a pattern consistent across lakes of different origin and limnological characteristics (Smol et al. 2005; Smol and Douglas 2007). Of these lakes, 67% have experienced directional ecosystem shifts since approximately AD 1850, and 87% of these are north of the Arctic Circle. Other areas of the Arctic have yet to experience these marked changes; for example some lakes on parts of northern Québec and Labrador and the west coast of Greenland have shown little directional ecosystem change since 1850 (Perren et al. 2006).

Another example of a large-scale synthesis study is the Circum-Arctic PaleoEnvironments (CAPE) project that integrated paleovegetation data (pollen and macrofossils) from across the Arctic to generate 1000-year time-slice maps of the Holocene and compared these with modeled estimates of past climate and vegetation. Such studies are valuable as they can establish a quantitative baseline against which to assess future changes (CAPE Project Members 2001). Meanwhile, Kaufman et al. (2004) examined 140 sites across the western Arctic to track the development of the northern-hemisphere Holocene thermal maximum along a latitude gradient. A variety of paleoenvironmental techniques were used including pollen and plant macrofossil analyses and chironomid-based temperature reconstructions.

In the Antarctic, the Scientific Committee for Antarctic Research (SCAR) has successfully drawn together research for a number of years, but has yet to carry out an integrated study of Antarctic paleolimnology.

3.5 Prospects

In addition to the studies so far described in this chapter, there are a number of promising new areas of research opening up to high-latitude paleolimnologists. Three of these are described below.

3.5.1 Paleolimnology of subglacial lakes

The sediments of subglacial lakes offer the potential to explore the origin of subglacial basins and the possibility that life might exist deep under the polar ice sheets. Detailed plans have been published outlining how this might be achieved (Siegert et al. 2006). At present the main focus is on the Antarctic lakes Vostok and Ellsworth, but both of these present the great challenges of drilling through several thousand meters of ice before reaching the lakes, and ensuring a non-contaminating sampling regime. However, an alternative approach is to drill into the sediments of former subglacial lake basins that have emerged from under the ice sheet during postglacial ice recession. For example, seismic surveys of Christie Bay in Great Slave Lake, the deepest lake basin in North America in the vicinity of the Keewatin Ice Dome during the Last Glacial Maximum, have revealed up to an approximately 150-m-thick layer of lake sediments, consisting of about 1 m of Holocene sediments, approximately 20 m of late-glacial varves, and up to 130 m of deep lake sediments. These deep lake sediments have been attributed to deposition during the last glacial period in a subglacial setting (S. Tulaczyk, personal communication). If correct, accessing subglacial lake sediments within the next few years is a distinct possibility.

3.5.2 Paleolimnology of earlier interglacial periods

Although we now live in a no-analog atmosphere (due to the rapid increases of greenhouse gases) there is still much that can be learned from the paleolimnological record of earlier interglacial periods. The last interglacial (MIS 5e) is of particular interest, as many records show that temperatures were warmer than present and that global melting of the ice sheets contributed an additional 3–5 m to global sea level. Recovering lake sediments from MIS 5e environments, when warmer conditions persisted for several thousand years, presents scientists with a valuable opportunity to

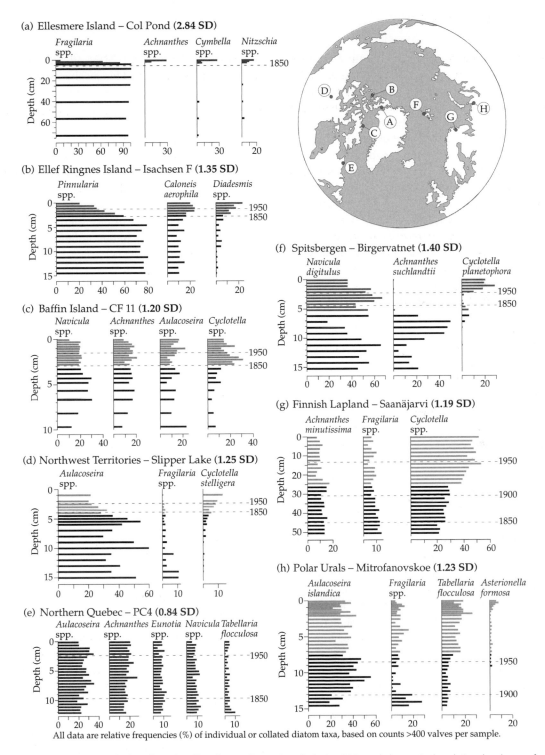

Figure 3.7 Representative paleolimnological profiles of recent (approximately the last 200 years) changes in the relative abundances of diatom assemblages from various regions of the circumpolar Arctic. The amount of assemblage change is further quantified by β diversity measurements (SD, standard deviation units). Many sites from the High Arctic recorded the greatest changes in assemblages, which were consistent with interpretation of recent warming, whereas other regions have shown less warming. Areas such as Northern Québec and Labrador have recorded little change. See Plate 15 for the colour version; from Smol *et al.* (2005).

determine how that particular location and eco-system responded to warmer conditions in the past, and is therefore a key tool in predicting the responses to the climate changes simulated by climate models for our immediate future (+1.3–6.3°C by 2100; Houghton 2001; see Plate 16). Remarkably, at both the northern and southern high latitudes, there are sites that escaped erosive action of the last glaciation and preserve in their sediments a record of one or more past interglacials. In the Arctic, the Clyde Foreland, Baffin Island, is yielding remarkable records of past interglacial diatom, chironomid, and pollen assemblages

(Figure 3.8) (Briner *et al.* 2007). In the Antarctic, analyses of biological and biogeochemical markers in the Larsemann Hills show a more productive biological community and greater habitat diversity during MIS 5e, suggesting higher temperatures than those being experienced in the present interglacial. From the composition of these MIS 5e sediments it is safe to predict that future elevated temperatures will allow new taxa to colonize and establish self-maintaining populations, and that continental Antarctica will face the re-invasion of the sub- and maritime Antarctic flora in the near future (Hodgson *et al.* 2006c).

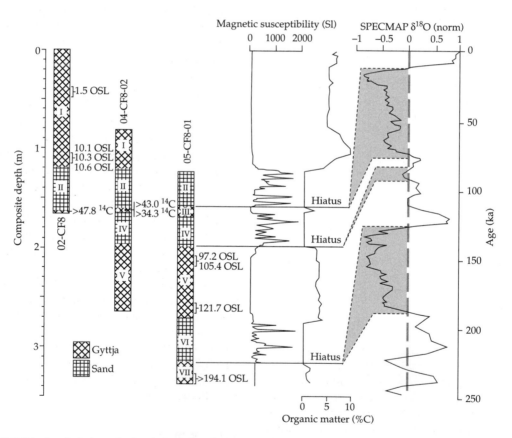

Figure 3.8 Paleolimnological records of environmental conditions experienced during previous interglacials have been found in the Clyde Foreland Region of Baffin Island. The left-hand side shows the stratigraphic and magnetic susceptibility logs from Lake CF8 which demonstrate the presence of two intercalated pre-Holocene organic sediment packages, believed from ^{14}C, optically stimulated luminescence (OSL) ages, and comparison with a normalized oceanic δ^{18}O global ice volume proxy, to represent deposition during the last (MIS 5) and penultimate (MIS 7) interglacials. Lake CF8 is the longest lake-sediment record thus far recovered from within the limits of continental glaciation and challenges the assumption that only sites distal to glacial margins are useful for reconstructing pre-Holocene environmental changes.

3.5.3 Paleolimnology of extreme environments

Studies of high-latitude limnology and paleolimnology can provide valuable insights into our understanding of life at its environmental limits. This has led to a number of research projects aimed at defining the boundaries of those conditions essential to various forms of life, and comparing this with what is known about water bodies that may exist elsewhere in our solar system, notably Mars and Europa (Doran *et al.* 1998, 2004). Antarctica has a number of lake habitats that approach extremes of salinity (e.g. Don Juan Pond), temperature (in saline lakes temperatures can fall below −10°C), thickness of perennial lake ice cover, pressure (subglacial lakes), light limitation, and ultra-oligotrophy. One example of life persisting at extremes can be found in Lake Vida, a closed-basin endorheic lake in the McMurdo Dry Valleys. The ice cover of Lake Vida is 19 m thick and forms an ice seal over a brine that is seven times as saline as sea water and has an average temperature of −10°C. This perennial lake ice has isolated the saline lake water from external air and water for thousands of years, but despite this scientists have been able to collect 2800-year-old desiccated fialmentous cyanobacteria from within the lake ice and reanimate it simply by providing liquid water and light (Doran *et al.* 2003). The ice, water, and sediments of high-latitude lakes such as Lake Vida offer the promise of further insights into the biochemistry and ecological organization of life under extreme conditions.

3.6 Conclusions

Given the dearth of long-term instrumental measurements and observations from polar regions, paleolimnological approaches offer important avenues to reconstruct these missing data-sets. Lake and pond sediments archive a surprisingly diverse array of physical, chemical, and biological information on environmental changes occurring within lakes as well as in lake catchments, and at regional to global scales via atmospheric transport. Thus far, most studies have focused on tracking recent (e.g. in the last two centuries) environmental changes; however, the wealth of information preserved in sedimentary profiles can be used to address many other questions. Despite clear signs of marked recent environmental changes in the high latitudes, we as yet have only a limited perspective on how Arctic and Antarctic climates and environmental conditions have varied in the more distant past.

Although sedimentary reconstructions cannot match the information content available in standard lake-monitoring programs, paleolimnologists have the major advantage that they can choose the time scale and, to some extent, temporal resolution of their analyses. Such approaches are especially important in polar regions, where other sources of proxy data, such as tree rings, may not be available. Moreover, a major strength of paleolimological records is that they can be used to identify regional patterns of climatic and other environmental changes, whereas other proxy records (such as ice cores) may provide information on a mixture of global and regional environmental signals (Anderson *et al.* 2004).

As more paleolimnological studies are completed, regional syntheses will become possible, and meta-analyses of these data can be used to address a wide spectrum of pressing environmental questions. However, as paleolimnological interpretations are completely dependent on our understanding of processes and species distributions in present-day limnological systems, close collaboration between limnologists working on a variety of time scales will be required. Although significant progress has been made in providing defendable interpretations of environmental changes, a better understanding is required of the limnology of polar lakes and the extent to which climate and other environmental changes can influence the sedimentary proxies. Much work remains, but it is evident that paleolimnological methods will continue to play a leading role in developing our understanding of past environmental changes and helping us to anticipate the magnitude, nature, and direction of future change.

Acknowledgments

Polar research in our laboratories is funded primarily by the Natural Environment Research Council UK, the Belgian Science Policy Office HOLANT project, the Natural Sciences and Engineering Research Council of Canada, and Polar Shelf

Canada. We thank the speakers who contributed to our special session on rapid environmental changes at high latitudes at the American Society of Limnology and Oceanography summer meeting in Victoria, Canada, 2006, who stimulated the development of some of our ideas. We are especially indebted to colleagues who have worked with us in the polar regions and who have shared with us their enthusiasm, knowledge, and friendship. We thank the editors, N. Michelutti, K. Rühland, and two anonymous reviewers for their constructive comments on this chapter.

References

Anderson, N.J. and Leng, M.J. (2004). Increased aridity during the early Holocene in West Greenland inferred from stable isotopes in laminated lake sediments. *Quaternary Science Reviews* **23**, 841–849.

Anderson, N.J., Ryves, D.B., Grauert, M., and McGowan, S. (2004). Holocene paleolimnology of Greenland and the North Atlantic Islands (North of 60 N). In Pienitz, R., Douglas, M.S.V., and Smol, J.P. (eds), *Long-term Environmental Change in Arctic and Antarctic Lakes*, pp. 319–347. Springer, Dordrecht.

Antoniades, D. and Douglas, M.S.V. (2002). Characterization of high Arctic stream diatom assemblages from Cornwallis Island, Nunavut, Canada. *Canadian Journal of Botany* **80**, 50–58.

Appleby, P.G., Jones, V.J., and Ellis-Evans, J.C. (1995). Radiometric dating of lake-sediments from Signy Island (Maritime Antarctic) – evidence of recent climatic-change. *Journal of Paleolimnology* **13**, 179–191.

Barnes, D.K.A., Hodgson, D.A., Convey, P., Allen, C.S., and Clarke, A. (2006). Incursion and excursion of Antarctic biota: past, present and future. *Global Ecology and Biogeography* **15**, 121–142.

Bennike, O. (2000). Palaeoecological studies of Holocene lake sediments from west Greenland. *Palaeogeography, Palaeoclimatology, Palaeoecology* **155**, 285–304.

Bentley, M.J., Hodgson, D.A., Smith, J.A., and Cox, N.J. (2005a). Relative sea level curves for the South Shetland Islands and Marguerite Bay, Antarctic Peninsula. *Quaternary Science Reviews* **24**, 1203–1216.

Bentley, M.J. *et al.* (2005b). Early Holocene retreat of the George VI Ice Shelf, Antarctic Peninsula. *Geology* **33**, 173–176.

Bernhard, G., Booth, C.R., and Ehramjian, J.C. (2000). *Changes in Antarctic UV levels in Relation to Ozone Hole Characteristics.* www.biospherical.com/NSF/default.asp. Biospherical Instruments, San Diego, CA.

Bissett, A.P., Gibson, J.A.E., Jarman, S. *et al.* (2005). Isolation, amplification, and identification of ancient copepod DNA from lake sediments. *Limnology and Oceanography Methods* **3**, 533–542.

Björck, S., Håkansson, H., Olsson, S., Barnekow, L., and Janssens, J. (1993). Palaeoclimatic studies in South Shetland Islands, Antarctica, based on numerous stratigraphic variables in lake sediments. *Journal of Paleolimnology* **8**, 233–272.

Björck, S. *et al.* (1996). Late Holocene palaeoclimatic records from lake sediments on James Ross Island, Antarctica. *Palaeogeography, Palaeoclimatology, Palaeoecology* **121**, 195–220.

Blais, J.M. *et al.* (2005). Arctic seabirds transport marine-derived contaminants. *Science* **309**, 445.

Boyle, J.F. (2007). Simulating loss of primary silicate minerals from soil due to long-term weathering using Allogen: comparison with soil chronosequence, lake sediment and river solute flux data. *Geomorphology* **83**, 121–135.

Briner, J.P., Axford, Y., Forman, S.L., Miller, G.H., and Wolfe, A.P. (2007). Multiple generations of interglacial lake sediment preserved beneath the Laurentide Ice Sheet. *Geology* **35**, 887–890.

Brock, C.P. *et al.* (2006). Spatial variability of stable isotopes and fossil pigments in surface sediments of Alaskan coastal lakes: constraints on quantitative estimates of past salmon abundance. *Limnology and Oceanography* **51**, 1637–1647.

CAPE Project Members (2001). Holocene paleoclimate data from the Arctic: testing models of global climate change. *Quaternary Science Reviews* **20**, 1275–1287.

Christoffersen, K.S. (2006). The influence of resting geese on the water chemistry of Arctic lakes. In *American Society of Limnology and Oceanography Summer Meeting Abstracts*, p. 23. ASLO, Victoria, BC.

Clarke, A., Barnes, D.K.A., and Hodgson, D.A. (2005). How isolated is Antarctica? *Trends in Ecology and Evolution* **20**, 1–3.

Cohen, A.S. (2003). *Paleolimnology: the History and Evolution of Lake Systems.* Oxford University Press, Oxford.

Convey, P. and Stevens, M.I. (2007). Antarctic biodiversity. *Science* **317**, 1877–1878.

Cook, A.J., Fox, A.J., Vaughan, D.G, and Ferrigno, J.G. (2005). Retreating glacier fronts on the Antarctic Peninsula over the past half-century. *Science* **308**, 541–544.

Coolen, M.J.L. *et al.* (2004). Combined DNA and lipid analyses of sediments reveal changes in Holocene haptophyte and diatom populations in an Antarctic lake. *Earth and Planetary Science Letters* **223**, 225–239.

Cromer, L., Gibson, J.A.E., Swadling, K., and Ritz, D.A. (2005). Faunal microfossils: indicators of Holocene ecological change in a saline Antarctic lake.

Palaeogeography, Palaeoclimatology, Palaeoecology **221**, 83–97.

Doran, P.T., Wharton, R.A., Des Marais, D.J., and McKay, C.P. (1998). Antarctic paleolake sediments and the search for extinct life on Mars. *Journal of Geophysical Research-Planets* **103**, 28481–28493.

Doran, P.T., Fritsen, C.H., McKay, C.P., Priscu, J.C., and Adams, E.E (2003). Formation and character of an ancient 19-m ice cover and underlying trapped brine in an "ice-sealed" east Antarctic lake. *Proceedings of the National Academy of Sciences USA* **100**, 26–31.

Doran, P.T., McKnight, D.M., and Mckay, C.P. (2006). Lake level response to extreme summer conditions in the McMurdo Dry Valleys, and paleoclimatic significance. In *American Society of Limnology and Oceanography Summer Meeting Abstracts*, p. 32. ASLO, Victoria, BC.

Doran, P.T. *et al.* (2004). Paleolimnology of extreme cold terrestrial and extraterrestrial environments. In Pienitz, R., Douglas, M.S.V., and Smol, J.P. (eds), *Developments in Paleoenvironmental Research*, vol. 8, *Long-term Environmental Change in Arctic and Antarctic Lakes*, pp. 475–507. Springer, Dordrecht.

Douglas, M.S.V. and Smol, J.P. (1999). Freshwater diatoms as indicators of environmental change in the high Arctic. In Stoermer, E.F. and Smol, J.P. (eds), *The Diatoms: Applications for the Environmental and Earth Sciences*, pp. 227–244. Cambridge University Press, Cambridge.

Douglas, M.S.V. and Smol, J.P. (2000). Eutrophication and recovery in the High Arctic: Meretta Lake revisited. *Hydrobiologia* **431**, 193–204.

Douglas, M.S.V., Smol, J.P., and Blake, Jr, W. (1994). Marked post-18th century environmental change in high Arctic ecosystems. *Science* **266**, 416–419.

Douglas, M.S.V., Smol, J.P., Savelle, J.M., and Blais, J.M. (2004). Prehistoric Inuit whalers affected Arctic freshwater ecosystems. *Proceedings of the National Academy of Sciences USA* **101**, 1613–1617.

Edwards, T.W.D., Wolfe, B.B., Gibson, J.J., and Hammarlund, D. (2004). Use of water isotope tracers in high latitude hydrology and paleohydrology. In Pienitz, R., Douglas, M.S.V., and Smol, J.P. (eds), *Long-term Environmental Change in Arctic and Antarctic Lakes*, pp. 187–207. Springer, Dordrecht.

Ellis-Evans, J.C., Laybourn-Parry, J., Bayliss, P.R., and Perriss, S.J. (1997). Human impact on an oligotrophic lake in the Larsemann Hills. In Battaglia, B., Valencia, J., and Walton, D.W.H. (eds), pp. 396–404. *Antarctic Communities: Species, Structure and Survival*. Cambridge University Press, Cambridge.

Engstrom, D.R., Fritz, S.C., Almendinger, J.E., and Juggins, S. (2000). Chemical and biological trends during lake evolution in recently deglaciated terrain. *Nature* **408**, 161–166.

Finney, B.P., Gregory Eaves, I., Douglas, M.S.V., and Smol, J.P. (2002). Fisheries productivity in the northeastern Pacific Ocean over the past 2,200 years. *Nature* **416**, 729–733.

Fisher, T.G., Smith, D.G., and Andrews, J.T. (2002). Preboreal oscillation caused by a glacial Lake Agassiz flood. *Quaternary Science Reviews* **21**, 873–878.

Fitzgerald, W.F. *et al.* (2004). Modern and historic atmospheric mercury fluxes in northern Alaska: global sources and Arctic depletion. *Environmental Science & Technology* **39**, 557–568.

Fredskild, B. (1983). The Holocene development of some low and high arctic Greenland lakes. *Hydrobiologia* **103**, 217–224.

Gregory-Eaves, I., Smol, J.P., Douglas, M.S.V., and Finney, B.P. (2003). Diatoms and sockeye salmon (*Oncorhynchus nerka*) population dynamics: reconstructions of salmon-derived nutrients in two lakes from Kodiak Island, Alaska. *Journal of Paleolimnology* **30**, 35–53.

Hambley, G.W. and Lamoureux, S.F. (2006). Recent summer climate recorded in complex varved sediments, Nicolay Lake, Cornwall Island, Nunavut, Canada. *Journal of Paleolimnology* **35**, 629–640.

Hendy, C.H. (2000). Late Quaternary lakes in the McMurdo Sound region of Antarctica. *Geografiska Annaler* **82A**, 411–432.

Hodgson, D.A. and Johnston, N.M. (1997). Inferring seal populations from lake sediments. *Nature* **387**, 30–31.

Hodgson, D.A. and Convey, P. (2005). A 7000-year record of oribatid mite communities on a maritime-Antarctic island: responses to climate change. *Arctic, Antarctic and Alpine Research* **37**, 239–245.

Hodgson, D.A., Johnston, N.M., Caulkett, A.P., and Jones, V.J. (1998). Palaeolimnology of Antarctic fur seal *Arctocephalus gazella* populations and implications for Antarctic management. *Biological Conservation* **83**, 145–154.

Hodgson, D.A., Doran, P.T., Roberts, D., and McMinn, A. (2004a). Paleolimnological studies from the Antarctic and subantarctic islands. In Pienitz, R., Douglas, M.S.V., and Smol, J.P. (eds), *Developments in Paleoenvironmental Research*, vol. 8, *Long-term Environmental Change in Arctic and Antarctic Lakes*, pp. 419–474. Springer, Dordrecht.

Hodgson, D.A. *et al.* (2004b). Environmental factors influencing the pigment composition of *in situ* benthic microbial communities in east Antarctic lakes. *Aquatic Microbial Ecology* **37**, 247–263.

Hodgson, D.A. *et al.* (2005a). Late Quaternary climate-driven environmental change in the Larsemann Hills, East Antarctica, multi-proxy evidence from a lake sediment core. *Quaternary Research* **64**, 83–99.

Hodgson, D.A. *et al.* (2005b). Late Pleistocene record of elevated UV radiation in an Antarctic lake. *Earth and Planetary Science Letters* **236**, 765–772.

Hodgson, D.A. *et al.* (2006a). Examining Holocene stability of Antarctic Peninsula Ice Shelves. *Eos Transactions, American Geophysical Union* **87**, 305–312.

Hodgson, D.A. *et al.* (2006b). Recent rapid salinity rise in three East Antarctic lakes. *Journal of Paleolimnology* **36**, 385–406.

Hodgson, D.A. *et al.* (2006c). Interglacial environments of coastal east Antarctica: comparison of MIS 1 (Holocene) and MIS 5e (Last Interglacial) lake-sediment records. *Quaternary Science Reviews* **25**, 179–197.

Houghton, J.T. (2001). *Climate Change 2001: The Scientific Basis. Intergovernmental Panel of Climate Change, 3rd Assessment Report.* Cambridge University Press, Cambridge.

Johannessen, O.M. *et al.* (2005). *Remote Sensing of Sea Ice in the Northern Sea Route: Studies and Applications*: Springer Praxis Books, UK.

Jones, V.J., Hodgson, D.A., and Chepstow-Lusty, A. (2000). Palaeolimnological evidence for marked Holocene environmental changes on Signy Island, Antarctica. *The Holocene* **10**, 43–60.

Kaufman, D.S. *et al.* (2004). Holocene thermal maximum in the Western Arctic (0° to 180° W). *Quaternary Science Reviews* **23**, 529–560.

Kokelj, S.V., Jenkins, R.E., Milburn, D., Burn, C.R., and Snow, N. (2005). The influence of thermokarst disturbance on the water quality of small upland lakes, Mackenzie Delta region, Northwest Territories, Canada. *Permafrost and Periglacial Processes* **16**, 343–353.

Krümmel, E. *et al.* (2003). Delivery of pollutants by spawning salmon. *Nature* **425**, 255–256.

Lambeck, K. and Chappell, J. (2001). Sea level change through the last glacial cycle. *Science* **292**, 679–686.

Leavitt, P.R. and Hodgson, D.A. (2001). Sedimentary pigments. In Smol, J.P., Birks, H.J.B., and Last, W.M. (eds), *Developments in Paleoenvironmental Research*, vol. 3, *Tracking Environmental Changes using Lake Sediments: Terrestrial Algal and Siliceous Indicators*, pp. 295–325. Kluwer Academic Publishers, Dordrecht.

Leavitt, P.R. *et al.* (2003). Climatic control of ultraviolet radiation effects on lakes. *Limnology and Oceanography* **48**, 2062–2069.

Long, A.J., Roberts, D.H., and Rasch, M. (2003). The deglacial and relative sea-level history of Innaarsuit, Disko Bugt, West Greenland. *Quaternary Research* **60**, 162–171.

Ludlam, S., Feeny, S., and Douglas, M.S.V. (1996). Changes in the importance of lotic and littoral diatoms in a high arctic lake over the last 191 years. *Journal of Paleolimnology* **16**, 184–204.

McGowan, S., Juhler, R.K., and Anderson, N.J. (2007). Autotrophic response to lake age, conductivity and temperature in two West Greenland lakes. *Journal of Paleolimnology* **39**, 301–317.

Michelutti, N., Wolfe, A.P., Vinebrooke, R.D., Rivard, B., and Briner, J.P. (2005). Recent primary production increases in arctic lakes. *Geophysical Research Letters* **32**, L19715, doi:10.1029/2005GL023693.

Mueller, D.R., Jeffries, M.O., and Vincent, W.F. (2003). Break-up of the largest Arctic ice shelf and associated loss of an epishelf lake. *Geophysical Research Letters* **30**, 2031, doi:10.1029/2003GL017931.

Muir, D. and Rose, N.L. (2004). Lake sediments as records of Arctic and Antarctic pollution. In Pienitz, R., Douglas, M.S.V., and Smol, J.P. (eds), *Long-term Environmental Change in Arctic and Antarctic Lakes*, pp. 209–239. Springer, Dordrecht.

Perren, B.B., Anderson, N.J., and Douglas, M. (2006). Diatoms reveal complex spatial and temporal patterns of recent limnological change in West Greenland. In *American Society of Limnology and Oceanography Summer Meeting*, p. 93. ASLO, Victoria, BC.

Pienitz, R., Douglas, M.S.V., and Smol, J.P. (eds) (2004). *Long-term Environmental Change in Arctic and Antarctic Lakes*. Springer, Dordrecht.

Pienitz, R. and Smol, J.P. (1993). Diatom assemblages and their relationship to environmental variables in lakes from the boreal forest-tundra ecotone near Yellowknife, Northwest Territories, Canada. *Hydrobiologia* **269/270**, 391–404.

Pienitz, R. and Vincent, W.F. (2000). Effect of climate change relative to ozone depletion on UV exposure in subarctic lakes. *Nature* **404**, 484–487.

Pienitz, R., Smol, J.P., Last, W., Leavitt, P.R., and Cumming, B. (2000). Multiproxy Holocene palaeoclimatic record from a saline lake in the Canadian Subarctic. *The Holocene* **10**, 673–686.

Quinlan, R., Douglas, M.S.V., and Smol, J.P. (2005). Food web changes in Arctic ecosystems related to climate warming. *Global Change Biology* **11**, 1381–1386.

Roberts, D., McMinn, A., Cremer, H., Gore, D., and Melles, M. (2004). The Holocene evolution and palaeosalinity history of Beall Lake, Windmill Islands (East Antarctica) using an expanded diatom-based weighted averaging model. *Palaeogeography, Palaeoclimatology, Palaeoecology* **208**, 121–140.

Roberts, S.J. *et al.* (2008). The Holocene history of George VI Ice Shelf, Antarctic Peninsula from clast-provenance analysis of epishelf lake sediments. *Palaeogeography Palaeoclimatology Palaeoecology* **259**, 258–283.

Rosqvist, G.C., Rietti-Shati, M., and Shemesh, A. (1999). Late glacial to middle Holocene climatic record of lacustrine biogenic silica oxygen isotopes from a Southern Ocean island. *Geology* **27**, 967–970.

Rühland, K. and Smol, J.P. (2005). Diatom shifts as evidence for recent Subarctic warming in a remote tundra lake, NWT, Canada. *Palaeogeography, Palaeoclimatology, Palaeoecology* **226**, 1–16.

Rühland, K., Priesnitz, A., and Smol, J.P. (2003). Paleolimnological evidence from diatoms for recent environmental changes in 50 lakes across Canadian Arctic treeline. *Arctic, Antarctic and Alpine Research* **35**, 110–123.

Ryves, D.B., McGowan, S., and Anderson, N.J. (2002). Development and evaluation of a diatom-conductivity model from lakes in West Greenland. *Freshwater Biology* **47**, 995–1014.

Sabbe, K., Verleyen, E., Hodgson, D.A., and Vyverman, W. (2003). Benthic diatom flora of freshwater and saline lakes in the Larsemann Hills and Rauer Islands, East Antarctica. *Antarctic Science* **15**, 227–248.

Schellnhuber, H.J. *et al.* (2006). *Avoiding Dangerous Climate Change*. Cambridge University Press, Cambridge.

Schindler, D.W., Curtis, P.J., Parker, B.R., and Stainton, M.P. (1996). Consequences of climatic warming and lake acidification for UV-B penetration in North American boreal lakes. *Nature* **379**, 705–708.

Schindler, D.W. and Smol, J.P. (2006). Cumulative effects of climate warming and other human activities of freshwaters of arctic and subarctic North America. *Ambio* **35**, 160–167.

Schwab, M. (1998). *Reconstruction of the Late Quaternary Climatic and Environmental History of the Schirmacher Ousis and the Wohlthat Massif (East Antarctica)*. Alfred Wegener Institute for Polar and Marine Research, Bremerhaven.

Siegert, M.J. (2001). *Ice Sheets and Late Quaternary Environmental Change*. Wiley, Chichester.

Siegert, M.J. *et al.* (2006). Exploration of Ellsworth Subglacial Lake: a concept paper on the development, organisation and execution of an experiment to explore, measure and sample the environment of a West Antarctic subglacial lake. *Reviews in Environmental Science and Bio/Technology* **6**, 161–179.

Smith, J.A. *et al.* (2006). Limnology of two Antarctic epishelf lakes and their potential to record periods of ice shelf loss. *Journal of Paleolimnology* **35**, 373–394.

Smith, J.A. *et al.* (2007). Oceanic and atmospheric forcing of early Holocene ice shelf retreat, George VI Ice Shelf, Antarctica Peninsula. *Quaternary Science Reviews* **26**, 500–516.

Smol, J.P. (1988). Paleoclimate proxy data from freshwater arctic diatoms. *Verhandlungen Internationale Vereingung der Limnologie* **23**, 837–844.

Smol, J.P. (2008). *Pollution of Lakes and Rivers: a Paleoenvironmental Perspective*. 2nd ed. Blackwell Publishing Oxford.

Smol, J.P. and Douglas, M.S.V. (2007). From controversy to consensus: making the case for recent climate change in the Arctic using lake sediments. *Frontiers in Ecology and the Environment* **5**, 466–474.

Smol, J.P. *et al.* (2005). Climate-driven regime shifts in the biological communities of arctic lakes. *Proceedings of the National Academy of Sciences USA* **102**, 4397–4402.

Sorvari, S., Korhola, A., and Thompson, R. (2002). Lake diatom response to recent Arctic warming in Finnish Lapland. *Global Change Biology* **8**, 171–181.

Squier, A.H., Hodgson, D.A., and Keely, B.J. (2005). Evidence of late Quaternary environmental change in a continental east Antarctic lake from lacustrine sedimentary pigment distributions. Antarctic Science. *Antarctic Science* **17**, 361–376.

Sun, L.G. and Xie, Z.Q. (2001). Changes in lead concentration in Antarctic penguin droppings during the past 3,000 years. *Environmental Geology* **40**, 1205–1208.

Sun, L.G., Xie, Z.Q., and Zhao, J.L. (2000). Palaeoecology—a 3,000-year record of penguin populations. *Nature* **407**, 858–858.

Tanabe, Y., Kudoh, S., Imura, S., and Fukuchi, M. (2008). Phytoplankton blooms under dim and cold conditions in freshwater lakes of East Antarctica. *Polar Biology*, **31**, 199–208.

Taton, A. *et al.* (2006). Contrasted geographical distributions and ecological ranges of cyanobacterial diversity from microbial mats in lakes of Eastern Antarctica. *FEMS Microbial Ecology* **57**, 272–289.

Verleyen, E., Hodgson, D.A., Sabbe, K., and Vyverman, W. (2004). Late Quaternary deglaciation and climate history of the Larsemann Hills (East Antarctica). *Journal of Quaternary Science* **19**, 361–375.

Verleyen, E., Hodgson, D.A., Milne, G.A. Sabbe, K., and Vyverman, W. (2005a). Relative sea level history from the Lambert Glacier region (East Antarctica) and its relation to deglaciation and Holocene glacier re-advance. *Quaternary Research* **63**, 45–52.

Verleyen, E., Hodgson, D.A., Sabbe, K., and Vyverman, W. (2005b). Late Holocene changes in ultraviolet radiation penetration recorded in an East Antarctic lake. *Journal of Paleolimnology* **34**, 191–202.

Vinebrooke, R.D. and Leavitt, P.R. (1996). Effects of ultraviolet radiation on periphyton in an alpine lake. *Limnology and Oceanography* **41**, 1035–1040.

Vyverman, W. *et al.* (2007). Historical processes constrain patterns in global diatom diversity. *Ecology* **88**, 1924–1931.

Wagner, B. and Melles, M. (2001). A Holocene seabird record from Raffles Sø, East Greenland, in response to climatic and oceanic changes. *Boreas* **30**, 228–239.

Wagner, B., Cremer, H., Hultzsch, N., Gore, D.B., and Melles, M. (2004). Late Pleistocene and Holocene history of Lake Terrasovoje, Amery Oasis, East Antarctica, and its climatic and environmental implications. *Journal of Paleolimnology* **32**, 321–339.

Wagner, B. *et al.* (2006). Glacial and postglacial sedimentation in the Fryxell basin, Taylor Valley, southern Victoria Land, Antarctica. *Palaeography, Palaeoclimatology, Palaeoecology* **241**, 320–337.

Weckström, J., Snyder, J.A., Korhola, A., Laing, T.E., and Macdonald, G.M. (2003). Diatom inferred acidity history of 32 lakes on the Kola Peninsula, Russia. *Water, Air, and Soil Pollution* **149**, 339–361.

Weller, G. (1998). Regional impact of climate change in the Arctic and Antarctic. *Annals of Glaciology* **27**, 543–552.

Wolfe, A.P. *et al.* (2004). Geochronology of high latitude lake sediments. In Pienitz, R., Douglas, M.S.V., and Smol, J.P. (eds), *Long-term Environmental Change in Arctic and Antarctic Lakes*, pp. 19–52. Springer, Dordrecht.

Zale, R. (1994). Changes in size of the Hope Bay Adelie penguin rookery as inferred from Lake Boeckella sediment. *Ecography* **17**, 297–304.

CHAPTER 4

The physical limnology of high-latitude lakes

Warwick F. Vincent, Sally MacIntyre, Robert H. Spigel, and Isabelle Laurion

Outline

The physical environment of high-latitude lakes exerts a wide-ranging influence on their aquatic ecology. Certain features of Arctic and Antarctic lakes, such as prolonged ice cover and stratification, persistent low temperatures, and low concentrations of light-absorbing materials, make them highly sensitive to climate change and to other environmental perturbations. These features also make such lakes attractive sites for limnological research and for developing models of broader application. In this review, we first describe the snow and ice dynamics of high-latitude lakes, and the factors that control their dates of freeze-up and break-up. We then examine the effects of ice cover and optical variables on the penetration of solar radiation that in turn influences heating and stratification, convective mixing, photochemical reactions, and photobiological processes including primary production. This is followed by a description of stratification and mixing regimes including meromixis in stratified saline lakes, and a summary of water budgets, currents, flux pathways and mixing processes that operate in ice-covered waters. We illustrate some of these physical limnological features of high-latitude lakes by detailed studies from two long-term ecological research sites, the Toolik Lake site in Arctic Alaska, and the McMurdo Dry Valleys site in the Ross Sea sector of Antarctica.

4.1 Introduction

The key physical characteristics that distinguish Arctic and Antarctic lakes are prolonged ice cover, persistently cold water temperatures, and the extreme seasonal fluctuation in energy supply, from continuous winter darkness to continuous sunlight in summer. These features are shared by high-latitude lakes in general, but there are also some limnological differences between the two polar regions, including differences in ice-cover thickness and duration, snow cover, mixing regimes, and the degree of underwater control of spectral irradiance by dissolved organic matter.

The more severe climate regime of Antarctica means that many of its lakes are covered by perennial ice: an ice cap up to several meters thick that has built up over many years and that remains in place even throughout summer. Lakes of the McMurdo Dry Valleys (latitude 77°30′S; Plate 4) have ice-cap thicknesses typically in the range of 3–6 m (Chinn 1993), with Lake Vida holding the record of 19 m of ice. Perennially ice-capped lakes are known to occur in the north polar region, but are relatively rare; for example, Ward Hunt Lake at latitude 83°N in High Arctic Canada (Plate 4) contains up to 4.1 m of ice (Bonilla *et al.* 2005) and Anguissaq Lake in northwest Greenland has up to 3.4 m of ice (Hobbie 1984).

Vast bodies of fresh water occur in both regions (for locations, see Figures 1.1 and 1.2) but in Antarctica the largest lakes are capped by up to

4 km of the overlying ice sheet (Plate 1). The most well known of these is Lake Vostok (latitude 77°S, 14 000 km²), and more than 100 other lakes as well as streams and thinner layers of water have now been detected beneath the Antarctic ice sheet (National Research Council 2007; see Chapter 7 in this volume). In contrast, the largest freshwater bodies in the Arctic are ice-free for up to several weeks each year, for example Great Bear Lake, Canada (latitude 65–67°N, 114 717 km²), Nettilling Lake, Canada (latitude 66.5°N, 5066 km²), Lake Taimyr, Russia (latitude 74.1°N, 4560 km²), and Lake Hazen, Canada (latitude 81.8°N, 542 km²; Plate 5).

In addition to their diverse freshwater lakes, both polar regions contain saline waterbodies, and these have some surprising physical properties. In the most extreme cases, the lakes are so hypersaline that they never freeze, even during winter darkness. For example, Deep Lake in the Vestfold Hills, Antarctica, has a salinity of 270 parts per thousand (ppt; sea water is about 35 ppt), and the water remains liquid and free of ice throughout the year, reaching a minimum temperature of –18°C. When ice freezes, it expels most of its dissolved materials into the remaining water, and only small quantities of the original salts, nutrients, and organic materials are retained within the ice. The almost complete freeze-up of shallow ponds in both polar regions thus gives rise to a highly concentrated mixture of solutes in their remaining bottom waters. This in turn depresses the freezing point and can allow liquid water conditions to persist throughout winter, at temperatures well below 0°C (for example, –12°C in Ross Island ponds, Antarctica; Schmidt *et al.* 1991). For the organisms that live in these ponds, these aqueous conditions of extreme cold and high salinities likely pose a severe challenge for winter survival. In deeper saline lakes, the salt exclusion during winter ice production can result in brine-rich density plumes that sink to the bottom of the lake. In this way, the deep waters of such lakes can accumulate salt and have much greater densities than their overlying surface waters. Finally, in saline lakes that are overlaid by lighter, fresher water and that are protected from wind-induced mixing by ice through most or all of the year, solar energy passing through the ice cap in summer can gradually heat the dense,

saline deeper waters to temperatures that are well above those in the overlying atmosphere (Vincent *et al.* 2008). These extreme solar-heated lakes are known from both polar regions, with deep-water temperatures that reach up to 14°C (Lake C1 in the Canadian High Arctic, Van Hove *et al.* 2006) and 22°C (Lake Vanda in the McMurdo Dry Valleys, Antarctica; Spigel and Priscu 1998).

4.2 Snow and ice dynamics

The dates of freeze-up and break-up of high-latitude lakes mark abrupt transition points in their seasonal limnology, and also provide a sensitive index of regional climate change. The seasonal pattern of lake ice dynamics has been simulated by applying a one-dimensional heat model (Duguay *et al.* 2003; Figure 4.1). This model gives a good estimation of the duration and thickness of ice cover for shallow lakes at high latitude, and predicts the ice-on and ice-off dates to within 2 days. For a shallow lake at Barrow, Alaska (latitude 71°N) the ice cover extends from mid-September to late June (up to 80% of the year) while for a similar site at Churchill, Manitoba (59°N), ice covers the lake from mid-October to early June (up to 65% of the year). In addition to local climate, two features play a major role in affecting ice thickness and dynamics. First, thick snow cover acts as an effective insulator, slowing the loss of heat by conduction to the atmosphere above. Snow-covered lakes, for example in protected valleys or in regions of higher snowfall, tend to have thinner ice than wind-swept lakes in the tundra, and their date of break-up is earlier as a result of less ice to melt (Figure 4.1). The second variable is lake depth. In deeper lakes, the water column acts as a heat sink in which much of the thermal energy is lost to greater depth and is unavailable for melting the ice. Hence, ice-off occurs later in the year in such lakes. In addition, deep lakes have a greater heat capacity per unit area and therefore take longer to cool in autumn before ice production begins. Great Slave Lake (latitude 62°N), for example, does not fully freeze up until mid-December to early January, some 6 weeks later than shallower lakes in the region (Oswald and Rouse 2004).

The duration of ice cover is sensitive to variations in climate, and therefore the ice-on and ice-out dates

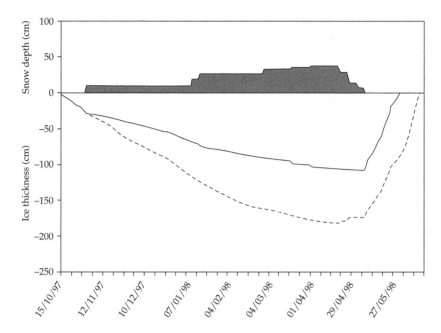

Figure 4.1 Application of a lake-ice model to lakes at Churchill, Canada, with (solid) and without (dashed line) snow cover. From Duguay *et al.* (2003).

as well as ice thickness can show large fluctuations from year to year and in response to long-term climate trends. In both the Arctic and Antarctica, lake ice cover has been identified as a sentinel of climate change, as well as a pivotal feature that amplifies the effects of climate change (Quesada *et al.* 2006). In the McMurdo Dry Valleys, a systematic increase in lake ice thickness accompanied a cooling trend in this region of Antarctica, and resulted in several limnological responses, including decreased phytoplankton production (Doran *et al.* 2002). In contrast, the Antarctic Peninsula region has experienced the most rapid warming in the southern hemisphere, resulting in decreased duration of ice cover; a variety of limnological shifts have been attributed to this effect, including the persistence of phytoplankton production over a longer growing season (Quayle *et al.* 2002).

In the northern hemisphere, long-term climate change has led to earlier dates of ice break-up and the onset of summer ice-free conditions in lakes that in the past have been covered by perennial ice (Van Hove *et al.* 2006 and references therein). The

date of ice-off is particularly critical for the energy and thermal regimes of lakes (Rouse *et al.* 2003) as well as for light availability for photosynthesis and biological UV exposure. In the 250 000-year paleolimnological record from Lake El'Gygytgyn, an ancient crater lake at latitude 67.5°N in Siberia, periods of highest paleoproductivity have been attributed to warm, ice-free summer conditions, whereas the lowest production rates have been attributed to cold, moist conditions in which the ice cover is perennial and coated by snow (Melles *et al.* 2007).

4.3 Underwater radiation

The supply and spectral composition of solar energy penetrating into water columns of high-latitude lakes depend on incident radiation and the nature of the overlying snow and ice cover. Once within the water column, the spectral transmission of the solar radiation is dictated by the absorption and scattering of photons by four categories of 'optically active' substances (materials that absorb and scatter light): pigment-containing algal cells (the phytoplankton), non-algal particles (especially detritus), colored dissolved organic matter (CDOM), and water itself.

The total annual solar radiation reaching the surface of the Earth's atmosphere drops substantially from the temperate zone to the polar regions, for example by 40% from latitudes 40–70°S. This effect is compounded by the high albedo (reflection of sunlight) of snow and ice that greatly reduces the amount of solar energy available for absorption and transfer to heat in the waters below. Mid-summer daily total irradiance can be as high in the polar regions as in the temperate zone ($\approx$40 MJ m^{-2} day^{-1} at the top of the atmosphere, with a daily maximum of approximately 800 W m^{-2} at the lake surface). At latitudes beyond the Arctic and Antarctic Circles, there are 24 h of daylight during summer, although with large variations in incident irradiance over the diurnal cycle (for example, there is a factor-of-5 change during mid-summer cloud-free days in the McMurdo Dry Valleys, with additional effects of topography depending on the site; Dana et al. 1998). Daily irradiance drops to zero during the two or more months of winter darkness, resulting in cold water temperatures and prolonged snow and ice cover. This seasonal irradiance cycle causes extreme fluctuations in the supply of energy for heating, melting, and for other processes that depend on solar radiation, such as photochemistry and photosynthesis.

From the 1980s onwards there has been much concern about the spectral composition of incident solar radiation in the polar regions, specifically the enhancement of UV-B irradiance associated with stratospheric ozone depletion. UV-B radiation (280–320 nm) is the photochemically and photobiologically most reactive waveband of the ground-level solar spectrum, and levels in spring have increased in the Arctic, as well as Antarctica. Changes in cloud cover can also affect UV radiation. At McMurdo Station, Antarctica, between 1989 and 2004 clouds reduced UV irradiance at 345 nm on average by 10% compared with clear-sky levels. Cloud-dependent reductions varied substantially by month and year, but rarely exceeded 60%, and generally increased with wavelength. Between September and November, the variability in UV caused by changes in total ozone was about twice as high as the UV variability due to clouds (Bernhard et al. 2006).

The reflective and attenuating properties of snow have a major effect on the underwater light regime of high-latitude lakes. For example, in early June in meromictic Lake A on Ellesmere Island (see Chapter 1), the maximum photosynthetically available irradiance (PAR; 400–700 nm) under 2 m of ice and 0.5 m of snow was about 5 μmol photons m^{-2} s^{-1}. This was equivalent to only 0.45% of incident PAR and was likely to severely limit primary production. Removal of snow from a 12 m^2 area on this lake increased PAR under the ice by a factor of 13 and biologically effective UV radiation by a factor of 16 (Belzile et al. 2001). Snow cover is highly variable and influenced directly by climate change, and modeling analyses show that such changes in snow cover would have a much greater effect on underwater UV exposure than moderate stratospheric ozone depletion (Vincent et al. 2007).

Lake ice is also an attenuator of solar radiation, but to an extent that depends on its gas content. White, bubble-containing ice is a highly scattering medium and strongly reduces the penetration of light to the waters beneath (Howard-Williams et al. 1998). However, lake ice that is largely free of bubbles can be fairly transparent to solar radiation, in the UV as well as PAR wavebands. This is because many of the impurities found in the waters beneath are excluded from the ice during freeze-up, including CDOM that strongly absorbs solar radiation in the blue and UV wavebands (Belzile et al. 2002). Spectral measurements at Lake Vanda, Antarctica, showed that about 10% of PAR passed through the 3.5-m-thick layer of clear perennial ice into the lake water beneath, and sufficient UV radiation penetrated into the upper water column to potentially cause photobiological effects (Vincent et al. 1998). A compilation of lake-ice attenuation data for Dry Valley lakes is given in Howard-Williams et al. (1998), and shows diffuse attenuation coefficients (the rate of exponential decline in irradiance with depth) for PAR in Lake Vanda and Lake Hoare of 0.49 to 1.33 m^{-1}, respectively.

CDOM is often one of the main constituents that attenuates light through the water column of high-latitude lakes and rivers (Laurion et al. 1997).

It is composed largely of humic materials derived from catchment vegetation and soils, and CDOM concentrations therefore tend to be much higher in forest-tundra and tundra lakes relative to alpine and polar desert lakes, where a greater proportion of the organic carbon is generated within the water body and is less colored. Some of the clearest lakes in the world are therefore to be found in the unvegetated catchments of Antarctica. In Lake Vanda, for example, the low inputs of CDOM, particles, and nutrients combine to result in crystal-clear blue waters in which maximum photosynthesis can occur some 60 m below the ice cap (see Chapter 9 in this volume). CDOM absorption of light increases with decreasing wavelength, and therefore even in the particle-rich waters of large Arctic rivers, this component controls the underwater attenuation of blue and UV radiation (Figure 4.2). Paleolimnological approaches have been used to infer long-term changes in the past in UV or PAR exposure in Arctic lakes (e.g. Pienitz and Vincent 2000) and Antarctic lakes (e.g. Verleyen *et al.* 2005) that result from changes in vegetation and thus CDOM concentrations, or from changes in ice and snow cover, lake depth, or other factors (see Chapter 3).

The influence of algal particles on underwater light is generally small in high-latitude lakes given the ultra-oligotrophic or oligotrophic nature of these waters. In a study of 204 lakes in the Canadian Arctic Archipelago, the planktonic chlorophyll *a* concentration ranged between less than $0.1–6 \, \mu g \, L^{-1}$ (Hamilton *et al.* 2001). However, in a study of 81 lakes and ponds spanning the Russian arctic treeline, chlorophyll *a* concentration was slightly higher, especially in the region of Pechora River where it reached $79 \, \mu g \, L^{-1}$ and averaged $3.5 \, \mu g \, L^{-1}$. In these lakes, secchi depth was correlated with particulates (chlorophyll *a*, particulate organic carbon, or particulate organic nitrogen) and less significantly with dissolved organic carbon. Where chlorophyll peaks do occur, they are often deep in the water column well below the ice, as is also found in highly stratified coastal Arctic seas (Figure 4.2).

In some high-latitude waters, non-algal particulates can also contribute substantially to the underwater attenuation of light, especially in the PAR waveband. These include waters that receive glacial flour (ground rock dust) from alpine glaciers, and river systems that receive particles from eroding permafrost soils. The influence of glacier-derived particles on lake primary production was investigated in lakes of the region studied by the Arctic Long-Term Ecological Research (LTER)

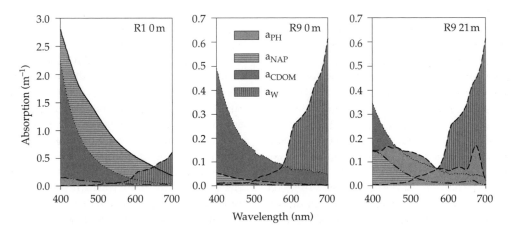

Figure 4.2 Spectral light absorption in an Arctic river and estuary. The curves are for each of the light-absorbing components (phytoplankton, a_{PH}; non-algal particles, a_{NAP}; colored dissolved organic matter, a_{CDOM}; and water itself, a_W) in samples from the Mackenzie River (station R1 0 m) and the adjacent coastal ocean (R9) at the surface (0 m) and in the deep chlorophyll maximum (21 m). From Retamal *et al.* (2008); the brown turbid water of this river (because of high a_{NAP} and a_{CDOM}) is shown in Plate 7.

program (Whalen *et al.* 2006). In one lake where high turbidity was measured due to resuspension of glacial silt from the lake bed, the average percentage of total lake volume and sediment surface located in the euphotic zone was reduced substantially compared with a lake situated on coarser glacial drift showing higher clarity (25 and 38% loss in lake and sediment volumes, respectively). This effect was translated into reduced primary production rates for the turbid lake ecosystem. The light regime of large Arctic rivers such as the Mackenzie River (Figure 4.2; Plate 8; Retamal *et al.* 2008) and its associated flood-plain lakes are very strongly influenced by suspended particles, with a secondary contribution by CDOM. Optical analyses of a series of lakes in the Mackenzie Delta showed that during the spring flood peak, suspended sediments were the major attenuator among most lakes (from 15 to 40% of total attenuation), other than those with high sills or positioned far from channel connection points (Squires and Lesack 2003). During the summer period of open water, however, CDOM appeared to be the most important light attenuator among almost all of the lakes in the central delta, with any algal effects of only minor importance for underwater light. The river was a source of CDOM, while within-lake processes contributed to non-colored dissolved organic carbon (DOC), leading to a CDOM–DOC inversion influence with increasing distance (increasing transparency) from the channel connection. Thaw or thermokarst lakes that form by permafrost erosion often show large lake-to-lake differences in color and other optical properties (Plates 8 and 9), and these in part are controlled by their suspended sediment load.

4.4 Stratification regimes

High-latitude lakes show many deviations from the dimictic pattern (two periods of mixing per year as found in warmer parts of the Arctic) that is especially common in north temperate regions. In lakes of the High Arctic and Antarctica that briefly lose their ice cover in summer, there may be insufficient time to warm the water above 3.98°C (the temperature of maximum density), and the water column remains cold, unstable, and mixing until refreezing and inverse stratification resumes

the next winter (Figure 4.3; see also Figure 1.3). Such lakes are termed 'cold monomictic' and their water-column circulation during summer is caused by convective mixing (their surface waters are warmed in the sun, become more dense, and therefore sink) in combination with wind-induced mixing. For lakes that warm only a little above the 3.98°C threshold, there is only weak resistance to wind-induced mixing, and the water bodies may stratify intermittently during calm sunny periods, and then mix to the bottom during storm events. This 'polymixis' is especially favored in polar environments because the change in density of water per degree C ($\Delta\rho/\Delta T$) is small at cold temperatures and rises nonlinearly with increasing temperature. For example, the shift from 4 to 5°C and from 29 to 30°C requires the same input of heat, but the resultant $\Delta\rho/\Delta T$ (and increase in stability) is 40 times higher for the latter.

The extent of water-column heating during summer is affected by incident solar radiation, the duration of snow and ice cover and air temperatures. It is also a function of water depth. Shallow lakes have less volume to heat for the same radiative and conductive input of energy across their surface, and they therefore melt out more rapidly. This feature also contributes to the more rapid melting of ice around the shallow edges of lakes. This produces a moat of open water encircling the lake, and the expanse of inshore ice-free water is often enlarged by stream inflows. A striking example of the effects of depth on stratification regime is seen in Lac à l'Eau Claire (Clearwater Lake), a large, multi-basin impact crater lake in the forest-tundra zone of Sub-Arctic Québec (Plate 9). This lake remains ice-covered for 9–10 months each year, and given differences in water depth, can exhibit cold mono-mixis, dimixis, and polymixis, all within the same lake but in different basins (Figure 4.4). The differential heating of lakes associated with variation in water depth can also result in variations across the same basin, with warm inshore water separated from cooler offshore water. The transition can be abrupt and associated with a sharp frontal region called a thermal bar. This phenomenon is well known from the North American Great Lakes, and has been observed in several large Arctic lakes such as Lake Taimyr (Robarts *et al.* 1999).

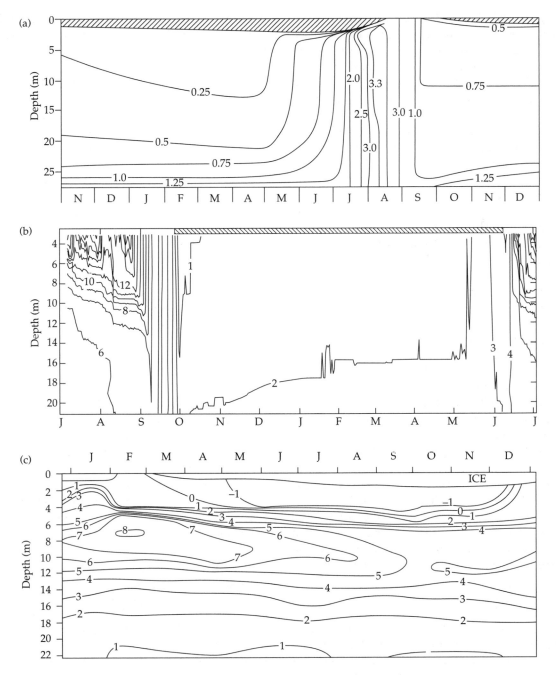

Figure 4.3 Stratification and mixing regimes in three polar lakes. Each panel shows temperatures in degrees Celsius. (a) Cold monomictic Char Lake, Canadian High Arctic (redrawn from Schindler *et al.* 1974). (b) Dimictic Toolik Lake, Alaska, from July 2004 to July 2005 (S. MacIntyre *et al.*, unpublished results). Inverse temperature gradient near ice-off is stabilized by increases in specific conductance (salt content) with depth. (c) Meromictic Ace Lake, Antarctica (redrawn from Burch 1988).

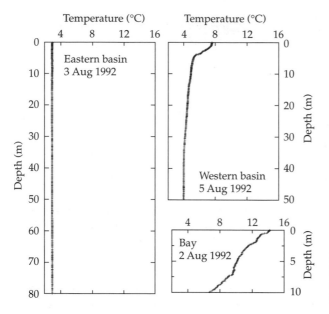

Figure 4.4 Three different stratification regimes in Lac à l'Eau Claire, northern Québec (Plate 9), during summer: cold **monomictic** (Eastern basin), **polymictic** (Western basin), and **dimictic** (Bay). Redrawn from Vincent and Hobbie (2000).

Another set of stratification and mixing regimes is found in the saline waters of the High Arctic and Antarctica. For the lakes with an ice cover that melts out each year, the surface waters may stratify and mix, while the bottom waters remain largely disconnected from the surface by a strong salinity-dependent density gradient, the pycnocline, that occurs between the two strata. Such lakes are termed meromictic (incompletely mixed), whereas those that are perennially capped by ice have been termed amictic (never mixed). The latter term, however, is inappropriate because several mixing processes are known to occur under ice including convective mixing, inflow-driven density currents, and thermohaline circulation.

The stratification regimes described above refer to the seasonal pattern of circulation and mixing; however, much shorter-term dynamics are also known to play a role in the structure and functioning of lake ecosystems. Lakes at all latitudes can be subject to short-term stratification caused by diurnal or multi-day warming that gives rise to temporary thermoclines (or diurnal thermoclines) near their surface. This may be sufficient to dampen moderate amounts of wind-induced mixing, in turn restricting the circulation of materials to the deeper waters of the epilimnion and prolonging

the exposure of near-surface biota and chemical constituents to bright solar radiation. These temporary thermoclines have been identified as a mechanism contributing toward UV-damage of phytoplankton (Milot-Roy and Vincent 1994) and UV-photodegradation of CDOM (Gibson *et al.* 2001). The thermoclines are likely to be less important in cold waters because of the nonlinear relationship between water density and temperature, but may be of greatest importance in colored waters that strongly absorb solar radiation, such as Toolik Lake, Alaska, and Sub-Arctic lakes.

4.5 Hydrological balance and flow pathways under the ice

High-latitude lakes have a number of distinctive features that influence the transport of materials from land to water, and that affect the pathways of transport once within the lake. The annual hydrological balance of a lake is the net sum of all water inputs (Q_i) minus all outputs (Q_o), giving a net change in volume (ΔV) of water stored in the lake:

$$\Delta V = Q_i - Q_o = Q_{is} + Q_{ig} + Q_{ip} - Q_{oo} - Q_{oe}$$

where the Q_i terms respectively are the annual inputs from streams, groundwater (typically low because of the permafrost soils), and direct precipitation on the lake, and the Q_o terms respectively are the annual water losses due to outflow and evaporation from the lake, all in cubic meters.

The Q_{is} term differs from many lakes at temperate latitudes in that it is highly seasonal, falling to zero during freeze-up in winter and remaining negligible until the late-spring melt season begins. This stream inflow is dictated primarily by melting glaciers (e.g. lakes of the McMurdo Dry Valleys and proglacial lakes in Greenland), or more commonly throughout the polar regions, by the melting of the winter snowpack. Both result in a sharp peak of discharge in late spring or early summer while the lakes are still ice-covered. At warmer locations, rainfall events during summer may also make a significant contribution to the annual water balance. Another striking difference relative to temperate and tropical systems is the effect of ice cover through most of the year. This results in evaporative losses that are generally low, although water is lost to the atmosphere directly from the ice, particularly in dry, windy climates. For lakes of the McMurdo Dry Valleys, for example, this ablation loss has been estimated at about 350 mm of ice per year.

The initial meltwaters from the snowpack or glaciers that accompany or immediately precede the peak discharge are typically enriched in DOC and other solutes including contaminants, in part because of interactions with the underlying soil and hyporheic zone (see Chapter 5), but also because of freeze-concentration and other fractionation effects within the snowpack. In ice-covered lakes, this initial, solute-rich cold water discharge is cooler and less dense than most of the inverse-stratified water column (see above). It therefore produces a flowing-water layer that is constrained to a thin zone immediately under the ice, and it may exit the lake rapidly via the outflow.

These hydrological and flow pathway effects are well illustrated by a 3-year study of mercury dynamics in Amituk Lake (latitude 75°N) on Cornwallis Island in the Canadian High Arctic (Semkin et al. 2005). The early melt of the snowpack was enriched in mercury as well as DOC and organic contaminants, and most of the input of these materials to the lake occurred during the spring snowmelt over

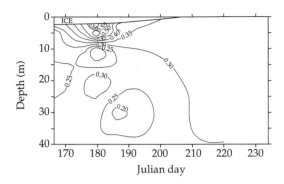

Figure 4.5 The through-flow of mercury (ng L^{-1}) under the ice in Lake Amituk in the Canadian High Arctic, during the period of peak spring inflow. Reproduced from Semkin et al. (2005), with permission from Elsevier.

a period of about 3 weeks. However, the lake was ice-covered and stratified during this time, and 59% of the annual mercury load traversed the lake as a mercury-rich bouyant current immediately beneath the ice (Figure 4.5), ultimately leaving the lake via its outflow. Only later in the summer when the ice had disappeared and the water column had warmed to become isothermal did the inflowing mercury and other solutes mix into the lake water. These observations indicate the critical importance of ice cover for the biogeochemical dynamics of high-latitude lakes.

4.6 Mixing and circulation beneath the ice

Although the presence of an ice cover isolates high-latitude lakes from direct wind-induced mixing through most of the year, a variety of other hydrodynamic mechanisms can occur, resulting in some circulation both horizontally and vertically. Inflowing density currents caused by cold river inputs, as described in the preceding section, provide one such mechanism for horizontal transport, and may lead to some turbulent mixing caused by the shear stress of the flowing water. There may be spatial variations in the heating and cooling of the lakewater, for example through differences in lake depth, ice thickness, and snow cover. These in turn will create density differences that could drive horizontal transport (Matthews and Heaney 1987).

Density currents can also be generated by local-ized high solute and particle concentrations. Salts and other materials are partially excluded from the ice during freeze-up (e.g. Belzile *et al.* 2002) and this can give rise to dense water at the ice–water inter-face that then sinks deeper into the water column. This mechanism also contributes to the formation of hypersaline bottom waters in meromictic high-latitude lakes that experience large quantities of ice production each year; examples include Ace Lake in the Vestfold Hills, Antarctica (Gibson 1999) and Romulus Lake in the Canadian High Arctic (Van Hove *et al.* 2006).

Another hydrodynamic mechanism that has generated considerable interest in the perennially ice-covered McMurdo Dry Valley lakes is the devel-opment of double diffusion cells caused by the dif-ference of two orders of magnitude in molecular diffusion rates for salt compared with heat (con-duction). These can expand and extend over con-siderable depth. In one of the lakes, Lake Vanda, a 'staircase' series of isohaline, isothermal convec-tion cells up to 20 m thick occurs in the upper water column, each sandwiched between layers of tem-perature and salinity gradients (Spigel and Priscu 1998). Radiotracer experiments on the largest of these cells showed that there were strong horizon-tal currents with velocities up to 1 cm s[-1] accom-panied by vertical turbulent mixing (Ragotzkie and Likens 1964).

There is considerable interest in the circulation and mixing dynamics of the subglacial lakes that lie beneath several kilometers of ice in Antarctica, and many of the mechanisms described above have been identified as potentially operating in such waters (National Research Council 2007). Some of these effects are amplified, however, by the unusual physical setting of these lakes. For example, at Lake Vostok, there is a 460-m difference in ice thickness between the north and south ends of the lake, leading to an estimated 0.31°C differ-ence in upper water temperatures along its 230-km length. Wüest and Carmack (2000) calculated that this would lead to pressure-driven horizontal flows of 0.3–0.6 mm s[-1] (up to about 20 km year[-1]), with the strength of this horizontal circulation dic-tated by heat fluxes at the ice-water interface, the extent of tilting of the ice and Coriolis force. These

calculations have been extended using numerical, three-dimensional ocean-circulation models, and predict flows of about 4500 m³ s[-1] in the northern and southern part of the lake and a cyclonic gyre in its center with velocities of around 0.1 cm s[-1] (Mayer *et al.* 2003). The predicted flow regime, however, was found to be highly sensitive to salinity, which at present is unknown.

In most freshwater lakes, the heat stored in the sediments in autumn will warm the overlying water, leading to a decrease in density and small-scale, buoyant turbulent plumes. As temperatures in these lakes decrease below 3.98°C, further geo-thermal heating of the bottom water will stabilize the bottom waters. Due to the great pressure on Lake Vostok and other lakes covered with thick ice, geothermal heating can continue to induce buoy-ant plumes when water temperatures are below 3.98°C. This is because the temperature of freezing (T_F; water at the ice–water interface should be close to this temperature) and the temperature of the maximum density of water (T_{MD}) are both functions of pressure. At atmospheric pressure, T_{MD} (3.98°C) is above T_F (0°C), and therefore small amounts of geothermal heating at the sediments will increase the density of the surrounding water and reinforce the inverse stratification. However, for fresh water at pressures above 28.4 MPa (280 atmospheres pressure; equivalent to ice 3170 m thick, assum-ing an ice density of 913 kg m[-3]), T_{MD} is less than T_F. Geothermal heating of the bottom water to above T_F will cause a decrease in its density and result in its upwards convection. For a freshwater column in Lake Vostok, Wüest and Carmack (2000) calculate that the vertical velocities within these convective plumes would be around 0.3 mm s[-1], which could completely mix the water column over a period of days.

4.7 Mixing and flow paths during ice-off and open-water conditions: Alaskan lakes

Detailed hydrodynamic studies have been under-taken at the Arctic LTER site, in particular during the open-water period in which temperatures can rise well above 3.98°C. Lakes at this site include shallow lakes that are polymictic during the ice-free period but which tend to freeze to the bottom during winter. Other lakes are deep enough that they do not freeze to the bottom in winter and thus, as long as oxygen concentrations are high enough, provide a refuge for fish in the winter. In summer, these lakes stratify and may be dimictic in contrast to the cold monomictic waterbodies at cooler high-latitude sites (see above). Mixing at ice-off in lakes in which temperatures have been measured under the ice is initially induced by convection. That is, instabilities are generated as water temperatures increase due to heating under the ice. The instabilities generally increase in size as the heating process continues. However, in Alaskan Arctic lakes, full overturn does not always occur under the ice. Overturn is inhibited by increases in salt content over the winter, particularly in the lower water column. These increases are likely due to remineralization as described by Mortimer (1971), and the lakes may be stable at ice-off (e.g. Figure 4.3b). Lake size determines whether these lakes mix subsequently. The smaller ones may be meromictic, or meromictic in some years; in consequence, deeper waters are often anoxic or hypoxic after ice-off. For larger lakes such as Toolik Lake (see Chapter 1 for a description of this site), wind forcing causes nonlinear internal waves that induce mixing shortly after ice-off, with eddy diffusivities (the parameter describing rates of turbulent diffusion) of 10^{-4}–10^{-3} m^{2}s^{-1}. Despite the high mixing, a chlorophyll maximum forms, attesting to the rapid growth rates of phytoplankton in early summer.

Stratification sets up quickly due to the relatively high attenuation from CDOM (diffuse attentuation coefficients for PAR in the range 0.5–1 m^{-1}), the low water temperatures, and consequently low evaporation rates, and the much warmer air than water temperatures at this time of year. The upper mixed layer is initially only 1–3 m deep and deepens slowly over the summer (Figure 4.6a). The rapid onset of stratification initially protects the newly formed chlorophyll maximum from being dispersed by wind. The stratification supports internal waves (Figure 4.6), which are intensified by persistent, high winds during transitions between warm and cold fronts. The degree of tilting of the thermocline, the amplitude of internal waves, and the onset of nonlinear internal wave breaking can be predicted

from the Wedderburn (W) and related Lake (L$_N$) numbers (Imberger and Patterson 1990). When these indices are above 10, the thermocline does not tilt. Partial upwelling occurs for values between 1 and 10, and upwelling to the surface occurs when these numbers decrease below 1. Because of the reduced temperature gradient across the thermocline and the lack of sheltering by trees or topography relative to similar sized lakes in the temperate zone, L$_N$ drops below critical values ($\approx$3) several times each summer, and internal waves in Arctic lakes even as small as 11 ha become nonlinear. Due to the resulting turbulence, eddy diffusivities in the metalimnion and hypolimnion respectively reach and sometimes exceed 10^{-6}–10^{-5} m^{2}s^{-1} (MacIntyre et al. 2006; S. MacIntyre, unpublished results).

The maximum heat content of lakes at the Arctic LTER site tends to occur in late July. Subsequently, as cold fronts move into the region, wind speeds and cooling associated with the fronts are intensified. Mixing increases in the upper water column. Depending on their depth, the lakes may mix fully, causing the surface entrainment of nutrients from bottom waters. For deeper lakes such as Toolik Lake, the mixing penetrates to depths of 8–12 m. However, the internal wave field is energized and becomes nonlinear (e.g. 7 August, Figure 4.6a, L$_N$ = 1). Entrainment of regenerated nutrients, such as ammonium, from the hypolimnion into the metalimnion is intensified at these times. Despite typical patterns of heat gain until mid-July and subsequent heat loss, there is considerable interannual variation in overall heating and in timing and rate of heat exchange. Maximum surface temperatures in July range from approximately 15 to 23°C. The resulting differences in stratification affect the rate of transport of nutrients between the hypolimnion and epilimnion. For instance, transport occurs on time scales of 1–4 days in cool summers when mixing in both the hypolimnion and epilimnion is intensified, but it takes a week or longer in warm summers when deepening of the epilimnion is less and the metalimnion, which acts as a barrier to transport, is thicker. Climate-induced changes in these exchange rates have the potential to affect mid-summer primary production. In warm summers with higher L$_N$, nutrients introduced directly into the upper mixed layer will be retained and

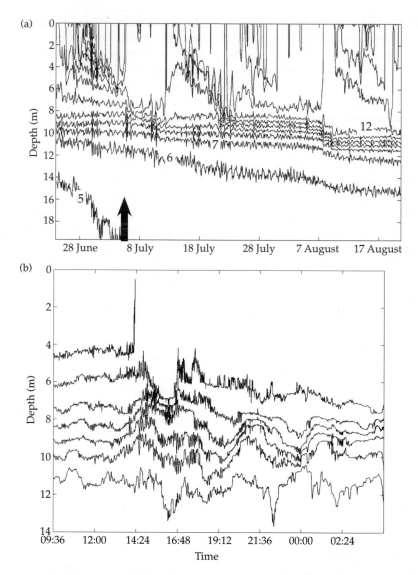

Figure 4.6 Short-term dynamics in Toolik Lake, Alaska. (a) Ten-minute-averaged isotherms during summer 2005 showing the strong, stable stratification in the thermocline after ice-off and periods with heating interspersed with periods of cooling. Diurnal winds with magnitudes of approximately 5 m s^{-1} induce internal waves that cause the up and down movement of isotherms with amplitude of 0.5 m. Four periods with higher winds and $1 \le L_N \le 3$ occurred between 28 June and 18 July. During these events, the frequency and amplitude of internal waves increased with concurrent increases in turbulence. (b) Expansion of the 4 July 2005 record showing the increased internal wave activity during an event with $L_N = 1$. Winds increased to 10 m s^{-1} by noon and remained high until 14:24 h. Isotherms in the thermocline slightly upwelled at that time but downwelled abruptly when winds ceased at 14:24 h. Upwelling occurred when winds increased again at 16:48 h. Relaxation of the wind at 19:12 h was followed by second-vertical-mode waves from 7 to 12 m (waves that expand and contract in the upper and lower **thermocline** at the same time). Throughout the event, high-frequency temperature fluctuations accompanied the larger-scale waves. Some of the fluctuations are high-frequency waves; others are turbulent fluctuations. The turbulence associated with internal waves at this time of year disperses the chlorophyll maximum; later in the season it results in vertical transport of regenerated nutrients. The data were obtained with Brancker TR-1050 temperature loggers sampling at 10-s intervals (S. MacIntyre et al., unpublished results).

will support growth. However, fluxes from deeper in the water column will be reduced. The converse occurs in cooler or windier summers with L_N more frequently decreasing to critical values. The formation of nonlinear waves which facilitate vertical transport depends on rates of heating, and on the magnitude and persistence of winds. Changes in the frequency of events with low L_N will thus be harbingers of change in Arctic ecosystems.

The flow path of incoming streams in Arctic lakes varies with temperature of the stream waters, the initial mixing of the stream water with the lake water, and the ambient stratification within the lake. Streams may be an important source of nutrients for phytoplankton in summer. One event with a discharge of $16\,m^3\,s^{-1}$ contributed 34% of the annual loading relative to an entire summer with low flows (MacIntyre *et al.* 2006). As the storm hydrograph begins to rise, intrusions tend to flow into the upper metalimnion. However, because temperatures of the streams decrease as discharge increases and because of the increased mixing near the inlet, intrusions during peak discharge occur deeper in the metalimnion. Whether the incoming nutrients are mixed into the euphotic zone and immediately available for growth depends upon the intensity of surface mixing that accompanies or follows the storm. Again, the efficacy of wind forcing and cooling in inducing fluxes of incoming nutrients is dependent on the overall stratification of the lake, and varies between warm and cool summers.

4.8 Stratification and mixing beneath perennial ice: McMurdo Dry Valley lakes

The snow- and ice-free valleys in the Trans-Antarctic Mountains west of McMurdo Sound contain several lakes that retain their ice cover year round, and that present a striking hydrodynamic contrast to the lakes of Arctic Alaska. Ice thickness varies from lake to lake, but generally is in the range of 3–6 m, with a narrow band around the edges of the lakes that melts out in most summers to form moats separating the ice from the shoreline. Perhaps the best known and most intensively studied among these are Lake Vanda in the Wright

Valley and lakes Fryxell, Hoare, and Bonney (Plate 4) in the Taylor Valley, all of which are currently sites of interdisciplinary research in the McMurdo Dry Valleys LTER program (see Chapter 1 in this volume for details).

Mixing and circulation times in the Dry Valley lakes are very long, with substantial modifications to density structure occurring mainly as a result of climate change, in contrast to the annual seasonal cycle of winter deep mixing and summer stratification that is typical of lakes that are ice-free for part of the year (Spigel and Priscu 1998). With the exception of water movement inside thermohaline convection cells in Lake Vanda (see above), turbulence is a rare occurrence in the lakes of the Dry Valleys, occurring only locally and intermittently. An example is the weak turbulence in the exchange flow in the narrows separating the two lobes of Lake Bonney, as inferred from temperature and conductivity microstructure measurements. Turbulence was restricted to boundary layers next to the bottom of the ice cover and next to the channel bed. Turbulence must also occur in wind-stirred surface water in moats, in inflows during episodes of prolonged or intense melting, and in buoyant meltwater rising along faces of submerged glaciers. But otherwise the interiors of these lakes are quiet.

Most of the McMurdo Dry Valley lakes are meromictic, with dilute surface freshwater overlying saline bottom water, for example up to 150 ppt in Lake Bonney. Like the saline Arctic lakes of Ellesmere Island, these highly stratified waters have gradually accumulated solar radiative heat over timescales of decades to millennia, and they have temperature profiles that do not resemble those normally found in freshwater lakes. Lake Bonney (Figure 4.7) and Lake Hoare have temperature maxima in the upper third of their water columns, whereas the maximum for Lake Fryxell (the shallowest of the lakes) is around mid-depth. Temperatures then decline steadily to minimum values at the bottom of the lakes. These minima are below 0°C in both lobes of Lake Bonney, but still well above the freezing point for the very saline waters found at the bottoms of these lakes. Temperature profiles, although retaining their general shape from year to year, show much greater variability than do conductivity profiles, as temperatures

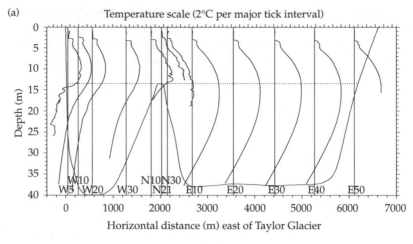

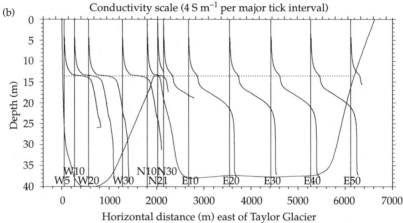

Figure 4.7 Horizontal variations in temperature and conductivity profiles in Lake Bonney, McMurdo Dry Valleys, Antarctica. Temperature (a) and conductivity (at 15°C; b) profiles measured on 10–11 January 1991 in Lake Bonney. Vertical lines mark positions of profiles as well as forming axes for 0°C and 0 S m^{-1}; the lake bottom is the curved line in the lower part of the graphs. Stations prefixed with a W are in the west lobe, E in the east lobe, and N in the narrows separating the two basins. The level of the top of the sill in the narrows is marked by the horizontal dotted line. From Spigel and Priscu (1998), with permission from the American Geophysical Union.

respond to seasonal and annual variations in heat transfer between the ice cover and the atmosphere. Values for maximum temperatures measured during the summer of 1990–1991 are indicative of typical values: 6.4°C for Lake Bonney east lobe, 3.2°C for Lake Bonney west lobe, 1.2°C for Lake Hoare, and 3.5°C for Lake Fryxell. Lake Hoare and the west lobe of Lake Bonney contain submerged faces of glaciers at one end of their basins: the Taylor Glacier in Lake Bonney west lobe (Plate 4), and the Canada Glacier in Lake Hoare. Although salinity profiles in the lakes are smooth, temperature

profiles exhibit more variability. For Lake Bonney's west lobe this is most noticeable in profiles made near the glacier and in the narrows (Figure 4.7). The fine-scale irregularities in temperature profiles are probably due to meltwater intrusions propagating into the lakes from the glacier faces and in the narrows of Lake Bonney, intrusions associated with exchange flows between the two basins. The glaciers at the ends of Lake Hoare and the west lobe of Lake Bonney are largely responsible for the cooler temperatures of these lakes compared with Lake Bonney east lobe. The cooler temperatures of Lake

Fryxell are due to its proximity to the coast and its exposure to a cooler climate.

In Lake Bonney, meltwater inflow is dominated by the Taylor Glacier and smaller mountain glaciers located at the western end of the west lobe (H. House, personal communication). Some features of the temperature and conductivity profiles are best explained with reference to the resulting net flow from the west lobe to the east lobe. Most of the meltwater inflows are probably relatively fresh and enter the lake just under the ice. As these inflows make their way east, they carry the salinity structure of the west lobe with them; it can be seen that conductivity profiles in the two lobes are virtually identical above the level of the narrows' sill (see the dotted horizontal line in Figure 4.7). Temperatures change along the way, however, as heat is exchanged with the atmosphere through the ice cover. Some meltwater inflows are saline, and form intrusions at depth in the west lobe. Water above such intrusions will be displaced upward, resulting in some leakage of salty water over the sill into the east lobe, where it will sink until it encounters the even higher salinities in the east lobe below the sill. Spigel and Priscu (1998) argued that the distinctive 'shoulder' in the east lobe conductivity profiles just below sill level is the signature of salty west-lobe water that is washed over the sill and sinks until it reaches its level of neutral density on top of the east lobe chemocline. The net east–west flow must also be partly responsible for maintaining the sharpness of the salinity gradient between 12 and 13 m in the west lobe by sweeping away salt that diffuses upward from below this level.

The unusual water-column properties of the McMurdo Dry Valleys lakes are intimately connected with the geology, geochemistry, and past climates of their surroundings. The extreme stability of the water columns, the lack of turbulence in the interiors of the lakes, and the very long time scales for mixing and circulation, mean that any localized mixing or horizontal variability can give rise to basin-scale motions that influence lake-wide water-column structure. Miller and Aiken (1996) have pointed out in their work on Lake Fryxell that horizontal redistribution of surface water on the scale of the entire lake basin can be caused by

freezing and melting processes on the underside of the lake's permanent ice cover. They showed how weak density currents generated by salt exclusion during moat freezing could explain a lens of relatively 'young' water found at depth in Lake Fryxell. Undoubtedly further examples of such basin-scale flows await discovery in these extreme polar lakes.

4.9 Conclusions

The most important physical characteristics of high-latitude lakes are their strong seasonal fluctuations in incident solar energy, the prevalence of low temperature conditions, and the presence of ice cover through much or all of the year. Ice cover has a pervasive influence on all aspects of polar lake ecosystems by directly affecting many key limnological properties and processes: solar energy supply, heat content of the water column, attenuation of light for photobiological and photochemical processes, transfer of kinetic energy from wind to water, exchange of gases with the atmosphere, the fate of inflowing materials, and the duration of stratification and mixing. Ice cover also provides the support for overlying snow, which itself has a major influence on spectral irradiance and the heat budget of the water column below. The process of ice-cover formation affects lake water chemistry and hydrodynamics by causing the freeze-concentration of solutes in the underlying water, and the production of density plumes.

High-latitude lakes are providing diverse, model systems for physical limnologists to explore general principles of lake structure and dynamics, as well as to discover novel processes found only in the Arctic or Antarctica, for example the hydrodynamics of subglacial lakes. For some questions, high-latitude waters are useful end members that extend the environmental range of key variables, for example for the effects of CDOM, ice thickness, and ice duration. All of these systems are highly vulnerable to ongoing climate change. Their persistent snow and ice cover is strongly dependent on the ambient climate regime, and loss of this cover has an abrupt impact on all ecosystem properties. The water temperatures of such lakes are often in the range where $\Delta\rho/\Delta T$ is small, and once exposed to the wind their water columns are unstable and sensitive to variations in local meteorology. The ongoing study of the physical limnology of high-latitude lakes will continue to generate new insights into the functioning of lake ecosystems in general, and will increasingly draw attention to their value as integrators and sentinels of global change.

Acknowledgments

Our limnological research on high-latitude lakes has been supported by the US National Science Foundation, the Natural Sciences and Engineering Research Council of Canada, the Network of Centres of Excellence programme Arctic Net, Fonds québécois de la recherche sur la nature et les technologies, Polar Shelf Canada, and the Canada Research Chair program. We thank Bernard Laval (University of British Columbia) and Claude Duguay (University of Waterloo) for their valuable review comments on the manuscript.

References

Belzile C., Vincent, W.F., Gibson, J.A.E., and Van Hove, P. (2001). Bio-optical characteristics of the snow, ice and water column of a perennially ice-covered lake in the high Arctic. *Canadian Journal of Fisheries and Aquatic Sciences* **58**, 2405–2418.

Belzile, C., Gibson, J.A.E., and Vincent, W.F. (2002). Colored dissolved organic matter and dissolved organic carbon exclusion from lake ice: implications for irradiance transmission and carbon cycling. *Limnology and Oceanography* **47**, 1283–1293.

Bernhard, G., Booth, C.R., Ehramjian, J.C., and Nichol, S.E. (2006). UV climatology at McMurdo Station, Antarctica, based on version 2 data of the National Science Foundation's Ultraviolet Radiation Monitoring Network. *Journal of Geophysical Research* **111**, D11201, doi:10.1029/2005JD005857.

Bonilla, S., Villeneuve, V., and Vincent, W.F. (2005). Benthic and planktonic algal communities in a high arctic lake: pigment structure and contrasting responses to nutrient enrichment. *Journal of Phycology* **41**, 1120–1130.

Burch, M.D. (1988). Annual cycle of phytoplankton in Ace Lake, an ice covered, saline meromictic lake. *Hydrobiologia* **165**, 59–75.

Chinn, T.J. (1993). Physical hydrology of the dry valley lakes. In Green, W.J. and Friedmann, E.I. (eds), *Physical*

and Biogeochemical Processes in Antarctic Lakes, pp. 1–51. American Geophysical Union, Washington DC.

Dana, G.L., Wharton, R.A., and Dubayah R. (1998). Solar radiation in the McMurdo Dry Valleys, Antarctica. In Priscu, J.C. (ed.), *Ecosystem Dynamics in a Polar Desert: The McMurdo Dry Valleys, Antarctica*, Antarctic Research Series, vol. 72, pp. 39–64. American Geophysical Union, Washington DC.

Doran, P.T. *et al.* (2002). Antarctic climate cooling and terrestrial ecosystem response. *Nature* **415**, 517–520.

Duguay, C.R.*et al.* (2003). Ice cover variability on shallow lakes at high latitudes: model simulations and observations. *Hydrological Processes* **17**, 3465–3483.

Gibson, J.A.E. (1999). The meromictic lakes and stratified marine basins of the Vestfold Hills, East Antarctica. *Antarctic Science* **11**, 175–192.

Gibson, J.A.E., Vincent, W.F., and Pienitz, R. (2001). Hydrologic control and diurnal photobleaching of CDOM in a subarctic lake. *Archives für Hydrobiologie* **152**, 143–159.

Hamilton, P.B., Gajewski, K., Atkinson, D.E., and Lean, D.R.S. (2001). Physical and chemical limnology of 204 lakes from the Canadian Arctic Archipelago. *Hydrobiologia* **457**, 133–148.

Hobbie, J.E. (1984). Polar limnology. In Taub, F.B. (ed.), *Lakes and Reservoirs. Ecosystems of the World*, vol. 23, pp. 63–104. Elsevier, Amsterdam.

Howard-Williams, C., Schwarz A-M, Hawes, I., and Priscu, J.C. (1998). Optical properties of the McMurdo Dry Valley lakes. In Priscu, J.C. (ed.), *Ecosystem Dynamics in a Polar Desert: The McMurdo Dry Valleys, Antarctica*, Antarctic Research Series, vol. 72, pp. 189–203. American Geophysical Union, Washington DC.

Imberger, J. and Patterson, J. (1990). Physical limnology. *Advances in Applied Mechanics* **27**, 303–475.

Laurion, I., Vincent, W.F., and Lean, D.R.S. (1997). Underwater ultraviolet radiation: development of spectral models for northern high latitude lakes. *Photochemistry and Photobiology* **65**, 107–114.

MacIntyre, S., Sickman, J.O., Goldthwait, S.A., and Kling, G.W. (2006). Physical pathways of nutrient supply in a small, ultra-oligotrophic arctic lake during summer stratification. *Limnology and Oceanography* **51**, 1107–1124.

Matthews, P.C. and Heaney, S.I. (1987). Solar heating and its influence on mixing in ice-covered lakes. *Freshwater Biology* **18**, 135–149.

Mayer, C., Grosfeld, K., and Siegert, M.J. (2003). Salinity impact on water flow and lake ice in Lake Vostok, Antarctica. *Geophysical Research Letters* **30**, 1767, doi:10.1029/2003GL017380.

Melles, M. *et al.* (2007). Sedimentary geochemistry of core PG1351 from Lake El'gygytgyn – a sensitive record of climate variability in the East Siberian Arctic during the past three glacial–interglacial cycles. *Journal of Paleolimnology* **37**, 89–104.

Miller, L.G. and Aiken, G.R. (1996). Effects of glacial meltwater inflow and moat freezing on mixing in an Antarctic lake as interpreted from stable isotope and tritium distributions. *Limnology and Oceanography* **41**, 66–76.

Milot-Roy, V. and Vincent, W.F. (1994). Ultraviolet radiation effects on photosynthesis: the importance of near-surface thermoclines in a subarctic lake. *Archives für Hydrobiologie Ergebnisse* **43**, 171–184.

Mortimer, C.H. (1971). Chemical exchanges between sediments and water in great lakes – speculations on probable regulatory mechanisms. *Limnology and Oceanography* **16**, 387–404.

National Research Council. (2007). *Exploration of Antarctic Subglacial Aquatic Environments: Environmental and Scientific Stewardship*. National Academies Press, Washington DC.

Oswald, C.J. and Rouse, W.R. (2004). Thermal characteristics and energy balance of various-size Canadian Shield lakes in the Mackenzie River Basin. *Journal of Hydrometeorology* **5**, 129–144.

Pienitz, R. and Vincent, W.F. (2000). Effect of climate change relative to ozone depletion on UV exposure in subarctic lakes. *Nature* **404**, 484–487.

Quayle, W.C., Peck, L.S., Peat, H., Ellis-Evans, J.C., and Harrigan, P.R. (2002). Extreme responses to climate change in Antarctic lakes. *Science* **295**, 645.

Quesada, A. *et al.* (2006). Landscape control of high latitude lakes in a changing climate. In Bergstrom, D., Convey, P., and Huiskes, A. (eds), *Trends in Antarctic Terrestrial and Limnetic Ecosystems*, pp. 221–252. Springer, Dordrecht.

Ragotzkie, R.A. and Likens, G.E. (1964). The heat balance of two Antarctic Lakes. *Limnology and Oceanography* **9**, 412–425.

Retamal, L., Bonilla, S., and Vincent, W.F. (2008). Optical gradients and phytoplankton production in the Mackenzie River and coastal Beaufort Sea. *Polar Biology* **31**, 363–379.

Robarts, R.D. *et al.* (1999). The biogeography and limnology of the Taimyr lake-wetland system of the Russian Arctic: an ecological synthesis. *Archive für Hydrobiologie* **121**, 159–200.

Rouse, W.R. *et al.* (2003). Interannual and seasonal variability of the surface energy balance and temperature of central Great Slave Lake. *Journal of Hydrometeorology* **4**, 720–730.

Schindler, D.W., Welch, H.E., Kalff, J., Brunskil, G.J., and Kritsch N. (1974). Physical and chemical limnology of

Char Lake, Cornwallis Island (75°N lat). *Journal of the Fisheries Research Board of Canada* **31**, 585–607.

Schmidt, S., Moskall, W., De Mora, S.J.D., Howard-Williams, C., and Vincent, W.F. (1991). Limnological properties of Antarctic ponds during winter freezing. *Antarctic Science* **3**, 379–388.

Semkin, R.G., Mierle, G., and Neureuther, R.J. (2005). Hydrochemistry and mercury cycling in a high arctic watershed. *Science of the Total Environment* **342**, 199–221.

Spigel, R.H. and Priscu, J.C. (1998). Physical limnology of McMurdo Dry Valley lakes. In Priscu, J.C. (ed.), *Ecosystem Dynamics in a Polar Desert: The McMurdo Dry Valleys, Antarctica*, Antarctic Research Series, vol. 72, pp. 153–187. American Geophysical Union, Washington DC.

Squires, M.M. and Lesack, L.F.W. (2003). Spatial and temporal patterns of light attenuation among lakes of the Mackenzie Delta. *Freshwater Biology* **48**, 1–20.

Van Hove, P., Belzile, C., Gibson, J.A.E., and Vincent, W.F. (2006). Coupled landscape-lake evolution in the Canadian High Arctic. *Canadian Journal of Earth Sciences* **43**, 533–546.

Verleyen, E., Hodgson, D.A., Sabbe, K., and Vyverman, W. (2005). Late Holocene changes in ultraviolet radiation penetration recorded in an east Antarctic lake. *Journal of Paleolimnology* **34**, 191–202.

Vincent, W.F., Rae, R., Laurion, I., Howard-Williams, C., and Priscu, J.C. (1998). Transparency of Antarctic lakes to solar ultraviolet radiation. *Limnology and Oceanography* **43**, 618–624.

Vincent, W.F. and Hobbie, J.E. (2000). Ecology of Arctic lakes and rivers. In Nuttall, M. and Callaghan, T.V. (eds), *The Arctic: Environment, People, Policies*, pp. 197–231. Harwood Academic Publishers, London.

Vincent, W.F., Rautio, M., and Pienitz, R. (2007). Climate control of underwater UV exposure in polar and alpine aquatic ecosystems. In Orbaek, J.B. *et al.* (eds), *Arctic Alpine Ecosystems and People in a Changing Environment*, pp. 227–249. Springer, Berlin.

Vincent, A., Mueller, D.R., and Vincent, W.F. (2008). Simulated heat storage in a perennially ice-covered high Arctic lake: sensitivity to climate change. *Journal of Geophysical Research* **113**, C04036.

Whalen, S.C., Chalfant, B.B., Fischer, E.N., Fortino, K.A., and Hershey, A.E. (2006). Comparative influence of resuspended glacial sediment on physicochemical characteristics and primary production in two Arctic lakes. *Aquatic Sciences* **68**, 65–77.

Wuest, A. and Carmack, E. (2000). A priori estimates of mixing and circulation in the hard-to-reach water body of Lake Vostok. *Ocean Modelling* **2**, 29–43.

CHAPTER 5

High-latitude rivers and streams

**Diane M. McKnight, Michael N. Gooseff, Warwick F. Vincent,
and Bruce J. Peterson**

Outline

Flowing-water ecosystems occur in the desert oases around the margins of Antarctica, and are common throughout the Arctic. In this review of high-latitude rivers and streams, we first describe the limnological properties of Antarctic streams by way of examples from the McMurdo Dry Valleys, including observations and experiments in the Taylor Valley Long-Term Ecological Research (LTER) site. These studies have drawn attention to the importance of microbial mats and the hyporheic zone in controling stream biogeochemistry, which in turn strongly influences the lake ecosystems that receive these inflows. We introduce the limnological characteristics of Arctic rivers and streams, and their dramatic response to nutrient enrichment as observed in a long-term experiment on the Kuparuk River, in the Toolik Lake, Alaska LTER site. Large rivers are an important feature of the Arctic, and discharge globally significant quantities of fresh water, dissolved organic carbon, and other materials into the Arctic Ocean. We review the rapidly increasing knowledge base on the limnology of these waters, and their associated lakes. The latter include the abundant lakes and ponds over their delta flood-plains, and the large stamukhi lakes that form behind thick sea ice near the river mouth. Both of these lake types experience extreme seasonal variations in all of their limnological properties, including water storage, and both influence the biogeochemical characteristics of the river water that is ultimately discharged to the sea. We conclude the review with a comparison of Arctic and Antarctic flowing waters and their associated flood-plains.

5.1 Introduction

Streams and rivers are important, dynamic ecosystems in high-latitude landscapes and their ecology reflects the extreme conditions of the polar environment. In Antarctica, surface streams (for subglacial flowing waters, see Chapter 7 in this volume) are restricted to the ice-free desert oases that are located near the coast. These are mostly small streams fed by glacial meltwater, with flow restricted to just a few weeks each year. Nonetheless, with the arrival of flow in summer, these streams become biogeochemically active ecosystems, influencing the flux of nutrients into the lakes in the valley bottoms. The longest river in the Antarctic, the Onyx River in the McMurdo Dry Valleys, is only 32 km in length. Furthermore, Antarctic streams are not significant sources of fresh water to the coastal marine environment, especially in comparison to the direct inputs from calving of glaciers. In contrast, the Arctic contains a diversity of stream and river ecosystems, ranging from low-order, spring-fed streams with perennial flow to large rivers that carry large, globally significant fluxes of fresh water to Arctic seas and the World Ocean. During the snowmelt period, the high flows carried by many Arctic streams and rivers act to flush flood-plain lakes and inundate wetlands, which are important habitats for wildlife. During the summer, the connectivity of the Arctic streams and rivers to the tundra landscape can vary dramatically with the thawing of the active layer.

Polar streams and rivers represent extreme or end-member ecosystems compared to streams typically found in temperate regions. Understanding hydrologic and ecological processes in these systems can provide insight into processes occurring in temperate systems. For example, Antarctic streams are excellent sites for studying processes occurring in the hyporheic zones of streams. The hyporheic zone is an important feature of all stream ecosystems and is the area underlying and adjacent to the stream channel where there is a continuous exchange of water. In temperate regions, the hyporheic zone is masked by the overlying riparian vegetation, but the lack of vegetation in the desert oases of the Antarctic allows the extent of the hyporheic zone to be directly seen as a damp area adjacent to the stream. Studies of these simplified and accessible stream systems have provided insightful models for wider understanding of hyporheic zone processes. Similarly, experiments in a well-studied Arctic river, the Kuparuk River on the North Slope of Alaska, have provided valuable understanding of the relationship between nutrient availability and cycling and changes in dominant phototrophs in the river (Plates 6 and 7).

Another driver for the study of polar streams is the accelerated change in polar regions due to changing global climate. These changes are driven by increases in greenhouse gases in the atmosphere, as well as depletion of ozone in the atmosphere over polar regions associated with reactions involving chlorofluorocarbons. Polar streams are dynamic hotspots on the landscape, and therefore responses to climate change in polar stream systems may be more immediate and evident than in terrestrial ecosystems of the Arctic or the massive glaciers of the Antarctic. Streams integrate the fluxes of water, sediment, and solutes from the surrounding landscape and monitoring of changes in these fluxes over time has the potential to provide insight into changes in terrestrial ecosystems. Because of the global significance of the fresh-water inflow to the Arctic Ocean, these fluxes of sediment and solutes carried by Arctic rivers are of great relevance for predicting change at the global scale.

5.2 Antarctic streams

Of the streams on the coastal regions of Antarctica, those in the McMurdo Dry Valleys have received the most study and are representative of streams in other coastal desert oases where the landscape is barren of vegetation (Vincent 1988). The investigations of dry valley streams by the New Zealand Antarctic Program began in the late 1960s provide an invaluable legacy. Specifically, the measurement of the flow in the Onyx River beginning in 1969 (Chinn 1993) provides the basis for the longest environmental record in the region. This record has been used to understand the warming trend that caused the lake levels to increase from 1903, when the lakes were first discovered by Robert Falcon Scott's expedition, until the late 1980s (Chinn 1993; Webster et al. 1996; Bomblies et al. 2001). Since that time, a decrease in lake levels has been associated with a cooling trend (Doran et al. 2002), which has been found to be associated with the development of the ozone hole.

Dry valley streams originate at the base of alpine, terminal, and piedmont glaciers or from proglacial lakes and shallow pools at the base of the glacier (Plate 6). The initiation of meltwater from these glaciers is driven by the warming of the surface layer of the glacier through a solid-state greenhouse effect and can begin when air temperatures are $-5°C$ and lower. Some streams travel short distances at the base of the glaciers before discharging into lakes on the valley floors. Other streams that drain higher-elevation glaciers flow for many kilometers through unconsolidated, highly permeable alluvium. In the small, first-order streams, discharge can vary as much as 10-fold during a day as a result of the changing insolation and sun angle (Vincent 1988; Conovitz et al. 1998). Stream temperatures also vary diurnally with changing insolation, reaching temperatures as high as 15°C on sunny days (Cozzetto et al. 2006).

Because streamflow depends upon solar insolation and the occurrence of temperatures above about $-5°C$, interannual variation in flow is great. For small streams in the McMurdo Dry Valleys, maximum discharge approached $1\,m^3\,s^{-1}$ during the warmest summer on record (2002) and the discharge in the Onyx River reached $30\,m^3\,s^{-1}$.

In contrast, during the cold summers more typical of a recent cooling period, many small streams exhibited low peak summer flows (less than $0.01\,m^3\,s^{-1}$) or no measurable flow (www.mcmlter.org). Dry valley streams typically have a substantial hyporheic zone, because of the high permeability of the alluvium, with rapid rates of exchange of water between the hyporheic zone and the main channel (Runkel *et al.* 1998; Figure 5.1). The hyporheic zone of dry-valley streams can be observed directly as a damp area adjacent to the stream. Further, because the source glaciers are effectively point sources of water and the surrounding landscape is extremely dry, these streams have no lateral inflow along the channel. The lower extent of the hyporheic zone is controlled by two factors: (1) the depth to permafrost, which increases progressively from 5 to 50 cm during the austral summer (Conovitz *et al.* 2006), and (2) the erosion of this frozen boundary by advection of warm water from the stream into the hyporheic zone (Cozzetto *et al.* 2006). As a result, an additional process limiting streamflow during cold summers is the loss of water from the channel as water soaks into the expanding hyporheic zone, which can delay the arrival of flow at the lake and be a significant component of the total annual flow for longer streams. This stored hyporheic water may allow

the overlying microbial mats to be physiologically active, as well as maintaining the hyporheic microbes, even during periods of limited surface flow during cold summers.

The discharge of dry valley streams is low, with peak summer flows typically less than $1\,m^3\,s^{-1}$, and therefore the streamflow is only sufficient to erode and transport appreciable quantities of the surficial material (primarily drift; Hall *et al.* 2000) in relatively steep reaches (McKnight *et al.* 2007). Thus in these glacially carved valleys, the channels of these streams are well-defined in the steep upper reaches. However, the stream channels are poorly defined and frequently divided into a multi-channel network through the flatter reaches, typically at lower elevations in zones that are submerged when lake levels rise. The quantity of sand and gravel transported by these streams is limited and substantial transport occurs infrequently, so the configuration of channel may be stable for decades or more, with or without regular summer flows.

In dry valley streams, the influence of dissolution of marine aerosols and in-stream weathering, on stream-water chemistry is reflected in the greater concentrations of major ions in longer streams and the progressive increase of ionic concentrations with stream length (Lyons *et al.* 1998; see also Chapter 8 in this volume). These solutes are generated through weathering reactions in the hyporheic zone and then brought to the stream through hyporheic exchange (Gooseff *et al.* 2002; Maurice *et al.* 2002; Figure 5.1). In addition, ion exchange with hyporheic and streambed sediments has also been shown to control dissolved solute loads of stream waters, particularly during enrichment additions to streams (Gooseff *et al.* 2004b). Nutrient sources to these stream ecosystems include atmospheric deposition of nitrogen species (Howard-Williams *et al.* 1997) and weathering of apatite as the main source of phosphate (Green *et al.* 1988).

At the landscape scale, biogeochemical processes occurring in the microbial mats (see Plate 10) and in the underlying hyporheic zone in the streams are important in regulating nutrient flux and the stoichiometric ratios of nutrients in the lakes (Howard-Williams *et al.* 1986, 1989a, 1989b; Vincent and Howard-Williams 1986, 1989; McKnight *et al.* 2004; Barrett *et al.* 2007). Howard-Williams *et al.*

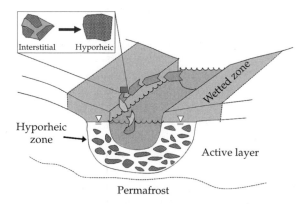

Figure 5.1 Processes occurring in steam hyporheic zones and in microbial mats. Stream water exchanges with the hyporheic zone where nutrients are taken up and where grain-surface weathering of hyporheic sediments increases dissolved silica and potassium concentrations in the hyporheic zone and stream.

(1989b) studied nitrogen cycling in dry valley streams and concluded that despite the cold temperatures, short periods of flow, and low rates of nitrogen fixation, the interannual persistence of microbial mats with high biomass results in rates of nitrogen cycling comparable with rates in many temperate streams. Hyporheic processes also control nutrient transport and cycling. Specifically, the hyporheic zone is an effective sink of nitrate due to either denitrification or assimilation by hyporheic microbial communities (Maurice *et al.* 2002; Gooseff *et al.* 2004a; McKnight *et al.* 2004), as well as a sink for soluble reactive phosphorus (Gooseff 2001; McKnight *et al.* 2004). Experimental nutrient additions have further quantified nutrient uptake and cycling in these benthic algal communities. For example, results from a nutrient addition experiment conducted at low flow in Green Creek, a small stream in Taylor Valley, showed that microbial mats controlled nitrate and phosphate concentrations not only by direct nutrient uptake but that dissimilatory nitrate reduction within the mats also influenced nitrogen cycling (Figure 5.1). Nitrogen redox processes are also important in the hyporheic zone, although the original source of nitrate in these systems is atmospheric deposition rather than weathering. Identification of denitrifying bacteria growing in the hyporheic zone (Maurice *et al.* 2002) indicated that nitrate may be transformed through hyporheic

biogeochemical processes. Experimental additions of nitrate and phosphate to Green Creek confirmed that hyporheic nutrient cycling was significant at the scale of the entire stream (Gooseff *et al.* 2004a; McKnight *et al.* 2004). Nutrient concentrations can be high during the first few hours and days of streamflow (Figure 5.2), reflecting the high nutrient release from glaciers during first melt, the initial flush of nutrients from newly wetted ground, and the reduced nutrient-stripping capacity of Antarctic stream ecosystems prior to full seasonal activation of their hyporheic and microbial mat communities.

The microbial mats and mosses are essentially perennial and remain in a freeze-dried state through the winter (Vincent and Howard-Williams 1986). The nature of the algal mats ranges from extensive, blanket-like coverage of the streambed to sparse patches occurring at the margins of the active channel or on the underside of rocks (Vincent 1988; Alger *et al.* 1997). The abundance of microbial mats is determined by physical features of the stream channel and the legacy of hydrologic events, such as episodes of high flow and prolonged periods of low flow. Microbial mats are typically abundant in many moderate gradient reaches where rocks occur in a stone pavement which provides a habitat with limited sediment mobility and scour (McKnight *et al.* 1998). In contrast, mats are sparse in steep stream sections

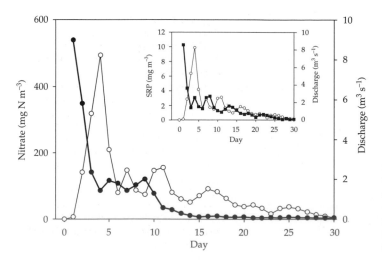

Figure 5.2 Discharge and nutrients in the Onyx River, Antarctica, over its 1983/1984 flow cycle, showing peak concentrations during the earliest period of seasonal flow. Open circles, river discharge; closed symbols, nitrate or soluble reactive phosphorus (SRP). Redrawn from the data in Vincent and Howard-Williams (1986).

where the rocks are uneven and jumbled or where the streambed is composed of moving sand. The microbial mats are composed chiefly of filamentous cyanobacteria of the family Oscillatoriaceae, with mats of *Phormidium* spp. abundant near the thalweg and mats of *Nostoc* spp. abundant nearer the stream margins (Broady 1980; McKnight *et al.* 1998). The *Phormidium* mats typically are composed of an upper layer, which is orange and enriched in accessory pigments, and an under layer with greater chlorophyll *a* content and greater photosynthetic activity (Vincent *et al.* 1993a, 1993b). Hawes *et al.* (1992) showed that *Nostoc* mats became dehydrated rapidly (within 5 h of exposure) and that their rates of photosynthesis and respiration recovered within only 10 min after rewetting. In similar experiments, *Phormidium* mats did not fully recover from drying even after many days; however, viable diaspores of *Phormidium* were found after 3 years of exposure (Hawes *et al.* 1992).

The microbial mats also contain a diverse assemblage of pennate diatoms from primarily aerophilic genera (Esposito *et al.* 2006). Of the 40 diatom species found in Taylor Valley, 24 have only been found in the Antarctic, indicating a high degree of endemism. The percentage of these Antarctic diatom species increased with decreasing annual streamflow and increasing harshness of the stream habitat associated with cessation of flow during the summer (Figure 5.3). The species diversity of assemblages reached a maximum when the Antarctic species accounted for 40–60% of relative diatom abundance. Limited numbers of tardigrades and nematodes are also present in the mats (Alger *et al.* 1997; Treonis *et al.* 1999), but grazing does not seem to control mat biomass. Mosses occur above the *Nostoc* mats on the stream banks, where they are wetted for short periods when flows are high, or in flat areas that become damp but lack open-water flow (Alger *et al.* 1997).

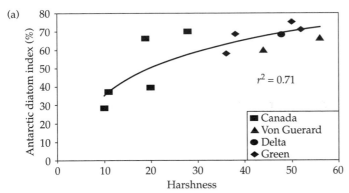

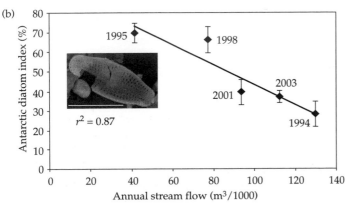

Figure 5.3 Diatom diversity in McMurdo Dry Valley streams as a function of stream flow and variability. (a) Average percentage abundance of Antarctic **diatoms** (relative to the total abundance of all **diatoms**, including cosmopolitan species) as a function of 'harshness', an index based on flow maxima, means, and variability. (b) Average percentage of Antarctic diatoms in the Canada Stream, Taylor Valley, as a function of total annual streamflow (in thousands of cubic meters). The strong negative relationship is shown with the trendline and r^2 value. Each point represents a yearly value, as indicated. Modified from Esposito *et al.* (2006), with permission from the American Geophysical Union. Insert: an endemic diatom from the McMurdo Dry Valleys, *Luticola muticopsis fo. evoluta*, from http://huey.colorado.edu/diatoms/about/index.php.

5.3 Case study: effect of flow restoration on microbial mats and ecosystem processes

The presence of salt crusts in lake sediments in the McMurdo Dry Valleys indicates that cold, dry climatic conditions have occurred repeatedly over the past 20 000 years and low streamflows have caused the lakes to become dry (Hendy *et al.* 1979). In studying these Antarctic stream ecosystems during the present period of generally increasing streamflow and rising lake levels during this century (Chinn *et al.* 1993; Webster *et al.* 1996; Bomblies *et al.* 2001) it is important to keep in mind that the cyanobacteria, diatoms, microbes, and other organisms comprising these ecosystems somehow persisted through recent colder periods of greatly reduced flow. With the projected increase in temperatures in Antarctica, temperatures may soon be warm enough to allow consistent generation of streamflow from glaciers at higher elevations, potentially reactivating stream ecosystems that have been dormant for centuries. To examine ecosystem processes important in the long-term survival of microbial mats under the extremely cold and dry conditions of the McMurdo Dry Valleys, flow was experimentally diverted to an abandoned channel, which had not received substantial flow for approximately two decades or more. This experiment was designed to address the time lag between the renewal of regular summer flow and the establishment of a functioning stream ecosystem.

The experiment was conducted in the upper reach of Von Guerard Stream, an inflow to Lake Fryxell in the Taylor Valley. Examination of aerial photographs of the stream taken over the previous four decades indicated that the limited water in the abandoned channel had been due to melting of accumulated snow. The measurement of ^{14}C in dried algal mats from the abandoned channel showed high values consistent with growth during the peak of the bomb ^{14}C spike in the atmosphere in the 1960s. The experimental diversion was carried out by constructing a low wall of sandbags filled with alluvium at the base of the steep upper reach of Von Guerard Stream to route a portion of the flow away from the eastern channel towards the previously abandoned western channel.

The surprising initial observation was that cyanobacterial mats were abundant in the reactivated channel within the first week, and within 2 weeks the black and orange mats were clearly visible as extensive areas of microbial life contrasting with the barren landscape on either side of the channel. These results suggest that the mats had been preserved in a cryptobiotic state in the channel and that colonization was not the controling process in the renewal of stream ecosystem function. Over the next few years, these mats had high rates of productivity and nitrogen fixation compared with mats from other streams in the same valley.

Monitoring of specific conductance and solute concentrations showed that for the first three years after the diversion, solute concentrations were greater in the reactivated channel than in most other dry-valley streams. Furthermore, as streamflow advanced downstream during the first several summers of the experiment, the concentrations of inorganic solutes increased in the downstream direction with the mobilization of soluble material that had accumulated over time. This downstream increase was not observed for nutrients, probably reflecting uptake by the rapidly growing microbial mats. Experiments in which mats from the reactivated channel and another stream were incubated in water from both streams indicated that the greater solute concentrations in the reactivated channel stimulated net primary productivity of the mats (Figure 5.4). Further, the microbial mats in the reactivated channel exhibited greater rates of nitrogen fixation than mats from other streams.

These stream-scale experimental results indicate that the cryptobiotic preservation of cyanobacterial mats in abandoned channels in the dry valleys allows for rapid response of these stream ecosystems to climatic and geomorphological change, similar to rapid responses observed in other arid zone stream ecosystems. The implications of this one experiment is that there is a potential for rapid stream ecosystem response to greater meltwater generation from glaciers at higher elevations and more inland in the McMurdo Dry Valleys and in other polar desert oases on the Antarctic coast.

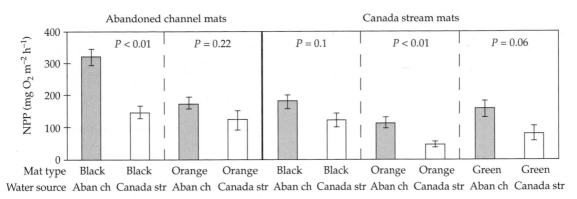

Figure 5.4 Effect of water from the reactivated channel (Abandoned channel, Aban ch) and Canada stream (Canada str) on net primary productivity (NPP) of *Phormidium* (orange) and *Nostoc* (black) mats from each stream. *P* values indicate the significance of *t* tests comparing two values. Reproduced from McKnight *et al.* (2007), with permission from Elsevier.

5.4 Arctic streams

One characteristic aspect of Arctic streams is the seasonal dynamics of ice in the stream, either as ice cover or as grounded ice. In response to the pronounced seasonal fluctuation in irradiance and air temperature, stream freeze–thaw cycles affect the annual hydrology patterns and stream ecosystem processes (Prowse 2001a, 2001b). In headwater stream systems, water columns freeze from top to bottom, and this grounded ice within the channel protects the bed from scour during spring snowmelt. Downstream, generally in larger rivers, discharge is perennial, but during the winter, an ice cover forms over the flowing water (Best *et al.* 2005). Although the ecology of ice-covered rivers is poorly understood, it is clear that the processes occurring in one season have an influence that carries forward to the next season and to subsequent years, thereby determining the overall resilience of Arctic stream ecosystems.

Ice also plays a role in the relationship between Arctic streams and their surrounding watershed. Most Arctic streams flow through tundra landscapes, underlain by permafrost. The shallow active layers provide little buffering of snowmelt or rain event waters, resulting in 'flashy' hydrologic stream response: rapid increases in stream discharge from melt and rain events. Event water is forced through the shallow active layer that is

generally dominated by organic-rich tundra soils, so lateral inflows to streams may provide large quantities of dissolved organic matter to tundra streams (Michaelson *et al.* 1998). In headwater streams this fresh dissolved organic matter is highly reactive and is enriched in aromatic moieties (Cory *et al.* 2007).

Arctic streams are mostly underlain by permafrost, and the potential for hyporheic exchange (the exchange of stream water between the open channel and the bed sediments) is controlled by the seasonal thaw depths underneath these channels. Brosten *et al.* (2006) documented the seasonal dynamic of thaw depths under several tundra streams in northern Alaska, noting that peak depths for first- and second-order alluvial channels can reach as much as 2 m, whereas peak depths under peat channels are generally less than 1 m deep. Despite the limited size of such 'aquifers', Edwardson *et al.* (2003) has demonstrated that hyporheic zones of Arctic streams are very important to nutrient processing in these channels, noting rapid nitrification of mineralized organic N and production of ammonium, phosphate, and carbon dioxide in the hyporheic zone.

The three main sources of water to first-order Arctic streams are channel heads, springs, and glaciers. Channel heads are characterized by the culmination of source area and/or water tracks such that surface flow is focused into a well-defined channel. Springs are generally perennial sources of water, and as a result spring head streams provide an over-winter fluvial habitat which may be ecologically important. Glacier melt streams are characterized by high turbidity as they tend to carry high loads of glacial flour: very fine sediment from the glacial source areas.

There are three common types of tundra stream channels: alluvial, peat, and combination alluvial/peat streams. Alluvial streams are composed of cobbles and gravels, and have mild to steep slopes, depending on landscape position. Except in glacial-fed streams, the sources of fine sediment are limited across the tundra landscape. However, bank material may provide fine-fraction material during high flow events. Peat streams are characterized by very low gradients, and many exhibit a beaded morphology: a series of deep pools connected by short, narrow runs. The bed sediments in the pools typically have high organic matter content, and a fairly fine size fraction. These channels are rarely scoured, and their structure does not promote transport of larger suspended material. Alluvial-peat channels share characteristics of both channel types. The morphology tends to be beaded, but with more rapid flow through larger connectors between pools. These beads, or pools, are less effective at slowing discharge than in the peat channels. Because of the moderately high-energy environment, alluvial/peat channels have bed materials that are a mix of gravel, cobble, and peat where lower velocities occur.

5.5 Case study: long-term effect of nutrient enrichment

The Kuparuk River (Plate 6) has been studied intensively since 1978 and is representative of the rivers that flow across the tundra of the North Slope of Alaska. The Kuparuk originates in the foothills of the Brooks Range and flows north-northeast to the Arctic Ocean, draining an area of 8107 km². It is classified as a clear-water tundra river (Craig and McCart 1975; Huryn et al. 2005) because it has no input from glaciers and little input from springs. The river is underlain by permafrost and is frozen solid from October to May; the open-water season is approximately 4 months, and summer water

temperatures average 8–13°C. Maximum precipitation occurs in July and August (Kane et al. 1989); however, the peak river discharge usually occurs in late May or early June during snowmelt. Mean annual precipitation in the region is approximately 15–30 cm/year, 30–40% of which falls as snow (McNamara et al. 1997).

Peat from eroding banks is the major source of particulate organic matter loading to the river, approximately 200–300 g C m^{-2} year^{-1} (Peterson et al. 1986). Dissolved organic carbon (DOC) comprises over 90% of the total organic carbon (mean, 6.8 mg C L^{-1}) export from the watershed (Peterson et al. 1986). Benthic net primary production is low compared with temperate streams, ranging from 15 to 30 g C m^{-2} year^{-1} (Peterson et al. 1986; Bowden et al. 1992). Soluble reactive phosphorus and ammonium levels are usually near the limits of detection (<0.05 μmol L^{-1}). Nitrate plus nitrite levels range from 0.1 to more than 10 μmol L^{-1}. Mean conductivity values (45 μS cm^{-1}) are typical for tundra streams of the North Slope and are relatively low compared with glacial and spring-fed rivers of the region. Mean pH is 7.0 and alkalinity averages about 250 μEq L^{-1}.

The reach of the Kuparuk that has been most intensely studied is located on the meandering, fourth-order stream section where the river straddles the Dalton Highway (68°38'N, 149°24'W). At this point the river drains an area of 143 km² of alpine and moist tundra communities. This case study focuses on this section of the river (Kriet et al. 1992; Hershey et al. 1997). Dwarf birch and willows line segments of the stream bank but do not shade the channel because of their low stature. The river channel has a mean width of about 20 m and is characterized by alternating riffles (20–30 m long and 5–30 cm deep) and pools (10–50 m long and 1–2 m deep) at the average summer discharge of 2–3 m³ s^{-1}. The riffle and pool substrate ranges from cobble to boulder. Average channel slope within the study reach is 0.6%.

Although bryophytes and filamentous algae are present in the upper Kuparuk River, epilithic diatoms dominate the autotrophic community. The dominant diatom species are *Hannaea arcus*, *Cymbella minuta*, *Achnanthes linearis*, *Achnanthes minutissima*, *Synedra ulna*, and *Tabellaria flocculosa*

(Miller *et al.* 1992). The macro-algal community consists primarily of *Tetraspora, Ulothrix, Stigeoclonium, Batrachospermum, Lemanea,* and *Phormidium*. The dominant bryophyte in the main river channel is the moss *Schistidium [Grimmia] agassizii*. The mosses *Hygrohypnum alpestre, H. ochraceum, Fontinalis neomexicana,* and *Solenostoma* sp. (a liverwort) are also present but are most frequently found in nutrient-rich spring tributaries.

The benthic invertebrate community includes filter-feeding black flies, *Stegopterna mutata* and *Prosimulium martini* (Hiltner 1985), the abundant mayflies *Baetis* spp. and *Acentrella* spp, and the less abundant mayfly *Ephemerella* sp. There are several species of chironomids and the tube builder *Orthocladius rivulorum* is one of the largest (Hershey *et al.* 1988). Also present is the caddisfly *Brachycentrus* (Hershey and Hiltner 1988). Arctic grayling (*Thymallus arcticus*) is the only fish species in this section of the river. Adult grayling reside in the river from the June spawning period through to mid-August, and then migrate upstream to overwinter in headwater lakes. Young-of-the-year (YOY; age 0) grayling are found primarily in pools (Deegan and Peterson 1992).

The first years of the Kuparuk River ecosystem study focused on baseline studies of hydrology, nutrient biogeochemistry, primary productivity, and organic matter budgets. In 1980, nutrient-enrichment bioassays showed that the low levels of dissolved phosphorus were limiting to primary production of the epilithic algal community (Peterson *et al.* 1981). This result and the very low level of primary production in the presence of 24-hour summer daylight raised the question of what would happen if nutrients were not limiting. In 1983 a long-term nutrient-fertilization study was started that continues until the present (Slavik *et al.* 2004). These studies have provided insights into how Arctic tundra streams function and respond to perturbation.

The whole-stream fertilization of the river with a constant drip addition of $10\,\mu g\,P\,L^{-1}$ stimulated algal production by about 1.5–2-fold on average (Peterson *et al.* 1993). The fertilization also increased bacterial activity, rates of respiration of the epilithic biofilm, and rates of decomposition of recalcitrant substrates such as lignin and lignocellulose (Peterson

et al. 1985). The bottom-up impacts of phosphorus addition increased the abundance of insects such as *Brachycentrus* and *Baetis*. In the first few summers of fertilization there was evidence that at times insect grazers limited chlorophyll accumulation on the stream bottom rocks. Although this can occur when mayfly and especially chironomid populations are at their maxima, subsequent studies with tracers (Wollheim *et al.* 1999) indicated that energy flow through grazers was on average far less than required to control algal biomass. However, the increased insect abundance stimulated the growth of both adults and YOY graylings in the fertilized reach relative to the reference reach (Deegan and Peterson 1992).

In spite of consistent responses to fertilization, the populations of algae and insects and the rates of fish growth proved to be highly variable (approximately 10-fold) from year to year both in the reference reach and in the fertilized reach (Deegan and Peterson 1992; Hershey *et al.* 1997; Slavik *et al.* 2004). Much of this variability was correlated with year-to-year variation in discharge. Under high-discharge conditions (and presumably high scouring) algal biomass was only half as great as under low-discharge conditions. Black fly abundance was also affected by flow (Hershey *et al.* 1997). Adult fish thrived when discharge was high but the YOY grew poorly under high-flow conditions, which are associated with low temperatures (Deegan *et al.* 1999). These impacts of discharge appear to be equally strong in the reference and fertilized reaches although the higher biomass and productivity in the phosphorus-fertilized reach made the effects easier to measure.

At the time when the major controls on stream ecosystem function appeared to have been resolved, an amazing transition occurred (Figure 5.5). After nearly a decade of continuous summer fertilization, the riffle habitat for several kilometers downstream of the phosphorus-addition point became covered with a carpet of two species of the moss *Hygrohypnum*: *Hygrohypnum alpestre* and *Hygrohypnum ochraceum* (Bowden *et al.* 1994; Slavik *et al.* 2004). This was a surprise because these species were not observed in these reaches of the Kuparuk prior to fertilization. A slow-growing species, *S. [Grimmia] agassizii*, was always present

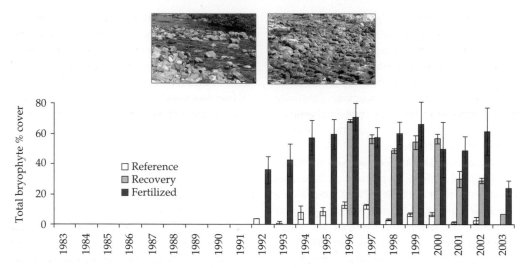

Figure 5.5 Striking long-term response to nutrient enrichment in the Kuparuk River, Alaska. The photographs illustrate the differences in stream-bottom coverage between the diatom-dominated stream bottom (reference reach; left) and the moss-dominated stream bottom (fertilized reach; right). Fertilization started in 1983 and terminated in 1996. The pictures were taken during a drought when the stream bottom was largely exposed but with average summer discharge these sites are submerged. The graph shows the temporal changes in moss cover in the reference, recovery, and fertilized reaches of the Kuparuk River. Original figure from B.J. Peterson; a colour version is shown in Plate 7.

in relatively low abundance but has not responded to the added phosphorus. Instead, the fast-growing *Hygrohypnum* spp. that are present in nutrient-rich seeps and springs far up river invaded the phosphorus-enriched reach. The impact of this moss invasion on the physics, biogeochemistry, and biota of the Kuparuk River can only be described as profound (Stream Bryophyte Group 1999).

Hygrohypnum provides a large amount of surface area for algal epiphytes and the moss community (moss plus epiphytic algae) is several-fold more productive than the epilithic biofilm in the reference reach (Arscott *et al.* 1998). The biomass of moss dwarfs by orders of magnitude the biomass of epilithic algae in the reference reach (Slavik *et al.* 2004). The moss fronds create a matrix containing slow-moving water compared with the swift currents above the moss carpet. This matrix traps and stores a large amount of fine particulate matter and may serve as an important locus of nutrient recycling. The moss matrix habitat is host to an insect community quite different in abundance and species composition than found in the rocky-bottom reference reach (Lee and Hershey 2000; Slavik *et al.*

2004). Chironomids, *Brachycentrus*, and a large mayfly (*Ephemerella*) are more abundant by an order of magnitude in the mossy fertilized reach. In contrast, other common insects including *Baetis*, black flies, and *Orthocladius* are less abundant in the mossy reach than in the reference reach. In spite of higher insect biomass and secondary production in the fertilized reach, the growth of adult drift-feeding grayling is no longer significantly greater than in the reference reach. Some aspect of the link between insect production and fish growth has been changed by the phosphorus-induced moss invasion but further work is required to understand this paradox.

During the almost 30 years of research on the Kuparuk River, the Arctic climate has been changing in ways that affect both the ecology of the river and how we conduct our long-term fertilization experiment (Hobbie *et al.* 1999). For example, the open-water season is now longer on average than it was in the 1970s and 1980s. Mean annual temperatures are increasing and summer precipitation and runoff have on average increased. In response, the start date of the annual nutrient drip has been advanced by a week. Warmer and wetter summer seasons appear to be increasing the concentrations and the export of inorganic nitrogen in the form of nitrate from the upper Kuparuk catchment, especially in the last 15 years (McClelland *et al.* 2007). It is possible that phosphorus input to the river is increasing but a change in dissolved phosphorus concentrations in the stream water is not anticipated because any available phosphorus inputs are rapidly assimilated by the phosphorus-limited algal and bryophyte communities. Finally there is the possibility that thermokarsts (collapse of the land surface due to subsurface ice and permafrost melt) may become more numerous as the climate continues to warm (Bowden *et al.* 2008). Thermokarsts impact tundra streams like the Kuparuk by loading them with sediments and with nutrients leached and weathered from newly thawed and exposed minerals and soils. How these changes in the Arctic climate and landscape will affect streams and rivers as climate warming is an important question for the future.

5.6 Large Arctic rivers

A striking limnological feature of the Arctic that has no parallel in Antarctica is the presence of large river ecosystems. These rivers drain vast catchments that begin at north temperate latitudes and that extend all the way to the Arctic coastline (Plates 7 and 8; Figure 1.2). Apart from their intrinsic value as a major class of aquatic ecosystems and as migratory routes for fish, birds, and other animals, these rivers provide important transport arteries for human settlements, by boat during the few months of open water and by wheeled and track vehicles after freeze-up. They also provide water supplies for drinking, fishing, agriculture, and industrial activities such as mining, and in their upper reaches most are harnessed to some extent for hydroelectricity. In their lower reaches, the rivers flow across extensive flood-plain deltas. These contain tens of thousands of lakes and ponds, providing a major habitat for aquatic wildlife. The rivers experience high-amplitude peak flows during the period of summer snowmelt, and during this time of year many of the flood-plain lakes are flushed out and replenished with river water.

Arctic rivers discharge huge quantities of fresh water, heat, solutes, and particulate materials into the Arctic Ocean, and there is intense interest in how these globally significant inputs may shift in response to ongoing climate change in the Arctic.

The fresh-water input to the Arctic Ocean has been estimated as 3299 km^3 per year, and much of this is via the five largest Siberian rivers (Yenisei, Lena, Ob, Pechora, and Kolyma) and the Mackenzie River in Canada. This input accounts for 11% of the total fresh-water drainage into the World Ocean, and given that the Arctic Ocean represents only 1% of the total ocean, Arctic seas are among the most freshwater-influenced in the world (Rachold *et al.* 2004). There is evidence that in concert with the regional warming trend there has been an increase in Arctic river discharge, and this may be a factor contributing towards the observed freshening of the North Atlantic Ocean, with implications for global ocean circulation (Peterson *et al.* 2006). Arctic rivers are also unusually rich in organic carbon, both particulate organic matter and dissolved materials (Dittmar and Kattner 2003). These are largely of terrestrial origin and undergo some biogeochemical processing and loss in the coastal zone of Arctic seas, including via ice-dammed lagoons, so-called stamukhi lakes that retain the river waters inshore for much of the year (Galand *et al.* 2008). A large portion of the dissolved organic matter, however, may be advected out over the Arctic shelf where it is removed by microbial activity (Garneau *et al.* 2006), photochemical processes (Bélanger *et al.* 2006), or by further advection including downward transport by density currents into the deep ocean (Dittmar 2004). The limnology of large Arctic rivers is therefore of broad significance to the structure and functioning of marine systems, from Arctic coastal waters to the global ocean.

The strong seasonality of the flow regime of large Arctic rivers is controlled by snow melting and by the presence and rupture of ice dams. Peak discharge is an extremely difficult (and dangerous) period to accurately gauge and sample because the waters are spread over large expanses and multiple channels, and are filled with fast-moving ice blocks. However, increasing data are becoming available for this critical part of the seasonal hydrograph. These observations show that some solute concentrations also rise with increasing discharge. In the Kolyma River, the concentration of DOC increases by almost a factor of 10 during peak discharge (Figure 5.6), and this brief period of spring snow melt contributes about half of the total annual DOC flux (Finlay *et al.* 2006). Carbon-14 dating of the DOC in this Arctic river during peak discharge showed that it was dominated by materials that were young in age, consistent with intense leaching of surface soils during the spring thaw. Later in the season, however, the ^{14}C age of the DOC increased substantially, suggesting mobilization from deeper soil horizons at that time (Neff *et al.* 2006). The DOC at peak flow is also highly colored, and this may have wide-ranging ecological effects via the absorption of UV and photosynthetically available solar radiation (Retamal *et al.* 2007, 2008).

The microbial and phytoplankton ecology of large Arctic rivers is still in an early stage of observation, and no seasonal records are available to date. These rivers are fed in part by lakes and flow long distances from these source waters to the sea. For example, the Mackenzie River flows 1800 km from Great Slave Lake to the coast and along the way receives input from many adjacent waterbodies including Great Bear Lake. Given this long transit time the waters can heat up considerably en route, for example to 15°C at the Mackenzie River mouth in June when air temperatures are still cool and the adjacent seas are only +1°C (Galand *et al.*

2008). There is also considerable opportunity for the inoculation and development of a riverine flora. Chlorophyll *a* measurements indicate moderate biomass concentrations of phytoplankton in Arctic rivers in summer, in the range of 1–5 mg m^{-3}, with dominance by diatoms and phytoflagellates such as cryptophytes (Meon and Amon 2004; Garneau *et al.* 2006; Galand *et al.* 2008). However, phytoplankton production rates are likely to be low given the deep mixing and strong absorption of light by colored dissolved and particulate organic material (Retamal *et al.* 2007, 2008).

Molecular and microscopy analyses for the Mackenzie River have shown the presence of an active microbial food web, high concentrations of picocyanobacteria and heterotrophic bacteria, and a surprisingly diverse community of Archaea (the third domain of microbial life, along with Bacteria and Eukarya) that likely reflects the heterogeneous source populations as well as diverse substrates for their metabolism and growth (Galand *et al.* 2006; Garneau *et al.* 2006). The Mackenzie River is especially rich in suspended particulate material (up to 100 g m^{-3}) and most of the heterotrophic production appears to be associated with bacteria attached to suspended particles (Garneau *et al.* 2006; Vallières *et al.* 2008).

The large Arctic rivers likely contain relatively low concentrations of zooplankton and benthic invertebrates given their fast turbulent flows and

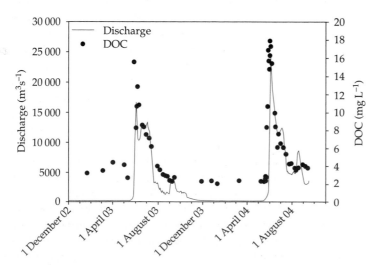

Figure 5.6 Seasonal variations in discharge and dissolved organic carbon (DOC) in the Kolyma River, 2003–2004. From Finlay *et al.* (2006), with permission from the American Geophysical Union.

low primary productivity. However, they provide migration conduits for many species of fish that may access the more abundant food sources in adjacent lakes and tributaries, as well as anadromous fish species that migrate via the river corridor to the ocean. The Mackenzie River, for example, contains several anadromous coregonid (whitefish) species, including one – *Stenodus leucichthys* – that was named, perhaps by early French trappers, the *inconnu* (meaning 'unknown' in French). Tagging and recapture of this species, which also occurs in northern Alaska and northern Eurasia, show that it can migrate more than 1500 km downstream to the Beaufort Sea (Stephenson *et al.* 2005). These and other coregonids have long been a traditional part of the local diet of the Mackenzie Inuit (Morrison 2000). In the large Russian Arctic rivers, a species that highly sought after for its meat and for caviar is the Siberian sturgeon *Acipenser baerii*. This species appears to be in decline, and two of its subspecies are threatened by extinction through a variety of causes that may include hydroelectric development and severe water pollution (Ruban 1997).

5.7 Arctic flood-plain lakes

Arctic rivers flow across extensive fluvial deltas that are underlain by permafrost. The uneven thawing of these regions results in thousands of lake basins that become filled with snow melt and with ice-jammed flood waters from the river. This aspect of the cryosphere distinguishes large Arctic river systems from their counterparts at lower latitudes. For example, the Amazon and Mississippi flood-plains also contain lakes, but these are a result of other geomorphological processes and in nowhere near the same abundance. The Arctic flood-plain waters are referred to as thaw lakes or thermokarst lakes, and they provide a rich habitat for aquatic plants, animals, and microbes. They also play a key role in the biogeochemical processing of river water before its discharge to the Arctic Ocean. The lakes are typically supersaturated in carbon dioxide and methane, and there is therefore much interest in their microbial ecology and photochemistry, and their abilities to convert organic carbon derived from the surrounding tundra catchments to greenhouse gases that may be released to the atmosphere.

Some 45 000 lakes larger than 0.14 ha have been inventoried in the Mackenzie River delta, in the western Canadian Arctic, along with another 4046 smaller waterbodies (Emmerton *et al.* 2007; Plate 8). The lakes vary in depth from 0.5 to 4.5 m, with a mean depth of about 1.5 m, although this depends on the season and flood year. Like other large Arctic rivers, the Mackenzie River flows from a relatively warm climate in the south to cold coastal Arctic conditions in the north, and during peak flow in spring–early summer the flows encounter ice jams in the downstream flood-plain delta. These cause the water to spread out over 11 200 km^2 of lakes and flooded vegetation, and at this time of year almost half the total annual discharge is stored temporarily in the delta. Most of the lakes are flooded, to an extent that depends on their elevation and distance from the river channels (see Figure 1.4), and this variable flooding controls their nutrient regime and other limnological characteristics (Lesack *et al.* 1998). This temporary storage on the delta provides sufficient time for biogeochemical and photochemical processing of the flood waters, which alters the composition of the riverine input of dissolved and particulate materials to the coastal Arctic Ocean (Emmerton *et al.* 2008). The connectivity between Mackenzie flood-plain lakes and their riverine source waters appears to be changing in complex ways as a result of climate warming in the region (Lesack and Marsh 2007).

Like thermokarst lakes in general, the shallow Arctic flood-plain waters often contain high concentrations of zooplankton that feed in part on seston, but also on mats of cyanobacteria and microalgae that coat the sediments (Rautio and Vincent 2006). The benthic mats can achieve high-biomass stocks and dominate overall primary production, whereas the phytoplankton populations tend to be relatively small. Aquatic plants (macrophytes) may also contribute substantially to the primary production of these waters, particularly in the less flooded lakes that have lower nutrient concentrations but are more transparent. For the lakes of the Mackenzie River delta, the frequency and duration of flooding also influences bacterial production, zooplankton community structure, and the presence of fish and birds (L. Lesack, unpublished results).

It has long been recognized that carbon dioxide and methane occur in high concentration in the waters that flow through the active layer of permafrost soils, and in the surface stream channels and basins that they discharge into (Kling *et al.* 1991). Additional greenhouse gas production may occur by photochemical degradation of organic matter in the flood-plain lakes and rivers, and by microbial activities in their water columns and sediments. In the abundant thaw lakes of the Kolyma River system, Siberia, huge quantities of methane are released as bubbles along the permafrost-lake margins. These thaw lake emissions appear to have increased by 58% in the final decades of the twentieth century as a result of increased melting of the permafrost, constituting an important feedback process that is likely to accelerate global warming (Walter *et al.* 2006).

5.8 Comparison of Arctic and Antarctic fluvial ecosystems

Polar streams and river networks are subject to flow regulation by seasonal climate variability. Thus, streams in both the north and the south polar regions are sensitive to local climate changes. These changes are likely to be considerably different in each region, though both are expected to experience generally warmer conditions. In both regions, streams are underlain by permafrost, although substream thaw depths may range from tens of centimeters to several meters. The presence of permafrost imposes a critical control on these streams because they interact with fairly small aquifers. These aquifers are of limited size and therefore do not, in general, provide a reservoir of water to contribute to streamflow. River networks of the Arctic are much more extensive than those in the Antarctic, largely because there is more ice-free land in the Arctic regions (the pan-Arctic area is approximately 22.4×10^6 km^2), and therefore Arctic stream water sources are farther from their oceanic termini. Thus, in the Arctic, there is much more geomorphologic evolution of the river. In the dry valleys of Antarctica, stream lengths are relatively short (<40 km) and stream networks are simple (second-order).

5.8.1 Stream discharge

In the Antarctic dry valleys, streams originate from glacier melt. Dry valley streamflow is obviously dependent upon a positive energy balance at glacier surfaces to generate melt from approximately November to February. Because of this dependence upon climate conditions, streamflow can cease even in the middle of the flow season, particularly if a snow event occurs. Fresh snow on the glacier surface will increase its albedo, and decrease the amount of melt energy. Additionally, streamflow has a distinct diurnal pattern with peaks and lulls controlled by the timing of maximum solar exposure on the glacier contributing area. The flow season ends from the top down; that is, meltwater generation generally ends and water remaining in the streambeds may drain to the field capacity of the substrate, or streambed drainage may be limited by the onset of freezing temperatures.

Arctic tundra streams originate from glacier melt, springs, and channel heads, although they are similar to dry valley streams in that they flow seasonally, generally from late May to September, except for the largest of Arctic rivers that flow year round. The flow season in the Arctic begins with melt of the annual snowpack (Kane *et al.* 1991), and later responds to rain and snow events throughout the summer. In headwater streams, streamflow generally ends with the onset of cold weather that begins to freeze stream water from the surface down toward the bed. In deep peat-lined channels, liquid water may remain under the ice cover late into the winter. Further downstream, large Arctic rivers flow perennially. In the winter, the rivers are ice-covered and flow is sustained by contributions from springs.

Along their lengths, dry valley streams generally lose water to evaporation or saturating the thaw depth beneath them. In low-flow seasons, the surrounding thawed streambed may accommodate several times the realized streamflow, in particular at the end of long reaches (Bomblies 1998). Thus these thawed sediments represent an important threshold reservoir that must be satisfied before streamflow can continue at the surface. Because most Antarctic lands do not receive rainfall, there is generally no lateral contribution of discharge to streams.

In the Arctic, however, during the height of the austral summer, snow and rain events can both contribute to streamflow. Thus tundra streams generally receive diffuse lateral inflows, which may also include contributions from seeps or springs. Because Arctic streams are underlain by permafrost, deeper groundwater additions (besides springs) are limited. Lakes and ponds are common in Arctic stream networks and may contribute to downstream discharge during times of low flow.

Under warmer conditions, permafrost may recede, and flow seasons may extend. In the Arctic, because of consistent summer rain events, such changes will likely result in more rapid geomorphological evolution of watersheds. In the Antarctic, warming may extend flow seasons and possibly glacial melt rates. The more regular occurrence of rain in the Antarctic would have substantial impacts on streams, landscapes, and glaciers, for example in the McMurdo Dry Valleys.

5.8.2 Stream ecosystems

Antarctic streams represent an important control of water and solutes, especially nutrients, as they link glaciers and closed-basin lakes. Dry valley streams generally contain low concentrations of nitrogen, phosphorus, and carbon (McKnight et al. 1999, 2004). The source of nitrogen species in dry valley streams is generally atmospheric deposition, whereas phosphorus is typically derived from weathering of local sediments, including that which is deposited on glaciers during strong katabatic wind events during the winter. Organic-matter loads in Antarctic streams are generally completely autochthonous, as there is little to no other organic matter across the landscape to contribute. Antarctic stream ecosystems are generally composed of cyanobacteria algae, diatoms, making up extensive benthic microbial mats, where conditions are appropriate, generally stable substrate composed of desert pavement (McKnight et al. 1999). Antarctic streams do not contain any higher-order plants, nor do they accommodate any fauna. These conditions are largely attributed to the fact that flow regimes are low enough in magnitude

that extensive channel degradation and shear are generally restricted.

Under a warmer climate, it is expected that nutrients and organic matter inputs to Antarctic streams would increase because of enhanced connections between streams and their surrounding landscapes. It is further expected that current stream ecosystem composition and function would shift toward fewer cold-tolerant microbial and diatom species, and enhanced hyporheic nutrient cycling due to greater depths to permafrost under and around streams. Warmer temperatures may cause more meltwater generation from higher-elevation glaciers, creating new, so-called reactivated stream ecosystems.

Arctic stream ecosystems, on the other hand, are more complex than those of the Antarctic, as they host a full suite of benthic fauna, including bryophytes (Stream Bryophyte Group 1999), insects (e.g. Huryn et al. 2005), and fish populations that generally over-winter in lakes. Tundra streams are subject to solute and nutrient inputs from their laterally contributing watershed. This delivery is especially important to the character and timing of carbon fluxes in headwater Arctic networks (Michaelson et al. 1998). Leaf litter from tundra plants also finds its way into streams adding additional allochthonous organic material.

Climate change in the Arctic is likely to lead to enhanced precipitation, and warmer temperatures, particularly in the winter. These changes will likely lead to modified vegetation across many Arctic watersheds, enhanced permafrost degradation, and extended flow seasons, thus modifying watershed flowpaths, biogeochemical cycling, and ultimately stream ecosystem structure and function. In the High Arctic polar desert environment, the loss of perennial snow banks may have a strong influence on high Arctic streams and wetlands that depend on them for a continuous supply of water in summer (Woo and Young 2006).

5.9 Conclusions

Rivers of both polar regions are integrators of processes that occur within contributing watersheds, and of cumulative climates. In the Antarctic, the suite of processes that are integrated is much smaller than in Arctic watersheds. However, it is the Antarctic regions that have the potential to experience the greatest modifications because of the potential for a much larger collection of drivers to which streams will respond. To date, much important research has been accomplished in rivers and flood-plains of polar regions, although our understanding is not yet extensive enough to warrant well-defined predictions of how these systems will respond to future climate change.

The expected warming in the polar regions will likely force significant changes to their lotic (flowing-water) ecosystems. Despite these changes, polar streams, rivers, and flood-plains will remain unique, compared with their temperate counterparts, partly because of the seasonality of energy input. Regardless of climate warming, and regardless of which seasonal distribution of warming, seasonal solar input patterns will not change, although cloud cover may. Thus, changes to polar lotic ecosystems must remain resilient to extended periods of both light and dark. The structure and function of these future systems will provide further opportunities to explore and test our fundamental theories of ecology and limnology. In the Antarctic, the dark austral winter reduces the energy available for glacier melt, so these streams will remain annually intermittent. Furthermore, these relatively simple stream systems may extend further up valley walls as streams follow the retreat of glaciers. Simultaneously, their mouths may follow upslope, though at a slower rate, as closed-basin lakes fill faster than sublimation/evaporation of their surfaces remove water. In the Arctic, where more persistent discharge occurs, even under the cover of winter ice in many places, warming will likely increase the distribution of flowing conditions during the winter, as less intense freezing leaves more water in channels and streambeds unfrozen. This is likely to result in increased annual heterotrophic processing of nutrients in these systems, which may affect the timing and magnitude of biogeochemical fluxes to coastal marine environments. Changes in ice dam conditions are likely to have wide-ranging effects through decreased water levels and changes in the timing of peak discharge.

Current research in polar streams, rivers, and flood-plains continues to expand our understanding of how these systems function. Manipulation experiments such as the fertilization experiment on the Kuparuk River in Alaska and the re-activation of a channel in the Lake Fryxell basin in Antarctica provide insight into potential changes to these systems under a warmer climate. These are careful experiments that seek to elucidate specific responses to specific perturbations; however, they do not account for the suite of changes, interactions, and feedback processes that are likely to occur across these landscapes in response to climate change. Continued vigorous research on these systems is necessary to better understand these flowing waters as ecological features of the changing polar landscape, and as biogeochemically active conduits that connect terrestrial environments to downstream lake and marine ecosystems.

Acknowledgments

We thank Lance Lesack and Craig Emmerton (Simon Fraser University) for helpful comments on the manuscript. Our research on high-latitude flowing water ecosystems is supported by the National Science Foundation (USA), ArcticNet in the Network of Centers of Excellence program (Canada), Polar Shelf Canada and the Natural Sciences and Engineering Research Council of Canada.

References

Alger, A.S. *et al.* (1997). *Ecological processes in a cold desert ecosystem: the abundance and species distribution of algal mats in glacial meltwater streams in Taylor Valley, Antarctica*. Occasional paper no. 51. Institute of Arctic and Alpine Research, Boulder, CO.

Arscott, D.B., Bowden, W.B., and Finlay, J.C. (1998). Comparison of epilithic algal and bryophyte metabolism in an arctic tundra stream, Alaska. *Journal of the North American Benthological Society* **17**, 210–227.

Barrett, J.E. *et al.* (2007). Biogeochemical stoichiometry of Antarctic Dry Valley ecosystems. *Journal of Geophysical Research–Biogeosciences* **112**, G01010, doi:10.1029/2005JG000141.

Bélanger, S. *et al.* (2006). Photomineralization of terrigenous dissolved organic matter in Arctic coastal waters from 1979 to 2003: interannual variability and implications of climate change. *Global Biogeochemical Cycles* **20**, GB4005.

Best, H., McNamara, J.P., and Liberty, L. (2005). Association of ice and river channel morphology determined using ground penetrating radar in the Kuparuk River, Alaska. *Arctic, Antarctic and Alpine Research* **37**, 157–162.

Bomblies, A., McKnight, D.M., and Andrews, E.D. (2001). Retrospective simulation of lake level rise in Lake Bonney based on recent 21-year record: indication of recent climate change in the McMurdo Dry Valleys. *Journal of Paleolimnology* **25**, 477–492.

Bomblies, A. (1998). *Climatic Controls on Streamflow Generation from Antarctic Glaciers.* Master's thesis, University of Colorado, Boulder, CO.

Bowden, W.B., Peterson, B.J., Finlay, J.C., and Tucker, J. (1992). Epilithic chlorophyll *a*, photosynthesis, and respiration in control and fertilized reaches of a tundra stream. *Hydrobiologia* **240**, 121–131.

Bowden, W.B., Finlay, J.C., and Maloney P.E. (1994). Long-term effects of PO_4 fertilization on the distribution of bryophytes in an arctic river. *Freshwater Biology* **32**, 445–454.

Bowden, W.B. *et al.* (2008). Potential effects of increased thermokarst activity on ecosystem dynamics in streams draining the foothills of the North Slope, Alaska. *Journal of Geophysical Research – Biogeosciences* **113** (in press).

Broady, P.A. (1982). Taxonomy and ecology of algae in a freshwater stream in Taylor Valley, Victorialand, Antarctica. *Archives für Hydrobiologie Supplement (Algological Studies)* **63**, 331–349.

Brosten, T. *et al.* (2006). Profiles of temporal thaw depths beneath two arctic stream types using ground-penetrating radar. *Permafrost and Periglacial Processes* **17**, 341–355.

Chinn, T.H. (1993). Physical hydrology of the dry valley lakes. In Green, W.J. and Friedmann, E.I. (eds), *Physical and Biogeochemical Processes in Antarctic Lakes*, pp. 1–51. Antarctic Research Series vol. 59. American Geophysical Union, Washington DC.

Conovitz, P.A., McKnight, D.M., MacDonald, L.H., Fountain, A.G., and House, H.R. (1998). Hydrologic processes influencing streamflow variation in Fryxell Basin, Antarctica. In Priscu, J.C. (ed.), *Ecosystem Dynamics in a Polar Desert: The McMurdo Dry Valleys*, pp 93–108. American Geophysical Union, Washington DC.

Conovitz, P.A., MacDonald, L.H., and McKnight, D.M. (2006). Spatial and temporal active layer dynamics along three glacial meltwater streams in the McMurdo Dry Valleys, Antarctica, *Arctic, Antarctic and Alpine Research* **38**, 42–53.

Cory, R.M., McKnight, D.M., Chin, Y.-P., Miller, P., and Jaros, C.L. (2007). Chemical characteristics of fulvic acids from Arctic surface waters: microbial contributions and photochemical transformations. *Journal of Geophysical Research* **112**, G04S51, doi:10.1029/2006JG000343.

Cozzetto, K., McKnight, D.M., Nylen, T., and Fountain A.G. (2006). Experimental investigations into processes controlling stream and hyporheic temperatures, Fryxell Basin, Antarctica. *Advances in Water Resources* **29**, 117–382.

Craig, P.C. and McCart, P.J. (1975). Classification of stream types in Beaufort Sea drainages between Prudhoe Bay, Alaska and the MacKenzie Delta, NWT. *Arctic and Alpine Research* **7**, 183–198.

Deegan, L.A. and Peterson, B.J. (1992). Whole-river fertilization stimulates fish production in an arctic tundra river. *Canadian Journal of Fisheries and Aquatic Sciences* **49**, 1890–1901.

Deegan, L.A., Golden, H.E., Harvey, C.J., and Peterson, B.J. (1999). Influence of environmental variability on the growth of age-0 and adult arctic grayling. *Transactions of the American Fisheries Society* **128**, 1163–1175.

Dittmar, T. (2004). Evidence for terrigenous dissolved organic nitrogen in the Arctic deep-sea. *Limnology and Oceanography* **49**, 148–156.

Dittmar, T. and Kattner, G. (2003). The biogeochemistry of the river and shelf ecosystem of the Arctic Ocean: a review. *Marine Chemistry* **83**, 103–120.

Doran, P.T. *et al.* (2002). Antarctic climate cooling and terrestrial ecosystem response. *Nature* **415**, 517–520.

Edwardson, K.J., Bowden, W.B., Dahm, C., and Morrice, J. (2003). The hydraulic characteristics and geochemistry of hyporheic and parafluvial zones in arctic tundra streams, North Slope, Alaska. *Advances in Water Resources* **26**, 907–923.

Emmerton, C.A., Lesack, L.F.W., and Marsh, P. (2007). Lake abundance, potential water storage and habitat distribution in the Mackenzie River Delta, western Canadian Arctic. *Water Resources Research*, W05419, doi:10.1029/2006WR005139.

Emmerton, C.A., Lesack, L.F.W., and Vincent, W.F. (2008). Mackenzie River nutrient fluxes to the Arctic Ocean and effects of the Mackenzie River delta. *Journal of Marine Systems* doi:10.1016/j.jmarsys.2007.10.001.

Esposito, R.M.M. *et al.* (2006). Antarctic climate cooling and response of diatoms in glacial meltwater streams, *Geophysical Research Letters* **33**, L07406, doi:10.1029/2006GL025903.

Finlay, J., Neff, J., Zimov, S., Davydova, A., and Davydov, S. (2006). Snowmelt dominance of dissolved organic carbon in high-latitude watersheds: Implications for characterization and flux of river DOC. *Geophysical Research Letters* **33**, L10401, doi:10.1029/2006GL025754.

Galand, P.E., Lovejoy, C., and Vincent, W.F. (2006). Remarkably diverse and contrasting archaeal communities in a large arctic river and the coastal Arctic Ocean. *Aquatic Microbial Ecology* **44**, 115–126.

Galand, P.E., Lovejoy, C., Pouliot, J., Garneau, M.-E., and Vincent, W.F. (2008). Microbial community diversity and heterotrophic production in a coastal arctic ecosystem: a stamukhi lake and its source waters. *Limnology and Oceanography* **53**, 813–823.

Garneau, M.-È., Vincent, W.F., Alonso-Sáez, L., Gratton, Y., and Lovejoy, C. (2006). Prokaryotic community structure and heterotrophic production in a river-influenced coastal arctic ecosystem. *Aquatic Microbial Ecology* **42**, 27–40.

Gooseff, M.N. (2001). *Modeling Hyporheic Exchange Influences on Biogeochemical Processes in Dry Valley Streams, Antarctica.* PhD dissertation, University of Colorado, Boulder, CO.

Gooseff, M.N., McKnight, D.M., Lyons, W.B., and Blum, A.E. (2002). Weathering reactions and hyporheic exchange controls on stream water chemistry in a glacial meltwater stream in the McMurdo Dry Valleys. *Water Resources Research* **38**, 1279, doi:10.1029/2001WR000834.

Gooseff, M.N., McKnight, D.M., Runkel, R.L., and Duff, J.H. (2004a). Denitrification and hydrologic transient storage in a glacial meltwater stream, McMurdo Dry Valleys, Antarctica. *Limnology and Oceanography* **49**, 1884–1895.

Gooseff, M.N., McKnight, D.M., and Runkel, R.L. (2004b). Reach-scale cation exchange controls on major ion chemistry of an Antarctic glacial meltwater stream. *Aquatic Geochemistry* **10**, 221–238.

Green, W.J., Angle, M.P., and Chave, K.E. (1988). The geochemistry of Antarctic streams and their role in the evolution of four lakes in the McMurdo Dry Valleys. *Geochimica Cosmochima Acta* **52**, 1265–1274.

Hall, B.L., Denton, G.H., and Hendy, C.H. (2000) Evidence from Taylor Valley for a grounded ice sheet in the Ross Sea, Antarctica. *Geografiska Annaler* **82A(2–3)**, 275–303.

Hawes, I., Howard-Williams, C., and Vincent, WF (1992). Desiccation and recovery of Antarctic cyanobacterial mats. *Polar Biology* **12**, 587–594.

Hendy, C.H., Healy, T.R., Raymer, E.M., Shaw, J., and Wilson, A.T. (1979). Late Pleistocene glacial chronology of the Taylor Valley, Antarctica and the global climate. *Quaternary Research* **11**, 172–184.

Hershey, A.E. and Hiltner, A.L. (1988). Effect of a caddisfly on black fly density: interspecific interactions limit black flies in an arctic river. *Journal of the North American Benthological Society* **7**, 188–196.

Hershey, A.E. et al. (1988). Nutrient influence on a stream grazer: *Orthocladius* microcommunities respond to nutrient input. *Ecology* **69**, 1383–1392.

Hershey, A.E. et al. (1997). The Kuparuk River: a long-term study of biological and chemical processes in an arctic river. In Milner, A.M. and Oswood, M.W. (eds), *Freshwaters of Alaska*, pp. 107–129. Springer-Verlag, New York.

Hiltner, A.L. (1985). *Response of Black Fly Species (Diptera: Simuliidae) to Phosphorus Enrichment of an Arctic Tundra Stream.* Thesis, University of Wisconsin, Madison, WI.

Hobbie, J.E. et al. (1999). Impact of global climate change on the biogeochemistry and ecology of an Arctic freshwater system. *Polar Research* **18**, 207–214.

Howard-Williams, C., Vincent, C.L., Broady, P.A., and Vincent, W.F. (1986). Antarctic stream ecosystems: variability in environmental properties and algal community structure. *Internationale Revue gesamten Hydrobiologie* **71**, 511–544.

Howard-Williams, C. and Vincent, W.F. (1989a). Microbial ecology of southern Victoria Land streams (Antarctica) I. Photosynthesis. *Hydrobiologia* **172**, 27–38.

Howard-Williams, C., Priscu, J.C., and Vincent, W.F. (1989b). Nitrogen dynamics in two Antarctic streams. *Hydrobiologia* **172**, 51–61.

Howard-Williams, C., Hawes, I., and Schwarz A-M (1997). Sources and sinks of nutrients in a polar desert stream, the Onyx River, Antarctica. In Lyons, W.B., Howard-Williams, C., and Hawes, I. (eds), *Ecosystem Processes in Antarctic Ice-Free Landscapes*, pp. 155–170, Balkema Press, Amsterdam.

Huryn, A.D. et al. (2005). Landscape heterogeneity and the biodiversity of arctic stream communities: a habitat template analysis. *Canadian Journal of Fisheries and Aquatic Sciences* **62**, 1905–1919.

Kane, D., Hinzman, L., Benson, C., and Liston, G. (1991), Snow hydrology of a headwater Arctic Basin 1. Physical measurements and process studies. *Water Resources Research* **27**, 1099–1109.

Kane, D.L., Hinzman, L.D., Benson, C.S., and Everett, K.R. (1989). Hydrology of Imnavait Creek, and arctic watershed. *Holarctic Ecology* **12**, 262–269.

Kling, G.W., Kipphut, G.W., and Miller, M.C. (1991). Arctic lakes and streams as gas conduits to the atmosphere: implications for tundra carbon budgets. *Science* **251**, 298–301.

Kriet, K., Peterson, B.J., and Corliss, T.L. (1992). Water and sediment export of the upper Kuparuk River drainage of the North Slope of Alaska. *Hydrobiologia* **240**, 71–81.

Lee, J.O. and Hershey, A.E. (2000). Effects of aquatic bryophytes and long-term fertilization on arctic stream insects. *Journal of the North American Benthological Society* **19**, 697–708.

Lesack, L.F.W., Marsh, P., and Hecky, R.E. (1998). Spatial and temporal dynamics of major solute chemistry among Mackenzie Delta lakes. *Limnology and Oceanography* **43**, 1530–1543.

Lesack, L.F.W. and Marsh, P. (2007). Lengthening plus shortening of river-to-lake connection times in the Mackenzie River Delta respectively via two global change mechanisms along the arctic coast. *Geophysical Research Letters* **34**, L23404.

Lyons, W.B. *et al.* (1998). Geochemical linkages among glaciers, streams and lakes within Taylor Valley, Antarctica. In Priscu, J.C. (ed.), *Ecosystem Processes in a Polar Desert: The McMurdo Dry Valleys, Antarctica*, pp. 77–91. Antarctic Research Series vol. 72. American Geophysical Union, Washington DC.

Maurice, P.A., McKnight, D.M., Leff, L., Fulghum, J.E., and Gooseff, M.N. (2002). Direct observations of aluminosilicate weathering in the hyporheic zone of an Antarctic Dry Valley stream. *Geochimica Cosmochimica Acta* **66**, 1335–1347.

McClelland, J.W., Stieglitz, M., Pan, F., Holmes, R.M., and Peterson, B.J. (2007). Recent changes in nitrate and dissolved organic carbon export from the Upper Kuparuk River, North Slope, Alaska. *Journal of Geophysical Research* **112**, G04S60, doi:10.1029/2006JG000371.

McKnight, D.M., Alger, A., and Tate, C.M. (1998). Longitudinal patterns in algal abundance and species distribution in meltwater streams in Taylor Valley, Southern Victoria Land, Antarctica. In Priscu, J.C. (ed.), *Ecosystem Processes in a Polar Desert: The McMurdo Dry Valleys, Antarctica*, pp. 93–108. Antarctic Research Series vol. 72. American Geophysical Union, Washington DC.

McKnight, D.M., Runkel, R.L., Tate, C.M., Duff, J.H., and Moorhead, D.L. (2004). Inorganic N and P dynamics of Antarctic glacial meltwater streams as controlled by hyporheic exchange and benthic autotrophic communities. *Journal of the North American Benthological Society* **23**, 171–188.

McKnight, D.M. *et al.* (1999). Dry Valley streams in Antarctica: ecosystems waiting for water. *Bioscience* **49**, 985–995.

McKnight, D.M. *et al.* (2007). Reactivation of a cryptobiotic stream ecosystem in the McMurdo Dry Valleys, Antarctica: a long-term geomorphological experiment. *Geomorphology* **89**, 186–204.

McNamara, J.P., Kane, D.L., and Hinzman, L.D. (1997). Hydrograph separations in an Arctic watershed using mixing model and graphical techniques. *Water Resources Research* **33**, 1707–1719.

Meon, B. and Amon, R.M.W. (2004). Heterotrophic bacterial activity and fluxes of dissolved free amino acids (DFAA) and glucose in the Arctic Rivers Ob, Yenisei and the adjacent Kara Sea. *Aquatic Microbial Ecology* **37**, 121–135.

Michaelson, G.J., Ping, C.L., Kling, G.W., and Hobbie, J.E. (1998). The character and bioactivity of dissolved organic matter at thaw and in the spring runoff waters of the arctic tundra north slope, Alaska. *Journal of Geophysical Research* **103(D22)**, **28**, 939–946.

Miller, M.C., DeOliveira, P., and Gibeau, G.G. (1992). Epilithic diatom community response to years of PO_4 fertilization: Kuparuk River, Alaska (68 N Lat.). *Hydrobiologia* **240**, 103–119.

Morrison, D. (2000). Inuvialuit fishing and the Gutchiak site. *Arctic Anthopology* **37**, 1–42.

Neff, J.C. *et al.* (2006). Seasonal changes in the age and structure of dissolved organic carbon in Siberian rivers and streams. *Geophysical Research Letters* **33**, L23401, doi: 10.1029/2006GL028222.

Peterson, B.J., Hobbie, J.E., Corliss, T.L., and Kriet, K. (1981). A continuous-flow periphyton bioassay: tests of nutrient limitation in a tundra stream. *Limnology and Oceanography* **28**, 583–591.

Peterson, B.J., Hobbie, J.E., and Corliss, T.L. (1986). Carbon flow in a tundra stream ecosystem. *Canadian Journal of Fisheries and Aquatic Sciences* **43**, 1259–1270.

Peterson, B.J. *et al.* (1985). Transformation of a tundra river from heterotrophy to autotrophy by addition of phosphorus. *Science* **229**, 1383–1386.

Peterson, B.J. *et al.* (1993). Biological responses of a tundra river to fertilization. *Ecology* **74**, 653–672.

Peterson, B.J. *et al.* (2006). Trajectory shifts in the arctic and subarctic freshwater cycle. *Science* **313**, 1061–1066.

Prowse, T.D. (2001a). River-ice ecology. I: Hydrologic, geomorphic, and water-quality aspects. *Journal of Cold Regions Engineering* **15**, 1–16.

Prowse, T.D. (2001b). River-ice ecology. II: Biological aspects. *Journal of Cold Regions Engineering* **15**, 17–33.

Rachold, V. *et al.* (2004). Modern terrigenous organic carbon input to the Arctic Ocean. In Stein, R.S. and Macdonald, R.W. (eds), *The Organic Carbon Cycle in the Arctic Ocean*, pp. 33–55. Springer, New York.

Rautio, M. and Vincent, W.F. (2006). Benthic and pelagic food resources for zooplankton in shallow high-latitude lakes and ponds. *Freshwater Biology* **51**, 1038–1052.

Retamal, L., Vincent, W.F., Martineau, C., and Osburn, C.L. (2007). Comparison of the optical properties of dissolved organic matter in two river-influenced coastal regions of the Canadian Arctic. *Estuarine and Coastal Shelf Science* **72**, 261–272.

Retamal, L., Bonilla, S., and Vincent, W.F. (2008). Optical gradients and phytoplankton production in the Mackenzie River and coastal Beaufort Sea. *Polar Biology* **31**, 363–379.

Ruban, G.I. (1997). Species structure, contemporary distribution and status of the Siberian surgeon *Acipenser baerii. Environmental Biology of Fishes* **48**, 221–230.

Runkel, R.L., McKnight, D.M., and Andrews, E.D. (1998). Analysis of transient storage subject to unsteady flow: diel flow variation in an Antarctic stream. *Journal of the North American Benthological Society* **17**, 143–154.

Slavik, K. *et al.* (2004). Long-term responses of the Kuparuk River ecosystem to phosphorus fertilization. *Ecology* **85**, 939–954.

Stephenson, S.A., Burrows, J.A., and Babaluk, J.A. (2005). Long-distance migrations by inconnu (*Stenodus leucichthys*) in the Mackenzie River system. *Arctic* **58**, 21–25.

Stream Bryophyte Group (1999). Roles of bryophytes in stream ecosystems. *Journal of the North American Benthological Society* **18**, 151–184.

Treonis, A.M., Wall, D.H., and Virginia, R.A. (1999). Invertebrate biodiversity in Antarctic dry valley soils and sediments. *Ecosystems* **2**, 482–492.

Vallières, C., Retamal, L., Osburn, C.L., and Vincent, W.F. (2008). Bacterial production and microbial food web structure in a large arctic river and the coastal Arctic Ocean. *Journal of Marine Systems*, doi:10.1016/j.jmarsys.2007.12.002.

Vincent, W.F. (1988). *Microbial Ecosystems of Antarctica.* Cambridge University Press, Cambridge.

Vincent, W.F. and Howard-Williams, C. (1986). Antarctic stream ecosystems: physiological ecology of a blue-green algal epilithon. *Freshwater Biology* **16**, 219–233.

Vincent, W.F. and Howard-Williams, C. (1989). Microbial ecology of southern Victoria Land streams (Antarctica). II. The effects of low temperature. *Hydrobiologia* **172**, 39–50.

Vincent, W.F., Howard-Williams, C., and Broady, P.A. (1993a). Microbial communities and processes in Antarctic flowing waters. In Friedmann, I. (ed.), *Antarctic Microbiology*, pp. 543–569. Wiley-Liss, New York.

Vincent, W.F., Downes, M.T., Castenholz, R.W., and Howard-Williams, C. (1993b). Community structure and pigment organisation of cyanobacteria-dominated microbial mats in Antarctica. *European Journal of Phycology* **28**, 213–221.

Walter, K.M., Zimov, S.A., Chanton, J.P., Verbyla, D., and Chapin, F.S. (2006). Methane bubbling from Siberian thaw lakes as a positive feedback to climate warming. *Nature* **443**, 71–75.

Webster, J., Hawes, I., Downes, M., Timperley, M., and Howard-Williams, C. (1996). Evidence for regional climate change in the recent evolution of a high latitude pro-glacial lake. *Antarctic Science* **8**, 49–59.

Wollheim, W.M. *et al.* (1999). A coupled field and modeling approach for the analysis of nitrogen cycling in streams. *Journal of the North American Benthological Society* **18**, 199–221.

Woo, M.K. and Young, K.L. (2006). High Arctic wetlands: their occurrence, hydrological characteristics and sustainability. *Journal of Hydrology* **320**, 432–450.

CHAPTER 6

Ice-based freshwater ecosystems

Ian Hawes, Clive Howard-Williams, and Andrew G. Fountain

Outline

Ice is a dominant component of polar landscapes which melts seasonally where combinations of solar energy flux, ice albedo, and air temperature permit. The resulting liquid water can persist on, and in, ice structures and these can support various types of ice-based aquatic ecosystem. We discuss the physical, chemical, and biotic features of the main types of ice-based aquatic ecosystems. We describe how dark-colored sediments on and in ice enhance absorption of solar radiation, and promote melting and the formation of habitats of varying sizes and longevity. We show how these range from 'bubbles' within glacial and perennial lake ice ($\approx 10^{-2}$ m diameter), and cryoconite holes ($\approx 10^{-1}$–10^{0} m diameter) on ice surfaces to large melt lakes ($\approx 10^{1}$–10^{2} m diameter) and rivers on ice shelves and ice sheets. We describe the shared physical, chemical, and biological properties, notably those that relate to their alternating liquid–frozen state, and how similar species assemblages tend to result, often shared with neighboring terrestrial habitats where extreme conditions also occur. Although there are tantalizing suggestions that microbial metabolism may occur within polar glacial ice, for the most part development of ice-based aquatic ecosystems depends on liquid water. Communities are predominantly microbial, with cyanobacteria and algae dominating the phototrophs, whereas microinvertebrates with stress-tolerating strategies (rotifers, tardigrades, and nematodes) are also present. We argue that while most of these ice-based systems are oligotrophic, with low concentrations of salts and nutrients supporting low rates of production, they are found across large areas of otherwise barren ice in polar regions and that to date their contributions to polar ecology have been underestimated. We suggest that they represent important biodiversity elements within polar landscapes, and would have been essential refugia from which polar-region ecosystems would have recovered after periods of extended glaciation.

6.1 Introduction

Ice and snow, the dominant features of polar landscapes, contribute to aquatic ecosystems through seasonal melting. Although much meltwater ultimately feeds land-based aquatic ecosystems, liquid water can also persist on and in ice itself long enough for aquatic ecosystems to develop (Howard-Williams and Hawes 2007). That meltwater systems develop on ice is in part due to latent heat of fusion. Raising a cubic decimeter (10^{-3} m^3)

of ice from a winter temperature of $-30°C$ to $0°C$ requires acquisition of approximately 58 kJ of heat, but turning ice at $0°C$ to water at the same temperature requires a further heat gain of approximately 333 kJ. This energy can come from absorption of solar radiation or from conductive transfer from the overlying atmosphere; on local scales it is absorption of solar radiation that tends to play a major role in determining ice melt. The albedo of snow and ice is normally high, reflecting 70–90% of the incoming solar energy (Paterson 1994). It has, however, long been recognized that melting of ice is promoted where accumulations of dark-colored material reduce this albedo and thus increase the absorption of solar energy relative to the adjacent ice (e.g. Sharp 1949; Gribbon 1979).

The manner in which deposits of wind-blown or avalanche-derived sediments evolve to form cylindrical cryoconite holes in glacier ice is now well established (e.g. Wharton *et al.* 1985; Fountain *et al.* 2004). In addition, the low-gradient, undulating surfaces of ice shelves are particularly favorable for the accumulation of sediment and large, long-lived, pools can be found associated with surface sediments on suitable ice shelves in both the Arctic and Antarctic (Howard-Williams *et al.* 1989; Vincent *et al.* 2000). Similarly, where the ice covers of perennially frozen lakes incorporate dark-colored material, liquid water may form during the summer months either on the surface or within ice covers, and these pockets of liquid water, or 'ice bubbles' *sensu* Adams *et al.* (1998), in turn support microbial ecosystems (Pearl and Pinckney 1996; Fritsen and Priscu 1998).

A somewhat different type of ice-based aquatic habitat, not dependent on sediment absorption of radiation, occurs where melt generated on the surfaces of glaciers coalesces to form flowing streams and supraglacial pools on the ice itself (Heywood 1977). Gradations exist between these categories of ice-based freshwater systems, but in this chapter we recognize ice bubbles, cryoconite holes, supraglacial pools, and streams to describe

ice-bound freshwater habitats. The special category of ice-based lake and river systems that develop underneath ice sheets and glaciers are described in Chapter 7 of this volume.

6.2 Ecosystems on and in glacial ice

6.2.1 Types of glacier-based ecosystem

Our categorization of freshwater habitats on glaciers is based primarily on their hydrology; *cryoconite holes* tend to be small, more or less isolated water pockets that generate their liquid water internally, *supraglacial pools* are fed from a catchment larger then the pool itself and not necessarily dependent on sediment, while *streams* are channelized pathways of flowing water. We currently know more about cryoconite holes than about the large meltwater pools and streams that are now known to occur on many polar glacier surfaces.

6.2.2 Cryoconite holes

Cryoconite holes are vertical cylindrically shaped melt holes in the glacier surface, which have a thin layer of sediment at the bottom and are filled with water (Figure 6.1). Although the holes have been known by a variety of names, including dust wells, dust basins, sub-surface melt pools, and baignoire (Agassiz 1847; Hobbs 1911; Sharp 1949), cryoconite hole is the term in most common use today. Cryoconite holes are common to ice-surfaced (as

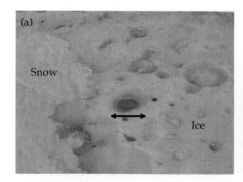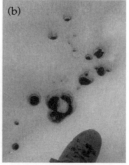

Figure 6.1 Cryoconite holes. (a) Holes with frozen lids on the surface of Canada Glacier, McMurdo Dry Valleys, Antarctica. The dark sediment patch is melting in, other holes are entirely frozen over. The scale arrow is about 20 cm. (b) Open holes on Storglaciaren, Sweden with a boot for scale.

opposed to snow-surfaced) zones of glaciers world-wide, including the Arctic (Von Drygalski 1897; Mueller *et al.* 2001; Säwström *et al.* 2002), temperate glaciers of the mid-latitudes (McIntyre 1984; Margesin *et al.* 2002), and the Antarctic (Wharton *et al.* 1985; Mueller *et al.* 2001; Fountain *et al.* 2004).

6.2.2.1 Physical processes in cryoconite holes

Cryoconite holes first form when small patches of sediment (wind-blown or avalanche-derived) accumulate in small depressions on the ice surface (Wharton *et al.* 1985; Fountain *et al.* 2004, 2008). The sediment absorbs solar radiation and initially melts into the surrounding ice faster than surrounding ice ablates, forming a water-filled hole (Figure 6.2). Cryoconite-hole width is probably determined by the size of the original sediment patch and can be a few centimeters to a meter or more in diameter. As the sediment melts deeper into the ice, the attenuation of radiation by absorption and scattering reduces heat gain and rate of melting and an

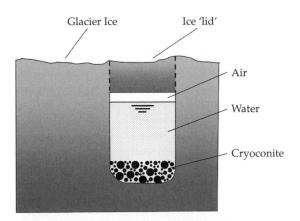

Figure 6.2 Cross-section schematic of a sealed cryoconite hole. Cryoconite holes freeze entirely during winter. The freezing process concentrates the solutes, possibly resulting in a delay of complete freezing of the entire hole as increasing salt concentration depresses the freezing point. The return of summer solar radiation again preferentially warms the sediment, which, if there has been winter or spring sublimation of surface ice, is now closer to the ice surface and again melts its way down through the ice according to the amount of heat it receives. In this way the sediment and associated biota may remain just beneath the ice surface, more or less isolated from the surrounding hydrologic system and the atmosphere for years and in some cases decades (Fountain *et al.* 2004).

equilibrium depth (typically 30–50 cm) is reached when the melt-rate in the hole matches the ablation rate of the ice surface (Gribbon 1979).

Two types of cryoconite hole are now recognized: those normally open to the atmosphere in summer, and those normally sealed with an ice lid. Closed holes are found where freezing air temperatures and persistent winds maintain a frozen ice surface over the liquid phase. In the Ross Sea sector of Antarctica, closed cryoconite holes occur at least as far north as 72°S (I. Hawes, personal observations). Closed holes have severely restricted exchange of water, solutes and gases and, unless surface conditions change dramatically, the cryoconite hole becomes entombed, freezing and melting beneath the glacier surface for years and capable of developing unusual biogeochemistries (Tranter *et al.* 2004). In some cases, however, apparently closed cryoconite holes can be hydrologically connected to a subsurface system of drainage passages (Fountain *et al.* 2004, 2008).

Open holes occur where the surface energy balance is close to melting and appear to be more common in the Arctic and temperate glaciers, where summers are warmer, than in the Antarctic (Mueller *et al.* 2001). Open holes tend to act like small ponds. Their water chemistries, while different from the host ice due to sediment dissolution and biological processes, are regularly diluted with fresh meltwater from the surface and exchange gases readily with the atmosphere.

6.2.2.2 Ecosystem processes in cryoconite holes

That cryoconite holes develop their own biota is not surprising, given that the snow that forms glacial ice encapsulates many propagules, including Archaea, bacteria, fungi, algae, and protozoa (Abyzov 1993; Christner *et al.* 2005; Castello and Rogers 2005) which can remain viable for long periods of time. Winds also carry such particles directly on to snow and ice surfaces as aeroplankton (Nkem *et al.* 2006). For open holes this is supplemented by surface meltwater and direct deposition into the holes. The microbial biota includes bacteria, virus-like particles, nanoflagellates, cyanobacteria, algae, and protozoa (Wharton *et al.* 1981, 1985; Mueller *et al.* 2001; Säwström *et al.* 2002; Christner *et al.* 2003). A metazoan census includes nematodes, tardigrades, rotifers (De Smet and Van Rompu 1994; Grongaard *et al.* 1999; Takeuchi *et al.* 2000; Christner *et al.* 2003; Porazinska *et al.* 2004), protozoans, copepods, and insect larvae (Kikuchi 1994; Kohshima 1984).

Studies of metabolic activity and nutrient analyses in cryoconite holes confirm the presence of active microbial consortia (Tranter *et al.* 2004; Foreman *et al.* 2007). Hodson *et al.* (2005) point to the loss of ammonium ions in the meltwaters on the glacier ice surfaces in Svalbard as evidence of microbial activity (photosynthesis) in the organic-rich sediments within the cryoconite holes. The water chemistry of the closed cryoconite holes of the McMurdo Dry Valleys also provides evidence of microbial activity. Photosynthesis in the holes, presumably by the green algae and cyanobacteria, is indicated by high saturations of oxygen (160%) and low values of pCO2 (Tranter *et al.* 2004). Values of net primary production, based on ^{14}C uptake, range from 67 to 146 mg C m^{-3} day^{-1} (C.M. Foreman, unpublished results) compared with 14–3768 mg C m^{-3} day^{-1} in open holes on Svalbard (Säwström *et al.* 2002).

In some closed holes, the chemistries can be extreme, with pH approaching 11 (Fountain *et al.* 2004), although Mueller *et al.* (2001) recorded an average pH of 8 for Canada Glacier cryoconite holes. This contrasts with the more dynamic, open holes on Arctic glaciers, where pH can be acidic (Mueller *et al.* 2001). High pH results first from hydrolysis of carbonates during the spring thaw that can increase pH to approximately 10.5, saturate the water with calcium carbonate and decrease pCO2. Photosynthesis further decreases pCO2 and increases pH to the observed extreme levels and can supersaturates the water with calcium carbonate. Ultimately, low pCO2 and high pH may limit photosynthesis. Thus in these poorly pH-buffered closed systems, photosynthesis may be constrained because of limitations on nutrients and new carbon dioxide sources. In-hole biogeochemical processes raise electrical conductivities in closed systems to several hundred µS cm^{-1}, substantially higher than melted ice.

Food webs in cryoconite holes are simple, with net flux of critical nutrients (inorganic carbon, nitrogen, and phosphorus) being dependent on external exchange and mineralization of sediment. Tranter *et al.* (2004) hypothesize that the cyanobacteria and algae in closed cryoconite holes must acquire nutrients scavenged from the sediments in the hole, although nitrogen fixation is a feature of several of the identified cyanobacterial taxa (Mueller *et al.* 2001). Phosphorus is almost certainly the limiting nutrient for growth. Heterotrophic bacteria found in the holes are comparable in abundance with those in high alpine lakes (Foreman *et al.* 2007), and these and fungi (Vincent *et al.* 2000) are together capable of exploiting different carbohydrates, amino acids, carboxylic acids, and aromatic compounds, suggesting that the cryoconite holes of dry-valley glaciers support metabolically diverse heterotrophic communities. The mixture of phototrophic and heterotrophic strategies, combined with the ability to survive in a frozen state through the dark winter months, produces a robust ecosystem capable of surviving many freeze–thaw cycles in isolation from the atmosphere and from subsurface water sources.

6.2.3 Supraglacial pools and streams

Aerial and satellite imagery shows that networks of supraglacial pools and interconnecting meltwater streams are widespread and common on ice shelves and on the Greenland Ice Sheet (e.g. Heywood 1977; Reynolds 1981; Maurette *et al.* 1986; Thomsen 1986; Säwström *et al.* 2002; Hodgson *et al.* 2004). Despite this abundance, little is known of these ecosystems. Whereas they are likely to share many features with cryoconite holes, their principal differences are hydrological and sediment-related. Pools will tend to have a much higher surface-area-to-volume ratio than cryoconite holes, a higher water-to-sediment ratio, and advective gains and loss of materials that do not occur in at least some cryoconite holes. In the 2005–2006 austral summer we sampled supraglacial ponds on the Tucker Glacier, Antarctica (72°S). Data from a typical pond are summarized in Table 6.1 and show an ultra-oligotrophic system, with even lower organic content and a lower pH than nearby cryoconite holes. Perhaps the best information on biological processes in supraglacial waters comes from the

Table 6.1 Characteristics of a typical supraglacial pond, and two nearby cryoconite holes (Cryo) on the Tucker Glacier Antarctica, January 2006. PC, particulate carbon; DOC, DON, DOP, dissolved organic carbon, nitrogen and phosphorus; DIN, dissolved inorganic nitrogen; DRP, dissolved reactive phosphorus. From I. Hawes, unpublished results

Site	Chlorophyll a (mg m^{-3})	PC (mg m^{-3})	pH	Conductivity (μS cm^{-1})	DON (mg m^{-3})	DOP (mg m^{-3})	DOC (g m^{-3})	DIN (mg m^{-3})	DRP (mg m^{-3})
Pond 1	0.04	63	8.3	7.8	31	2	2.9	6	1
Cryo 1	0.70	229	9.2	27	171	3	4.7	11	2
Cryo 2	0.49	285	9.3	26	284	3	3.1	17	3

Arctic (Säwström *et al.* 2002), where phytoplankton were found actively photosynthesizing at a rate of 7–8 mg C m^{-3} h^{-1}.

Studies of ponds on the George VI and Amery ice shelves provide insight into how critical snowfall and density of underlying firn are to meltwater accumulation. On those ice shelves, meltwater accumulates in surface irregularities where snowfall is below a threshold of 2×10^5 g m^{-2} year^{-1} and when firn density exceeds 820 kg m^{-3}, when it becomes impermeable to water (Reynolds 1981). Once a melt pool has formed, it becomes self-reinforcing in that the surface has a lower albedo and rapidly absorbs radiation penetrating overlying snow. The George VI Ice Shelf contains not only open pools, but also what appear to be ice-sealed englacial pools that lie within a meter of the glacier surface and can be tens of centimeters deep. All persist for at least several years (Reynolds 1981). As far as we are aware the biological communities and ecosystem processes in these waters remain unstudied.

6.3 Ecosystems on floating ice shelves

6.3.1 Types of ice-shelf ecosystem

The mass balance of the floating ice shelves of the Arctic and Antarctic regions is maintained by various processes. Some are fed from high altitude, with down-slope flow feeding ice to the lower, floating parts; others represent long-term accumulation of sea ice or are sustained by basal freezing of underlying waters. Ice is lost by calving, melting, and ablation. Ice shelves with substantial ablation zones include Antarctica's George VI, Amery, and

McMurdo ice shelves and the collection of remnant ice shelves along the northern coastline of Ellesmere Island in High Arctic Canada. Ablation zones allow sediment that has been deposited on, or incorporated into, the ice to migrate and accumulate towards the surface and thus promote the generation of surface meltwater that can in turn support aquatic ecosystems. Ice shelves support some of the most spectacular ice-based aquatic ecosystems on Earth (Plates 2 and 3).

The McMurdo Ice Shelf may be the most extensive surface-ablation area in Antarctica (Swithinbank 1970); it is virtually static, balanced by freezing of sea water beneath and ablation and melting on the surface. Basal freezing incorporates marine debris into the ice (Debenham 1920) and this is transported up through the ice as the surface ablates. Distribution of uplifted sediment on the surface varies and gives rise to a gradation of surfaces from almost fully covered with sediment that is 100 mm or more thick to a sparse sediment cover (Kellogg and Kellogg 1987). Differences in sediment cover determine the nature of the ice surface. Thick sediment cover ($\geq$100 mm; Isaac and Maslin 1991) insulates underlying ice and prevents melting, thus creating a very stable ice surface within which large, long-lived, sediment-lined ponds develop (Howard-Williams *et al.* 1989). Patchy layers of sediment typically are a mosaic of bare ice, thin sediments that enhance melting, and thicker sediments that retard melting; the net effect being a highly dynamic and irregular surface with mobile sediments and generation of copious meltwater (Debenham 1920).

The Ward Hunt Ice Shelf (83°N, 74°W) and nearby Markham Ice Shelf in the Canadian Arctic are ablation zones that also derive sediment from aeolian transported terrestrial sources and from the underlying seabed (Vincent *et al.* 2004a, 2004b; Mueller and Vincent 2006). As with the McMurdo Ice Shelf, the undulating surfaces of these ice shelves accumulate meltwater in a series of streams and pools (Figure 6.3). Satellite imagery (see Figure 17.4) and aerial photos of the Ward Hunt Ice Shelf (Plate 2) show an alignment of these ponds with the prevailing winds (Vincent *et al.* 2000). A feature of Arctic ice shelves is that they are shrinking. They underwent considerable break-up over the twentieth century, with further loss in recent years (Mueller *et al.* 2003). Since 2000, extensive fractures have appeared in the 3000-year-old Ward Hunt Ice Shelf (83°N, 75°W) and large sections have disintegrated (Figure 17.4; see also Chapter 1 in this volume). The nearby Ayles Ice Shelf broke out completely in August 2005. The recent 30-year period of accelerated warming is implicated, as it has also been in loss of ice shelves from western Antarctica. Notable among these losses was the Larsen B Ice Shelf, which was known to have large meltwater ponds on its surface in its final stages (Shepherd *et al.* 2003) but we know of no details on the limnological characteristics of these.

6.3.2 Physical processes in ice-shelf ponds

Pond freezing is a gradual process with the ice front gradually moving down through the pond water as heat accumulated over the summer is slowly lost (Figure 6.4). The delay in freezing at depth results in the liquid-water phase extending several months after surface ice forms (Mueller and Vincent 2006). In spring, ice-based systems can become liquid long before air temperatures are above freezing (Hawes *et al.* 1999; Figure 6.4) while during summer, pond temperatures can reach over 10°C (Figure 6.5). Thus liquid water, at moderate temperature, can be present for more than 3 months, whereas the ice-free period ranges from zero to a little over 1 month (Hawes *et al.* 1993, 1997, 1999).

The process of salt concentration during freezing described for cryoconite holes is important in ponds, where stratification verging on meromixis can develop (Wait *et al.* 2006; Figure 6.5). Exclusion of salts from the ice matrix results in formation of a brine in the pond bottom which may be so saline that it remains liquid below −20°C. During this process, sequential precipitation of different minerals, first mirabilite and then gypsum, results in a change in ionic ratios as the brine volume

Figure 6.3 Sampling a meltwater pond on the Markham Ice Shelf, Canadian High Arctic. Photograph: W.F. Vincent.

and temperature decrease. The following summer a layer of melted ice overlies the brine pool which, where the density difference is sufficient to resist mixing, may persist all summer (Wait *et al.* 2006; Figure 6.5) and even to allow an inverse thermal gradient to develop (Hawes *et al.* 1999). Summer photosynthesis results in the monimolinia providing supersaturation of dissolved oxygen and extreme high pH with a corresponding marked depletion in dissolved inorganic carbon (Hawes *et al.* 1997). This situation is reversed in winter when oxygen is depleted through respiration in the dark and the deep layers of these ponds provide a sequentially variable environment for microbial populations. Freeze-concentration and evaporative concentration of salts, and dynamic water levels, result in ponds with an array of salinities and ionic compositions, from near fresh to hypersaline.

6.3.3 Ecosystem processes in ice-shelf ponds

Where sediments are patchy, water bodies are small and dynamic and support communities of low biomass, dominated by fast-growing 'colonizers' such as diatoms, particularly *Pinnularia cymatopleura*, and coccoid chlorophytes (Howard-Williams *et al.* 1990). The biota of more stable and long-lived ice-shelf ponds is often dominated by mats of algae and cyanobacteria (Howard-Williams *et al.* 1989, 1990; Vincent *et al.* 2000, 2004a, 2004b; Mueller and Vincent 2006) with biomass of the cyanobacterial mats on the McMurdo Ice Shelf as high as 400 mg chlorophyll *a* m^{-2} (Howard-Williams *et al.* 1990; Hawes *et al.* 1993, 1997). The cyanobacteria that dominate the more long-lived ponds show a degree of variation in species between ponds of different salinities (Howard-Williams *et al.* 1990). Mats are mostly dominated by species of *Phormidium*, *Oscillatoria*, *Nostoc*, and *Nodularia*. *Nodularia* is particularly common in brackish waters, while *Oscillatoria* cf. *priestleyi* was the dominant in a highly saline pond (>70 mS cm^{-1}). Phylogenetic diversity is much higher than morphological diversity and salinity appears to be a major factor determining diversity structure (Jungblut *et al.* 2005). The diatoms *Pinnularia cymatopleura*, *Nitzchia antarctica*, and *Navicula* spp. were

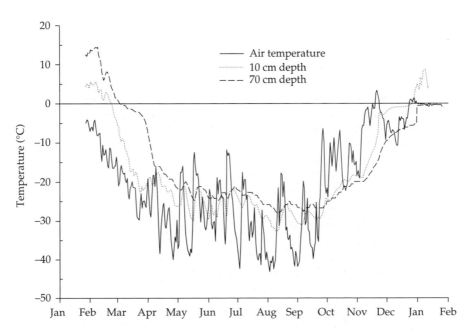

Figure 6.4 Thermistor records of temperatures in air and at two depths in a pond on the McMurdo Ice Shelf, during 1999. I. Hawes and C. Howard-Williams, unpublished results.

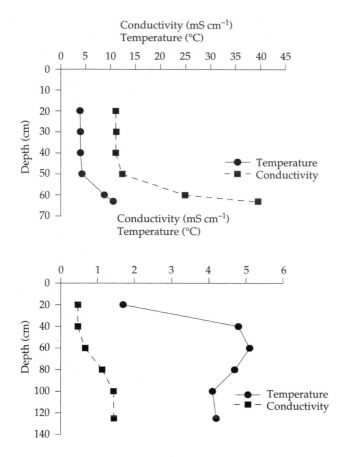

Figure 6.5 Vertical profiles of temperatures and conductivity in two ponds on the McMurdo Ice Shelf. The upper figure is for P70E, a pond which was ice-free at the time of sampling, and the lower figure is for Fresh Pond, which was frozen over all summer. Note the differences in scales of the *x* axes. I. Hawes and C. Howard-Williams, unpublished results.

associated with fresh waters, other naviculoids inculding *Navicula shackletonii* and *Navicula muticopsis* with fresh to brackish (11 mS cm⁻¹) waters and *Tropidoneis laevissima* and *Amphiprora* sp. with the most saline (Vincent and James 1996; Hawes *et al.* 1997). Benthic primary productivity is dealt with in detail in Chapter 10.

One species of tardigrade (probably a *Macrobiotus* sp.), three nematodes (three species of *Plectus*, *Plectus frigophilus* and two unidentified species) and seven rotifers (six bdelloids and one monogonontid) have been found associated with the cyanobacterial mats. Of the invertebrates, the rotifers were most abundant (4×10^5 m⁻²) and of these, a *Philodina* species, is most abundant by far, with dense aggregations often forming orange spots on microbial mats (Suren 1990). In addition to the benthic communities, ice shelf ponds support a planktonic community with phytoflagellates, *Chlamydomonas*,

Chroomonas, and *Ochromonas*; the latter dominating the most saline ponds (James *et al.* 1995). James *et al.* (1995) found that 'benthic' cyanobacteria were frequently entrained into the plankton. Twenty-two species of protozoa were also found in this survey, of which 15 could be considered truly planktonic. These fed primarily on bacteria-sized prey, with none showing any ability to consume particles of 5 μm diameter or greater.

Microbial mats dominate biomass and productivity in most ice-shelf ponds (Chapter 10). These typically take the form of cohesive layers of cyanobacteria overlying anoxic sediments. An orange-brown surface layer contains high concentrations of photoprotective pigment that shield lower, phycocyanin-rich blue-green layers (Vincent *et al.* 1993, 2004b) where both dissolved oxygen concentration and oxygen evolution reach their peaks (Hawes *et al.* 1997). A feature of ice-shelf ponds, as well as many other ice-based systems, is the high pH that can result from photosynthetic depletion of carbon dioxide and bicarbonate. Twelve ponds sampled on the McMurdo Ice shelf in 2002 had a mean pH of 9.7 (±0.5 S.D.). At such high a pH photosynthesis might be expected to be carbon-limited, but mat communities are well positioned to take advantage of any inorganic carbon/nutrients that are generated from their underlying sediments where pH is much lower and inorganic carbon and other nutrient concentrations are higher (Hawes *et al.* 1993). Physiologically, the mat-forming filamentous cyanobacteria consistently are primarily psychrotolerant, with temperature optima above 20°C (Hawes *et al.* 1997; Mueller and Vincent 2006). Phylogenetic analysis using molecular techniques suggests that the psychrotolerant trait occurs widely in the cyanobacterial lineage and that polar strains are often closely related to temperate ones. Psychrophily, in contrast, is confined to a narrow branch of the cyanobacteria, which includes bipolar strains (Nadeau *et al.* 2001).

A full suite of microbiological processes occurs within the ice-shelf ponds, allowing effective recyling of carbon, nitrogen, sulphur, and other elements within the semi-closed systems. Nutrient dynamics are exemplified by a nitrogen budget for a pond on the McMurdo Ice Shelf (Table 6.2). Autotrophic fixation of nitrogen (average 12.8 mg m^{-2} day^{-1} and similar to those at temperate latitudes) accounted for 67% of total nitrogen inputs whereas nitrogen release from the sediments as recycled ammonium/nitrogen, was 32%. Snow and ice melt provided the remaining 1%. Unlike nitrogen, other essential nutrients such as phosphorus cannot be biologically fixed and must come from weathering of minerals or atmospheric deposition.

Retention of nitrogen was high at 98%, much of it as biota, but retention as dissolved organic nitrogen (DON) appears to be very marked in pond waters. At the McMurdo Ice Shelf, DON concentrations can exceed 13 g m^{-3}, very high values for natural, unpolluted waters. Observations suggest that DON in these ponds is relatively refractory (Vincent and Howard-Williams 1994) and accumulates over time similarly to the major ions (Figure 6.6). The sources, sinks, and dynamics of DON in ice-based systems still need to be adequately addressed.

Table 6.2 Nitrogen budget for the water column of a pond on the McMurdo Ice Shelf. Data compiled from Hawes *et al.* (1997), Downes *et al.* (2000) and Fernandez-Valiente *et al.* (2001)

Process	Flux (mg m^{-2} day^{-1})	Percentage of inflow
In		
1. Precipitation	0.2	1
2. N fixation	12.8	67
3. Recycled N	6.0	32
Total in	19	100
Out		
4. Denitrification	0.4	2
5. Phytoplankton uptake	3.8	20
6. Benthic uptake	4.8	26
7. Storage*	10	52
Total out	19	100

*By difference.

Steep biogeochemical gradients are a feature of microbial mats. Oxygen concentration falls from supersaturated to anoxia a few millimeters into the microbial mats (Hawes *et al.* 1997). Mountfort *et al.* (2003) examined the anaerobic processes occurring in the underlying anoxic sediments. They found that, while nitrate reduction and methanogensis did occur, sulphate reduction was the dominant process. Interestingly these authors reported that all of these processes continued after freeze-up, with methanogenesis showing less sensitivity to subzero temperatures than sulphate reduction. This resulted in a gradual shift in dominant terminal electron acceptors with declining temperature.

6.4 Lake-ice ecosystems

6.4.1 Introduction

Although perennially ice-covered lakes are found through many regions of Antarctica and the colder parts of the Arctic, the only ice-covered lakes from which extensive information on ice-bound communities has been obtained are those from the McMurdo Dry Valleys. Living material within the ice of these lakes was first reported by Wilson (1965) and further investigated by Parker *et al.* (1982). Since

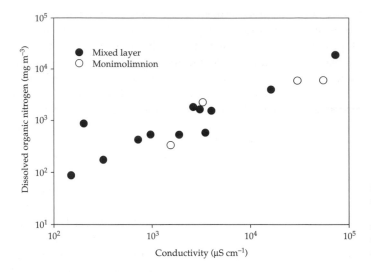

Figure 6.6 Dissolved organic nitrogen and conductivity in ponds on the McMurdo Ice Shelf. Note the logarithmic scales. Closed circles are the values from the upper mixed layer, open circles are from the lower layers when these were present at the time of sampling.

then investigations have revealed a variety of biota trapped within lake ice covers (Fritsen *et al.* 1998). As well as systems within lake ice, the sediment-rich surface of some perennially ice-covered lakes facilitates surface melting and the generation of freshwater systems on top of the ice, analogous to the sediment-lined surfaces of the McMurdo and Ward Hunt Ice Shelves.

6.4.2 Physical processes in lake ice

Ice temperatures vary with depth into the lake ice, being close to ambient air temperature close to the ice surface (surface pools will freeze to temperatures of −40°C), while at increasing depths into ice they are buffered from extremes. At 2 m into the ice, temperatures below −20°C are rare, whereas at 3 m depth −10°C is the lowest experienced each year (Priscu and Christner 2004).

The dynamics of perennial lake ice, with basal freezing balancing surface ablation, means that materials may enter lake ice from above, as with cryoconite holes on glaciers, but also from below, should materials be trapped in the ice cover during freezing, as with some ice shelves. Materials that freeze into the ice from above migrate downwards under the influence of solar heating and seasonal warming, leaving a liquid trail behind, until reaching a depth to which insufficient solar radiation

penetrates to allow further migration (Adams *et al.* 1998). Materials that freeze into the ice from below include fragments of the microbial mats that grow on the beds of these lakes and these will move upwards as ice ablates at the surface (Wilson 1965; Parker *et al.* 1982).

Regardless of direction of movement, dark materials in lake ice will tend to accumulate at the depth where melting rate balances surface ablation. The lower attenuation of radiation by lake ice compared to the firn ice of glaciers allows lake-ice ecosystems to sink meters rather than the decimeters of cryoconite holes into glacial ice. Thus lake ice communities form deeply entombed 'bubbles' of water, gas, and mineral and biotic inclusions within the ice. These undergo complex physico-chemical changes during the multiple seasonal freeze–thaw cycles that they are exposed to (Adams *et al.* 1998).

6.4.3 Ecosystem processes in lake ice

Communities within lake ice occupy a habitat similar to cryoconite holes in many ways (Priscu *et al.* 1998) but differ in that, while lake ice bubbles exist, they are never open to atmospheric exchange. Few measurements have been made, but it is apparent that communities within lake ice bubbles are photosynthetically active, although light-saturated rates of chlorophyll *a*-specific photosynthesis were 1–100-fold lower in the ice bubbles than in the lake water below (Fritsen and Priscu 1998). This may be due to the long period of subzero temperatures (at least −20°C) over the winter months, although the chemical conditions in the bubbles, after long periods (possibly years) of enclosure in repeatedly frozen and thawed lake ice, make these waters chemically extreme. Carbon dioxide limitation is almost certainly a constraint on photosynthesis, in waters where pH rises to 10 or more (Fritsen *et al.* 1998).

As with cryoconite holes (see Christner *et al.* 2003), molecular characterization of the diversity of bacteria and cyanobacteria in the ice bubbles has shown a wide variety of microbial groups (Gordon *et al.* 2000). 16S rDNA hybridization experiments demonstrated once again the degree of similarity between communities in ice-based and ice-derived communities in that dominant members of the diverse lake-ice microbial community are also found in adjacent microbial mats in melt streams, cryoconite holes, and other local habitats from which they probably originated.

6.5 The significance of ice-based systems

Despite their widespread distribution within ablation zones, the significance of ice-based systems in polar landscapes has received relatively little attention. Mueller *et al.* (2001) noted that cryoconite holes covered 12% of the surface of the lower ablation zone of White Glacier on Axel Heiberg Island, Nunavut, and up to 8% of that of the Canada Glacier in Antarctica. Fountain *et al.* (2004) found 3–15% coverage on four different glaciers in the dry valleys including Canada Glacier. Multiplying the total area of cryoconite holes by the average biomass within a hole, Mueller *et al.* estimated ice-based biomass (as loss on ignition) of the two glaciers to be 10.0 and 1.5 g Cm^{-2} respectively. Although these are low values, even in comparison with surrounding polar soils (216 g Cm^{-2} was measured in soils close to the Canada Glacier; Moorhead *et al.* 1999), there are several reasons why this pool of living material may have disproportionately significant ecosystem roles.

First, numerous authors have alluded to the potential of cryoconite holes to act as refugia for the biota within an otherwise hostile landscape, and point to the broad similarities between the flora and fauna of cryoconite holes to each other and to surrounding terrestrial and aquatic systems (e.g. Wharton *et al.* 1985). Recent molecular studies have confirmed the close relationships between organisms in these various locales (Priscu and Christner 2004). The supposition that aeolian and perhaps other transport mechanisms link these small elements of the watershed follows naturally from these observations and it becomes sensible to think of an the extensive, if fragmented, ice-based ecosystem as forming a continuum between distant, soil-based habitats. Aeolian transport of sediment and biota have been shown to be important processes in polar regions (Lancaster 2002; Nkem *et al.* 2006), and windblown dust has been shown to contain abundant propagules of cyanobacteria, eukarotic algae, and bryophytes (Hawes 1991).

As well as creating spatial continua, cryoconite-type habitats may also play a role in providing temporal continuity by acting as refugia through climatic excursions, including ice ages, facilitating rapid recolonization with cold-tolerant organisms when the ice retreats (Wharton *et al.* 1985). Recently there has been growing acceptance that during its early history the Earth went through one or a series of so-called snowball Earth phases, when almost the whole planet froze over (Hoffman *et al.* 1998). Vincent *et al.* (2000) argued that supraglacial habitats would have been among the few that would have provided suitable conditions for the survival of the microbial organisms and communities that are thought to have existed at the time of the largest event (600 million years before present, BP), including the cyanobacteria which still dominate the flora of ice-based ecosystems. Indeed, arguments have been advanced that the concentration of organisms into low-volume compartments within icy systems that would have accompanied the snowball Earth phases, coupled with the long-term stability of biota under cold conditions, may have provided ideal conditions for the development of symbiotic associations and eukaryotic development in the Precambrian (Vincent *et al.* 2000). Even in Quaternary glacial cycles, ice covered 11–18% of the Earth's surface and ice-based ecosystems must have been more extensive than at present (Hoffman and Schrag 2000).

The evidence for long-term viability of bacteria, viruses, fungi, and eukaryotes in polar ice is becoming increasingly strong (Castello and Rogers 2005). Whether these organisms are metabolically active at ambient temperatures or merely preserved is not yet clear, but it does seem likely that some organisms trapped in ice during its formation do retain viability and may form part of the early colonists of ice-based systems as they begin to melt. This is another way whereby ice-based systems may form an important intermediary for recolonization of terrestrial habitats during periods of glacial retreat.

Ice-based systems also play a role in the flux of matierals through polar landscapes. Hodson *et al.* (2005) showed how microbial activity within supraglacial freshwater systems on Arctic glaciers could make significant contributions to evolution of the chemistry of water draining glacial surfaces. They attributed the unexpectedly large fluxes of dissolved and particulate organic nitrogen from the glacier surfaces to nitrogen assimilation by microbial communities in cryoconite holes. The only measurement of the productivity of polar cryoconite holes of which we are aware (Säwström *et al.* 2002) indicates that rates of photosynthesis can be at least as high as water from nearby lakes, particularly in hole bottom water, and, whereas individual holes may have small volumes, the possibility that the large area occupied by cryoconite holes on some glaciers may result in substantial total carbon fixation requires further investigation. For the ice-shelf-based ecosystems there is no question that they represent substantial foci for biomass accumulation and productivity, far exceeding that of other habitats with the regions that they occupy.

6.6 Conclusions

In this review a surprising congruity between what are often treated as different types of ice-based system has emerged. We have shown how aquatic ecosystems can develop on and in any kind of perennial ice, most usually in ablation areas of relatively impermeable ice. In many cases, the existence of an aquatic ecosystem is triggered by an exposure of dark material on the ice surface that promotes melting through absorption of solar radiation; the type of system that develops, regardless of type of ice, is determined by the thickness, size, and stability of these sediment deposits. Indeed, there is a such an overlap between ice bubbles, ice-lidded and open cryoconites, ice-covered and ice-free ponds, and ice-bound lotic ecosystems as to make our splitting of these habitats a convenience rather than a reality. All share the seasonal fluctuations between frozen and unfrozen, and the common mechanism of ice formation means that the freeze-concentration effects in gas and salt contents are similar.

Whether ice, water, or land is the 'natural' abode of the organisms that are shared and exchanged with nearby terrestrial, ice-based, and aquatic ecosystems is perhaps an unanswerable and irrelevant question. The dominance of psychrotrophs over psychrophiles (Nadeau *et al.* 2001) supports the view that the ability to tolerate a varied range of habitats rather than to specialize in ice-bound systems is an important trait among the ice community. Our tendency to think of ice-based aquatic ecosystems, even in polar regions, as suboptimal oddities is perhaps more a reflection of our bias as warm-zone, terrestrial animals than a realistic assessment of their niche. It is better to think of these biota as a dynamic group of extremotrophs that are able to occupy whatever habitats are available at any given time. At both poles, and in most kinds of habitat, they are dominated by cyanobacteria, although a wide range of other organisms – including algae, fungi, protists, rotifers, nematodes, and tardigrades – is present. The required characteristics of successful colonists include tolerance of freezing and desiccation (Hawes *et al.* 1992), of high irradiance during summer (usually through synthesis of photoprotective pigments; Vincent *et al.* 1993, 2004a) and in some cases tolerance to high salinities (Hawes *et al.* 1997) and, wherever gas exchange is restricted, high pH (e.g. Fritsen *et al.* 1998). Therefore, similarities among biota between Arctic and Antarctic, glacier, and lake-ice habitats should not be surprising. Cold conditions are neither new nor rare on Earth. In the past they have been more and less widespread, but the ability of organisms to both tolerate the extremes of cold environments and extract sufficient resources from an inhospitable and often isolated habitat to grow and reproduce have clearly enabled them to persist.

This picture of dynamic interaction between land and ice becomes yet more logical when the relative timescales of biological evolution and habitat change are superimposed. Over evolutionary (millennial) time scales the relative abundance of ice and land in both polar regions has undergone substantial change. An example sequence from Victoria Land (Doran *et al.* 1994) is shown below.

• 15 000–10 000 years BP: during the last glaciation there were few if any ice-free areas. The main refuges for microbial life were ice-bound ecosystems (cryoconites, glacial ice ponds, and ice shelves).
• 10 000–5000 years BP: in the early interglacial during glacial retreat from small parts of the continent the ice-free areas would have been colonized from the ice-bound ecosystems by washout from **cryoconite** holes and aeolian dispersal from ice shelves.
• 5000 years BP–present: in mid to late interglacials the above process may be balanced by a 'reverse' colonization of glaciers from aeolian dispersal from ecosystems that have subsequently developed on ice-free land.

Over long time periods we may expect a mixing back and forth of communities between ice-bound systems and ice-free land. A major ecosystem stressor over millennia are the glaciations and it is the ice-bound systems that provide the genetic pools of microbial consortia that allow for ecosystem biogeochemical pathways, and presumably the soil-based systems when the ice retreats. This existing balance in favor of ice-bound systems may be changing rapidly with warming of the polar regions and glacial retreat (e.g. Schiermeier 2004) which will allow a greater diversity of organisms to colonize the ice-free lands and marked changes to existing ecosystems. The consequences for ice-based habitats will be complex and potentially severe, perhaps resulting in elimination from some areas. However, if our hypothesis that the ice biota is less a specialized one and more part of an assemblage of stress-tolerant opportunists

which cross-colonize between and across ice- and soil-based systems prove correct, we can expect ice-bound ecosystems to return as soon as ice does. Testing these hypotheses may be significant to our full understanding of whether polar climate amelioration threatens unique ice-based ecosystems or whether there is sufficient long-term resilience to cope with long-term change.

Acknowledgments

Thanks are due to those people, too many to list by name, with whom we have shared experiences of ice-based aquatic ecosystems and have contributed through their support, ideas and stimulating discussions over the years. Many agencies have supported our research; we would particularly acknowledge the New Zealand Foundation for Research, Science and Technology (CO1X0306) and the US National Science Foundation (ANT-0423595). Logistic support has come from the New Zealand and US Antarctic Research Programs and Polar Shelf Canada.

References

Abyzov, S.S. (1993). Microorganisms in Antarctic ice. In Friedmann, E.I. (ed.), *Antarctic Microbiology*, pp. 265–296. Wiley, New York.

Adams, E.E., Priscu, J.C., Fritsen, C., Smith, S.R., and Brackman, S.L. (1998). Permanent ice covers of the McMurdo Dry Valley Lakes, Antarctica: Bubble formation and metamorphism. In Priscu, J.C. (ed.), *Ecosystem Dynamics in a Polar Desert: the McMurdo Dry Valleys, Antarctica*, pp. 281–295. Antarctic Research Series vol. 72. American Geophysical Union, Washington DC.

Agassiz, L. (1847). *Nouvelles études et expériences sur les glaciers actuels leur structure leur progression, et leur action physique sur le sol*. Victor Masson, Paris.

Castello, J.D. and Rogers, S.O. (2005). *Life in Ancient Ice*. Princeton University Press, Princeton, NJ.

Christner, B.C., Kvitko, B.H., and Reeve, J.N. (2003). Molecular identification of Bacteria and Eukarya inhabiting an Antarctic cryoconite hole. *Extremophiles* **7**, 177–183.

Christner, B.C., Mosley-Thompson, E., Thompson, L.G., and Reeve, J.N. (2005). Classification of bacteria from polar and non-polar glacial ice. In Castello, J.D. and Rogers, S.O. (eds), *Life in Ancient Ice*, pp. 227–239. Princeton University Press, Princeton, NJ.

Debenham, F. (1920). A new mode of transportation by ice: the raised marine muds of South Victoria Land (Antarctica). *Quarterly Journal of the Geological Society of London* **75**, 51–76.

De Smet, W.H. and Van Rompu, E.A. (1994). Rotifera and Tardigrada from some Cryoconite holes on a Spitsbergen (Svalbard) glacier. *Belgian Journal of Zoology* **124**, 27–37.

Doran, P.T., Wharton, Jr, R.A., and Lyons, W.B. (1994) Paleolimnology of the McMurdo Dry Valleys, Antarctica. *Journal of Paleolimnology* **10**, 85–114.

Downes, M.T., Howard-Williams, C., Hawes, I., and Schwarz, A.-M.J. (2000). Nitrogen dynamics in a tidal lagoon at Bratina Island, McMurdo Ice Shelf, Antarctica. In Davison, W., Howard-Williams, C., and Broady, P. (eds), *Antarctic Ecosystems: Models for Wider Ecological Understanding.* Proceedings of the VII SCAR Biology Symposium, pp. 19–25. Caxton Press, Christchurch.

Fernandez-Valiente, E., Quesada, A., Howard-Williams, C., and Hawes, I. (2001). N_2-Fixation in cyanobacterial mats from ponds in the McMurdo Ice Shelf, Antarctica. *Microbial Ecology* **42**, 338–349.

Foreman, C.M., Sattler, B., Mikucki, J., Porazinska, D.L., and Priscu JC. (2007). Metabolic activity and diversity of cryoconites in the Taylor Valley, Antarctica. *Journal of Geophysical Research* **112**, G04S32, doi:10.1029/2006JG000358.

Fountain, A.G., Tranter, M., Nylen, T.H., Lewis, K.J., and Mueller, D.R. (2004) Evolution of cryoconite holes and their contributions to meltwater runoff from glaciers in the McMurdo Dry Valleys, Antarctica. *Journal of Glaciology* **50**, 35–45.

Fountain, A.G., Nylen, T.H., Tranter, M., and Bagshaw, E. (2008). Temporal variations in physical and chemical features of cryoconite holes on Canada Glacier, McMurdo DryValleys, Antarctica. *Journal of Geophysical Research* **113**, G01S92, doi:10.1029/2007JG000430.

Fritsen, C.H. and Priscu, J.C. (1998). Cyanobacterial assemblages in permanent ice covers on Antarctic lakes: distribution, growth rate and temperature response to photosynthesis. *Journal of Phycology* **34**, 587–597.

Fritsen, C.H., Adams, E.E., McKay, C.P., and Priscu, J.C. (1998), Permanent ice covers of the McMurdo dry Valleys lakes, Antarctica: Liquid water contents. In Priscu, J.C. (ed.), *Ecosystem Dynamics in a Polar Desert: the McMurdo Dry Valleys, Antarctica*, pp. 269–280. Antarctic Research Series vol. 72. American Geophysical Union, Washington DC.

Gordon, D.A., Priscu, J.C., and Giovannoni, S. (2000). Origin and phylogeny of microbes living in permanent Antarctic lake ice. *Microbial Ecology* **39**, 197–202.

Gribbon, P.W. (1979). Cryoconite holes on Sermikaysak, West Greenland, *Journal of Glaciology* **22**, 177–181.

Grongaard, A., Pugh, P.J.A., and McInnes SJ. 1999. Tardigrades and other cryoconite biota on the Greenland Ice Sheet. *Zoologischer Anzeiger* **238**, 211–214.

Hawes, I. (1991). Airborne propagule sampling on the McMurdo Ice Shelf, 1990/91 – a comparison of samplers. *BIOTAS Newsletter* **6**, 7–10.

Hawes, I., Howard-Williams, C., and Vincent, W.F. (1992). Desiccation and recovery of Antarctic cyanobacterial mats. *Polar Biology* **12**, 587–594.

Hawes, I., Howard-Williams, C., and Pridmore, R.D. (1993). Environmental control of microbial biomass in the ponds of the McMurdo Ice Shelf, Antarctica. *Archiv für Hydrobiologie* **127**, 271–287.

Hawes, I., Howard-Williams, C., Schwarz, A.-M.J., and Downes, M.T. (1997). Environment and microbial communities in a tidal lagoon at Bratina Island, McMurdo Ice Shelf, Antarctica. In Battaglia, B., Valencia, J., and Walton, D. (eds), *Antarctic Communities: Species, Structure, and Survival*, pp. 170–177. Cambridge University Press, Cambridge.

Hawes, I., Smith, R., Howard-Williams, C., and Schwarz, A.-M.J. (1999). Environmental conditions during freezing, and response of microbial mats in ponds of the McMurdo Ice Shelf, Antarctica. *Antarctic Science* **11**, 198–208.

Heywood, R.B. (1977). Limnological survey of the Ablation Point area, Alexander Island, Antarctica. *Philosophical Transactions of the Royal Society of London Series B Biological Sciences* **279**, 39–54.

Hobbs, W.H. (1911). *Characteristics of Existing Glaciers.* Macmillian Co., New York.

Hodgson, D.A., Doran, P.T., Roberts, D., and McMinn, A. (2004). Paleolimnological studies from the Antarctic and sub antarctic islands. In Pienitz, R., Douglas, M.S.V., and Smol, J.P. (eds), *Long-term Environmental Change in Arctic and Antarctic Lakes*, pp. 419–474. Springer, Dordrecht.

Hodson, A.J., Mumford, P.N., Kohler, J., and Wynn, P.M. (2005). The high Arctic glacial ecosystem: new insights from nutrient budgets. *Biogeochemistry* **72**, 233–256.

Hoffman, P.F. and Schrag, D.P. 2000. Snowball Earth. *Scientific American*, **68**, 68–75.

Hoffman, P.F., Kaufman, A.J., Halverson, G.P., and Schrag, D.P. (1998). A Neoproterozoic snowball Earth. *Nature* **281**, 1342–1346.

Howard-Williams, C. and Hawes I. (2007) Ecological processes in Antarctic inland waters: interactions between physical processes and the nitrogen cycle. *Antarctic Science* **19**, 205–217.

Howard-Williams, C., Pridmore, R., Downes, M.T., and Vincent, W.F. (1989). Microbial biomass, photosynthesis and chlorophyll a related pigments in the ponds of the McMurdo Ice Shelf, Antarctica. *Antarctic Science* **1**, 125–131.

Howard-Williams, C., Pridmore, R.D., Broady, P.A., and Vincent, W.F. (1990). Environmental and biological variability in the McMurdo Ice Shelf ecosystem. In Kerry, K.R. and Hempel, G. (eds), *Antarctic Ecosystems: Ecological Change and Conservation*, pp. 23–31. Springer Verlag, Berlin.

Isaac, P. and Masli, D. (1991). *Physical Hydrology of the McMurdo Ice Shelf*. Science Report to the New Zealand Antarctic Research Programme. Antarctica New Zealand, Christchurch.

James, M.R., Pridmore, R.D., and Cummings, V.J. (1995). Planktonic communities of melt pools on the McMurdo Ice Shelf, Antarctica. *Polar Biology* **15**, 555–567.

Jungblut, A.-D. *et al.* (2005). Diversity of cyanobacterial mat communities in three meltwater ponds of different salinities, McMurdo Ice Shelf, Antarctica. *Environmental Microbiology* **7**, 519–529

Kellogg, D.E. and Kellogg, T.B. (1987). Diatoms of the McMurdo Ice Shelf, Antarctica: implications for sediment and biotic reworking. *Palaeogeography, Palaeoclimatology and Palaeoecology* **60**, 77–96.

Kikuchi, Y. (1994). *Glaciella*, a new genus of freshwater *Canthyocamyidae* (*Copepoda Harpacticoida*) from a glacier in Nepal, Himalayas. *Hydrobiologia* **292/293**, 59–66.

Kohshima, S. (1984). A novel cold-tolerant insect found in a Himalayan glacier. *Nature* **310**, 225–227.

Lancaster, N. (2002). Flux of eolian sediment in the McMurdo Dry Valleys, Antarctica: a preliminary assessment. *Arctic, Antarctic and Alpine Research* **34**, 318–323.

Margesin, R., Zacke, G., and Schinner, F. (2002). Characterization of heterotrophic microorganisms in alpine glacier cryoconite. *Arctic, Antarctic and Alpine Research* **34**, 88–93.

Maurette, M., Hammer, C., Brownlee, D.E., Reeh, N., and Thomsen, H.H. (1986). Traces of cosmic dust in the blue ice lakes of Greenland. *Science* **233**, 869–872.

McIntyre, N.F. (1984). Cryoconite hole thermodynamics. *Canadian Journal of Earth Sciences* **21**, 152–156.

Moorhead, D.L. *et al.* (1999). Ecological legacies: Impacts on ecosystems of the McMurdo Dry Valleys. *Bioscience* **49**, 1009–1019.

Mountfort, D.O., Kaspar, H.F., Asher, R.A., and Sutherland, D. (2003). Influences of pond geochemistry, temperature and freeze-thaw on terminal anaerobic processes occurring in sediments of six ponds of the McMurdo Ice Shelf, near Bratina Island, Antarctica. *Applied and Environmental Microbiology* **69**, 583–592.

Mueller, D., Vincent, W.F., Pollard, W.H., and Fritsen, C.H. (2001). Glacial cryoconite ecosystems: A bipolar comparison of algal communities and habitats. *Nova Hedwigia* **123**, 173–197.

Mueller, D.R., Vincent, W.F., and Jeffries, M.O. (2003). Ice shelf break-up and ecosystem loss in the Canadian high Arctic. *Eos* **84**, 548–552.

Mueller, D.R. and Vincent, W.F. (2006) Microbial habitat dynamics and ablation control on the Ward Hunt Ice Shelf. *Hydrological Processes* **20**, 857–876.

Nadeau, T., Millbrandt, E.C., and Castenholz, R.W. (2001). Evolutionary relationships between cultivated Antarctic Oscillatorians (cyanobacteria). *Journal of Phycology* **37**, 650–654.

Nkem, J.N. *et al.* (2006). Wind dispersal of soil invertebrates in the McMurdo Dry Valleys, Antarctica. *Polar Biology* **29**, 346–352.

Parker, B.C., Simmons, G.M., Wharton, R.A., Seaburg, K.G., and Love, F.G. (1982). Removal of organic and inorganic matter from Antarctic lakes by aerial escape of blue-green algal mats. *Journal of Phycology* **19**, 72–78.

Paterson, W.S.B. (1994). *The Physics of Glaciers*, 3rd edn. Pergamon, Oxford.

Pearl, H.W. and Pinckney, J.L. (1996). Ice aggregates as a microbial habitat in Lake Bonney, dry valley lakes, Antarctica: nutrient-rich microzones in an oligotrophic ecosystem. *Antarctic Journal of the United States* **31**, 220–222.

Priscu, J.C. and Christner, B.C. (2004). Earth's icy biosphere. In Bull, A. (ed.), *Microbial Diversity and Bioprospecting*, pp. 130–145. ASM Press. Washington DC.

Priscu, J.C. *et al.* (1998). Perennial Antarctic lake ice: an oasis for life in a polar desert. *Science* **280**, 2095–2098.

Reynolds, J.M. (1981) Lakes on George VI Ice Shelf, Antarctica. *Polar Record* **20**, 435–432.

Säwström, C., Mumford, P., Marshall, W., Hodson, A., and Laybourn-Parry, J. (2002). The microbial communities and primary productivity of cryoconite holes in an Arctic glacier (Svalbard 79° N), *Polar Biology* **25**, 591–596.

Schiermeier, Q. (2004). Greenland's climate, a rising tide. *Nature* **428**, 114–115.

Sharp, R.P. (1949). Studies of superglacial debris on valley glaciers, *American Journal of Science* **247**, 289–315.

Shepherd, A., Wingham, D., Payne, T., and Skvarca, P. (2003). Larsen Ice Shelf has progressively thinned. *Science* **302,** 856–859.

Suren, A. (1990). Microfauna associated with algal mats in melt pools of the Ross Ice Shelf. *Polar Biology* **10**, 329–335.

Swithinbank, C. (1970). Ice movement in the McMurdo Sound area of Antarctica. *Proceedings of the International*

Symposium on Antarctic Glaciological Exploration, pp. 472–486. International Association of Scientific Hydrology, Ghentbrugge.

Takeuchi, N., Kohshima, S., Yoshimura, Y., Seko, K., and Fujita, K. (2000). Characteristics of cryoconite holes on a Himalayan glacier, Yala Glacier Central Nepal. *Bulletin of Glaciological Research* **17**, 51–59.

Thomsen, H.H. (1986). Photogrammetric and satellite mapping of the margin of the inland ice. West Greenland. *Annals of Glaciology* **8**, 164–167.

Tranter, M. *et al.* (2004). Extreme hydrochemical conditions in natural microcosms entombed within Antarctic ice. *Hydrological Processes* **18**, 379–387.

Vincent, W.F., Downes, M.T., Castenholz, R.W., and Howard-Williams, C. (1993). Community structure and pigment organisation of cyanobacteria-dominated microbial mats in Antarctica. *European Journal of Phycology* **28**, 213–221.

Vincent, W.F. and Howard-Williams, C. (1994). Nitrate-rich inland waters of the Ross Ice Shelf region, Antarctica. *Antarctic Science* **6**, 339–346.

Vincent, W.F. and James, M.R. (1996). Biodiversity in extreme aquatic environments: lakes, ponds and streams of the Ross Sea sector, Antarctica. *Biodiversity and Conservation* **5**, 1451–1471.

Vincent, W.F., Mueller, D., and Bonilla, S. (2004a). Ecosystems on ice: the microbial ecology of the Markham Ice Shelf in the high Arctic. *Cryobiology* **48**, 103–112.

Vincent, W.F., Mueller, D., Van Hove, P., and Howard-Williams, C. (2004b). Glacial periods on early Earth and implications for the evolution of life. In Seckbach, J. (ed.), *Origins: Genesis, Evolution and Diversity of Life*, pp. 483–501. Kluwer Academic Publishers, Dordrecht.

Vincent, W.F. *et al.* (2000). Ice shelf microbial ecosystems in the high Arctic and implications for life on snowball Earth. *Naturwissenschaften* **87**, 137–141.

Von Drygalski, E. (1897). *Grønland-Expedition der Gesellschaft für Erdkunde zu Berlin 1891–1893*. W.H. Kuhl, Berlin.

Wait, B.R., Webster-Brown, J.G., Brown, K.R., Healy, M., and Hawes, I. (2006). Chemistry and stratification of Antarctic meltwater ponds I: Coastal ponds near Bratina Island, McMurdo Ice Shelf. *Antarctic Science* **18**, 515–524.

Wharton, Jr, R.A., Vinyard, W.C., Parker, B.C., Simmons, Jr, G.M., and Seaburg, K.G. (1981). Algae in cryoconite holes on Canada Glacier in Southern Victoria Land, Antarctica. *Phycologia* **20**, 208–211.

Wharton, R.A., McKay, C.P., Simmons, G.M., and Parker, B.C. (1985). Cryoconite holes on glaciers. *Bioscience* **35**, 499–503.

Wilson, A.T. (1965). Escape of algae from frozen lakes and ponds. *Ecology* **46**, 376.

CHAPTER 7

Antarctic subglacial water: origin, evolution, and ecology

John C. Priscu, Slawek Tulaczyk, Michael Studinger,
Mahlon C. Kennicutt II, Brent C. Christner, and
Christine M. Foreman

Outline

Recent discoveries in the polar regions have revealed that subglacial environments provide a habitat for life in a setting that was previously thought to be inhospitable. These habitats consist of large lakes, intermittently flowing rivers, wetlands, and subglacial aquifers. This chapter presents an overview of the geophysical, chemical, and biological properties of selected subglacial environments. The focus is on the large subglacial systems lying beneath Antarctic ice sheets where most of the subglacial water on our planet is thought to exist. Specifically, this chapter addresses the following topics: (1) the distribution, origin, and hydrology of Antarctic subglacial lakes; (2) Antarctic ice streams as regions of dynamic liquid-water movement that influence ice-sheet dynamics; and (3) subglacial environments as habitats for life and reservoirs of organic carbon.

7.1 Introduction

Over the last decade, interest in subglacial lakes and rivers has matured from a curiosity to a focus of scientific research. The earliest evidence of large subglacial lakes was provided by Russian aircraft pilots who flew missions over the Antarctic continent in the 1960s (Robinson 1964). Speculation about the presence of lakes was verified by airborne radio-echo soundings collected during the 1960s and 1970s (Drewry 1983) in which flat reflectors at the bottom of ice sheets were interpreted as indicating subglacial accumulations of liquid water (Oswald and de Robin 1973; de Robin et al. 1977). However, it was Kapitsa's description of subglacial Lake Vostok in Antarctica that convinced the scientific community of the existence of major reservoirs of water beneath thick ice sheets (Kapitsa et al. 1996). We now know that more than 150 lakes exist beneath the Antarctic ice sheets (Priscu et al. 2003, 2005; Siegert et al. 2005) and that many may be connected by networks of subglacial streams and rivers (Gray et al. 2005; Wingham et al. 2006; Fricker et al. 2007; see Plate 1). Recent evidence also indicates that subglacial lakes may initiate and maintain rapid ice flow and should be considered in ice-sheet mass balance assessments. Liquid water had previously been documented beneath Antarctic ice streams (Engelhardt et al. 1990), the Greenland ice sheet (Fahnestock et al. 2001; Andersen et al. 2004), and smaller continental glaciers (e.g. Mikucki et al. 2004; Bhatia et al. 2006). Estimates indicate that the volume of Antarctic subglacial lakes alone exceeds 10 000 km^3 (Dowdeswell and Siegert 1999), with Lake Vostok ($\approx$5400 km^3; Studinger et al. 2004) and Lake 90°E (1800 km^3; Bell et al. 2006) being the largest.

Subglacial environments were originally speculated to be devoid of life (e.g. Raiswell 1984). However, discoveries of microbial life in McMurdo Dry Valley lake ice (Priscu et al. 1998), accretion ice above Lake Vostok (Priscu et al. 1999; Christner et al.

2006), within Greenland and Antarctic glacial ice (e.g. Christner *et al.* 2006), at the beds of alpine (Sharp *et al.* 1999) and High Arctic glaciers (Skidmore *et al.* 2005), in subglacial volcanic calderas (Gaidos *et al.* 2004), and in outlet glaciers draining the polar plateau (Mikucki and Priscu 2007) have all provided information about the expected diversity and biogeochemical importance of biology in subglacial environments. It is now known that subglacial biology plays a role in geochemical processes, offering new insights into the evolution and biodiversity of life on our planet (Priscu and Christner 2004). The discovery of viable organisms in subglacial environments has extended the known limits of life on Earth, providing strong evidence that life has successfully radiated into virtually all aquatic habitats on Earth that contain 'free' liquid. Subglacial liquid environments are an exciting frontier in polar science and will provide an improved understanding of the coupling of geological, glaciological, and biological processes on our planet.

This chapter presents an overview of the geophysical, chemical, and biological properties of selected aquatic subglacial environments. The focus is on the large subglacial systems lying beneath Antarctic ice sheets where most of the subglacial water on our planet is thought to exist (e.g. Siegert *et al.* 2006; Priscu and Foreman, 2008). This chapter addresses the following specific topics: (1) the distribution, origin, and hydrology of Antarctic subglacial lakes; (2) Antarctic ice streams as regions of dynamic liquid-water movement that influence ice-sheet dynamics; and (3) subglacial environments as habitats for life and reservoirs of organic carbon. Surface ice-based ecosystems are discussed in Chapter 6.

7.2 Antarctic subglacial lakes and rivers: distribution, origin, and hydrology

7.2.1 Distribution

The analysis of airborne surveys collected between 1967 and 1979 (Siegert *et al.* 1996) initially revealed at least 70 subglacial lakes beneath the Antarctic ice sheet. Dowdeswell and Siegert (2002) categorized these subglacial lakes as: (1) lakes in subglacial basins in the ice-sheet interior; (2) lakes perched on

the flanks of subglacial mountains; and (3) lakes close to the onset of enhanced ice flow. Lakes in the first category are found mostly in and on the margins of subglacial basins. Lakes in this category can be divided into two subgroups. The first subgroup is located where subglacial topography is relatively subdued; the second subgroup of lakes occur in topographic depressions, often closer to subglacial basin margins, but still near the slow-flowing center of the Antarctic ice sheets. Where bed topography is subdued, deep subglacial lakes are unlikely to develop. Lake Vostok is the only subglacial lake that occupies an entire subglacial trough. Other troughs, such as the Adventure Subglacial Basin, contain several smaller lakes (e.g. Wingham *et al.* 2006). 'Perched' subglacial lakes are found mainly in the interior of the ice sheet, and on the flanks of subglacial mountain ranges. In several cases, small subglacial lakes (<10 km long) have been observed perched on the stoss face (i.e. facing the direction from which a glacier moves) of large (>300 m high), steep (gradient >0.1) subglacial hills. At least 16 subglacial lakes occur at locations close to the onset of enhanced ice flow, some hundreds of kilometers from the ice-sheet crest (Siegert and Bamber 2000). About 20 other lakes have also been reported by Popov *et al.* (2002), who analyzed radio-echo sounding data collected between 1987 and 1991 by Russian Antarctic Expeditions in central Antarctica between Enderby Land and 90°S. Radar surveys carried out in 1999 and 2001 by the Italian Antarctic Program over Aurora and Vincennes Basins and over Belgica Highlands revealed 14 subglacial lakes in addition to the original Siegert inventory (Tabacco *et al.* 2003). The Italian survey defined the boundary conditions of a relatively large lake (Subglacial Lake Concordia) located in the Vincennes Basin, at 74°06′S, 125°09′E. The ice thickness over Lake Concordia ranges from 3900 to 4100 m, the surface area is greater than 900 km^2, and the water depth is estimated to be about 500 m in the central basin (Tikku *et al.* 2005). The high density of lakes in the Dome C region suggests that they are hydrologically connected within a watershed, making them an important system for the study of subglacial hydrology, and biological and geochemical diversity. The recent discovery of four large lakes with surface areas of

3915, 4385, 1490, and 3540 km² near the onset of the fast-flowing Recovery Ice Stream, a catchment that compromises 8% of the East Antarctic Ice Sheet and contributes 58% of the flux into the Filchner Ice Shelf, has led to the suggestion that these lakes may be responsible for the initiation of the ice stream (Bell *et al.* 2007). If this is true, then basal hydrology should be taken into consideration in numerical models of ice-sheet motion.

Siegert *et al.* (2005) combined radar-sounding interpretations from Italian, Russian and US researchers to revise the number of lakes known to exist beneath the ice sheet from 70 to 145. Approximately 81% of the detected lakes lie at elevations less than a few hundred meters above sea level whereas the majority of the remaining lakes are 'perched' at higher elevations. Lake locations from the new subglacial lake inventory are shown in Figure 7.1 relative to local 'ice divides' calculated from the satellite-derived surface elevations of Vaughan *et al.* (1999) and their spatial relationship to subglacial elevation and hydraulic fluid potential. Most of the lakes identified (66%) lie within 50 km of a local ice divide and 88% lie within 100 km of a local divide. Even lakes located

far from the Dome C/Ridge B cluster and associated with very narrow catchments lie either on or within a few tens of kilometers of the local divide marked by the catchment boundary. The hydraulic potential reveals that some of the lakes along the divide could be hydraulically connected, whereas others located on either side of the divide may not be in communication as the divides tends to follow the line of maximum fluid potential.

7.2.2 Origin

The association of subglacial lakes with local ice divides and regions of high hydraulic fluid potential leads to a fundamental question concerning the evolution of subglacial lake environments: does the evolving ice sheet control the location of subglacial lakes or does the fixed lithospheric character necessary for lake formation (e.g. basal morphology, geothermal flux, or the nature of sub-ice aquifers) constrain the evolution of ice-sheet catchments? With the exception of central West Antarctica (e.g. Anandakrishnan *et al.* 2007) little is known about either the lithospheric character along these catchment boundaries or the history

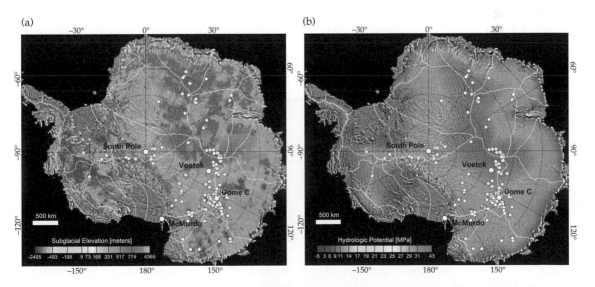

Figure 7.1 Distribution of Antarctic subglacial lakes in relation to the ice divides (light lines in both panels), subglacial elevation (a) and hydraulic fluid potential (b). The base map is a MODIS image mosaic. The water should flow from lakes with high fluid potential to lakes with lower potential, assuming that there is a connection. The larger white circles indicate the locations of major field stations. A colour version is shown in Plate 1.

of their migration as discerned from layering within the ice sheet.

Together with Lake Vostok (14 000 km^2; >800 m deep), the 90°E and Sovetskaya lakes define a province of major lakes on the flanks of the Gamburtsev Subglacial Mountains (Bell *et al.* 2006) (Figure 7.2). The estimated water depths of the 90°E (2000 km^2; ≈900 m deep) and Sovetskaya (1600 km^2; >800 m deep) lakes are similar to the maximum water depths deduced from seismic and gravity inversions over Lake Vostok and are also similar to other tectonically controlled lakes such as fault-bounded lakes including Tahoe, USA (501 m) and Issyk-kul, Kyrgyzstan (668 m), as well as rift lakes including Tanganyika, Africa (1479 m), Malawi, Africa (706 m), and Baikal, Siberia (1637 m) (Herdendorf 1982). With the exception of Great Slave Lake, Canada (624 m), glacially scoured lakes tend to have maximum water depths of less than 420 m (Herdendorf 1982). Whereas the majority of surface lakes are glacial in origin (75%) (Meybeck 1995), most (85%) of the deep lakes (>500 m) are tectonic in origin (Herdendorf 1982). The steep, rectilinear morphology of Lakes Vostok, 90°E, and Sovetskaya indicate a tectonic origin (see Figure 7.2 and Bell *et al.* 2006). Tectonic control of these basins is not indicative of active tectonics or elevated geothermal heat flow, but the basin-bounding faults may provide conduits of active fluid flow rich in dissolved minerals into the lakes (Studinger *et al.*

2003). These deep elongate basins probably pre-date the onset of Antarctic glaciation and were likely surface lakes before becoming overlain by glacial ice. The tectonically controlled depth of these lakes should provide relatively constant water depths through changing climatic conditions over the past 10–35 million years (Bell *et al.* 2006). Deep subglacial lakes are likely to have been stable through many glacial cycles and may have developed novel ecosystems, in contrast to the shallower lakes. This contention is corroborated by recent evidence from Great Slave Lake, Canada, which showed that the benthic sediments were undisturbed by the retreat of the Laurentian ice sheet across the lake basin (S. Tulaczyk, unpublished results).

Although arguments have been made for the tectonic origins of deep subglacial lakes (Studinger *et al.* 2003; Bell *et al.* 2006), there continues to be debate about whether subglacial lakes in Antarctica reside in active tectonic basins or along old inactive zones of structural weakness that once provided guidance for subglacial erosion. Much of East Antarctica, where the majority of subglacial lakes have been found so far (see Figure 7.1), is thought to have assembled between 500 and 800 million years ago. However, our knowledge of the interior of the continent, the distribution of major tectonic boundaries and old zones of structural weakness is limited due to a paucity of data. In regions where geophysical data, such as surface and airborne

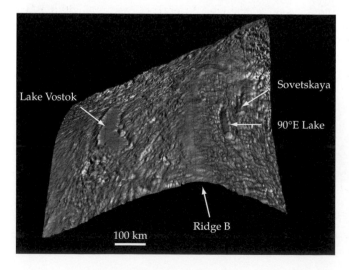

Figure 7.2 MODIS mosaic of Ridge B region including the 90°E and Sovetskaya lakes as well as Lake Vostok. Geographic north is to the bottom. See Bell *et al.* (2006) for details.

geophysics, and continent-wide geodetic networks can be combined, it is possible to discern whether tectonically controlled basins or zones of structural weakness are the preferred sites of subglacial lakes (SALE 2007). Views of the origins of structures that bisect the interior of the East Antarctic continent continue to evolve. Further knowledge based on high-resolution dating techniques will be needed to clarify whether subglacial lakes preferentially form along major tectonic boundaries when water is available.

7.2.3 Hydrology

For over 30 million years the Antarctic continent has had a hydrologic system in which redistribution of atmospheric precipitation is accomplished predominantly through flow of ice. When observed from the surface, Antarctica lives up to its reputation as a 'frozen continent'. However, recent scientific observations indicate the existence of a dynamic subglacial system of liquid-water generation, storage, and discharge in Antarctica, which is similar in some ways to river and lake systems on the other continents. Understanding the physical, chemical, and biological properties of this liquid subglacial water is currently one of the most exciting frontiers in Antarctic science and has the potential to change our basic understanding of the coupling of geological, glaciological, and biological processes in Antarctica.

Although the hydrology of certain alpine temperate glaciers have been studied for some time, knowledge of sub-ice-sheet hydrology in Antarctica is rudimentary. We can put only loose quantitative constraints on the dynamics of sources and sinks of water beneath the ice. To date, our understanding of the topology, geometry, and efficiency of subglacial drainage networks in Antarctica is based mostly on models. We know more about drainage channels on Mars than about liquid-water flow features in Antarctica. Subglacial Antarctic lakes provide the most impressive evidence for the presence and importance of subglacial water in Antarctica. Antarctic lakes may hold over 8% of all lacustrine fresh water on Earth, enough to cover the whole continent with a uniform water layer approximately 1 m deep. These estimates of water

volumes in subglacial lakes are surprising given the fact that rates of subglacial water production are 100–1000 times slower than mean effective precipitation rates on other continents (≈ 0.001 compared with ≈ 0.3 m year^{-1}, respectively). This means that average water residence time in the subglacial zone of Antarctica is equal to approximately 1000 years, which is likely a reflection of the slow rates of drainage of liquid water through subglacial environments. From other glacial drainage systems we know that water drainage often involves long periods of water accumulation, punctuated by dramatic flood events, known typically under their Icelandic name, *jökulhlaups*. Such floods could modulate ice flow rates, be significant agents of geomorphic change, and may release living organisms and organic carbon from subglacial lakes.

Despite the dearth of information on the subglacial hydrology in Antarctica, there is a growing appreciation that the origins and cycling of water in these systems plays an important role in continent-wide hydrological processes. Recently, Gray *et al.* (2005) and Wingham *et al.* (2006) presented evidence for large discharges of water (≈ 0.3 km^3, ≈ 5 m s^{-1}; and ≈ 1.8 km^3, ≈ 50 m s^{-1}, respectively) beneath ice sheets occurring over a period of months. These processes appear to extend over distances of tens or even hundreds of kilometers. Wingham *et al.* (2006) and Fricker *et al.* (2007) further hypothesized that subglacial basins may be flushed periodically. This has important ramifications for the residence time of water and water-circulation patterns in subglacial lakes. Mixing and transport processes within lakes established by *in situ* chemical, geothermal, and biogeochemical activities, or by pressure melting of ice, would be disrupted by the rapid throughput of water. Periodic discharges would also alleviate or reduce gas pressure build-up from the disassociation of gas hydrate during the melting of basal ice and exchange resident biota between subglacial lakes. If these hydrologic processes occur on a large scale, solute and microbial redistribution throughout the subglacial water environment may occur frequently. Understanding the parameters that control subglacial water balances will be a major challenge for future subglacial environment research.

Evidence is emerging that subglacial lakes may also drain catastrophically to the ocean. Lewis *et al.* (2006) demonstrated recently that the dramatic morphology of the Labyrinth in the Wright Valley and the drainages associated with the Convoy Range, both located in the McMurdo Dry Valleys along the Trans-Antarctic Mountain front, may be the result of large floods during the mid-Miocene era. The mid-Miocene was a period when the Antarctic climate cooled and the East Antarctic Ice Sheet experienced a major expansion, growing far beyond its present limits. The ice-free regions of southern Victoria Land exhibit extensive bedrock-channel networks most probably carved during catastrophic outbursts of subglacial lakes (Denton and Sugden 2005). The sudden and repeated drainage of the subglacial lake system through the Labyrinth/Convoy regions occurred between 12.4 and 14.4 million years ago when the East Antarctic Ice Sheet was larger and the melting-ice margins terminated in the Southern Ocean. During this period, a significant addition of freshwater from subglacial floods ($\approx$6000 km^3, $\approx$2$\times$10^6 m^3 s^{-1}, sea level rise $\approx$1.6 cm; Lewis *et al.* 2006) could have triggered alternate modes of ocean circulation within the Ross Sea and Southern Ocean (e.g. Mikolajewicz 1998). The results of recent climate modeling suggest that regional and global climate are sensitive to freshwater influx into the Ross Embayment. This flux can alter thermohaline circulation, disrupt deep-water formation, and impact sea-ice extent (Mikolaiewicz 1998; Lewis *et al.* 2006). Such changes, if they were rapid, could influence climate on a global scale.

7.3 Antarctic ice streams: regions of dynamic liquid water movement that influence ice-sheet dynamics

Antarctic **subglacial lakes** tend to be located in the central parts of the ice sheet, where ice becomes thick enough to provide sufficient thermal insulation for the basal thermal regime to melt ice in contact with the bedrock. As the ice sheet thins towards the edges, increasing basal shear heating in zones of fast ice sliding becomes the primary mechanism maintaining basal melting (Llubes *et al.* 2006). In these fringe regions, basal melting rates (>10 mm year^{-1}) may be several times greater than in the deep interior ($\approx$1 mm year^{-1}). Average basal

melting on the continental scale is approximately 2 mm year^{-1} (Vogel *et al.* 2003; Llubes *et al.* 2006).

The combination of relatively abundant fresh water associated with fine-grained sediments that typify the subglacial and basal zones of ice sheets may provide an important habitat for life in addition to subglacial lakes. Whereas the total area of subglacial lakes is estimated to cover less than 1% of Antarctica (Dowdeswell and Siegert 1999), the zone of saturated sediments is likely to extend over most of the continent forming what can be considered as our planet's largest wetland. It is unlikely that basal meltwater is confined only to the pore spaces of glacial sediments immediately underlying the ice base (typically 1–10 m; Alley *et al.* 1997). In North America, glacial meltwater penetrated the upper 100's of meters of rocks and sediments (e.g. McIntosh and Walter 2005). Even making the conservative assumption that basal meltwater infiltrated just the top approximately 0.1 km beneath the Antarctic ice sheet, the volume of subglacial groundwater is likely to fall in the range of 10^4–10^5 km^3 (assuming $\approx$10$\times$10^6 km^2 for the area and average porosity ranging between 1 and 10%). This volume would increase to 10^6 km^3 assuming basal meltwater infiltration of 1 km and a porosity of 10%, which is consistent with the formerly glaciated portions of North America (McIntosh and Walter 2005). Based on these estimates, the volume of subglacial groundwater beneath the Antarctic ice sheet would be 100 times greater than that estimated for subglacial lakes, an order of magnitude greater than all nonpolar surface fresh water (e.g. atmosphere, surface streams, and rivers), and would account for almost 0.1% of all water on our planet (Table 7.1).

Subglacial zones of West Antarctic ice streams represent an important example of potential freshwater-saturated subglacial wetlands. Their physical and chemical characteristics have been studied during recent decades through geophysical and glaciological investigations (e.g. Alley *et al.* 1987; Tulaczyk *et al.* 1998; Gray *et al.* 2005). Existing predictions of basal melting and freezing rates in the drainage basin of Ross ice streams indicate a predominance of melting in the interior with melting rates decreasing downstream. This region has an area of approximately 0.8$\times$10^6 km^2 and an average net

Table 7.1 The major water reservoirs on Earth. Data for the polar ice sheets and Antarctic groundwater are shown in bold. Compiled from Wetzel (2001) and Gleick (1996). Antarctic groundwater volume assumes a surface area of $10 \times 10^6 \, km^2$, depth of 1 km, and porosity of 10% (see text for details).

Reservoir	Volume ($km^3 \times 10^6$)	Percentage of total
Oceans	1370	96.94
Antarctic ice sheet	**30**	**2.12**
Groundwater	9.5	0.67
Greenland Ice Sheet	**2.6**	**0.18**
Antarctic 'groundwater'	**1**	**0.07**
Lakes	0.125	0.009
Soil moisture	0.065	0.005
Atmosphere	0.013	0.001
Antarctic subglacial lakes	**0.010**	**0.001**
Streams and rivers	0.0017	0.0001

melting rate of about $3 \, mm \, year^{-1}$ ($\approx 2.5 \, km^3 \, year^{-1}$) (Joughlin *et al.* 2004).

Recent observations indicate the presence of small subglacial lakes (≈ 5–10 km diameter) beneath ice streams that exchange water at rates of approximately $100 \, m^3 \, s^{-1}$ (Gray *et al.* 2005). These findings imply the presence of a highly dynamic subglacial water drainage system, which is consistent with borehole observations of temporally (and spatially) variable subglacial water pressure (Engelhardt and Kamb 1997). A borehole camera deployed on one of the ice streams revealed the presence of an approximately 1.6-m-deep cavity of liquid water, which may be part of a more widespread water-drainage system. In the same area, the camera imaged an approximately 10–15-m layer of refrozen water, which possesses a record of changes in subglacial hydrology over time scales ranging between decades and millennia (Vogel *et al.* 2005). In addition to basal water flow, the subglacial sediments themselves experience deformation and horizontal transport, associated with fast ice flow (Alley *et al.* 1987; Tulaczyk *et al.* 2001).

Samples collected from the subglacial environment beneath the West Antarctic ice streams have revealed the widespread presence of porous subglacial till (Tulaczyk *et al.* 2000a, 2000b). Subglacial core samples recovered from dozens of locations on three different ice streams permitted investigation of both the mineral component and pore water in these cores. The subglacial sediment is spatially uniform, presumably because it is generated by erosion of relatively homogeneous, widespread Tertiary glaciomarine sediments (Tulaczyk *et al.* 1998). This material contains many unstable minerals, which are susceptible to chemical weathering, such as pyrite and hornblende. The sediments are saturated by fresh water (dissolved solids $<0.5 \, g \, l^{-1}$), contain approximately 0.2% (by weight) of organic carbon, and contain ions that can act as redox couples capable of supporting chemotrophic life (Vogel *et al.* 2003).

7.4 Subglacial environments as habitats for life and reservoirs of organic carbon

As discussed in previous sections of this chapter, there is a diverse range of subglacial environments on our planet ranging from the relatively small liquid-water habitats that exist beneath polar and temperate glaciers to the systems we now know are present beneath the Antarctic ice sheet. Life in all of these must proceed without immediate input from the atmosphere and in the absence of light. Space constraints do not allow us to discuss all of these environments in detail so we have chosen to focus on what is known about Lake Vostok, the largest subglacial lake, located beneath the East Antarctic Ice Sheet.

7.4.1 Lake Vostok

Much attention is currently focused on the exciting possibility that the subglacial environments of Antarctica may harbor microbial ecosystems under thousands of meters of ice, which have been isolated from the atmosphere for as long as the continent has been glaciated (20–25 million years, Naish *et al.* 2001). The discovery during the early 1970s and subsequent inventory of subglacial lakes in Antarctica (Kapitsa *et al.* 1996; Siegert *et al.* 1996) rarely mentioned their biological potential until Priscu *et al.* (1999) and Karl *et al.* (1999) showed the presence, diversity, and metabolic potential of bacteria in frozen lakewater (accreted ice) overlying the liquid waters of Lake Vostok. Owing to differences in the pressure melting point caused by the tilted ice ceiling, lakewater refreezes (accretes) at the base of the ice sheet in the central and southern regions of Lake Vostok, removing water from the lake (e.g. Studinger *et al.* 2004). Hence, constituents in the accretion ice should reflect those in the actual lakewater in a proportion equal to the partitioning that occurs when water freezes (Priscu *et al.* 1999; Siegert *et al.* 2001; Christner *et al.* 2006).

Profiles of prokaryotic cell abundance through the entire Vostok core reveal a 2–7-fold higher cell density in accretion ice than the overlying glacial ice, implying that Lake Vostok is a source of bacterial carbon beneath the ice sheet (Figure 7.3). Cell densities ranged from 34 to 380 cells ml^{-1} in the glacial ice between 171 and 3537 m and the concentration of total particles of more than 1 μm ranged from 4000 to 12 000 particles ml^{-1}, much (30–50%) of which was organic in origin (Royston-Bishop *et al.* 2005; Priscu *et al.* 2006; Christner *et al.* 2008). A 6-fold increase in bacterial cell density was detected in samples of ice core from depths of 3540 and 3572 m where glacial ice transitions to accretion ice. Measurements of membrane integrity indicated that that the majority of cells were viable in both the glacial and accretion ice (Christner *et al.* 2006). The accretion ice below 3572 m contained fewer particles than glacial ice and the deepest accretion ice (3622 m) had the lowest number of total particles of all accretion-ice samples. Bacterial density followed the same trend as the density of mineral particles within the ice core (Royston-Bishop *et al.* 2005). These results, in concert with geophysical data from the lake basin led Christner *et al.* (2006) to contend that a shallow embayment located in the southwestern portion of the lake supports higher densities of bacteria than the lake proper. Christner *et al.* (2006), using partitioning coefficients obtained from lakes in the McMurdo Dry Valleys (see also Priscu *et al.* 1999), estimated that the number of bacteria in the surface waters of the shallow embayment and the lake proper should be approximately 460 and 150 cells ml^{-1}, respectively. These concentrations are much lower than those found in the permanently ice-covered lakes in the McMurdo Dry Valleys (≈10^5 ml^{-1}; Takacs and Priscu 1998), indicating that Lake Vostok is a relatively unproductive system.

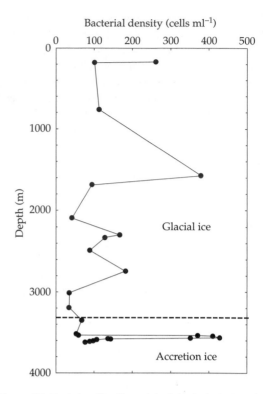

Figure 7.3 Vertical profile of bacterial cell density in the Vostok ice core. The horizontal dashed line denotes the transition from glacial ice to accretion ice. Cell density was determined on melted ice treated with the DNA stain SYBR Gold and counted by epifluorescence microscopy. Modified from Christner *et al.* (2006).

Sequence data obtained from DNA encoding for small-subunit ribosomal RNA (16S rDNA) revealed phylotypes that were most closely related to extant members of the alpha-, beta-, and gamma-Proteobacteria, Firmicutes, Bacteroidetes, and Actinobacteria. If the accreted ice microbes are representative of the lake microbiota, these data imply that microbes within Lake Vostok do not represent an evolutionarily distinct subglacial biota (Christner *et al.* 2008). The time scale of isolation within Lake Vostok (>15 × 10^6 years) is not long in terms of prokaryotic evolution compared with their 3.7 × 10^9-year history on Earth, and studies of species divergence of other prokaryotes have shown that species-level divergence may take approximately 100 million years (Lawrence and Ochman 1998). However, other mechanisms of genetic change (such as recombination) could allow more rapid alteration of organism phenotype allowing for adaptation to conditions within Lake Vostok (Page and Holmes 1998), which would not be reflected in evolutionary changes in the 16S rRNA gene. An alternative scenario is that glacial meltwater entering the lake forms a lens overlying the Vostok water column. If so, the microbes discovered within accretion ice would likely have spent little time in the lake water itself (few, if any, cell divisions occurring) before being frozen into the accretion ice. The microbes within the main body of the lake below such a freshwater lens may have originated primarily from basal sediments and rocks and, if so, their period of isolation may be adequate for significant evolutionary divergence, particularly given the potential selection pressures that exist within subglacial environments. PCR-based analyses of the microbial diversity in Lake Vostok accretion ice based on 16S rRNA genes (Christner *et al.* 2001; Bulat *et al.* 2004) has revealed two phylotypes closely related to thermophilic bacteria. One of them is related to a facultative chemolithoautotroph identified previously in hot springs and capable of obtaining energy by oxidizing hydrogen sulphide at reduced oxygen tension. Evidence for the presence of hydrothermal input is supported by the recent interpretation of He^3/He^4 data from accretion ice (Petit *et al.* 2005), which implies that there may be extensive faulting beneath Lake Vostok, which could introduce geochemical energy sources to the southern part of the lake. If this emerging picture is correct, Lake Vostok could harbor a unique assemblage of organisms fueled by chemical energy. Although it seems inevitable that viable microorganisms from the overlying glacial ice, and in sediment scoured from bedrock adjacent to the lake, are regularly seeded into the lake, the question remains of whether these or pre-existing microorganisms have established an ecosystem in Lake Vostok. If a microbial ecosystem were found to exist within the water or sediment of these subsurface environments, it would be one of the most extreme and unusual ecosystems on Earth.

The 16S rRNA gene sequence data from the Vostok accretion ice allow comparisons to be made with physiologically well-characterized organisms that exist in public databases. In addition to the data of Bulat *et al.* (2004), suggesting the presence of thermophiles that may use hydrogen for energy and carbon dioxide as a carbon source, several of the phylotypes reported by Christner *et al.* (2006) are most closely related to aerobic and anaerobic species of bacteria with metabolisms dedicated to iron and sulphur respiration or oxidation. This similarity implies that these metals play a role in the bioenergetics of microorganisms that occur in Lake Vostok. As cautioned by Christner *et al.* (2006), these metabolic estimates were made on relatively distant phylogenetic relationships (<95% identity); hence, these conclusions remain tentative until the organisms are characterized physiologically. Given this caveat, Figure 7.4 shows the possible chemoautotrophic metabolic pathways that may occur in Lake Vostok. Importantly, the substrates involved can be supplied to the lake by physical glacial processes and do not require geothermal input. These pathways could supply organic carbon and support heterotrophic metabolisms that use O_2, NO_3^-, SO_4^{2-}, S^0, or Fe^{2+} as electron acceptors.

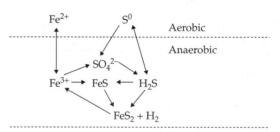

Chemoautotrophic (i.e. CO_2-based) microbial system.....

.....supplying organic C for use by heterotrophs:
$(CH_2O)n + Xoxidized \rightarrow CO_2 + Xreduced$
Where Xoxidized may be O_2, NO_3^-, SO_4^{2-}, S^0, Fe^{3+}

Figure 7.4 Diagram showing the potential biogeochemical pathways that may be important in the surface waters of Lake Vostok based on 16 S rDNA sequence data. This pathway involves chemoautotrophic fixation of carbon dioxide using iron and sulphur as both electron donors and electron acceptors. The pathway also shows the potential transformations that may occur across an aerobic/**anaerobic** boundary (the sequence data revealed the presence of both aerobic and **anaerobic** bacteria). Organic carbon produced by chemoautotrophic metabolism could then fuel heterotrophic metabolism within the lake.

7.4.2 Microbial ecology of icy environments

A major question that arises in the study of sub-glacial and other icy environments concerns the role of microorganisms in terms of ecosystem structure and function. Stated more succinctly: are the organisms we observe 'freeloaders' or are they actively involved in ecosystem processes? If the latter, we would expect to observe distinct bio-geographical distributions of organisms within icy environments, much like those observed in cyano-bacteria in hot springs (Papke and Ward 2004). Microbiological surveys of icy environments have identified common bacterial genera from global locations (e.g. Priscu and Christner 2004; Christner *et al.* 2008). We have used data from these surveys to show that more than 60% of these isolates group into only six genera (Figure 7.5), implying that icy habitats are indeed selecting for specific groups of microorganisms and that these organisms may be actively growing within icy environments. Caution should be taken when applying the results sum-marized in Figure 7.5 to subglacial lakes because

there is an emerging view (discussed previously in this chapter) that many of the lakes are connected by advective flow, which could disperse gene pools within and perhaps across watersheds. The balance between growth rates and rates of advection must be known before subglacial biogeography can be understood. We must further know if the micro-organisms within the subglacial environments were derived solely from the overlying glacial ice or from the sedimentary environment underlying the ice sheet. This has important implications for the evolutionary times scales involved. If derived from the overlying ice sheet, the organisms would only have about 1 million years to evolve, a time scale that is highly unlikely to lead to evolution, particularly at the slow growth rates expected. If the organisms were derived from the sediment environment, their lineages could have evolved over perhaps 30 million years, which approximates the time when the first major glaciations were thought to have occurred in Antarctica.

Before we can understand unequivocally the biology and selection pressures within subglacial

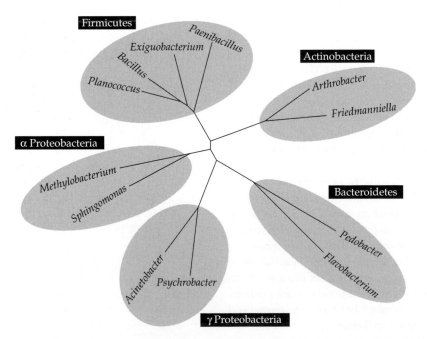

Figure 7.5 The most frequently recovered genera from glacial ice and subglacial environments based on the phylogeny of the 16 S rRNA molecule . The various source environments in which these genera have been documented are quite diverse and share little in common except that all are permanently frozen. See Priscu and Christner (2004) and Christner *et al.* (2008) for details.

environments, the habitat must be defined precisely. For example, most of what we know about Lake Vostok comes from accretion ice, which formed from near-surface lake water. Physical limnology models that assess water circulation in Lake Vostok have been reviewed recently by Siegert *et al.* (2005). Lake Vostok circulation is predicted to be a consequence of differences in the pressure melting point between the north and south ends of the lake. In a freshwater (i.e. low ionic strength) lake, geothermal heating will warm bottom waters to a temperature higher than that of the upper layers. The water density will decrease with increasing temperature, resulting in an unstable water column leading to convective circulation where cold meltwater sinks and water warmed by geothermal heat ascends. Conversely, if the lake is slightly saline, the fresh glacier meltwater will be buoyant relative to bulk lake water, and the northern meltwater would spread southward and upward. If the horizontal salinity gradient is great enough to compensate for geothermal warming, water would move into regions of progressively lower pressure, displacing lake water in the south. The cold northern water would eventually refreeze onto the ice sheet base some distance from where it first melted. In this case, a conveyor of fresh cool meltwater would migrate from north to south beneath the ice sheet, causing displacement of warmer dense lake water from the south to the north. If the lakes are vertically chemically stratified, an upper layer of cold fresh water can circulate over a deep layer of warm saline water. The heat in the deep saline water may originate from biogenic processes as well as geothermal heating.

Air hydrate (a naturally occurring solid composed of crystallized water (ice) molecules, forming a rigid lattice of cages (a clathrate) with most of the cages containing a molecule of air) is suspected of playing a role in establishing the physical, chemical, and biological characteristics of subglacial lakes. Atmospheric air, captured in ice sheets, occurs exclusively as gas hydrate at an ice thickness of a few kilometers (Hondoh 1996). In large subglacial lakes, such as Lake Vostok, with a geometry that favors the establishment of a melt–freeze cycle, the melting of the ice sheet would release air hydrate to the water. Accretion ice is nearly gas-free relative

to overlying glacial ice due to the exclusion of gas during refreezing (Jouzel *et al.* 1999). This exclusion would lead to increased dissolved gas concentrations in Lake Vostok (Lipenkov and Istomin 2001; McKay *et al.* 2003). Dissolved oxygen concentrations are predicted to be as much as 50 times higher than air-equilibrated water (McKay *et al.* 2003). Within 400 000 years ($\approx$29 residence times), gas concentrations in the lake are predicted to reach levels that favor the formation of air hydrate. Air hydrate formed in the lake would float to the surface of the lake if formed from air or sink if it contains more than 10% carbon dioxide. The estimates of Lake Vostok water dissolved oxygen concentrations do not consider the effects of biological removal processes or hydrologic recharge (McKay *et al.* 2003). High oxygen levels can react with intracellular molecules to produce superoxides and hydrogen peroxide, the latter of which can generate free radicals following the reaction with certain intracellular metals (e.g. iron). These strong oxidants can degrade macromolecules and damage membranes. To combat the effects of these oxidants, we would expect the microorganisms within the lakes to contain antioxidants and other detoxifying agents such as superoxide dismutase and catalase.

Dissolved organic carbon (DOC) plays a key role in ecosystem carbon cycling owing to its role as an energy and carbon source for heterotrophic organisms. DOC concentrations in Lake Vostok accretion ice are low (<70 µM; Christner *et al.* 2006), implying that DOC concentrations in surface lake waters are low as well. Priscu *et al.* (1999) and Christner *et al.* (2006), using partitioning coefficients of DOC from McMurdo Dry Valley lakes, estimated DOC levels in the surface waters of Lake Vostok between 86 and 100 µM (160 µM for the shallow embayment). Although these DOC levels can support heterotrophic growth, molecular evidence from Lake Vostok accretion ice implies that carbon dioxide fixation by chemolithoautotrophs may also be an important source of new organic carbon in Lake Vostok (Christner *et al.* 2006). The new carbon produced by this process can then provide a substrate to support heterotrophic activity.

One must also consider the sediments that exist within subglacial environments in any discussion of metabolic activity, and distribution and evolution of microorganisms. Sediments have been shown to be hotspots for biological diversity and activity in most surface lakes on our planet, and they should have a significant role in all subglacial microbial processes. Lake sediments are generally organic, nutrient-rich, and support a wide range of reduction/oxidation conditions over relatively small spatial scales (e.g. Wetzel 2001). Consequently, sedimentary habitats can support diverse metabolic lifestyles. Data from Lake Bonney (McMurdo Dry Valleys) show that sediment bacterial diversity and metabolic activity exceed those in the overlying water column (J.C. Priscu, unpublished results), and preliminary data from the subglacial sediment environment beneath the ice streams of West Antarctica reveal that microbial density exceeds 10^7 cells g sediment^{-1} (B. Lanoil, unpublished results). Biogeochemical transformations produce chemical gradients within the sediments, which affect the properties of the overlying water column. For example, organisms utilizing oxygen as a terminal electron acceptor can produce anoxic or suboxic conditions within the sediments and the lower portion of the water column. Sediment biogeochemistry can also produce compounds (e.g. CH_4, H_2S, NH_4^+, PO_4^{3-}) that diffuse into the water column, supplying redox couples and nutrients to water column bacteria. As with surface lakes, sediments in subglacial lakes should play an important role in the lake ecosystem.

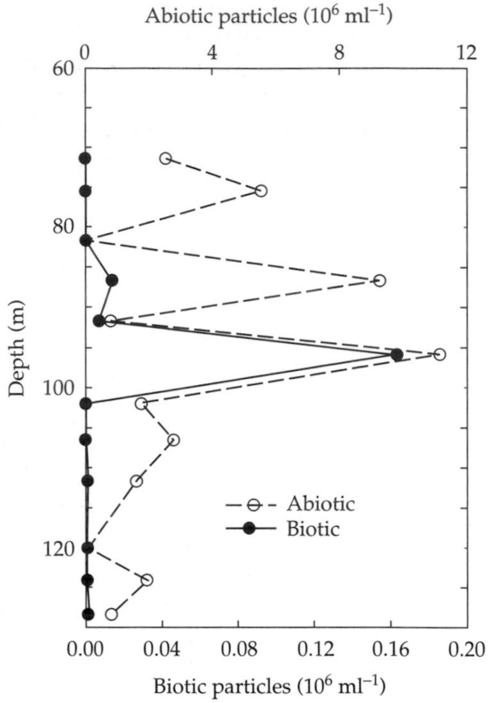

Figure 7.6 Vertical profiles of biotic and abiotic particles in the 30–130-m (below the surface) glacial ice from the West Antarctic Ice Sheet divide core site. The data were obtained on melted samples processed with a flow cytometer. Biotic particles were determined by particle fluorescence following application of the DNA stain SYTO 60.

7.4.3 Subglacial environments as reservoirs of organic carbon

Snowfall typically accumulates on the Antarctic ice sheets at a rate of 2–10 cm year^{-1} (accumulation of ≈22 cm year^{-1} can occur near ice divides (e.g. www.waisdivide.unh.edu/) and carries with it a record of gases, ions, and particles present in the atmosphere at the time of snowfall (e.g. Priscu *et al.* 2006). An initial analysis of particle data collected with a flow cytometer from 70 to 130 m (which represents ≈1000 years of accumulation) at an ice divide in the West Antarctic Ice Sheet revealed total densities ranging from 2×10^4 to 1×10^7 particles ml^{-1}, with the highest densities occurring 94 m beneath the surface (Figure 7.6). Concentrations of bacterial-sized particles containing DNA (determined by staining samples with SYTO 60, a DNA-binding fluorescent dye) ranged from 2.1×10^{-2} to 1.6×10^5 ml^{-1}, with the highest levels at 94 m, the depth with the highest abiotic particle density. Although these data from the West Antarctic Ice Sheet site represent only the near-surface ice, they do imply that biotic and abiotic particle densities at the West Antarctic Ice Sheet site are higher than in the Vostok core collected from the East Antarctic Ice Sheet (Christner *et al.* 2006). These ice entombed bacteria will eventually be released to the subglacial environment following a period of transit (hundreds of thousands of years). Owing to the large volume of Antarctic glacial ice (see Table 7.1), these ice-bound bacteria may represent an important reservoir of global organic carbon. Priscu and Christner (2004) first estimated the organic carbon reservoirs associated with the Antarctic ice sheet and associated

Table 7.2 Summary of the prokaryotic cell number, prokaryotic carbon (Cell carbon), and dissolved organic carbon (DOC) computed for Antarctic subglacial lakes, the ice sheet, and the subglacial aquifer. Carbon concentrations are in Petagrams (10^{15} g). Global estimates of cell number in freshwater lakes and rivers, the open ocean, the terrestrial subsurface and soils are from Whitman et al. (1998; Tables 1 and 5). Global cellular carbon reservoirs for the terrestrial subsurface and soils are also from Whitman et al. (1998); a carbon content of 11 fg C cell^{-1} was used to obtain the carbon reservoirs in the freshwater lakes and rivers, and open ocean. DOC estimates for fresh waters, the open ocean, and the terrestrial subsurface assume average values of 2.2, 0.5, and 0.7 mg l^{-1} for these systems (Thurman 1985) and volumes of 2.31 × 10^5, 1.37 × 10^9, and 6.00 × 10^7 km^3, respectively (Wetzel 2001). Cellular and DOC pools in subglacial lakes were estimated using values projected for the main body of Lake Vostok (1.50 × 10^8 cells m^{-3} and 1.03 × 10^3 mg m^{-3}, respectively (Christner et al. 2006), in concert with the volume for all subglacial lakes of 10 000 km^3 (Dowdeswell and Siegert 1999). Cellular and DOC data from ice cores collected below Vostok Station (Christner et al. 2006) were depth-weighted to yield average values of 1.34 × 10^8 cells m^{-3} and 3.09 × 10^2 mg m^{-3}. These values were used with an ice-sheet volume of 3.01 × 10^7 km^3 (IPCC 1995) to compute the cellular reservoirs within the ice sheet. A subglacial aquifer volume of 10^7 km^3 was computed assuming that depth equals 1 km and surface area equals 1.0 × 10^7 km^2. The aquifer volume was used with a cellular density obtained from samples collected beneath the Kamb Ice Stream (2 × 10^7 cells g^{-1}) and a sediment density of 2 g cm^{-3} to compute the number of total cells in the aquifer (B. Lanoil, unpublished results). The method used to determine cellular abundance includes cells attached to sediment particles and those residing within the pore waters. Hence, our results provide a bulk estimate of cells within the entire aquifer, not just the pore water. The DOC reservoir in subglacial groundwater was computed using a subglacial aquifer porosity of 10% and an average value for groundwater DOC of 700 mg m^{-3} (Thurman 1985; assuming all DOC is in pore water). Cellular carbon in all Antarctic habitats was computed using a conversion factor of 11 fg C cell^{-1}, determined for bacteria from a permanently ice-covered lake in the McMurdo Dry Valleys, Antarctica (Takacs and Priscu 1998). NA, not computed owing to the large variation in published values.

| | Antarctica | | | | Greenland | Both poles | Global | | | |
	Lakes	Ice sheet	Subglacial aquifer	Total	Ice sheet	Lakes, ice sheet and subglacial water	Fresh water	Open ocean	Terrestrial subsurface	Soils
Cell number	1.50 (×10^{21})	4.03 (×10^{24})	4.00 (×10^{29})	4.00 (×10^{29})	3.51 (×10^{23})	4.00 (×10^{29})	1.31 (×10^{26})	1.20 (×10^{29})	2.50 (×10^{30})	2.60 (×10^{29})
Cell carbon (Pg)	1.65 (×10^{-8})	4.43 (×10^{-5})	4.40 (×10^9)	4.40 (×10^9)	3.86 (×10^{-6})	4.40 (×10^9)	1.44 (×10^{-3})	1.32 (×10^9)	2.15 (×10^2)	2.60 (×10^1)
DOC (Pg)	1.03 (×10^{-2})	9.29 (×10^9)	7.00 (×10^{-1})	1.00 (×10^1)	9.50 (×10^{-1})	1.10 (×10^1)	5.08 (×10^{-1})	6.85 (×10^2)	4.20 (×10^1)	NA
Cell carbon+DOC (Pg)	1.03 (×10^{-2})	9.29 (×10^9)	5.10 (×10^9)	1.44 (×10^1)	9.50 (×10^{-1})	1.54 (×10^1)	5.10 (×10^{-1})	6.86 (×10^2)	2.57 (×10^2)	NA

subglacial lakes, but they had relatively little data to work with. The recent publication by Christner *et al.* (2006) now provides the most complete dataset with which to estimate carbon reservoirs in the Antarctic ice sheet and subglacial lakes. Using these data, plus new values on subglacial groundwater and bacterial density in Antarctica as well as new data on DOC in Greenland ice, the density of bacterial cells, bacterial carbon, and DOC were estimated in (1) Antarctic subglacial lakes, (2) Antarctic and Greenland ice sheets, and (3) Antarctic and Greenland subglacial groundwater (Table 7.2). These calculations reveal that the largest bacterial carbon pool (>99%) occurs in subglacial groundwater, whereas the largest pool of DOC (93%) occurs within the glacial ice. These results reflect the relatively high prokaryotic cell densities used for the groundwater estimates (2×10^7 cell g^{-1}) and the immense volume of the Antarctic ice sheet (3×10^7 km^3), respectively. All carbon pools estimated for Greenland ice are about an order of magnitude below those for the Antarctic equivalents. The prokaryotic carbon pool from both poles exceeds that estimated in all surface fresh waters (rivers and lakes) by more than two orders of magnitude; this is equivalent to that in the open ocean but is 49 and six times lower than that in terrestrial groundwater and soils, respectively.

DOC levels within polar regions exceed those in surface fresh water by 20-fold, but are 63 and 38 times lower than DOC in the open ocean and terrestrial groundwater, respectively (global DOC pools were not computed for soils owing to the extreme site-specific variability observed in the literature). These comparisons indicate that Earth's polar regions contain a significant reservoir of prokaryotic carbon and DOC. This result should not be surprising given that more than 70% of our planet's fresh water resides in polar regions as ice and subglacial water. The estimates in Table 7.2 will continue to be refined as we learn more about the geophysical, chemical and biological properties on our polar regions. Data in Table 7.2 support the contention of Priscu and Christner (2004) that polar ice, particularly Antarctic ice, contains an organic carbon reservoir that should be considered when addressing issues concerning global carbon dynamics.

Acknowledgments

We thank all members of the Scientific Research Program SALE (Scientific Committee on Antarctic Research; SCAR)) for their input over the years. Many of the ideas incorporated into this chapter were derived directly from information presented and synthesized at an advanced science and technology planning workshop focusing on subglacial Antarctic lake environments in the International Polar Year 2007–2008 held in Grenoble, France (April 2006). J.C.P. was supported by National Science Foundation grants OPP0432595, OPP0440943, OPP0631494, and MCB0237335 during the preparation of this chapter.

References

Alley, R.B., Blankenship, D.D., Bentley, C.R., and Rooney, S.T. (1987). Till beneath ice stream B. 3. Till deformation: evidence and implications. *Journal of Geophysical Research* **92**, 8921–8930.

Alley, R.B. *et al.* (1997). How glaciers entrain and transport basal sediment: physical constraints. *Quaternary Science Reviews* **16**, 1017–1038.

Anandakrishnan, S., Catania, G.A., Alley, R.B., and Horgan, H.J. (2007). Discovery of till deposition at the grounding line of Whillans Ice Stream. *Science* **315**, 1835–1838.

Andersen, K.K. *et al.* (2004). High-resolution record of Northern Hemisphere climate extending into the last interglacial period. *Nature* **431**, 147–151.

Bell, R.E., Studinger, M., Fahnestock, M.A., and Shuman, C.A. (2006). Tectonically controlled subglacial lakes on the flanks of the Gamburtsev Subglacial Mountains, East Antarctica *Geophysical Research Letters* **33**, L02504, doi:10.1029/2005GL025207.

Bell, R.E., Studinger, M., Shuman, C.A., Fahnestock, M.A., and Joughin, I. (2007). Large subglacial lakes in East Antarctica at the onset of fast-flowing ice streams. *Nature* **445**, 904–907.

Bhatia, M., Sharp, M.J., and Foght, J.M. (2006). Distinct bacterial communities exist beneath a High Arctic polythermal glacier. *Applied and Environmental Microbiology* **72**, 5838–5845.

Bulat, S.A., Alekhina, I.A., Blot, M. *et al.* (2004). DNA signature of thermophilic bacteria from the aged accretion ice of Lake Vostok, Antarctica: implications for searching for life in extreme icy environments. *International Journal of Astrobiology* **3**, 1–12.

Christner, B.C., Mosley-Thompson, E., Thompson, L.G., and Reeve, J.N. (2001). Isolation of bacteria and 16S rDNAs from Lake Vostok accretion ice. *Environmental Microbiology* **3**, 570–577.

Christner, B.C. *et al.* (2006). Limnological conditions in subglacial Lake Vostok, Antarctica. *Limnology and Oceanography* **51**, 2485–2501.

Christner, B.C., Skidmore, M.L., Priscu, J.C., Tranter, M., and Foreman, C. (2008). Bacteria in subglacial environments. In Margesin, R., Schinner, F., Marx, J.-C., and Gerday, C. (ed.), *Psychrophiles: from Biodiversity to Biotechnology*, pp. 51–71. Springer Publishers, Berlin.

Denton, G.H. and Sugden, D.E. (2005). Meltwater features that suggest Miocene ice-sheet overriding of the Transantarctic Mountains in Victoria Land, Antarctica. *Geografiska Annaler* **87**, 67–85.

de Robin, G.Q., Drewry, D.J., and Meldrum, D.T. (1977). International studies of ice sheet and bedrock. *Philosophical Transactions of the Royal Society of London Series B Biological Sciences* **279**, 185–196.

Dowdeswell, J.A. and Siegert, M.J. (1999). The dimensions and topographic setting of Antarctic subglacial lakes and implications for large-scale water storage beneath continental ice sheets. *Geological Society of America Bulletin* **111**, 254–263.

Dowdeswell, J.A. and Siegert, M.J. (2002). The physiography of modern Antarctic subglacial lakes. *Global and Planetary Change* **35**, 221–236.

Drewry, D.J. (1983). *Antarctica: Glaciological and Geophysical Folio.* Scott Polar Research Institute, University of Cambridge, Cambridge.

Engelhardt, H., Humphrey, N., Kamb, B., and Fahnestock, M. (1990). Physical conditions at the base of a fast moving Antarctic ice stream. *Science* **248**, 57–59.

Engelhardt, H. and Kamb, B. (1997). Basal hydraulic system of a west antarctic ice stream: Constraints from borehole observations. *Journal of Glaciology* **43**, 207–230.

Fahnestock, M, Abdalati, W., Joughin, I., Brozena, J., and Gogineni, P. (2001). High geothermal heat flow, basal melt, and the origin of rapid ice flow in central Greenland. *Science* **294**, 2338–2342.

Fricker, H.A., Scambos, T., Bindschadler, R., and Padman, L. (2007). An active subglacial water system in West Antarctica mapped from space. *Science* **315**, 1544–1548.

Gaidos, E. *et al.* (2004). A viable microbial community in a subglacial volcanic crater lake, Iceland. *Astrobiology* **4**, 327–344.

Gleick, P.H. (1996). Water resources. In Schneider, S.H. (ed.), *Encyclopedia of Climate and Weather*, vol. 2, pp. 817–823. Oxford University Press, New York.

Gray, L. *et al.* (2005). Evidence for subglacial water transport in the West Antarctic Ice Sheet through three-dimensional satellite radar interferometry. *Geophysical Research Letters* **32**, L03501, doi:10.1029/2004GL021387.

Herdendorf, C.E. (1982). Large lakes of the world. *Journal of Great Lakes Research* **8**, 379–412.

Hondoh, T. (1996). Clathrate hydrates in polar ice sheets. *Proceedings of the 2nd International Conference on Natural Gas Hydrates*, pp. 131–138. Toulouse.

IPCC (1995). *Climate Change, Impacts. Adaptations and Mitigation of Climate Change: Scientific-technical Analysis*, Watson, R.T., Zinyowera, M.C., Moss, R.H., and Dokken, D.J. (eds). Cambridge University Press, Cambridge.

Joughin, I., Tulaczyk, S., MacAyeal, D.R., and Engelhardt, H. (2004). Melting and freezing beneath the Ross ice streams, Antarctica. *Journal of Glaciology* **50**, 96–108.

Jouzel, J. *et al.* (1999). More than 200 meters of lake ice above subglacial Lake Vostok, Antarctica. *Science* **286**, 2138–2141.

Kapitsa, A.P., Ridley, J.K., Robin, G. de Q., Siegert, M.J., and Zotikov, I.A. (1996). A large deep freshwater lake beneath the ice of central East Antarctica. *Nature* **381**, 684–686.

Karl, D.M. *et al.* (1999). Microorganisms in the accreted ice of Lake Vostok, Antarctica. *Science* **286**, 2144–2147.

Lawrence, J.G. and Ochman, H. (1998). Molecular archaeology of bacterial genomes. *Proceedings of the National Academy of Sciences USA* **95**, 9413–9417.

Lewis, A.R., Marchant, D.R., Kowalewski, D.E., Baldwin, S.L., and Webb, L.E. (2006). The age and origin of the Labyrinth, western Dry Valleys, Antarctica: evidence for extensive middle Miocene subglacial floods and freshwater discharge to the Southern Ocean. *Geology* **34**, 513–516.

Lipenkov, V.Y. and Istomin, V.A. (2001). On the stability of air clathrate-hydrate crystals in subglacial Lake Vostok, Antarctica. *Materialy Glyatsiologicheskikh Issledovanii* **91**, 129–133, 138–149.

Llubes, M., Lanseau, C., and Rémy, F. (2006). Relations between basal condition, subglacial hydrological networks and geothermal flux in Antarctica. *Earth and Planetary Science Letters* **241**, 655–662.

McIntosh, J.C. and Walter, L.M. (2005). Volumetrically significant recharge of Pleistocene glacial meltwaters into epicratonic basins: constraints imposed by solute mass balances. *Chemical Geology* **222**, 292–309.

McKay, C.P., Hand, K.P., Doran, P.T., Anderson, D.T., and Priscu, J.C. (2003). Clathrate formation and the fate of noble and biologically useful gases in Lake Vostok, Antarctica. *Geophysical Research Letters* **30**, 1702, doi:10.1029/2003GL017490.

Meybeck, M. (1995). Global distribution of lakes. In Lerman, A., Imboden, D.M., and Gat, J.R. (eds), *Physics* and Chemistry of Lakes, pp. 1–35. Springer-Verlag, Berlin.

Mikolajewicz, U. (1998). Effect of meltwater input from the Antarctic ice sheet on the thermohaline circulation. Annals of Glaciology **27**, 311–315.

Mikucki, J.A., Foreman, C.M., Sattler, B., Lyons, W.B., and Priscu, J.C. (2004). Geomicrobiology of Blood Falls: an iron-rich saline discharge at the terminus of the Taylor Glacier, Antarctica. Aquatic Geochemistry **10**, 199–220.

Mikucki, J.A. and Priscu, J.C. (2007). Bacterial diversity associated with Blood Falls, a subglacial outflow from the Taylor Glacier, Antarctica. Applied and Environmental Microbiology **73**, 4029–4039.

Naish, T.R. et al. (2001). Orbitally induced oscillations in the East Antarctic Ice Sheet at the Oligocene/Miocene boundary. Nature **413**, 719–723.

Oswald, G.K.A. and de Robin, G.Q. (1973). Lakes beneath the Antarctic Ice Sheet. Nature **245**, 251–254.

Page, R.D.M. and Holmes, E.C. (1998). Molecular Evolution: a Phylogenetic Approach. Blackwell Science, Oxford.

Papke, R.T. and Ward, D.M. (2004). The importance of physical isolation to microbial diversification. FEMS Microbiology Ecology **48**, 293–303.

Petit, J.R., Alekhina, I., and Bulat, S. (2005). Lake Vostok, Antarctica: exploring a subglacial lake and searching for life in an extreme environment. In Gargaud, M., Barbier, B., Martin, H., and Reisse, J. (eds), Lectures in Astrobiology, vol. I, Advances in Astrobiology and Biogeophysics, pp. 227–288. Springer, Berlin/ Heidelberg.

Popov, S.V., Masolov, V.N., Lukin, V.V., and Sheremetiev, A.N. (2002). Central part of East Antarctica: bedrock topography and subglacial lakes (abstract). Scientific Conference: Investigation and Environmental Protection of Antarctica, pp. 84–85. Arctic and Antarctic Research Institute (AARI), St. Petersburg.

Priscu, J.C., Adams, E.E., Lyons, W.B. et al. (1999). Geomicrobiology of subglacial ice above Lake Vostok, Antarctica. Science **286**, 2141–2144.

Priscu, J.C. and Christner, B.C. (2004). Earth's icy biosphere. In Bull, A.T. (ed.), Microbial Biodiversity and Bioprospecting, pp. 130–145. American Society for Microbiology Press, Washington DC.

Priscu, J.C., Christner, B.C., Foreman, C.M., and Royston-Bishop, G. (2006). Biological material in ice cores. In Elias, S.A. (ed.), Encyclopedia of Quaternary Sciences, vol. 2, pp. 1156–1166. Elsevier, Oxford.

Priscu, J.C. and Foreman, C.M. (2008). Lakes of Antarctica. In Likens, G.E. (ed.), Encyclopedia of Inland Waters. Elsevier, Oxford, in press.

Priscu, J.C. et al. (1998). Perennial Antarctic lake ice: an oasis for life in a polar desert. Science **280**, 2095–2098.

Priscu, J.C. et al. (2003). An international plan for Antarctic subglacial lake exploration. Polar Geography **27**, 69–83.

Priscu, J.C. et al. (2005). Exploring subglacial Antarctic Lake environments. EOS, Transactions of the American Geophysical Union **86**, 193–197.

Raiswell, R. (1984). Chemical models of solute acquisition in glacial meltwaters. Journal of Glaciology **30**, 49–57.

Robinson, R.V. (1964). Experiment in visual orientation during flights in the Antarctic. Soviet Antarctic Expedition Information Bulletin **2**, 233–234.

Royston-Bishop, G. et al. (2005). Incorporation of particulates into accreted ice above subglacial Lake Vostok, Antarctica. Annals of Glaciology **40**, 145–150.

SALE (2007). Subglacial Antarctic Lake Environments (SALE) in the International Polar Year 2007–08: Advanced Science and Technology Planning Workshop, 24–26 April 2006, Grenoble, France. http://salepo.tamu.edu/ saleworkshop.

Sharp, M. et al. (1999). Widespread bacterial populations at glacier beds and their relationship to rock weathering and carbon cycling. Geology **27**, 107–110.

Siegert, M.J., Dowdeswell, J.A., Gorman, M.R., and McIntyre, N.F. (1996). An inventory of Antarctic subglacial lakes. Antarctic Science **8**, 281–286.

Siegert, M.J. and Bamber, J.L. (2000). Subglacial water at the heads of Antarctic ice-stream tributaries. Journal of Glaciology **46**, 702–703.

Siegert, M.J., Carter, S., Tabacco, I., Popov, S., and Blankenship, D.D. (2005). A revised inventory of Antarctic subglacial lakes. Antarctic Science **17**, 453–460.

Siegert, M.J. et al. (2006). Exploration of Ellsworth Subglacial Lake: a concept paper on the development, organisation and execution of an experiment to explore, measure and sample the environment of a West Antarctic subglacial lake. Reviews in Environmental Science and Biotechnology **6**, 161–179.

Siegert, M.J. et al. (2001). Physical, chemical and biological processes in Lake Vostok and other Antarctic subglacial lakes. Nature **414**, 603–609.

Skidmore, M., Anderson, S.P., Sharp, M., Foght, J., and Lanoil, B.D. (2005). Comparison of microbial community composition in two subglacial environments reveals a possible role for microbes in chemical weathering processes. Applied and Environmental Microbiology **71**, 6986–6997.

Studinger, M., Bell, R.E., and Tikku, A.A. (2004). Estimating the depth and shape of subglacial Lake Vostok's water cavity from aerogravity data. Geophysical Research Letters **31**, L12401, doi:10.1029/2004GL019801.

Studinger, M. *et al.* (2003). Geophysical models for the tectonic framework of the Lake Vostok region, East Antarctica. *Earth and Planetary Science Letters* **216**, 663–677.

Tabacco, I.E. *et al.* (2003). Evidence of 14 new subglacial lakes in Dome C-Vostok area. *Terra Antarctic Reports* **8**, 175–179.

Takacs, C.D. and Priscu, J.C. (1998). Bacterioplankton dynamics in the McMurdo Dry Valley lakes: Production and biomass loss over four seasons. *Microbial Ecology* **36**, 239–250.

Thurman, E.M. (1985). *Organic Geochemistry of Natural Waters.* Nijhoff and Junk Publishers, Dordrecht.

Tikku, A.A. *et al.* (2005). Influx of meltwater to subglacial Lake Concordia, East Antarctica. *Journal of Glaciology* **51**, 96–104.

Tulaczyk, S.M., Kamb, B., Scherer, R.P., and Engelhardt, H.F. (1998). Sedimentary processes at the base of a West Antarctic ice stream; constraints from textural and compositional properties of subglacial debris. *Journal of Sedimentary Research* **68**, 487–496.

Tulaczyk, S., Kamb, W.B., and Engelhardt, H.F. (2000a). Basal mechanics of Ice Stream, B., West Antarctica 1. Till mechanics. *Journal of Geophysical Research* **105**, 463–481.

Tulaczyk, S., Kamb, W.B., and Engelhardt, H.F. (2000b). Basal mechanics of Ice Stream, B., West Antarctica 2. Undrained plastic bed model. *Journal of Geophysical Research* **105**, 483–494.

Tulaczyk, S., Kamb, B., and Engelhardt, H.F. (2001). Estimates of effective stress beneath a modern West Antarctic ice stream from till preconsolidation and void ratio. *Boreas* **30**, 101–114.

Vaughan, D.G., Bamber, J.L., Giovinetto, M., Russell, J., and Cooper, A.P.R. (1999). Reassessment of net surface mass balance in Antarctica. *Journal of Climate* **12**, 933–946.

Vogel, S.W., Tulaczyk, S., and Joughin, I.R. (2003). Distribution of basal melting and freezing beneath tributaries of Ice Stream C: Implication for the Holocene decay of the West Antarctic Ice Sheet. *Annals of Glaciology* **36**, 273–282.

Vogel, S.W. *et al.* (2005). Subglacial conditions during and after stoppage of an Antarctic Ice Stream: Is reactivation imminent? *Geophysical Research Letters* **32**, L14502, doi:10.1029/2005GL022563.

Wetzel, R.G. (2001). *Limnology: Lake and River Ecosystems.* Academic Press, San Diego, CA.

Whitman, W.B., Coleman, D.C., and Wiebe, W.J. (1998). Prokayotes: the unseen majority. *Proceedings of the National Academy of Sciences USA* **95**, 6578–6583.

Wingham, D.J., Siegert, M.J., Shepherd, A., and Muir, A.S. (2006). Rapid discharge connects Antarctic subglacial lakes. *Nature* **440**, 1033–1036.

CHAPTER 8

Biogeochemical processes in high-latitude lakes and rivers

W. Berry Lyons and Jacques C. Finlay

Outline

Polar aquatic ecosystems are excellent laboratories for biogeochemical research. The polar regions are among the least modified by human activities, so there are opportunities to study biogeochemical processes in the absence of overwhelming anthropogenic influences. In addition, there are abundant freshwater ecosystems in which comparative or experimental work can be conducted, and increasing evidence for environmental change is driving a rapid expansion in polar research. Extensive surveys of lake chemistry now exist, and in this chapter we draw upon these data to summarize the biogeochemical composition of polar lakes, and to illustrate the growing potential for cross-system comparisons. We describe the general features of biogeochemical cycles in polar aquatic environments, and the important and sometimes unique controls over biogeochemical processes. In addition we briefly address key studies, new literature, expanding areas of research, and geochemical linkages in polar systems. Finally, we end the review with an analysis of climate change scenarios for the Arctic and Antarctic.

8.1 Introduction

There has been little attempt in the past to compare and contrast Arctic and Antarctic aquatic systems, let alone their biogeochemistries. There are similarities as well as differences in aquatic environments between the two polar regions (Grémillet and LeMaho 2003; Laybourn-Parry 2003; see also Chapter 1 in this volume). Both are dominated by cryospheric processes and have common temperature, radiation, and humidity regimes (Laybourn-Parry 2003; Pienitz *et al.* 2004), whereas differences include the period of melt and, hence flow, in streams and the amount and type of lake ice cover (Doran *et al.* 1997). Variations in the quality and quantity of landscape vegetation also greatly influence biogeochemical cycles and processes in these high-latitude aquatic environments (Lyons *et al.* 2001). In general, the higher the latitude, the shorter the melt and flow cycle, the longer period

for lake ice and lesser the amount of terrestrial organic soils and vegetation. All of these factors serve to reduce the influence of terrestrial biogeochemistry. For example, the McMurdo Dry Valley aquatic systems at 78°S sustain stream flows from 4 to 10 weeks per year (Plate 6), have perennial lake ice covers (Plate 4), and are little affected by inputs of terrestrial organic carbon.

Antarctic ecosystems have been termed 'unique and sensitive', containing organisms with an 'extremely high level of adaptation to the most drastic abiotic conditions on Earth' (Grémillet and LeMaho, 2003). Laybourn-Parry (2003) has pointed out that both polar regions contain lakes of various depths, sizes, and salinities, but that hypersaline systems are more common in the Antarctic. Antarctica also has a much greater number of epishelf lakes, freshwater lakes sitting on sea water, which thereby have direct contact with the ocean (Laybourn-Parry 2003). Many permanently ice-covered lakes exist in the Antarctic, including the McMurdo Dry Valley lakes, Beaver Lake (MacRobertson Land), and Untersee (Queen Maud Land), where detailed limnological investigations have taken place. There are also lakes that are frozen to the bottom or almost so where hypersaline brines exist under more than 10 m of ice (Doran *et al.* 2003). Small lakes and ponds are also abundant in the ice-free regions of Antarctica and in the McMurdo Dry Valley region their biogeochemistries have been extensively documented (Healy *et al.* 2006). In addition to these 'surface' lakes and ponds, over 100 subglacial lakes of various depths and sizes occur beneath the Antarctic ice sheets (Siegert 2005; Plate 1).

Some of these features important to Antarctic lakes also influence the coldest Arctic lakes. For example, permanent ice cover may be found in lakes of northern Canada and Greenland; relict sea water is present in some lakes, resulting in meromictic conditions (Van Hove *et al.* 2006); evaporative concentration is a significant factor in biogeochemical composition of some Greenland lakes (Anderson *et al.* 2001). These systems constitute a small percentage of the overall number of Arctic lakes, however. Instead, the diversity of biogeochemical characteristics in Arctic lakes is more strongly linked to the nature of their contact with terrestrial ecosystems, geological heterogeneity, and climate. These factors may result in greater overall complexity and biogeochemical diversity of Arctic aquatic ecosystems compared with their Antarctic counterparts (Lyons *et al.* 2001). Access to some parts of the Arctic is comparatively easier, however, so there are more long-term and seasonally representative data-sets available than for Antarctica.

Most Antarctic lakes are found around the coastal 'oasis' of the continent where either glacier retreat or isostatic rebound have created ice-free regions where water can accumulate (Hodgson *et al.* 2004). In contrast, lakes and permanent rivers are abundant throughout much of the pan-Arctic, and hydrological transport and processing of elements in freshwater ecosystems play a major role in the biogeochemistry of both inland and marine waters. A wide diversity of lake types exist in the Antarctic, ranging from closed-basin

systems, some thought to be Pleistocene in age, to epishelf lakes formed in marine embayments or dammed by ice shelves, to epiglacial lakes formed against melting glaciers (Laybourn-Parry 2003; Hodgson *et al.* 2004). A similarly wide diversity of Arctic lakes exists, with additional influences of thermokarst (permafrost thaw), glacial scour, and wind action on lake formation (see Chapter 2 for further details on lake formation). The diversity of age and hydrology of these different types of lakes lead to large variations in physical, chemical, and biological processes.

Polar lakes exist and have been investigated over a broad range of latitudes, from lakes on Signy Island at 60°S to Lake Wilson at 80°S in the Antarctic, and from thaw lakes on discontinuous permafrost at 55°N (Plates 8 and 9) to Ward Hunt Lake at 83°N in the Arctic (Plate 4). This latitudinal span in each polar region corresponds to large variations in melt season, ice-cover duration, and landscape vegetation. Other important differences among the Antarctic lake and stream systems are the influence of past climatic change on the current system. This concept of ecological carry over or 'legacy' is extremely important in controlling biogeochemical processes in McMurdo Dry Valley aquatic systems. In the Arctic, effects of landscape age, geology, permafrost, and vegetation exert strong controls over biogeochemical processes and composition through effects on weathering, inputs of organic matter, and hydrologic processes.

River and stream systems in the Antarctic are small by Arctic standards (see Plates 6 and 7). The Onyx River, Wright Valley (McMurdo Dry Valleys) and the Algae River, Bunger Hills being the longest rivers on the continent at 28 and 25 km in length, respectively. The amount of literature on the biogeochemistry of Antarctic lakes is much greater than that on Antarctic streams. In contrast, the Arctic contains some of Earth's largest rivers, and some large lakes. The amount of literature on Arctic rivers is comparable with that on lakes, in part because of the great importance of water and solute fluxes to the Arctic Ocean.

The sources of solutes, gases, and solids to the streams and lakes in the Antarctic are influenced by similar processes: seasonality and source of melt, ice cover, landscape vegetation, groundwater–surface water interactions, and legacy effects. Solutes in the streams and lakes come from a number of sources including glacier/snowmelt dominated by marine aerosols, chemical weathering in the stream channels and hyporheic zones, and in some cases remnant sea water (e.g. Green *et al.* 2005). The quantity and quality of the solute input varies with landscape position, elevation, and proximity to the ocean. Stream gradient has an important influence on hyporheic exchange and hence chemical weathering and nutrient dynamics (Gooseff *et al.* 2002) (see Chapter 5). Supraglacial processes can also greatly influence stream and lake chemistry through particle dissolution and/or biochemical reactions. Gas geochemistry in the lakes is controlled in great part by the extent and duration of the ice cover with extreme supersaturations of oxygen, nitrogen, argon, neon, and helium occurring in the permanently ice-covered lakes due to lack of atmospheric exchange (e.g. Craig *et al.* 1992). Legacy effects, such as the past cryocentration of lake waters during cooler, drier climatic regimes when ice covers were actually lost for relatively long periods of time, can also have great significance in the overall biogeochemical dynamics of Antarctic lakes (Priscu 1995). These episodes of drawdown and cryoconcentration of solutes have created a number of meromictic lakes with hypersaline, anoxic monimolimnia where nutrients are highly concentrated. Diffusion of these nutrients into the upper, oxic, nutrient-depleted waters is extremely important to lake production (Priscu 1995; Laybourn-Parry 2003), with the development of deep chlorophyll maxima (see Chapter 9 in this volume). The lack of terrestrial input of organic carbon into McMurdo Dry Valley lakes and streams and their similar chemical characteristics indicate that the source of this carbon is derived from microorganisms only (McKnight *et al.* 1993).

8.2 Carbon cycle

Studies of carbon cycling are of particular interest because of the environmental sensitivity of polar freshwater carbon cycles, and their role in responses of terrestrial ecosystems to climate change. Surveys of carbon distributions are most common, but detailed carbon dynamics have been

described in a number of Antarctic lakes, including Lake Bonney, McMurdo Dry Valleys, and Beaver Lake, an epishelf lake at 70°48′S, 68°15′E (Priscu *et al.* 1999; Laybourn-Parry *et al.* 2006), and a number of lakes and rivers across the Arctic, including Toolik Lake, Alaska, and Char Lake, in high Arctic Canada (see Chapter 11).

8.2.1 Inorganic carbon dynamics

Inorganic carbon derived from glacier ice and from the atmosphere is a major source of total inorganic carbon (ΣCO_2; i.e. the sum of CO_2, HCO_3^- and CO_3^{2-}) for the streams and lakes in Antarctica. Direct glacier melt into supraglacial lakes can have very old ΣCO_2 (i.e. low ^{14}C values), whereas stream ΣCO_2 is usually equilibrated with the atmosphere (Takahashi *et al.* 1999). McMurdo Dry Valley streams are supersaturated with respect to atmospheric carbon dioxide and therefore the streams are undersaturated with respect to calcium carbonate (Neumann *et al.* 2001), leading to the potential dissolution of carbonate minerals in the stream channels and hyporheic zones. Annual fluxes of $3\,mmol\,m^{-2}$ of CO_2 have been measured entering Lake Fryxell, McMurdo Dry Valleys, from streams (Neumann *et al.* 2001).

Unlike temperate lakes and, in fact, most other polar lakes, the permanent ice covers of the McMurdo Dry Valley lakes eliminate physical mixing and carbon dioxide exchange with the atmosphere, therefore greatly restricting the redistribution of ΣCO_2 (Neumann *et al.* 2004). The surface waters of the lakes, however, can be greatly depleted, relative to the atmosphere, in carbon dioxide, with values as low as $10^{-4.3}$ (*p*CO2) in Lake Bonney, McMurdo Dry Valleys. This depletion is thought to be due to biological uptake without replenishment via the atmosphere due to the ice cover (Neumann *et al.* 2001). Recent work using $\delta^{13}C$ supports the idea that carbon dioxide depletion via photosynthesis is a major seasonal occurrence, especially in the shallow water regimes where abundant benthic mats exist (Lawson *et al.*, 2004). A large upward flux from aphotic deeper waters can provide an important source of carbon dioxide for photosynthesis in some lakes (Neumann *et al.* 2001). All the McMurdo Dry Valley lakes have upper to mid-depth $\delta^{13}C$-CO_2 maxima, reflecting the preferential biological uptake of ^{12}C, whereas only a few have deep waters with negative values of $\delta^{13}C$, indicating extensive remineralization of organic carbon (Neumann et al. 2004).

Green et al. (1988) measured the solubility of calcium carbonate in the McMurdo Dry Valley lakes and found that Lakes Hoare, Miers and Joyce are supersaturated in the euphotic zone relative to calcite, but undersaturated at depth. Lake Fryxell was supersaturated throughout the water column. These findings suggest that only in Lake Fryxell would calcium carbonate derived pelagically be preserved in the sediments. The data suggest that the shallow, benthic portions of the lakes would also preserve calcium carbonate, especially where benthic mats are abundant and photosynthesis drives the pH and inorganic carbon system towards carbonate precipitation. These ideas are supported by the observation that paleo delta deposits throughout Taylor Valley contain both organic matter and calcium carbonate. Calcium carbonate precipitation has also been documented in some of the Vestfold Hills lakes. Organic Lake has both monohydrocalcite and aragonite, indicating both inorganic and biogenic precipitation of carbonate minerals (Bird et al. 1991). The $\delta^{13}C$ and $\delta^{18}O$ analysis of this calcium carbonate indicates that in the most recent history of the lake, calcium carbonate is primarily derived from inorganic processes, not biological activity. On the other hand, the very low ionic strength epishelf lakes are extremely undersaturated with respect to calcium carbonate minerals (Doran et al. 2000).

Permafrost and low temperature are also strong constraints on weathering in the Arctic, leading to low HCO_3^- concentrations in many rivers and lakes. Dissolved carbon dioxide is also a significant constituent of ΣCO_2, but, in contrast to Antarctic fresh waters, much of this carbon dioxide is derived from mineralization of transported terrestrial organic carbon (e.g. Jonsson et al. 2003). Arctic fresh waters are almost always net sources of carbon dioxide to the atmosphere, so consideration of gaseous as well as soluble forms of carbon are important to accurately measure aquatic and terrestrial net ecosystem production (Kling et al. 1991). In addition, methane production represents another dynamic pathway

for carbon transformation and transport in Arctic fresh waters. In Siberian lakes, thermokarst activity increases microbial access to Pleistocene-aged carbon, stimulating high rates of methane and carbon dioxide release (Walter et al. 2006). Lakes contribute significantly to high-latitude greenhouse gas fluxes, and may represent an important positive feedback to climate warming.

As for many other solutes in Arctic fresh waters, ΣCO_2 and methane concentrations and fluxes are seasonally dynamic and challenging to study. For example, although HCO_3^- concentrations are generally low due to the presence of permafrost, they typically rise during winter in rivers due to inputs from groundwater (e.g. Smedberg et al. 2006). Carbon dioxide and methane accumulate under ice in rivers and lakes, so degassing during ice-out is a major factor in annual carbon fluxes (e.g. Phelps et al. 1998). A final example is that fluxes of methane from lakes can occur largely via ebullition, requiring special sampling methods for accurate measurement (Walter et al. 2006).

Autotrophic production by phytoplankton in Arctic fresh waters is strongly nutrient-limited and generally low (see Section 8.3 on nutrients, below). As a consequence, photosynthetic effects on inorganic carbon dynamics are often minimal (Kling et al. 1992a; Jonsson et al. 2003) and calcium carbonate precipitation is uncommon. Significant precipitation has been observed in some areas such as shallow coastal lakes of northern Alaska and hypersaline lakes near the ice-sheet terminus in Greenland (Anderson et al. 2001).

8.2.2 Dissolved organic carbon dynamics

The dissolved organic carbon (DOC) concentrations in Antarctic streams are very low, ranging from 17 to 75 μM in the McMurdo Dry Valley streams (McKnight et al. 1993). Glacier ice where the meltwater originates has even lower values (≈8 μM; Lyons et al. 2007), suggesting that some DOC entering the lakes is derived from the algal communities in the streams. Maximum DOC values in the lakes range from 2090 and 2370 μM at depth, respectively, in Lakes Fryxell and Vanda (McMurdo Dry Valleys); however, the surface waters of these lakes contain less than 25 μM DOC. Maximum water-column values in Beaver Lake were 54 μM (Laybourn-Parry *et al.* 2006). Saline lakes in the Vestfold Hills have varying amounts of DOC from maximum season values of approximately 58 μM in Pendant Lake to values as high as 3330 μM in Rookery Lake (Laybourn-Parry *et al.* 2002). The lowest DOC values in these Vestfold Hills lakes occur in the late austral summer. This observation along with strong coincidence between DOC concentrations and bacterial growth suggests DOC consumption. The Signy Island lakes have higher DOC concentrations in the austral summer with concentrations as high as approximately 100 μM, but decrease to undetectable levels in the late austral winter (Butler 1999).

The major carbon fractions of both autochthonous (derived from biological processes within the lake) and allochthonous (derived from outside the lake) sources in the McMurdo Dry Valley lakes are similar (fulvic and hydrophilic acids), and reflect a similar source of microbially derived components (McKnight *et al.* 1993). The composition of minor components of the DOC vary between lakes and between their sediments and water column, however, reflecting differences in microbial communities and lake history. For example, fatty acids increase with depth until the anoxic layer in Lake Vanda, then decrease down to the sediments (Matsumoto 1993). The low-molecular-mass fraction of the DOC pool in the Taylor Valley lakes is thought to be labile and an important source of bacterial energy (McKnight *et al.* 1993; Takacs *et al.* 2001). Budgetary considerations suggest that in the Taylor Valley lakes, bacterial respiration is 1.25–2 times greater than the DOCsupply, indicating that either bulk DOC is being drawn down or that seasonal decomposition of particulate carbon is also a major DOC source (Takacs *et al.* 2001).

Methane production has been observed in a number of Antarctic lakes, including Lake Fryxell, McMurdo Dry Valleys, Lake Untersee, East Antarctica (Smith *et al.* 1993; Wand *et al.* 2006), and Ace Lake, Vestfold Hills (Coolen *et al.* 2004). Lake Untersee has one of the highest concentrations (≈21.8 mM) ever observed in a natural aquatic ecosystem. In Lake Fryxell methane is produced only in the sediments, whereas in Lake Untersee, water-column production was observed. In both

lakes, carbonate reduction was the predominant methanogenic pathway. Methane consumption in the anoxic water column of Lake Fryxell where sulphate reduction is occurring is a major carbon-processing mechanism in the lake (Smith *et al.* 1993).

DOC is of particular interest in Arctic fresh waters because it is often the dominant form of carbon, is sensitive to environmental change, is associated with transport of nutrients and trace elements, and can strongly affect underwater light (see Chapter 4). Concentrations of DOC in rivers are elevated compared with those of temperate rivers, and highly varied temporally. DOC concentrations in lakes are noticeably elevated compared with typical values for temperate lakes only in certain areas (see Table 8.2, below; see Gregory-Eaves *et al.* 2000). A prevalent feature of DOC in northern fresh waters is that isotopes and other tracers consistently show a dominant contribution by terrestrial organic matter (e.g. Neff *et al.* 2006). Despite the overall consistency of terrestrial organic carbon sources to Arctic fresh waters, there is substantial spatial variation in concentration and composition of DOC. Of the many physical and biological factors that influence DOC production and transport, effects of temperature appear to be very important in both laboratory studies and field surveys (Frey and Smith 2005). Some of the observed spatial variation may be related to the availability of soluble organic compounds associated with terrestrial vegetation, with greater concentrations generally found in soils and fresh waters with trees and shrubs (e.g. Pienitz *et al.* 1997). Other factors such as soil chemistry and acid deposition (Evans *et al.* 2005) have also been implicated in trends of increasing river DOC concentration and flux in north temperate watersheds but their importance in Arctic regions is unknown.

A second prominent feature of DOC in Arctic fresh waters is substantial temporal variation. Hydrologic fluxes are dominated by spring snowmelt for many regions, and these peak flows from uplands carry a disproportionately high load of DOC to rivers and lakes (Finlay *et al.* 2006). Concentrations of DOC and other organic compounds increase with flow in many watersheds (see Chapter 5). Interestingly, opposite patterns are observed for low-gradient, wetland watersheds (Laudon *et al.* 2004). DOC exported by rivers during snowmelt is of modern age, and more dominated by aromatic compounds compared with low-flow periods (e.g. Neff *et al.* 2006). Although terrestrial DOC is often recalcitrant in fresh water, spring snowmelt mobilization of labile materials can lead to relatively rapid metabolism and nutrient cycling despite the prevailing cold temperatures (e.g. Michaelson *et al.* 1998). Overall, spring periods are underrepresented in sampling efforts to characterize fluxes and composition of dissolved organic matter (DOM).

8.3 Nutrient cycling

Many Antarctic lakes are oligotrophic or even ultra-oligotrophic with photosynthesis rates of less than 2.5 μmol C L^{-1} day^{-1}. However, there are exceptions, such as lakes that have been influenced by bird rookeries or saline lakes in the Vestfold Hills that have been derived from relict sea water. Antarctic lakes vary considerably in their nutrient concentrations and in many cases phosphorus appears to be limiting for phytoplankton production (Laybourn-Parry 2003). Thus nutrient limitation is a major feature of most Antarctic limnetic systems. Although physical factors such as low light and temperature are important constraints on production in many lakes, nutrient limitation can also be significant. Dissolved inorganic nitrogen and phosphate measurements have been made in a vast majority of Antarctic lakes, and, where investigated, seasonal nutrient cycles occur, with the lowest concentrations occurring from November through the austral autumn (e.g. Henshaw and Laybourn-Parry 2002). A number of investigations in the McMurdo Dry Valley lakes indicate that nutrients are taken up rapidly in streams via algal mats and/or the hyporheic zone (Howard-Williams *et al.* 1986; McKnight *et al.* 2004). Howard-Williams *et al.* (1986) demonstrated great variability in nutrient concentrations between streams but also showed that the early summer discharge contained higher levels of dissolved nitrogen and phosphorus. Canfield and Green (1985) also showed early-season high fluxes in stream nutrients. McKnight *et al.* (2004) illustrated that nitrate uptake occurred in both stream

channels and the hyporheic zone but phosphate was only lost through the main channel. These data imply that organisms associated with the hyporheic zone obtain their phosphorus through *in-situ* weathering processes. Recent work in the Antarctic maritime indicates that microbes in both snowpack and ice-marginal environments can take up dissolved inorganic nitrogen and phosphate, thereby complicating the snowmelt and elution processes (Hodson 2006). It is clear from these works and others that the nutrient chemistry is modified greatly from melting glaciers through the streams into the lakes. Supraglacial processes can also impact the nutrient chemistry. Recent research in the Taylor Valley aquatic system strongly indicates that the C:N:P stoichiometry is modified greatly as water flows along the hydrological continuum from snow/glacier ice to the closed-basin lakes. The biota helps control the elemental ratios in these waters, although landscape age also has great influence (Barrett *et al.* 2007). Streams on younger surfaces provide more phosphorus relative to nitrogen, and streams with high abundances of algal mats have much lower N:P water ratios. Thus landscape age and history along with the gradient and geomorphology of the streams (which in large part help constrain the biological complexity of the stream channels and hyporheic zones) play an important role in regulating the nutrient input into the lakes.

Experimental work by Priscu (1995) demonstrated that in the McMurdo Dry Valleys, Lakes Bonney and Lake Vanda are phosphorus-deficient, whereas Lake Fryxell and Lake Hoare are nitrogen-deficient (see Chapter 9). Nitrate, soluble reactive phosphate, DOC, reactive silicate, plus chloride and sulphate profiles for these four major Taylor Valley lakes (McMurdo Dry Valleys) are shown in Figure 8.1. The deep-water enrichments of dissolved nitrogen and phosphorus in some of the McMurdo Dry Valley lakes and other lakes in Antarctica are related to legacy events tied to the hydrologic and climatic history of these lakes. This is definitely the case for Lakes Bonney, Fryxell, and Vanda (Priscu 1995). These past events show that cryoconcentrated solutes can greatly impact nutrient dynamics in the lakes (Laybourn-Parry 2003) and may even be related to trace-metal toxicity that

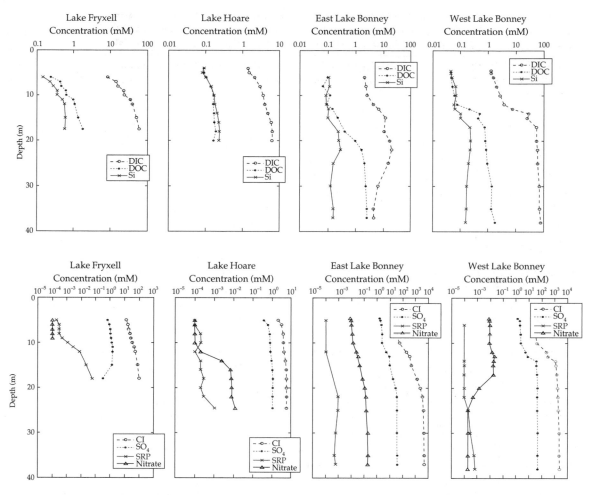

Figure 8.1 Depth profiles of dissolved inorganic carbon (DIC), dissolved organic carbon (DOC), reactive silicate (Si), chloride (Cl), sulfate (SO₄), soluble reactive phosphate (SRP), and nitrate from the four major Taylor Valley lakes, Southern Victoria Land, Antarctica. From www.mcmlter.org.

is thought to exist in the east lobe of Lake Bonney (Ward *et al.* 2003).

As in the Antarctic region, Arctic lakes are generally highly oligotrophic, leading to low rates of planktonic biomass (Table 8.1) and primary production. As a consequence, production in Arctic lakes is often dominated by benthic algae, so factors such as light limitation are also important in regulating primary production and nutrient uptake (e.g. Bonilla *et al.* 2005).

Nitrogen in Arctic streams and lakes is often overwhelmingly dominated by dissolved organic nitrogen (DON) (Dittmar and Kattner 2003). Some

exceptions include streams draining steep, barren ground and discontinuous permafrost watersheds (Jones *et al.* 2005; Schindler *et al.* 1974) and winter baseflow when NO₃⁻ makes a large contribution to nitrogen pools or transport. Elsewhere, dissolved inorganic nitrogen (DIN) concentrations are often low relative to DON. Dynamics of DON in Arctic fresh waters are similar to those of DOC in many respects, including peak lability, concentration, and flux during spring snowmelt (Stepanaukas *et al.* 2000).

Low DIN concentrations may account for the evidence for nitrogen limitation of phytoplankton

Table 8.1 Responses in McMurdo Dry Valley lakes (Taylor Valley, Antarctica) to a period of high stream discharges (flood year 2001–2002). Biomass: depth integrated phytoplankton chlorophyll a; PPR, phytoplankton production in the upper water column.

Lake	Biomass (next season, 2002–2003)	PPR (next season, January 2002, 2002–2003)	
West Bonney	≈50% increase	≈23% decrease	2× increase
East Bonney	≈150% increase	≈10% increase	2× increase
Hoare	≈20% increase	No change	No change
Fryxell	No change	5× increase	6× increase at 5 m

Source: Foreman *et al.* (2004).

growth in some Arctic lakes (e.g. Levine and Whalen 2001; Bergstrom *et al.* 2005). Less work has been done in rivers, which generally show high dissolved N:P, possibly indicating more consistent phosphorus deficiency. Even where phosphorus limitation is documented, demand for DIN is often relatively high (Wollheim *et al.* 2001), and colimitation by nitrogen and phosphorus is common (e.g. Bowden *et al.* 1992; Levine and Whalen 2001). Nitrogen-limited primary production is somewhat of a paradox considering the large terrestrial pools of nitrogen in soils of surrounding watersheds. Despite these large nitrogen pools, slow rates of decomposition and nitrogen mineralization on land lead to limited availability of DIN for terrestrial plants. Strong terrestrial demand for DIN coupled with low rates of atmospheric inputs typically leads to strong watershed retention of atmospheric nitrogen inputs (e.g. Kortelainen and Saukkonen 1998; but see Jones *et al.* 2005).

Concentrations of phosphorus in Arctic fresh waters are often extremely low (Table 8.2), leading to limitation or colimitation of autotrophic production in many lakes and streams. Modest elevation of phosphorus (sometimes nitrogen and phosphorus) leads to dramatic changes in ecosystem structure and processes. Long-term phosphorus addition experiments in a river (Slavik *et al.* 2004) and combined nitrogen and phosphorus addition to a lake in Alaska (O'Brien *et al.* 2005) demonstrate strong limitation to autotrophic production, with both effects on higher trophic levels and stimulation of nutrient cycling. Although biotic demand for PO$_4$ is extremely high, geochemical processes

can also exert strong effects on phosphorus availability. Surficial sediments are usually oxic in Arctic lakes, and have a large capacity to bind PO$_4$ (Whalen and Cornwell 1985). Thus these lakes may recovery rapidly from pulsed phosphorus inputs (O'Brien *et al.* 2005).

Other nutrients have generally received less attention in both Arctic and Antarctic regions. Relatively little systematic research has been done on silicate dynamics in Antarctic lakes. The one investigation suggests that there is little biological control on silicate concentrations due in part because of the low numbers of diatoms in the water column (Figure 8.1). Thermodynamic calculations using data from the McMurdo Dry Valley lakes indicate that amorphous or biogenic SiO$_2$ is undersaturated in all the lakes except for the deep waters of the east lobe of Lake Bonney. This suggests that any SiO$_2$·H$_2$O produced in the water column, such as diatom debris, should dissolve (Lyons *et al.* 1988). Diatoms are much more abundant in the stream and lake algal mats.

Sulphate reduction in both the anoxic hypolimnia and sediments occurs in many Antarctic lakes, especially the ones with strong chemoclines such as Lakes Fryxell, Vanda, and Hoare (Green *et al.* 1988), in the McMurdo Dry Valleys, Lake Untersee, and in many of the relict sea-water lakes of the Vestfold Hills (Franzmann *et al.* 1988). Sulphur-oxidizing bacteria have been recently documented in Lake Fryxell, suggesting a complex sulphur cycle in the lake (Figure 8.1). The identified bacteria are chemolithotrophs and exist in both the oxic and anoxic portions of the lake (Sattley and Madigan 2006),

Table 8.2 The chemical composition of lakes and ponds for selected Arctic regions. Mean values were assessed for lakes with maximum depth of over 2 m for recent surveys of lake chemistry. Analytical methods were approximately similar for most studies. PD, polar desert; T, tundra; FT, forest-tundra; TChla, total **chlorophyll a** uncorrected for phaeophytin; TP, total phosphorus; TN, total nitrogen. The first column gives the number of lakes, n.

n	Study	Region	Land cover	Year/period	Latitude (N)	TP (μg L^{-1})	TN (mg L^{-1})	TChla (μg L^{-1})	DOC (mg L^{-1})	ΣDIC (mg L^{-1})	Individual ions (mg L^{-1})					
											SO$_4$	Ca	Mg	Na	K	Cl
16	Ruhland and Smol (1998)	Central Canada	T	1998	65°24′	9.61	0.20	0.99	2.33	1.61	1.94	0.88	0.59	0.29	0.32	0.85
2	Brutemark et al. (2006)	SW Greenland	T		66°59′	1.20	0.73	0.10	65.50		40.83	23.28	87.86	178.5	64.4	233.5
6	Laperriere et al. (2003)	Alaska	FT	1990s	67°00′	4.17	0.30	1.38								
10	Laperriere et al. (2003)	Alaska	T	1990s	68°00′	6.95	0.28	1.51								
40	Jonsson et al. (2003)	Northern Sweden	T	2000	68°13′	6.82	0.23	0.62	4.18	1.63						
10–38	Kling et al. (1992b, 2000)	Alaska	T	1990s	68°23′			1.85	3.48	6.15	3.56	17.30	2.06	0.48	0.39	0.22
6	Gregory-Eaves et al. (2000)	Alaska	T	1996	68°24′	8.23	0.29	0.71	24.63	15.25	4.82	21.53	2.25	0.48	0.37	1.52
6	Pienitz et al. (1997)	NW Territories	FT	1990	69°00′	23.73	0.67	1.63	16.15	6.38	24.73	16.52		3.68	1.03	3.02
18	Pienitz et al. (1997)	NW Territories	T	1990	69°00′	10.33	0.39	1.21	8.86	14.18	8.28	18.83		4.99	1.31	7.94
23	Duff et al. (1998)	Taimyr Peninsula	FT	1993	69°14′	5.40		1.00	4.00	6.10	3.90	6.30	1.90	0.80	0.20	0.40
31	Duff et al. (1998)	Lena delta	T	1994	71°18′	6.50		1.40	4.10	4.00	1.90	5.30	1.60	0.80	0.40	0.60
9	Michelutti et al. (2002a)	Victoria Island	PD	1997	72°10′	1.35	0.18	0.54	1.19	16.29	2.50	19.98	5.38	0.44	0.22	1.03
17	Lim et al. (2005)	Banks Island	T	2000	72°57′	11.31	0.27	1.23	3.55	12.45	14.71	14.48	8.71	23.15	1.29	41.02
15	Lim et al. (2001)	Bathhurst Island	PD	1994	75°42′	13.57	0.46	0.68	2.82	19.37	6.39	31.56	4.66	3.01	0.49	5.78
13	Lim and Douglas (2003)	Devon Island	PD	2000	75°48′	3.89	0.11	0.68	1.95	16.12	4.55	19.49	5.79	1.28		3.08
1	Antoniades et al. (2003)	Ellef Ringnes Island	PD	1997	78°29′	6.30	0.08	1.30	1.00	2.40	17.30	4.50	2.30	1.60	0.50	0.90
12	Michelutti et al. (2002b)	Axel Heiberg Island	PD	1998	80°12′	4.10	0.25	0.64	3.51	10.66	39.85	21.68	6.67	9.48	2.03	14.04

and are thought to play an important role in the biogeochemical cycling of carbon. Photosynthetic sulphur bacteria are also known from Lake Fryxell, saline lakes in the Vestfold Hills, ice-shelf microbial mat communities, and certain Arctic lakes (e.g. Van Hove *et al.* 2006).

A number of the McMurdo Dry Valley lakes contain very high concentrations of N_2O, dimethyl sulphide (DMS), and dimethyl sulphoxide (DMSO). Vincent *et al.* (1981) first documented N_2O in Lake Vanda at concentrations exceeding $2 \mu M$. Their experimental work indicated that the N_2O was produced via nitrification in the oxic zone of the lake. Loss was due to both diffusion upward in the oxic zone and by denitrification in the anoxic zone below the chemocline (Figure 8.2). Recent work confirmed that N_2O is produced via nitrification and ammonium oxidation in all the McMurdo Dry Valley lakes with production rates, computed with a one-dimensional diffusion model, as high

as $5.3 \, nmol \, N \, day^{-1}$ (Priscu 1997). Both DMS and DMSO are abundant in the deeper waters of Lake Bonney with values of DMS reaching 317 nM in the west lobe, and DMSO values as high as approximately 200 nM in the east lobe of the lake (Lee *et al.* 2004). These high concentrations are extremely unusual as they occur in the absence of primary production. Lee *et al.* (2004) concluded that the reduction of DMSO to DMS in the west lobe, but not in the east lobe, could explain the differences in concentrations between these two species.

Concentrations of most major ions are dilute in Arctic waters compared with many other regions. For example, Table 8.2 shows concentrations of major ions such as Ca^{2+}, SO_4^{2-}, and HCO_3^- in lakes in most regions to be extremely low. Rivers are similarly dilute, on average, but less well characterized and more temporally varied. Low ionic strength is related primarily to limited weathering, resulting from low temperatures and permafrost, which

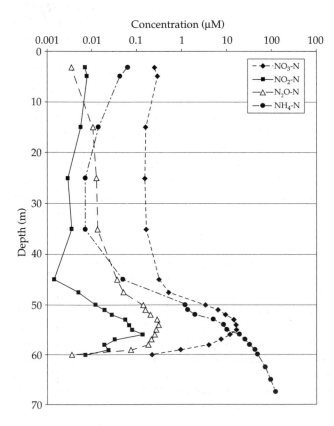

Figure 8.2 Depth profiles of nitrogen species from Lake Vanda, Wright Valley, Southern Victoria Land, Antarctica. Drawn from Vincent *et al.* (1981).

limits access to mineral soils. An additional factor is the low human population density and absence of urban and industrial development in most parts of the Arctic, leading to limited anthropogenic contributions to atmospheric deposition and point source pollution. An important exception is long-range transport of toxic compounds which has elevated concentrations even in some of the most remote streams and lakes (see Chapter 16). Further, local industrial sources are present in some areas (e.g. smelters on the Kola Peninsula and near Norilsk in Russia, and oil and gas fields at Prudhoe Bay, Alaska) which elevate local pollutant inputs (e.g. Gordeev *et al.* 2004).

Despite their generally dilute nature, there is substantial diversity in the chemical composition of Arctic fresh waters. Table 8.2 summarizes chemical composition of lakes by region from some recent surveys. Even at this broad scale, substantial regional variation in chemical composition can be seen in Arctic lakes. Considerably more variation is observed at smaller, local scales. Regional and local diversity may be partially explained by the following factors: climate and hydrologic variation, effects of marine inputs, geological heterogeneity, and presence or absence of permafrost. These factors are considered briefly below.

Climate influences ion composition via evaporative or runoff-dominated processes affecting water chemistry. Although evapotranspiration is generally low in polar regions, precipitation is also often low. Thus some regions, such as parts of Greenland, evaporative concentration of solutes is a primary control over lake chemistry (Anderson *et al.* 2001). Effects of marine inputs are prevalent near coastal areas, which are extensive. Here, marine inputs have a strong influence over ionic composition and concentrations, in particular elevating SO_4^{2-}, Na^+, and Cl^- concentrations (e.g. Pienitz *et al.* 1997). Geological heterogeneity, due to varied glacial, periglacial, and geothermal processes, also influence chemical composition of Arctic fresh waters. For example, varied glaciation of northern Alaska has produced landscapes that range in age from 10 000 to over 1 million years old. Despite the presence of continuous permafrost, landscape age influences chemistry of surrounding lakes through varied weathering and

DOC dynamics (Hobbie *et al.* 2002). For rivers, a classic example is the 1969 survey of the McKenzie River showing different composition of tributaries depending on geology and permafrost presence (Reeder *et al.* 1972). Where permafrost coverage is discontinuous, high rates of weathering may occur in permafrost free areas, and rates may be enhanced by cryogenic processes or recent exposure of permafrost.

8.4 Geochemical linkages

A series of seminal papers by Green and his coworkers were the first to approach the aquatic systems of Antarctica from a holistic, 'watershed' perspective. These papers compared stream element inputs with total lake concentrations of the respective elements of interest and determined elemental residence times, mass budgets of each element, and lake 'ages' (e.g. Green *et al.* 1988). This pioneering work was based on stream inputs for as little as one austral summer. With the establishment of the McMurdo Dry Valleys Long-Term Ecological Research (MCM-LTER) program in 1993 in Taylor Valley, and the season-to-season gauging of the Taylor Valley streams, it became clear that stream flows and, hence, geochemical fluxes to the lakes, can vary dramatically from year to year. Lyons *et al.* (1998) followed up on the earlier work of Green and coined the term 'geochemical continuum,' meaning the geochemical change of waters as they evolve from glacier melt to proglacial streams to closed basin lakes. This idea was later extended to include geochemical processes occurring in the glaciers as well as the hyporheic zones of the streams (Lyons *et al.* 2001). Recent work in the McMurdo Dry Valleys and on Signy Island has added to our overall knowledge of how aquatic systems in Antarctica evolve from a nutrient perspective (Hodson 2006; Barrett *et al.* 2007). These papers demonstrate that exchanges of liquid water and biological processes both play important roles in changing nutrient concentrations and speciation across Antarctic landscapes. Similar work in Alaska at Glacier Bay (Engstrom *et al.* 2000) and Toolik Lake (Hobbie *et al.* 2002) is providing similar information for recently glaciated Arctic landscapes.

Because of the seasonal and sometimes aperiodic variations of liquid water production in polar environments, geochemical fluxes are closely linked to hydrologic variations. In unvegetated watersheds, early-season stream flows may have very high solute concentrations as they flush previously precipitated salts, newly produced weathering products, and aeolian inputs into lakes and ponds. For watersheds with extensive vegetation and soils, snowmelt carries large fluxes of nutrients and trace elements in DOM. Changes in elevation and distance from the ocean also can greatly impact solution chemistry, as first noted by Wilson (1981). The geochemical variation is reflected in changes in the aerosol chemistry, changes in moisture, and changes in temperature. For example, the higher, drier elevations of the McMurdo Dry Valleys have soils with higher NO_3^- relative to Cl^- and SO_4^{2-}, whereas the lower elevations, closer to the marine source, have higher Cl^- concentrations.

Streamflow is highly variable in polar regions, and higher stream flows lead to larger inputs of solutes into lakes. For example, Foreman *et al.* (2004) demonstrated the significance of an extreme high-flow event on the biogeochemistry of the McMurdo Dry Valley lakes. This high flow of meltwater occurred because of an unseasonably warm mid-December to mid-January in 2001–2002 when record stream discharge lead to increased lake levels and a decrease in lake ice thickness. The initial flood event decreased primary production in one lake, but increased it in others (Table 8.1). The following summer both primary production and biomass increased (Table 8.1). These responses are thought to be related to the extremely high nutrient inputs by this flood event. For example, the phosphate flux to Lake Fryxell was four times greater than the longer-term annual inputs (Foreman *et al.* 2004). The warming event with its subsequent input of glacier melt served to recharge the system with water and nutrients, thereby greatly increasing production in the water column (Foreman *et al.* 2004). Flood events in the McMurdo Dry Valleys, such as the one of the 2001–2002 austral summer, occur aperiodically, but at what could be described as 'quasi-quintennial' to decadal. These sporadic events are probably a major long-term driver of production in these lakes.

8.5 Future responses to a warming climate

Lyons *et al.* (2006) recently put forth a scenario of how Antarctic aquatic systems might respond to warming climate. Similar conceptual models have been developed by other Antarctic limnologists (A. Camacho, personal communication). These scenarios were based on long-term monitoring of ecological, hydrological, and meteorological data from Signy Island, McMurdo Dry Valleys, and Livingston Island. With warming, especially warming in the austral summer, longer melt seasons will lead to larger volumes of liquid water flowing through streams into the lakes. Increased stream flow could lead to decreased algal mat abundances in the streams (McKnight *et al.* 1999). This would decrease nutrient uptake in the streams, thereby increasing the dissolved flux into the lakes. Lake volumes will increase and lake ice could thin and larger moat areas would occur in the austral summer. The latter would lead to increasing atmosphere exchange. In the lower latitudes of the continent and in the sub-Antarctic, ice cover would be lost for longer periods of time, thereby greatly enhancing atmospheric contact. This increase in flow of meltwater would increase the nutrient fluxes into the surface waters of the lakes.

The increased nutrient fluxes, greater atmospheric exchange enhancing carbon dioxide input, decreased ice covers (enhanced photosynthetically available radiation), and warmer temperatures would in turn increase primary production in the lakes. The higher volume of fresh glacier meltwater into the mixolimnia would enhance the physical stability of lakes whose bottom waters are brackish to hypersaline. The increased stability and primary production will enhance the organic carbon flux to the deeper waters, which in the most stably stratified lakes could produce suboxic to anoxic conditions. This increased eutrophy will influence the biological structure and function of the lakes.

Ongoing changes in hydrologic, physical, and biological conditions in fresh waters are evident across a broad swath of the Arctic. Recent reviews summarize these diverse changes (e.g. Rouse *et al.* 1997; Hobbie *et al.* 1999; Hinzman *et al.* 2005). Here we focus on biogeochemical responses to hydrologic, thermal, and permafrost changes that will affect the biogeochemical properties of Arctic fresh waters.

8.5.1 Hydrologic change

As discussed above, runoff volume and timing play a key role in determining concentrations and fluxes of solutes. Recent increases in precipitation, and perhaps permafrost melting, have contributed to a significant increase in water flux from large Arctic rivers (e.g. McClelland *et al.* 2006). Increasing river discharge may increase flux of many solutes, but the magnitude will depend on the specific mechanisms that govern their production and availability in soils for transport. Altered spatial and temporal distributions of precipitation are also likely, and could have important consequences for many regions which currently have very low summer rainfall.

Another hydrologic change with important biogeochemical consequences is the increase in groundwater infiltration that accompanies loss of permafrost and seasonally frozen surface layers. Permafrost watersheds have characteristically flashy hydrographs and little groundwater flow compared with permafrost-free watersheds. A decrease in permafrost or seasonally frozen surface soils will increase infiltration of water into soils (Figure 8.3), with manifold consequences for biogeochemical composition of runoff to fresh waters. An example of the effects of permafrost on hydrology and solute composition can be seen in the Sub-Arctic Caribou-Poker watershed in central Alaska. Watersheds ranging from 4 to 53% permafrost cover show large contrasts in hydrology and solute form and flux (Jones *et al.* 2005; Petrone *et al.* 2006). Streams in watersheds with less permafrost show dampened hydrologic variation, with higher inorganic and lower organic solute concentrations relative to those with extensive permafrost. Water infiltration through mineral soils allows greater retention and mineralization of organic matter, and increased transport of weathering products (Figure 8.3).

8.5.2 Direct effects of rising temperatures

Rising temperatures will have broad impacts on biogeochemical fluxes and cycling of elements in Arctic fresh waters. Warmer temperatures will have diverse physical and biological effects that could include increased primary production (Flanagan *et al.* 2003) and stratification in lakes, decreased snow and ice cover, increased weathering rates, and changes in vegetation within watersheds. All

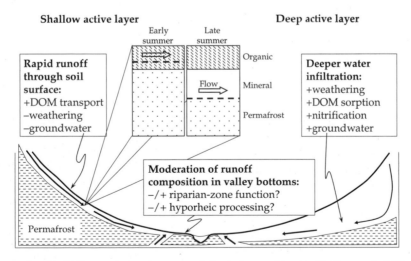

Figure 8.3 A conceptual model of influence of **permafrost** on runoff chemical composition. Modified from work of O'Donnell and Jones (2006) and Petrone *et al.* (2006).

of these changes could have major effects on bio-geochemical processes.

Perhaps the most notable direct effect of increased temperature on Arctic fresh waters will be changes in dynamics of DOM production and transport from watersheds. Increasing tempera-tures may have positive effects on generation of DOM in soils. Laboratory experiments consistently show temperature sensitivity of DOM production in soils (e.g. Neff and Hooper 2002). A recent sur-vey of wetland streams in central Siberia yields data that support an important role for temperature in regulating DOC mobilization to aquatic eco-systems (Figure 8.4). Streams in cold permafrost watersheds had modest DOC concentrations that were unrelated to the extent of wetlands within the watershed. In contrast, DOC was strongly related to wetlands in warmer regions to the south, sug-gesting a large mobilization of DOC as tempera-ture increases in northern wetlands. Because behavior of DON is in some respects similar to DOC in northern watersheds, DON concentrations may also rise as a result of continued warming,

and this has important implications for the Arctic carbon cycle since terrestrial vegetation is strongly nitrogen-limited.

Acting against stimulatory effects on DOM pro-duction are other temperature-dependent changes that may have opposing effects on DOM inputs to aquatic ecosystems. For example, microbial deg-radation of DOM to carbon dioxide and NH_4^+ is positively affected by temperature, possibly lead-ing to a reduction in DOC and DON transport with warming. As discussed below, greater expo-sure of mineral soils with permafrost thaw could decrease fluxes of organic materials via effects of sorption.

8.5.3 Permafrost thaw

An important indirect effect of rising temperatures is thawing of permafrost, a dominant feature of much of the Arctic. Permafrost temperatures are increasing and there are local examples of near-surface permafrost loss in many regions. A recent modeling study suggests substantial reductions

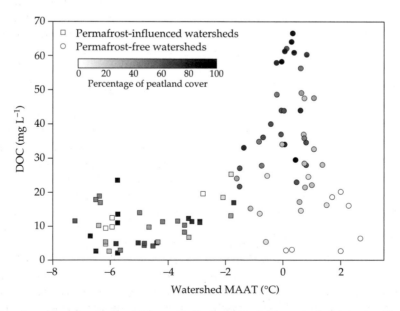

Figure 8.4 Temperature dependence of summertime DOC concentration in streams draining watersheds in western Siberia during 1999–2001. Squares and circles indicate cold **permafrost**-influenced and warm **permafrost**-free watersheds, respectively. Peatland cover varied similarly between both watershed types but there was no relationship between peatlands and DOC in cold **permafrost**-influenced streams. MAAT, mean annual air temperature. Redrawn from Frey and Smith (2005).

in permafrost in the upper 3 m of soils within the next 100 years (Lawrence and Slater 2005). Permafrost thaw will have strong direct effects on biogeochemical composition of fresh water in addition to those effects mediated through changes in hydrology, soil temperature, and weathering. Immediate effects of permafrost thaw would be through release of easily leached compounds from permafrost. Permafrost appears to contain large soluble pools of ions and these are likely to be removed rapidly and transported to streams and lakes (Kokelj *et al.* 2005). Permafrost is of diverse composition in the pan-Arctic, however, and little is known about the general composition of soluble organic and inorganic compounds in much of the permafrost regions that are sensitive to thaw.

Much less is known about nutrients. Permafrost may be relatively rich in organic carbon and nitrogen and phosphorus, and release of these solutes during permafrost thaw could be large (Zimov *et al.* 2006). There remains great uncertainty regarding concentrations of carbon, nitrogen, and phosphorus in near-surface permafrost in the Arctic region. Recent studies in Siberia indicate that soil- and permafrost-bound DOC and nutrients are apparently being released into rivers and lakes in northern Siberia. Radiocarbon dating of DOC indicates low but detectable fluxes of Pleistocene carbon to rivers (Neff *et al.* 2006). There is some evidence of mobilization of soil nutrients because stream nitrogen and phosphorus levels in the region are roughly 2–6-fold those observed in other parts of the Arctic (J.C. Finlay *et al.*, unpublished results). Thus it is possible that rapid warming of nutrient-rich permafrost could increase nitrogen and phosphorus fluxes to aquatic ecosystems.

We summarize the longer-term effects of permafrost thaw on flux and composition of Arctic fresh waters with a conceptual model (Figure 8.3). Extensive hydrologic contact with organic soils, and little interaction with mineral soils in permafrost watersheds with shallow active layers account for the high organic matter content and extremely dilute ionic composition of many streams and lakes. The exposure of poorly weathered soils during permafrost thaw coupled with increased water flux could provide a sustained increased supply of weathering products that could greatly alter the chemical composition of streams and lakes in permafrost regions. Evidence for such changes has been seen in an increase of weathering products in Toolik Lake over the past 30 years (Hinzman *et al.* 2005).

Increased hydrologic water contact with mineral soils is also likely to greatly reduce the export of organic material. Mineral soils can stabilize organic materials onto charged surfaces. Such stabilization can be for long periods, as part of soil development, or could allow greater microbial mineralization of DOM in soils, resulting in increases in oxidized forms of nitrogen and carbon in soils, streams, and lakes. There is little aquatic data to assess this hypothesis, but a recent study of Sub-Arctic watershed nitrogen budgets showed that low permafrost cover is associated with high rates of stream NO_3^- fluxes relative to DON. High NO_3^- losses resulted in greater stream export than atmospheric inputs, an unusual result for nitrogen retention for undisturbed watersheds (Jones *et al.* 2005). Similar effects may have been observed in the Yukon River, where comparison of DOC flux for two periods (1978–1980 and 2001–2003) shows a large recent decrease (Striegl *et al.* 2005). These studies illustrate the need to understand interactions between temperature and hydrologic change for understanding the complex, multiple effects of climate alterations on Arctic fresh waters.

8.6 Conclusions

The vast polar regions present both challenges and opportunities to the study of freshwater biogeochemistry. The challenges generally arise from landscape heterogeneity in a remote and variable environment. The Arctic and Antarctica contain streams, rivers, and lakes whose diversity rivals that of any other region on Earth. Unpredictable temporal variation in the environment, such as the timing and magnitude of spring snowmelt, have key roles in biogeochemical cycles yet are extremely difficult to study without sustained intensive monitoring programs, which are uncommon. As a consequence, the knowledge base for polar regions remains inadequate. For example, Arctic freshwater biogeochemical studies have primarily emphasized flux estimation (e.g. nitrogen and carbon), surveys of chemical compositions, and tracer studies using isotope, geochemical, or organic markers to determine sources of solutes and particulate matter. With some notable exceptions, studies of biogeochemical processes are relatively rare, and more confined spatially. Time-series data, and data for many important parts of the Arctic such as Siberia, are also uncommon. Most of the studies that provided much of the first modern scientific data for Arctic fresh waters decades ago have not been continued, leaving a limited basis for understanding ecosystem responses to climate change. Much of our current understanding of large river chemical fluxes comes from studies done decades ago (Dittmar and Kattner 2003), using widely varied approaches (see Holmes *et al.* 2000; Finlay *et al.* 2006).

The research cited in this review has clearly demonstrated that relatively subtle climate change could significantly impact the biogeochemistry of polar aquatic systems. Because of this, there is a need to integrate both biogeochemical observations and process studies into climatic response investigations as a much-needed link to understanding the ecological consequences of climate change in these sensitive and unique aquatic ecosystems.

Acknowledgments

Thanks to the McMurdo Dry Valleys Long-Term Ecological Research (MCM-LTER) investigators, scientists, and students who helped collect and analyze the Taylor Valley data presented within; special thanks in this regard to John C. Priscu. Both authors are very thankful to Johanna Laybourn-Parry and Warwick F. Vincent for asking them to write this chapter. We also thank Kathy Welch, Chris Gardner, Mike Mellesmoen, and Michele Larrimer for their help with manuscript preparation.

References

Anderson, N.J., Harriman, R., Ryves, D.B., and Patrick, S.T. (2001). Dominant factors controlling variability in the ionic composition of West Greenland Lakes. *Arctic, Antarctic and Alpine Research* **33**, 418–425.

Antoniades, D., Douglas, M.S.V., and Smol, J.P. (2003). The physical and chemical limnology of 24 ponds and one lake from Isachsen, Ellef Ringnes Island, Canadian High Arctic. *International Review of Hydrobiology* **88**, 519–538.

Barrett, J.E. *et al.* (2007). Biogeochemical stoichiometry of Antarctic Dry Valley ecosystems. *Journal of Geophysical Research–Biogeosciences* **112**, G01010, doi:10.1029/2005JG000141.

Bergstrom, A.K., Blomqvist, P., and Jansson, M. (2005). Effects of atmospheric nitrogen deposition on nutrient limitation and phytoplankton biomass in unproductive Swedish lakes. *Limnology and Oceanography* **50**, 987–994.

Bird, M.I., Chivas, A.R., Radnell, C.J., and Burton, H.R. (1991). Sedimentological and stable-isotope evolution of lakes in the Vestfold Hills, Antarctica. *Palaeogeography, Palaeoclimatology, Palaeoecology* **84**, 109–130.

Bonilla, S., Villeneuve, V., and Vincent, W.F. (2005). Benthic and planktonic algal communities in a high arctic lake: pigment structure and contrasting responses to nutrient enrichment. *Journal of Phycology* **41**, 1120–1130.

Bowden, W.B., Peterson, B.J., Finlay, J.C. *et al.* (1992). Epilithic chlorophyll a, phototsynthesis, and respiration in control and fertilized reaches of a tundra stream. *Hydrobiologia* **240**, 121–131.

Brutemark, A., Rengefors, K., and Anderson, N.J. (2006). An experimental investigation of phytoplankton nutrient limitation in two contrasting low Arctic lakes. *Polar Biology* **29**, 487–494.

Butler, H.G. (1999). Temporal plankton dynamics in a maritime Antarctic lake. *Archives of Hydrobiology* **146**, 311–339.

Canfield, D.E. and Green, W.J. (1985). The cycling of nutrients in a closed-basin Antarctic lake: Lake Vanda. *Biogeochemistry* **1**, 233–256.

Coolen, M.J.L. *et al.* (2004). Evolution of the methane cycle in Ace Lake (Antarctica) during the Holocene: response of methanogens and methanotrophs to environmental change. *Organic Geochemistry* **35**, 1151–1167.

Craig, H., Wharton, R.A., and McKay, C.P. (1992). Oxygen supersaturation in ice-covered Antarctic lakes: biological versus physical contributions. *Science* **255**, 318–321.

Dittmar, T., and Kattner, G. (2003). The biogeochemistry of the river and shelf ecosystem of the Arctic Ocean: a review. *Marine Chemistry* **83**, 103–120.

Doran, P., Adams, P., and Ecclestone, M. (1997). Arctic and Antarctic lakes: contrast or continuum? In Lewkowicz, A.G. (ed.), *Poles Apart: a Study in Contrasts*. Proceedings of an International Symposium on Arctic and Antarctic Issues, University of Ottawa, 25–27 September 1997. University of Ottawa Press, Ottawa.

Doran, P., Wharton, R.A., Lyons, W.B., Des Marais, D.J., and Andersen, D.T. (2000). Sedimentology and geochemistry of a perennially ice-covered epishelf lake in

Bunger Hills Oasis, East Antarctica. *Antarctic Science* **12(2)**, 131–140.

Doran, P., Fritsen, C.H., McKay, C.P., Priscu, J.P., and Adams, E. (2003). Formation and character of an ancient 19-m ice cover and underlying trapped brine in an "ice-sealed" east Antarctic lake. *Proceedings of the National Academy of Sciences USA* **100(1)**, 26–31.

Duff, K.E., Laing, T.E., Smol, J.P., and Lean, D.R.S.(1998). Limnological characteristics of lakes located across Arctic treeline in northern Russia. *Hydrobiologia* **391**, 205–222.

Engstrom, D.R. *et al.* (2000). Chemical and biological trends during lake evolution in recently deglaciated terrain. *Nature* **408**, 161–166.

Evans, C.D., Monteith, D.T., and Cooper, D.M. (2005). Long-term increases in surface water dissolved organic carbon: observations, possible causes and environmental impacts. *Environmental Pollution* **137**, 55–71.

Finlay, J., Neff, J., Zimov, S., Davydova, A., and Davydov, S. (2006). Snowmelt dominance of dissolved organic carbon in high-latitude watersheds: implications for characterization and flux of river DOC. *Geophysical Research Letters* **33**, L10401, doi:10.1029/2006GL025754.

Flanagan, K.M. *et al.* (2003). Climate change: the potential for latitudinal effects on algal biomass in aquatic ecosystems. *Canadian Journal of Fisheries and Aquatic Science* **60**, 635–639.

Foreman, C.M., Wolf, C.F., and Priscu, J.C. (2004). Impact of episodic warming events on the physical, chemical and biological relationships of lakes in the McMurdo Dry Valleys, Antarctica. *Aquatic Geochemistry* **10**, 239–268.

Franzmann, P.D., Skyring, G.W., Burton, H.R., and Deprez, P.P. (1988). Sulphate reduction rates and some aspects of the limnology of four lakes and a fjord in the Vestfold Hills, Antarctica. *Hydrobiologia* **165**, 25–33.

Frey, K.E. and Smith, L.C. (2005). Amplified carbon release from vast West Siberian peatlands by 2100. *Geophysical Research Letters* **32**, L09401, doi:10.1029/2004GL022025.

Gooseff, M.N., McKnight, D.M., Lyons, W.B., and Blum, A.E. (2002). Weathering reactions and hyporheic exchange controls on stream water chemistry in a glacial meltwater stream in the McMurdo Dry Valleys. *Water Resources Research* **38(12)**, 15:1–15:17.

Gordeev, V.V., Rachold, V., and Vlasova, I.E. (2004). Geochemical behaviour of major and trace elements in suspended particulate material of the Irtysh river, the main tributary of the Ob River, Siberia. *Applied Geochemistry* **19**, 593–610.

Green, W.J., Angle, M.P., and Chave, K.E. (1988). The geochemistry of Antarctic streams and their role in the evolution of four lakes of the McMurdo Dry Valleys. *Geochimica Cosmochimica Acta* **52**, 1265–1274.

Green, W.J. *et al.* (2005). Geochemical processes in the Onyx River, Wright Valley, Antarctica: major ions, nutrients, trace metals. *Geochimica Cosmochimica Acta* **69(4)**, 839–850.

Gregory-Eaves, I. *et al.* (2000). Characteristics and variation in lakes along a north-south transect in Alaska. *Archiv für Hydrobiologie* **147**, 193–223.

Grémillet, D. and LeMaho, Y. (2003). Arctic and Antarctic ecosystems: poles apart? In Huiskes, A.H.L. *et al.* (eds), *Antarctic Biology in a Global Context*, pp. 169–175. Backhuys Publishers, Leiden.

Healy, M., Webster-Brown, J.G., Brown, K.L., and Lane, V. (2006). Chemistry and stratification of Antarctic meltwater ponds II: Inland ponds in the McMurdo Dry Valleys, Victoria Land. *Antarctic Science* **18**, 525–534.

Henshaw, T. and Laybourn-Parry, J. (2002). The annual patterns of photosynthesis in two large, freshwater, ultra-oligotrophic Antarctic lakes. *Polar Biology* **25**, 744–752.

Hinzman, L.D. *et al.* (2005). Evidence and implications of recent climate change in northern Alaska and other Arctic regions. *Climatic Change* **72**, 251–298.

Hobbie, J.E. *et al.* (1999). Impact of global change on biogeochemistry and ecosystems of an Arctic freshwater system. *Polar Research* **18**, 207–214.

Hobbie, S.E., Miley, T.A., and Weiss, M.S. (2002). Carbon and nitrogen cycling in soils from acidic and nonacidic tundra with different glacial histories in Northern Alaska. *Ecosystems* **5**, 761–774.

Hodgson, D.A., Doran, P.T., Roberts, D., and McMinn, A. (2004). Paleolimnological studies from the Antarctic and subantarctic islands. In Pientiz, R., Douglas, M.S.V., and Smol, J.P. (eds), *Long-term Environmental Change in Arctic and Antarctic Lakes*, pp. 419–474. Developments in Paleoenvironmental Research, vol. 8. Springer, Dordrecht.

Hodson, A. (2006). Biogeochemistry of snowmelt in an Antarctic glacial ecosystem. *Water Resources Research* **42**, W11406, doi:10.1029/2005WR004311.

Holmes, R.M. *et al.* (2000). Flux of nutrients from Russian rivers to the Arctic Ocean: can we establish a baseline against which to judge future changes? *Water Resoures Research* **36**, 2309–2320.

Howard-Williams, C., Vincent, C.L., Broady, P., and Vincent, W.F. (1986). Antarctic stream ecosystems: variability in environmental properties and algal community structure. *Internationale Revue der gesamten Hydrobiolobiologie* **71**, 511–544.

Jones, J.B. *et al.* (2005). Nitrogen loss from watersheds of interior Alaska underlain with discontinuous permafrost. *Geophysical Research Letters* **32**, L02401, doi:10.1029/2004GL021734.

Jonsson, A., Karlsson, J., and Jansson, M. (2003). Sources of carbon dioxide supersaturation in clearwater and humic lakes in northern Sweden. *Ecosystems* **6**, 224–235.

Kling, G.W., Kipphut, G.W., and Miller, M.C. (1991). Arctic lakes and streams as gas conduits to the atmosphere: implications for tundra carbon budgets. *Science* **251**, 298–301.

Kling, G.W., Kipphut, G.W., and Miller, M.C. (1992a). The flux of CO_2 and CH_4 from lakes and rivers in Arctic Alaska. *Hydrobiologia* **240**, 23–36.

Kling, G.W. *et al.* (1992b). The biogeochemistry and zoogeography of lakes and rivers in Arctic Alaska. *Hydrobiologia* **240**, 1–14.

Kling, G.W. *et al.* (2000). Integration of lakes and streams in a landscape perspective: the importance of material processing on spatial patterns and temporal coherence. *Freshwater Biology* **43**, 477–497.

Kokelj, S.V., Jenkins, R.E., Milburn, D., Burn, C.R., and Snow, N. (2005). The influence of thermokarst disturbance on the water quality of small upland lakes, Mackenzie Delta Region, Northwest Territories, Canada. *Permafrost and Periglacial Processes* **16**, 343–353.

Kortelainen, P. and Saukkonen, S. (1998). Leaching of nutrients, organic carbon and iron from Finnish forestry land. *Water, Air, and Soil Pollution* **105**, 239–250.

Laperriere, J.D., Jones, J.R., and Swanson, D.K. (2003). Limnology of Lakes in Gates of the Arctic National Park and Preserve, Alaska. *Lake and Reservoir Management* **19**, 108–121.

Laudon, H., Kohler, S., and Buffam, I. (2004). Seasonal TOC export from seven boreal catchments in northern Sweden. *Aquatic Sciences* **66**, 223–230.

Lawrence, D.M. and Slater, A.G. (2005). A projection of severe near-surface permafrost degradation during the 21st century. *Geophysical Research Letters* **32**, L24401, doi:10.1029/2005GL025080.

Lawson, J., Doran, P.T., Kenig, F., Des Marais, D.J., and Priscu, J.C. (2004). Stable carbon and nitrogen isotopic composition of benthic and pelagic organic matter in lakes of the McMurdo Dry Valleys, Antarctica. *Aquatic Geochemistry* **10**, 269–301.

Laybourn-Parry, J. (2003). Polar limnology, the past, the present and the future. In Huiskes, A.H.L. *et al.* (eds), *Antarctic Biology in a Global Context*, pp. 329–375. Backhuys Publishers, Leiden.

Laybourn-Parry, J., Quayle, W., and Henshaw, T. (2002). The biology and evolution of Antarctic saline lakes in relation to salinity and trophy. *Polar Biology* **25**, 542–552.

Laybourn-Parry, J., Madan, N.J., Marshall, W.A., Marchant, H.J., and Wright, S.W. (2006). Carbon dynamics in an ultra-oligotrophic epishelf lake (Beaver Lake, Antarctica) in summer. *Freshwater Biology* **51**, 1116–1130.

Lee, P.A. *et al.* (2004). Thermodynamic constraints on microbially mediated processes in lakes of the McMurdo Dry Valleys, Antarctica. *Geomicrobiology Journal* **21**, 221–237.

Levine, M.A. and Whalen, S.C. (2001). Nutrient limitation of phytoplankton production in Alaskan Arctic foothill lakes. *Hydrobiologia* **455**, 189–201.

Lim, D.S.S. and Douglas, M.S.V. (2003). Limnological characteristics of 22 lakes and ponds in the Haughton Crater region of Devon Island, Nunavut, Canadian High Arctic. *Arctic, Antarctic and Alpine Research* **35**, 509–519.

Lim, D.S.S., Douglas, M.S.V., and Smol, J.P. (2005). Limnology of 46 lakes and ponds on Banks Island, NWT, Canadian Arctic Archipelago. *Hydrobiologia* **545**, 11–32.

Lim, D.S.S. *et al.* (2001). Physical and chemical limnological characteristics of 38 lakes and ponds on Bathurst Island, Nunavut, Canadian High Arctic. *International Review of Hydrobiology* **86**, 1–22.

Lyons, W.B., Laybourn-Parry, J., and Welch, K.A. (2006). Antarctic lake systems and climate change. In Bergstrom, D., Convey, P., and Huiskes, A. (eds), *Trends in Antarctic Terrestrial and Limnetic Ecosystems: Antarctica as a Global Indicator* , pp. 273–295. Kluwer Academic Publishers, Dordrecht.

Lyons, W.B., Welch, K.A., and Doggett, K. (2007). Organic carbon in Antarctic snow. *Geophysical Research Letters* **34**, L02501, doi:10.1029/2006GL028150.

Lyons, W.B. *et al.* (1998). Geochemical linkages among glaciers, streams, and lakes within the Taylor Valley, Antarctica. In Priscu, J.C. (ed.), *Ecosystem Dynamics in a Polar Desert: The McMurdo Dry Valleys Antarctica*. Antarctic Research Series, vol. 72. American Geophysical Union, Washington DC.

Lyons, W.B. *et al.* (2001). The McMurdo Dry Valleys Long-Term Ecological Research Program: new understanding of the biogeochemistry of the Dry Valley lakes: a review. *Polar Geography* **25(3)**, 202–217.

Matsumoto, G.I. (1993). Geochemical features of the McMurdo Dry Valley Lakes, Antarctica. In Green, W.J. and Friedmann, E.I. (eds), *Physical and Biogeochemical Processes in Antarctica Lakes*, pp. 95–118. Antarctic Research Series, vol. 59, American Geophysical Union, Washington DC.

McClelland, J.W. *et al.* (2006). A pan-Arctic evaluation of changes in river discharge during the latter half of the

20th century. *Geophysical Research Letters* **33**, L06715, doi:10.1029/2006GL025753.

McKnight, D.M., Aiken, G.R., Andrews, E.D., Bowles, E.C., and Harnish, R.A. (1993). Dissolved organic material in Dry Valley Lakes: a comparison of Lake Fryxell, Lake Hoare, and Lake Vanda. In Green, W.J. and Friedmann, E.I. (eds), *Physical and Biogeochemical Processes in Antarctica Lakes*, pp. 119–134. Antarctic Research Series, vol. 59. American Geophysical Union, Washington DC.

McKnight, D.M., Runkel, R.L., Tate, C.M., Duff, J.H., and Moorhead, D.L. (2004). Inorganic N and P dynamics of Antarctic glacial meltwater streams as controlled by hyporheic exchange and benthic autotrophic communities. *Journal of the North American Benthological Society* **23(2)**, 171–188.

McKnight, D.M. *et al.* (1999). Dry Valley Streams in Antarctica: ecosystems waiting for water. *Bioscience* **49(12)**, 985–995.

Michaelson, G.J., Ping, C.L., Kling, G.W., and Hobbie, J.E. (1998), The character and bioactivity of dissolved organic matter at thaw and in the spring runoff waters of the arctic tundra north slope, Alaska. *Journal of Geophysical Research* **103**, 28939–28946.

Michelutti, N., Douglas, M.S.V., Lean, D.R.S., and Smol, J.P. (2002a). Physical and chemical limnology of 34 ultra-oligotrophic lakes and ponds near Wynniatt Bay, Victoria Island, Arctic Canada. *Hydrobiologia* **482**, 1–13.

Michelutti, N., Douglas, M.S.V., Muir, D.C.G., Wang, X., and Smol, J.P. (2002b). Limnological characteristics of 38 lakes and ponds on Axel Heiberg Island, High Arctic Canada. *International Review of Hydrobiology* **87**, 385–399.

Neff, J.C. and Hooper, D.U. (2002). Vegetation and climate controls on potential CO_2, DOC and DON production in northern latitude soils. *Global Change Biology* **8**, 872–884.

Neff, J.C. *et al.* (2006). Seasonal changes in the age and structure of dissolved organic carbon in Siberian rivers and streams. *Geophysical Research Letters* **33**, L23401, doi:10.1029/2006GL028222.

Neumann, K., Lyons, W.B., Priscu, J.C., and Donahoe, R.J. (2001). CO_2 concentrations in perennially ice-covered lakes of Taylor Valley, Antarctica. *Biogeochemistry* **56**, 27–50.

Neumann, K., Lyons, W.B., Priscu, J.C., Desmarais, D.J., and Welch, K.A. (2004). The carbon isotopic composition of dissolved inorganic carbon in perennially ice-covered Antarctic lakes: Searching for a biogenic signature. *Annals of Glaciology* **39**, 518–524.

O'Brien, W.J. *et al.* (2005). Long-term response and recovery to nutrient addition of a partitioned Arctic lake. *Freshwater Biology* **50**, 731–741.

O'Donnell, J.A. and Jones, J.B. (2006). Nitrogen retention in the riparian zone of catchments underlain by discontinuous permafrost. *Freshwater Biology* **51**, 854–864.

Petrone, K.C., Jones, J.B., Hinzman, L.D., and Boone, R.D. (2006). Seasonal export of carbon, nitrogen, and major solutes from Alaskan catchments with discontinuous permafrost. *Journal of Geophysical Research-Biogeosciences* **111**, G02020, doi:10.1029/2005JG000055.

Phelps, A.R., Peterson, K.M., and Jeffries, M.O. (1998). Methane efflux from high-latitude lakes during spring ice melt. *Journal of Geophysical Research* **103**, 29029–29036.

Pienitz, R., Smol, J.P., and Lean, D.R.S. (1997). Physical and chemical limnology of 59 lakes located between the southern Yukon and the Tuktoyaktuk Peninsula, Northwest Territories (Canada). *Canadian Journal of Fisheries and Aquatic Sciences* **54**, 330–346.

Pienitz, R., Douglas, M.S.V., and Smol, J.P. (2004). Paleolimnological research in polar regions: An introduction. In Pienitz, R., Douglas, M.S.V., and Smol, J.P. (eds), *Environmental Change in Arctic and Antarctic Lakes*, pp. 1–17. Developments in Paleoenvironmental Research, vol. 8. Springer, Dordrecht.

Priscu, J.C. (1995). Phytoplankton nutrient deficiency in lakes of the McMurdo Dry Valleys, Antarctica. *Freshwater Biology* **34**, 215–227.

Priscu, J.C. (1997). The biogeochemistry of nitrous oxide in permanently ice-covered lakes of the McMurdo Dry Valleys, Antarctica. *Global Change Biology* **3**, 301–315.

Priscu, J.C. *et al.* (1999). Carbon transformations in a perennially ice-covered Antarctic lake. *Bioscience* **49(12)**, 997–1008.

Reeder, S.W., Hitchon, B., and Levinson, A.A. (1972). Hydrochemistry of the surface waters of the McKenzie River drainage basin, Canada- I. Factors controlling inorganic composition. *Geochimica Cosmochimica Acta* **36**, 825–865.

Rouse, W.R. *et al.* (1997). Effects of climate change on the fresh waters of Arctic and subarctic North America. *Hydrological Processes* **11**, 873–902.

Ruhland, K. and Smol, J.P. (1998). Limnological characteristics of 70 lakes spanning Arctic treeline from coronation gulf to Great Slave Lake in the Central Northwest Territories, Canada. *International Review of Hydrobiology* **83**, 183–203.

Sattley, W.M. and Madigan, M.T. (2006). Isolation, characterization, and ecology of cold-active chemolithotrophic, sulphur-oxidizing bacteria from perennially ice-covered Lake Fryxell, Antarctica. *Applied and Environmental Microbiology* **72(8)**, 5562–5568.

Schindler, D.W., Welch, H.E., Kalff, J., Brunskill, G.J., and Kritsch, N. (1974a). Physical and chemical limnology of

Char Lake, Cornwallis Island (75°N lat). *Journal of the Fisheries Research Board of Canada* **31**, 585–607.

Siegert, M.J. (2005). Lakes beneath the ice sheet: the occurrence, analysis, and future exploration of Lake Vostok and other Antarctic subglacial lakes. *Annual Reviews of Earth Planet Sciences* **33**, 215–245.

Slavik, K. *et al.* (2004). Long-term responses of the Kuparuk River ecosystem to phosphorus fertilization. *Ecology* **85**, 939–954.

Smedberg, E., Mörth, C.-M., Swaney, D.P., and Humborg, C. (2006). Modeling hydrology and silicon-carbon interactions in taiga and tundra biomes from a landscape perspective: implications for global warming feedbacks. *Global Biogeochemical Cycles* **20**, GB2014, doi: 10.1029/2005GB002567.

Smith, R.L., Miller, L.G., and Howes, B.L. (1993). The geochemistry of methane in Lake Fryxell, an amictic, permanently ice-covered, Antarctic lake. *Biogeochemistry* **21**, 95–115.

Stepanaukas, R., Laudon, H., and Jorgensen, N.O.G. (2000). High DON bioavailabilty in boreal streams during a spring flood. *Limnology and Oceanography* **45**, 1298–1307.

Striegl, R.G., Aiken, G.R., Dornblaser, M.M., Raymond, P.A., and Wickland, K.P. (2005). A decrease in discharge-normalized DOC export by the Yukon River during summer through autumn. *Geophysical Research Letters* **32**, L21413, doi:10.1029/2005GL024413.

Takacs, C.D., Priscu, J.C., and McKnight, D.M. (2001). Bacterial dissolved organic carbon demand in McMurdo Dry Valley lakes, Antarctica. *Limnology and Oceanography* **46(5)**, 1189–1194.

Takahashi, H.A., Wada, H., Nakamura, T., and Miura, H. (1999). ^{14}C anomaly of freshwater algae in Antarctic coastal ponds and lakes. *Polar Geoscience* **12**, 248–257.

Van Hove, P.C., Belzile, C., Gibson, J.A.E., and Vincent, W.F. (2006). Coupled landscape-lake evolution in High Arctic Canada. *Canadian Journal of Earth Sciences* **43**, 533–546.

Vincent, W.F., Downes, M.T., and Vincent, C.L. (1981). Nitrous oxide cycling in Lake Vanda, Antarctica. *Nature* **292**, 618–620.

Walter, K.M., Zimov, S.A., Chanton, J.P., Verbyla, D., and Chapin, F.S. (2006). Methane bubbling from Siberian thaw lakes as a positive feedback to climate warming. *Nature* **443**, 71–75.

Wand, U., Samarkin, V.A., Nitzsche, H.-M., and Hubberten, H.-W. (2006). Biogeochemistry of methane in the permanently ice-covered Lake Untersee, central Dronning Maud Land, East Antarctica. *Limnology and Oceanography* **51(2)**, 1180–1194.

Ward, B.B., Granger, J., Maldonado, M.T., and Wells, M.L. (2003). What limits bacterial production in the suboxic region of permanently ice-covered Lake Bonney, Antarctica? *Aquatic Microbial Ecology* **31**, 33–47.

Whalen, S.C. and Cornwell, J.C. (1985). Nitrogen, phosphorus and organic carbon cycling in an Arctic lake. *Canadian Journal of Fisheries and Aquatic Sciences* **42**, 797–808.

Wilson, A.T. (1981). A review of the geochemistry and lake physics of the Antarctic Dry Valleys. In McGinnis, L.D. (ed.), *Dry Valley Drilling Project*, pp. 185–192. Antarctic Research Series, vol. 33. American Geophysical Union, Washington DC.

Wollheim, W.M. *et al.* (2001). Influence of stream size on ammonium and suspended particulate nitrogen processing. *Limnology and Oceanography* **46**, 1–13.

Zimov, S.A. *et al.* (2006). Permafrost carbon: stock and decomposability of a globally significant carbon pool. *Geophysical Research Letters* **33**, L20502, doi:10.1029/2006GL027484.

CHAPTER 9

Phytoplankton and primary production

Michael P. Lizotte

Outline

Phytoplankton are microscopic organisms capable of contributing to primary production. These photosynthetic microbes include bacteria (oxygenic cyanobacteria and anoxygenic photosynthetic bacteria), many forms of eukaryotic algae, and other chloroplast-containing protists (e.g. ciliates). Biomass accumulation in polar lakes often occurs in layers, including deep chlorophyll maxima, especially beneath ice cover. The majority of polar lakes studied would be classified as ultra-oligotrophic based on maximum chlorophyll concentrations. However, few lakes have been studied seasonally to determine the annual peak in biomass, thus trophic status is probably underestimated; intensively studied lakes are mostly oligotrophic. Eutrophication has been noted in polar lakes, usually associated with colonies of marine birds and mammals, but also due to human activities. The seasonal progression of primary production is initiated and ended by the large seasonal changes in solar radiation. A springtime peak is evident in most polar lakes, with subsequent enhancement or restraint by changes in light (e.g. ice and snow cover; water clarity), nutrient availability (e.g. depletion by primary producers, influx from streams), or losses (e.g. grazing, disease, washout, sedimentation). Comparisons between Arctic and Antarctic lakes imply that there may be differences in biodiversity, lake trophic status, and primary production that can improve understanding of the how polar lake phytoplankton are influenced by climate, nutrient supply, biotic interactions, and their own capacity to acclimate to their environment.

9.1 Introduction

The phytoplankton are a critical foundation for food webs in polar lake ecosystems, and provide scientists with many examples of how microbial life survives and flourishes in extreme environments. The aim of this chapter is to review current knowledge about the diversity of microbes contributing to the primary production in the water column of polar lakes, the range of biomass accumulation and productivity in polar lakes, and the environmental factors that limit primary production. The number and diversity of studies has grown considerably in the past few decades, which allows for a substantial comparison between Arctic and Antarctic systems in each of these fundamental topics.

9.2 Photosynthetic plankton

The planktonic organisms associated with photosynthetic production in polar lakes include bacteria, eukaryotic algae, and ciliated protists (Table 9.1). Almost all taxonomic groups of eukaryotic algae and cyanobacteria have been recorded in both the Arctic and Antarctic. There have been fewer observations of anoxygenic photosynthetic bacteria and plastid-sequestering and symbiotic ciliates. Genomic data and taxonomic revisions have led to recognition of more major taxonomic groups. Thus

Table 9.1 Photosynthetic taxa observed in the plankton of polar lakes. The Prasinophyceae, Chrysophytes, Cryptophytes, Dinoflagellates, Euglenoids, and Haptophytes are classified here as Algae, but elsewhere in this volume are classified as Phytomastigophora or the phytoflagellates within the sub-Kingdom Protozoa in the Kingdom Protista; for a recent classification of all microbial eukaryotes see Adl *et al.* (2005).

Taxonomic group		Common genera, Arctic	Common genera, Antarctic
Bacteria			
Cyanobacteria	Chroococcales	*Aphanocapsa, Aphanothece, Chroococcus, Gloeocapsa, Microcystis, Synechococcus, Woronichinia*	*Aphanocapsa, Chroococcus, Gloeocapsa, Synechococcus, Synechocystis*
	Nostocales	*Anabaena*	*Anabaena*
	Oscillatoriales	*Oscillatoria, Phormidium, Planktothrix*	*Oscillatoria, Phormidium*
Green sulphur	Chlorobia	*Chlorobium, Pelodictyon, Prosthecochloris*	*Chlorobium*
Purple sulphur	Gammaproteobacteria	*Thiocapsa, Thiocystis*	*Chromatium, Thiocapsa*
Purple non-sulphur	Alphaproteobacteria		*Rhodospirillum, Rhodopseudomonas*
	Betaproteobacteria		*Rhodoferax*
Algae			
Chlorophytes	Chlorophyceae	*Ankyra, Botryococcus, Chlamydomonas, Oedogonium, Oocystis, Pediastrum, Scenedesmus*	*Ankyra, Chlamydomonas, Chloromonas, Dunaliella*
	Klebsormidiophyceae	*Stichococcus*	*Raphidonema, Stichococcus*
	Pedinophyceae	*Pedinomonas*	*Pedinomonas*
	Prasinophyceae	*Monomastix*	*Pyramimonas*
	Trebouxiophyceae	*Ankistrodesmus, Chlorella*	*Ankistrodesmus, Chlorella*
	Ulvophyceae	*Binuclearia*	*Gloetila*
	Zygnematophyceae	*Closterium, Cosmarium, Euastrum, Mougeotia, Staurastrum*	*Closterium, Cosmarium, Cylindrocystis, Mougeotia*
Chrysophytes	Chrysophyceae	*Chrysococcus, Chromulina, Dinobryon, Ochromonas, Pseudokephyrion, Uroglena*	*Chromulina, Ochromonas, Pseudokephyrion*
	Dictyophyceae	*Pseudopedinella*	*Pseudopedinella*
	Eustigmatophyceae	*Chlorobotrys*	
	Synurophyceae	*Mallomonas, Synura*	*Mallomonas*
	Xanthophyceae	*Tribonema*	*Tribonema*
Diatoms	Bacillariophyceae	*Navicula, Nitzschia, Pinnularia*	*Achnanthes, Navicula, Nitzschia, Pinnularia*
	Coscinodiscophyceae	*Aulacoseira, Cyclotella, Rhizosolenia, Stephanodiscus*	*Aulacoseira, Cyclotella, Stephanodiscus*
	Fragilariophyceae	*Asterionella, Diatoma, Fragilaria, Synedra, Tabellaria*	*Asterionella, Fragilaria, Tabularia*
Cryptophytes	Cryptophyceae	*Chroomonas, Cryptomonas, Rhodomonas*	*Chroomonas, Cryptomonas, Rhodomonas*
Dinoflagellates	Dinophyceae	*Amphidinium, Ceratium, Gymnodinium, Gyrodinium, Peridinium*	*Gonyaulax, Gymnodinium, Gyrodinium, Peridinium*
Euglenoids	Euglenophyceae	*Euglena, Phacus, Trachelomonas*	*Euglena*
Haptophytes	Prymnesiophyceae	*Chrysochromulina*	*Chrysochromulina*
Protozoa: Ciliates			
	Litostomatea	*Mesodinium*	*Mesodinium*
	Prostomatea		*Plagiocampa*
	Spirotrichea		*Strombidium, Euplotes*
	incertae sedis	*Limnostrombidium*	

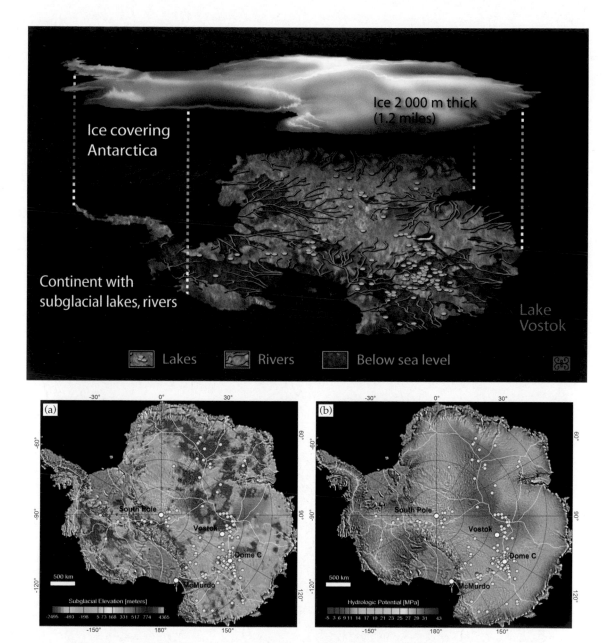

Plate 1 Subglacial lakes and rivers. Top: an artist's depiction of the lakes, rivers, and watersheds that lie beneath the thick Antarctic ice sheet. Credit: National Science Foundation (USA). Bottom: distribution of Antarctic subglacial lakes in relation to the ice divides (light lines in both panels), subglacial elevation (a) and hydraulic fluid potential (b). The base map is a MODIS image mosaic. The water should flow from lakes with high fluid potential to lakes with lower potential, assuming that there is a connection. The white circles indicate the locations of major field stations. Credit: J.C. Priscu.

Plate 2 Supraglacial lakes in the Canadian High Arctic. Top: elongate meltwater lakes on the Ward Hunt Ice Shelf. Bottom: microbial mats on the Markham Ice Shelf. Photographs: D. Sarrazin and W.F. Vincent.

Plate 3 Supraglacial lakes and biota on the McMurdo Ice Shelf, Antarctica. Top: surface of the ice. Middle: ponds on the undulating ice, with a surface layer of moraine, salt, and microbial mats. Bottom: epifluorescence micrograph of mat cyanobacteria (left) and the bdelloid rotifer *Philodina gregaria* that feeds on and accumulates carotenoid pigments from the mats (right; drawing from Murray 1910). Photographs: W.F. Vincent.

Plate 4 Perennially ice-covered lakes in polar desert catchments. Top: Ward Hunt Lake in the Canadian High Arctic. Photograph: D. Antoniades. Bottom: the west lobe of Lake Bonney in the McMurdo Dry Valleys, Antarctica. Photograph: H. Basagic.

Plate 5 Polar desert lakes. Top: Lake Hazen, a large deep lake in the Canadian High Arctic. Bottom: Ace Lake, a meromictic lake in the Vestfold Hills, Antarctica. Photographs: W.F. Vincent and I. Bayly.

Plate 6 Flowing-water ecosystems. Top: the glacier-fed Adams Stream in the McMurdo Dry Valleys, Antarctica. Bottom: the Kuparuk River, Alaska. Photographs: W.F. Vincent.

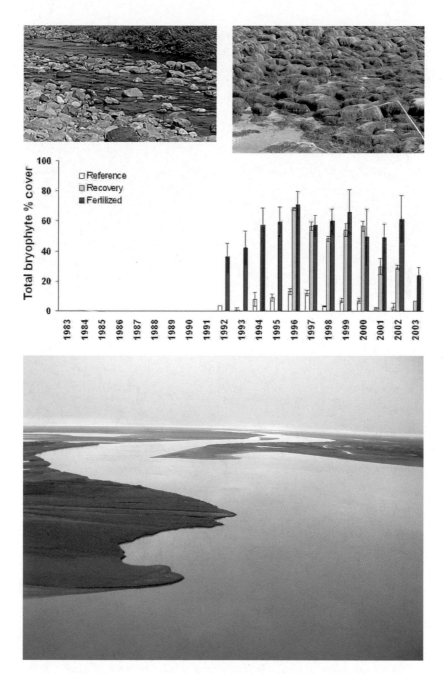

Plate 7 Flowing-water ecosystems. Top: stimulation of aquatic mosses (bryophytes) by long-term nutrient enrichment in the Kuparuk River. Bottom: one of the main channels of the Mackenzie River, Canada. Photographs: B.J. Peterson and W.F. Vincent; see Chapter 5.

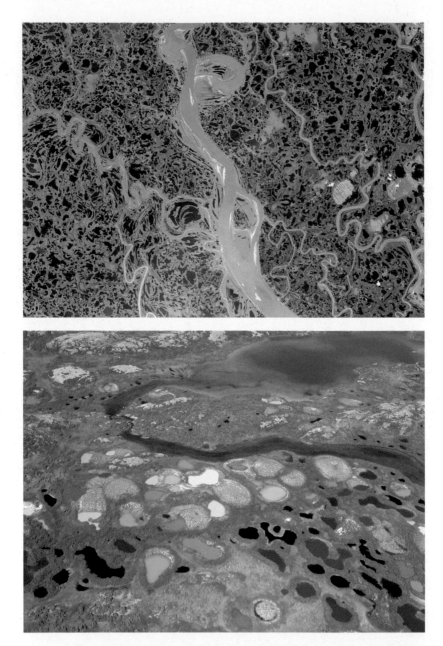

Plate 8 Arctic lakes and ponds. Top: satellite image of the Mackenzie River and its floodplain lakes. Photograph: NASA GSFC/METI/ERSDAC/ JAROS and the U.S./Japan ASTER Science Team. Bottom: multicolored, permafrost thaw lakes in northern Québec. Photograph: I. Laurion.

Plate 9 Lakes in the low Arctic, forest-tundra zone, Québec. Top: permafrost thaw lakes at the edge of the treeline. Photograph: I. Laurion. Bottom: the eastern basin of Lac à l'Eau Claire (Clearwater Lake), still ice-covered in mid-summer. Photograph: W.F. Vincent.

Plate 10 Benthic cyanobacterial communities. (a) Oscillatorian mat in a meltwater pond on the Koettlitz Glacier, Antarctica. (b) *Nostoc commune*, from a High Arctic pond, Bylot Island. (c) Cyanobacterial mat in Lake Vanda, Antarctica. (d) Benthic mat from a pond on the McMurdo Ice Shelf, Antarctica. (e) Ultrastructure of a cyanobacterial mat. (f) Migration of oscillatorian cyanobacteria to the surface of a pond mat in an ultraviolet-screening experiment. (g) Cyanobacterium-coated moss pillar in Syowa Lake, East Antarctica. Photographs: C. Howard-Williams (a); I. Laurion (b); I. Hawes (c); A. Quesada (d, f); A. De los Ríos and C. Ascaso (e); S. Imura (g).

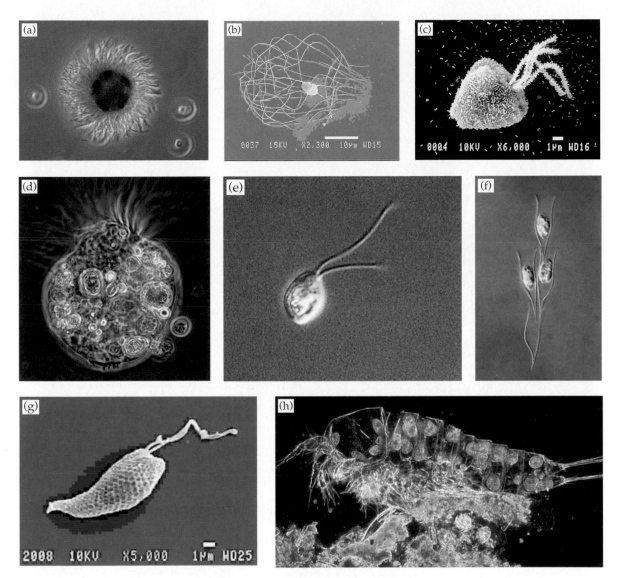

Plate 11 Protists in Arctic and Antarctic lakes. (a) *Mesodinum rubrum* from Ace Lake, stained with Lugol's iodine. (b) *Diaphanoeca grandis*, a choanoflagellate from Highway Lake, Vestfold Hills. (c) *Pyramimonas gelidicola*, a common phytoflagellate in the Vestfold Hills saline lakes. (d) An oligotrich ciliate containing sequestered plastids. (e) *Katablepharis*, a heterotrophic flagellate from Char Lake. (f) *Dinobryon*, a colonial mixotrophic phytoflagellate from Char Lake. (g) A cryptophyte, a common group of phytoflagellates. (h) *Tetrahymena* (hymenostome ciliates) inside the exoskeleton of a dead copepod. Photographs: J. Laybourn-Parry, G. Nash, P. Patterson, and W.F. Vincent.

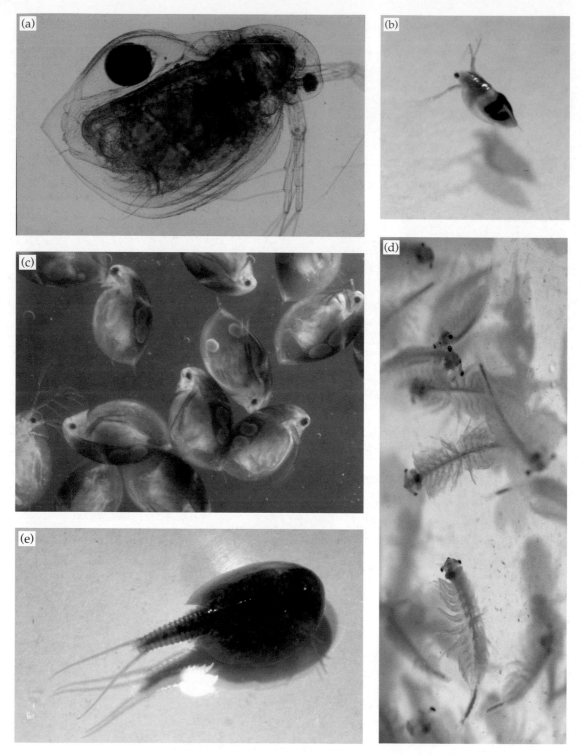

Plate 12 Pigmented zooplankton. (a) *Daphniopsis studeri*, a cladoceran from East Antarctica. (b) Melanized *Daphnia middendorfiana* carrying ephippium, from a lake at Resolute, Canadian High Arctic. (c) *Daphnia umbra* from a pond in Kilpisjärvi, Finnish Lapland. (d) *Artemiopsis stefanssoni* fairy shrimps from a pond in Resolute. (e) Tadpole shrimp *Lepidurus arcticus* from a pond near Lake Hazen, Ellesmere Island. Photographs: J. Laybourn-Parry (a); P. Van Hove (b, d, e); P. Junttila (c).

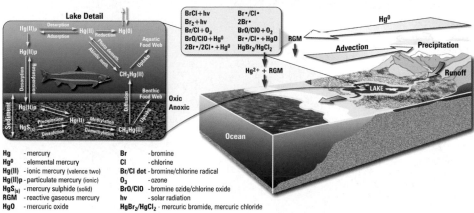

Hg	- mercury	Br	- bromine
Hg⁰	- elemental mercury	Cl	- chlorine
Hg(II)	- ionic mercury (valence two)	Br/Cl dot	- bromine/chlorine radical
Hg(II)p	- particulate mercury (ionic)	O₃	- ozone
HgS(s)	- mercury sulphide (solid)	BrO/ClO	- bromine ozide/chlorine oxide
RGM	- reactive gaseous mercury	hv	- solar radiation
HgO	- mercuric oxide	HgBr₂/HgCl₂	- mercuric bromide, mercuric chloride

Plate 13 Arctic fish and contaminant pathways. Top: large and small forms of Arctic char (*Salvelinus alpinus*) from Lake Hazen, Canadian High Arctic. Photograph: M. Power. Bottom: transport processes and reactions controlling mercury, which bioaccumulates through the food web to fish (from Wrona *et al.* 2006). Meth., methylation; Demeth., demethylation.

Plate 14 Top: Chemical and biological stratification in Ace Lake, a meromictic (permanently stratified) lake in the the Vestfold Hills, Antarctica. Bottom left: sediment sampling for paleolimnological studies of the lakes and fjords of Ellesmere Island, using a percussion sediment corer. Bottom right: a 1600-year-old lake sediment core from the Larsemann Hills, East Antarctica. Photographs: J. Laybourn-Parry, W.F. Vincent, and D. Hodgson.

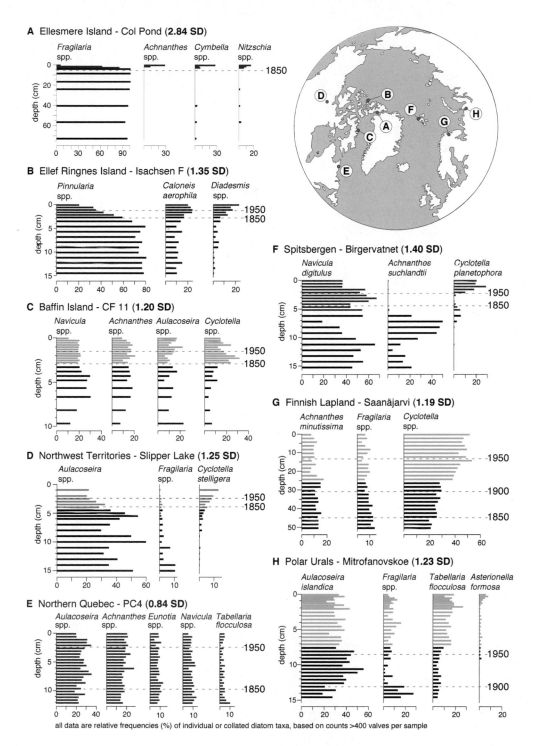

Plate 15 Paleolimnological profiles of recent (approximately the last 200 years) changes in the relative abundances of diatom assemblages from various regions of the circumpolar Arctic. The amount of assemblage change is further quantified by β diversity measurements (SD, standard deviation units). Many sites from the High Arctic (shown in red) recorded the greatest changes in assemblages, which were consistent with interpretation of recent warming. In regions with less warming, profiles are shown in orange and green. Meanwhile, areas such as Northern Québec and Labrador recorded little change (shown in blue). From Smol *et al.* (2005); see Chapter 3.

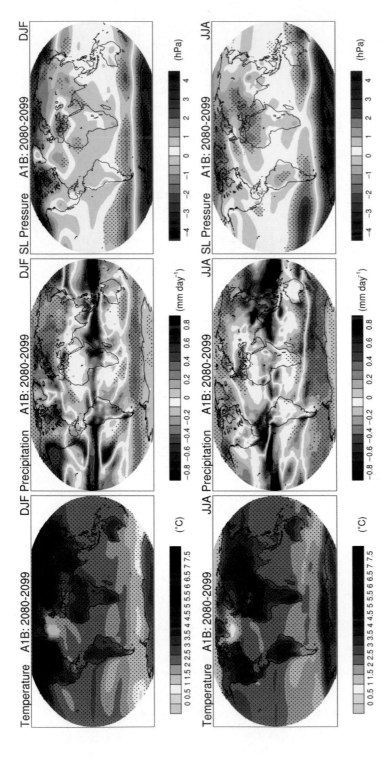

Plate 16 Predictions of major climate change ahead. Multimodel mean changes in surface air temperature (°C, left), precipitation (mm day^{-1}, middle) and sea-level pressure (hPa, right) for boreal winter (DJF, top) and summer (JJA, bottom). Changes are given for the period 2080–2099 relative to 1980–1999. Stippling denotes areas where the magnitude of the multimodel ensemble mean exceeds the intermodel standard deviation. Reproduced by permission of the Intergovernmental Panel on Climate Change. Meehl *et al.* (2007).

References

Meehl, G.A. *et al.* (2007). Global climate projections. In Solomon, S. *et al.* (eds), *Climate Change 2007: The Physical Science Basis. Contribution of Working Group I to the Fourth Assessment Report of the Intergovernmental Panel on Climate Change*, pp. 747–846. Cambridge University Press, Cambridge.

Murray, J. (1910) On collecting at Cape Royds. In Murray, J. (ed.), *British Antarctic Expedition 1907–9: Reports on the Scientific Investigations, Vol. I Biology*, pp. 1–15. William Heinemann, London.

Smol, J.P. *et al.* (2005). Climate-driven regime shifts in the biological communities of Arctic lakes. *Proceedings of the National Academy of Sciences USA* **102**, 4397–4402.

Wrona, F.J., Prowse, T.D., and Reist, J.D. (eds) (2006). Climate change impacts on Arctic freshwater ecosystems and fisheries. *Ambio* **35**, 325–415.

Table 9.1 is arranged by traditional common names (which have utility in understanding ecological relationships) and the class-level names in online taxonomic databases (e.g. www.catalogueoflife. org, http://sn2000.taxonomy.nl/Taxonomicon). The genera listed are taken directly from the literature cited (i.e. there has been no attempt to update for species re-designated into new or existing genera). However, the reader should be aware that elsewhere in this volume, phytoflagellates, here classified as Algae, are referred to as Protozoa, within the sub-phylum Phytomastigophora; for a recent classification of all eukaryotic microbes, see Adl *et al.* (2005). At the species and genus level, the validity of past taxonomic designations (e.g. use of temperate species names based on morphological similarity) and questions of endemism will continue to be assessed as genomic and advanced microscopic techniques provide new information (Vincent 2000a). The major groups of planktonic photosynthetic organisms will be discussed in terms of presence, absence (or not yet recorded), and which groups have been observed to dominate planktonic communities in polar lakes.

9.2.1 Photosynthetic bacteria

Cyanobacteria are the most commonly reported photosynthetic bacteria in lakes. Cyanobacteria contain chlorophyll *a* (Chl-*a*) as a major photosynthetic pigment, and conduct their most efficient photosynthesis using two photosystems, using water as an electron source and producing oxygen as a major waste product of photosynthesis. As photosynthetic producers, cyanobacteria usually operate much like the eukaryotic algae in the plankton; thus cyanobacteria and algae compete for resources and are part of the same food web. However, many species of cyanobacteria are also capable of operating anoxygenic photosynthesis, which is less efficient, running a cyclic pathway with only one photosystem and using different electron acceptors, such as hydrogen sulphide. Whereas anoxygenic photosynthesis may allow some cyanobacteria to colonize anaerobic environments to avoid aerobic competitors or grazers, it can place them in competition with the other types of photosynthetic bacteria.

Although rare in polar seas, cyanobacteria are common in high-latitude lakes (Vincent 2000b) and range from small single cells to large multicelled filaments and colonies. The order (called a subsection in the classification system used by bacteriologists) Nostocales includes mostly large colonies with specialized cells for carrying out nitrogen fixation, and includes *Anabaena* that is sometimes encountered in polar-lake plankton. Most of the cyanobacterial diversity and abundance is within the orders Chroococcales (coccoid cells, some in colonies) and Oscillatoriales (filamentous species without heterocysts). The latter includes *Oscillatoria* and *Phormidium*, which have been reported as dominant and bloom species in some polar lakes, particularly in deep chlorophyll layers. Cyanobacteria are often observed as the dominant component of benthic mats in lakes, shores and inflowing streams (see Chapter 10 in this volume), and the possibility exists that minor populations in the plankton have been advected to the lakes, or are from species that grow in both plankton and benthos (tychoplanktonic). Colonial planktonic forms include *Microcystis* in the Arctic and *Gloeocapsa* in the Antarctic. Single-celled genera often dominate the picophytoplankton, most notably *Synechococcus*, and are the most abundant photosynthetic cell types in many polar lakes (Vincent 2000b).

The other bacteria listed in Table 9.1 are generally referred to as anoxygenic ('not oxygen-producing') photosynthetic bacteria. They have traditionally been classified in terms of their major photosynthetic pigment and electron acceptors. 'Purple' bacteria contain either bacteriochlorophyll *a* or *b*, whereas the 'green' bacteria usually contain bacteriochlorophylls *c*, *d*, or *e*. So-called sulphur bacteria typically use reduced sulphur compounds (sulphide or elemental sulphur) as an electron source, whereas 'non-sulphur' bacteria typically use organic molecules or hydrogen (although some species can use sulphide at low concentrations). This matrix of pigments and electron sources defines four types of bacteria, but one, the green non-sulphur bacteria, has only been observed in benthic mats of polar lakes and not the plankton (Madigan 2003).

The green sulphur bacteria are obligate anaerobes; thus they are limited to environments high in reduced sulphur and lacking oxygen. The have been found in lakes with anaerobic hypolimnia with high sulphide concentrations (e.g. Burke and Burton 1988). *Chlorobium* is the most important genus, though a more diverse community of green sulphur bacteria is reported from the European Arctic (Lunina *et al.* 2005). Green sulphur bacteria form well-developed deep chlorophyll maxima in Arctic meromictic lakes and result in a high-turbidity scattering layer of sulphur particles (e.g. Van Hove *et al.* 2006).

Purple sulphur bacteria reported in polar lakes are members of the Gammaproteobacteria, and can tolerate some oxygen but will only conduct photosynthesis under anaerobic conditions. They have been reported from deeper lakes in Antarctica (e.g. Burke and Burton 1988; Madigan 2003). Recent studies of an Arctic lake dominated by purple sulphur bacteria (Lunina *et al.* 2005) showed that they can be responsible for a majority of the primary production (Ivanov *et al.* 2001). This group is not always found in lakes that would appear to have the expected growth conditions; for example, Lake Fryxell in Antarctica (Madigan 2003).

Purple non-sulphur bacteria include species from different bacterial classes, but have some common attributes. Photosynthesis occurs under anaerobic conditions, they can utilize organic compounds as electron sources for photosynthesis or directly for growth, and most are sensitive to high sulphide concentrations, and are thus limited to thin layers at the interface between sulphide-rich bottom waters and oxygenated surface waters. However, one of the Betaproteobacter species, a *Rhodoferax* strain cultured from Antarctic Lake Fryxell showed wide tolerance for sulphide and inhabited the entire sulphide-rich hypolimnion (Jung *et al.* 2004). The Alphaproteobacter group has been more widely observed in Antarctic lakes, including gas-vacuolate species that may dominate the primary production of the lake (Karr *et al.* 2003). Ellis-Evans (1996) suggests that anoxygenic photosynthetic bacteria might not be widespread because only certain lake types have good conditions for their growth (e.g. hypersaline hypolimnia), thus species of purple non-sulphur bacteria (with their trophic flexibility) might be the most widely distributed. No reports of this group in Arctic lakes have emerged.

9.2.2 Eukaryotic phytoplankton

Studies of algal and protistan diversity in polar lakes have been concentrated over the past half century, during which traditional groupings have been split. As a transition from the older literature and to help ecologists trying to proscribe certain functional groups (e.g. based on roles in biogeochemical cycles), the review that follows will discuss the traditional groupings such as chlorophytes, chrysophytes, cryptophytes, diatoms, and dinoflagellates. Only exceptional observations will be cited in this text, while the most common genera can be found repeatedly in taxonomic listings for Arctic lakes (e.g. Kalff 1967; Sheath and Hellebust 1978; Alexander *et al.* 1980; Wallen and Allen 1982; O'Brien *et al.* 1997; Vezina and Vincent 1997; Cremer and Wagner 2004) and Antarctic lakes (e.g. Light *et al.* 1981; Parker *et al.* 1982; Laybourn-Parry and Marchant 1992; Spaulding *et al.* 1994; Izaguirre *et al.* 1998; Unrein and Vinocur 1999; Allende and Izaguirre 2003). As with the cyanobacteria, there are questions about some species observed in the plankton that are usually associated with benthic habitats, particularly for diatoms and certain chlorophytes, prasinophytes, and xanthophytes.

The traditional group identified as chlorophytes or green algae that have been observed in polar lakes include members of several classes, with Chlorophyceae and Trebouxiophyceae having the most common species. Whereas all the classes have genera found at both poles, Arctic lakes appear to have much greater species diversity. The planktonic chlorophytes include common flagellates (e.g. *Chlamydomonas*), unicells, or small colonies (e.g. *Ankyra*, *Pediastrum*), and sometimes filamentous forms (e.g. *Oedogonium*). The trebouxiophytes are single or colonial cells that include the bloom-forming *Ankistrodesmus* and the commonly reported *Chlorella*. The klebsormidiophytes include bloom-forming *Stichococcus* and *Raphidonema*; the latter includes common algae from snow fields, thus it is possible that 'blooms' might be based on advected cells after snowmelt. Prasinophytes are mostly flagellates and include *Pyramimonas*, which is commonly observed in and sometimes dominates Antarctic lakes (e.g. Laybourn-Parry *et al.* 2002; the deep chlorophyll maximum of Lake Fryxell, Vincent 1981).

The planktonic Zygmatophyceae include a diverse group of desmids. The species diversity of desmids is much higher in the Arctic than the Antarctic (Coesel 1996); the few multispecies observations in the Antarctic are from the Antarctic Peninsula and South Atlantic islands (e.g. Izaguirre *et al.* 1998), suggesting that long-range migratory birds may be a vector for this group. The ulvophytes are a large class of filamentous and macro-algae, and smaller filamentous genera have made rare appearances in the plankton at both poles. The pedinophytes are a new group with few genera, represented by the flagellate *Pedinomonas* in polar lakes (e.g. Alexander *et al.* 1980; Light *et al.* 1981).

The chrysophytes, traditionally proscribed as yellow-brown algae exclusive of diatoms, include several major classes, but the Chrysophyceae retain the most important species. As with the green algae, the overall species diversity is much higher in the Arctic. The Chrysophyceae include common polar genera of the single-celled flagellates *Chromulina*, *Ochromonas*, and *Pseudokephyrion*. Arctic lakes frequently have colonial species such as *Uroglena* and *Dinobryon*. Dictyophytes are rare, but the genus *Pseudopedinella* has been observed at both poles (e.g. Sheath and Hellebust 1978; Butler *et al.* 2000). Eustigmatophytes have only been reported from the Arctic with the genus *Chlorobotrys* (Alexander *et al.* 1980). Synurophytes include the common flagellates *Synura* and *Mallomonas*; the species diversity is higher in the Arctic, whereas only *Mallomonas* has been reported, rarely, in the Antarctic (e.g. Spaulding *et al.* 1994; Butler *et al.* 2000). The xanthophytes are represented by the filamentous *Tribonema* (Alexander *et al.* 1980; Allende and Izaguirre 2003), which may have stream or benthic origins.

Cryptophytes have been reported repeatedly as dominant members of polar lake plankton, including the unicellular flagellates of *Chroomonas*, *Cryptomonas*, and *Rhodomonas*. They occur in coastal ponds as well as larger lakes, including meromictic systems (Vincent and Vincent 1982b).

Diatoms are unified as a functional group due to their high demand for silica which they use to make a cell wall. They are also of great interest because these cell walls preserve well as fossils and have great utility as paleolimnological indicators. All three diatom classes are diverse and well represented in polar lakes, although only a few genera have species that form blooms. The species diversity is much greater in the Arctic, as are diatom dominance or blooms. Bacillariophytes include the common genera of *Navicula*, *Nitzschia*, and *Pinnularia*; there are also reports of *Achnanthes* in small Antarctic lakes (e.g. Unrein and Vinocur 1999), but they occur by resuspension as this genus is usually attached. The coscinodiscophytes include the most common planktonic diatoms in polar lakes, including *Aulacoseira*, *Cyclotella*, and *Stephanodiscus*, although blooms have only been reported in the Arctic (e.g. Kalff *et al.* 1975; Cremer and Wagner 2004). The fragilariophytes include *Asterionella*, *Diatoma*, *Fragilaria*, and *Tabellaria*, with blooms reported in the Arctic.

Dinoflagellates, have been widely reported in polar regions with species of *Gymnodinium*, *Gyrodinium*, and *Peridinium* at both poles. In the Antarctic coastal lakes, the colonial *Gonyaulax* has been reported (Laybourn-Parry *et al.* 2002), possibly of marine origin. Dinoflagellate blooms, usually of *Gymnodinium* species, have been reported beneath the ice in early spring in the Arctic (e.g. Kalff *et al.* 1975; Sheath and Hellebust 1978) and Antarctica (Butler *et al.* 2000).

Other taxa of flagellated phytoplankton have been reported from the Euglenophyceae and Prymnesiophyceae, but not as dominant species. The euglenophytes are more common and diverse in the Arctic (e.g. Sheath and Hellebust 1978). Prymnesiophytes are rare in fresh waters, but *Chrysochromulina* has been reported from both poles (e.g. Light *et al.* 1981; O'Brien *et al.* 1997).

9.2.3 Ciliates

A contribution to photosynthesis can be made by ciliates that either harbor symbiotic algae or sequester captured algal chloroplasts, which are held for a substantial period (days to weeks) before digestion. These ciliates are considered mixotrophic, with ecological functions as both photosynthetic autotrophs and heterotrophic consumers. The observations may be more numerous and thorough in the Antarctic (Laybourn-Parry 2002), where a symbiotic species of *Mesodinium* and chloroplast-sequestering species of *Euplotes*, *Plagiocampa*, and *Strombidium* have been reported. All these ciliate genera have been observed in Arctic lakes, including *Mesodinium* in High Arctic Lake A (P. Van Hove *et al.*, unpublished results). For the Arctic, photosynthetic ciliates are *Limnostrombidium* (Laybourn-Parry and Marshall 2003).

9.3 Biomass

To follow changes in phytoplankton populations, various methods have been devised to estimate their biomass. Because biomass is usually dilute in natural waters, direct weighing of phytoplankton is not possible; it would require concentrating cells from many, perhaps hundreds of, liters of water, with the added difficulty of separating planktonic animals and other non-photosynthetic organisms that overlap in size. The most common approaches for measuring biomass have focused on counting organisms or measuring a common chemical constituent (photosynthetic pigments).

The main method for measuring biomass of phytoplankton has been through light microscopy. The phytoplankton cells are counted and the average biovolume of the cells is used to estimate the biovolume for each taxa. These biovolume estimates are often converted to carbon units based on carbon:biovolume ratios. The latter can be a limitation of the method, as carbon:biovolume ratios are subject to physiological variations, and can also differ greatly among taxa. The main advantage of the method is that one can derive population-level estimates (e.g. Sheath and Hellebust 1978; Alexander *et al.* 1980; Spaulding *et al.* 1994; Izaguirre *et al.* 2001) that help to understand competition among phytoplankton species, differential response to environmental conditions, grazing impacts, disease outbreaks, suspension of attached or benthic species, and many other important studies. The main drawback to the method is that it is labor-intensive. Some populations, particularly of small, single-celled organisms, can be studied using automated flow-cytometry systems that can sort some major phytoplankton groups based on pigment and fluorescent characteristics; however, this technique has seen limited use in polar lakes (e.g. Andreoli *et al.* 1992).

The most common method for estimating the biomass of phytoplankton is the measurement of chlorophyll *a* (Chl-*a*) concentrations. This method has several advantages over microscopy-based methods: it can be used as a measure of total phytoplankton biomass, it is easier and faster (allowing more numerous observations), and it can be based on the same cell concentration techniques (filtration) used in chemical and radiotracer studies. However, chlorophyll has limitations as a biomass estimate. Concentrations per cell are subject to acclimation of the photosynthetic system, nutrient availability, and other physiological conditions. Whereas Chl-*a* is an abundant pigment in cyanobacteria, eukaryotic algae, and photosynthetic ciliates, it will miss or severely underestimate bacteriochlorophyll-containing photosynthetic bacteria. Finally, results can vary among the different standard methods (trichromatic spectrophotometry, fluorometry, and high-performance liquid chromatography, or HPLC) due to different interference effects and pigment-extraction techniques. The most advanced methodology, albeit the most expensive and labor-intensive, is HPLC, which can also be used to derive information about other pigments that are specific to particular taxonomic groups. This chemotaxonomy approach can be used to derive biomass estimates (as a percentage of total Chl-*a* biomass) for several major groups (e.g. Lizotte and Priscu 1998; Bonilla *et al.* 2005).

As Chl-*a* biomass has been the most common method employed over the past 40 years, it provides the most useful measure for comparing lakes. Chl-*a* concentration has now been measured in hundreds of polar lakes in surveys and has been widely used in seasonal studies of phytoplankton dynamics. Chl-*a* concentrations in Arctic lakes range from below levels of detection (often less than $0.01\,\mathrm{mg\,m^{-3}}$, although this is subject to the volume filtered and the detector used) to over $100\,\mathrm{mg\,m^{-3}}$ (Table 9.2). Even higher concentrations have been reported in the Antarctic (Table 9.3). A data-set derived from the sources in these two tables was used to derive frequency distributions for polar lakes in each hemisphere. Only the highest value reported for a given lake was used, including single observations; if the study only reported a range for a number of lakes, only the upper value (representing a single

Table 9.2 Chlorophyll concentrations reported for plankton of Arctic lakes.

Region	Lake	Chlorophyll concentration (mg Chl-am^{-3})	Source
West N. America (110–170°W)	16 lakes	0.3–148	Kling *et al.* (1992)
	24 lakes	0–1.4	Pienitz *et al.* (1997)
	10 lakes	0.05–4.7	Kling *et al.* (2000)
	39 lakes	0.2–5.1	Levine and Whalen (2001)
	2 lakes	0.9–18.2	Whalen *et al.* (2006)
	N2	0.5–4.5	O'Brien *et al.* (2005)
	Oil Lake	0.1–3.0	Miller *et al.* (1978)
	Pond B	0.1–2.1	Alexander *et al.* (1980)
	Toolik	0.5–5.9	Whalen and Alexander (1986a, 1986b)
Canadian Shield and Archipelago (60–130°W)	32 lakes	0.52–6.8	Welch and Legault (1986)
	9 lakes	0.1–29.8	Vezina and Vincent (1997)
	7 lakes	0.04–1.24	Markager *et al.* (1999)
	6 lakes	0.22–1.02	Van Hove *et al.* (2006)
	Methane	0.75–3.98	Welch *et al.* (1980)
Greenland	74 lakes	0.2–5.4	Jeppesen *et al.* (2003)
Europe (0–70°E)	2 lakes	0.2–1.65	Laybourn-Parry and Marshall (2003)
	15 lakes	0.33–21	Ellis-Evans *et al.* (2001)
	Imandra	1–7	Moiseenko *et al.* (2001)

lake) was used in the frequency distribution. The results (Figure 9.1) showed that that most observations are in the lowest bins (<2 mg Chl-am^{-3}), but that one-quarter of the Antarctic data-set were in the higher ranges (>8 mg Chl-am^{-3}). For comparison, Padisak (2004) reviewed the classification of lakes into different trophic states based on the annual mean and maximum for Chl-a concentration (Table 9.4). This scale is useful as a guide to how nutrient-rich and biologically productive is a given lake ecosystem.

Based on regional surveys and seasonal studies of individual lakes, some general conclusions can be drawn. First, there have been very few observations of hypertrophic lakes in polar regions. In most cases, these systems have been associated with imports of animal wastes by natural causes, such as breeding colonies; for example, ponds and lakes near penguin colonies on Antarctica (Parker 1972; Spurr 1975; Izaguirre *et al.* 1998). In Pony Lake on Ross Island, Antarctica, the enrichment by an Adélie penguin colony has resulted in extreme phytoplankton cell densities, and a measured Chl-a concentration of 839 mg m^{-3} (Vincent and Vincent

1982b). Even eutrophic lakes are rare, but often associated with animal colonies (Ellis-Evans 1981b; Izaguirre *et al.* 2001; Laybourn-Parry *et al.* 2002; Toro *et al.* 2007) and, in the Arctic, with human sewage (Kalff and Welch 1974). Mesotrophic lakes have not been widely reported in the Arctic, though they are known in Alaska (e.g. O'Brien *et al.* 2005; Whalen *et al.* 2006), Svalbard (Ellis-Evans *et al.* 2001), and probably Russia (e.g. Moiseenko *et al.* 2001). A larger proportion of Antarctic lakes are in the mesotrophic category, including numerous lakes on the Antarctic Peninsula (e.g. Izaguirre *et al.* 1998; Unrein and Vinocur 1999; Allende and Izaguirre 2003), the Vestfold Hills (Heath 1988; Laybourn-Parry *et al.* 2002, 2005), Signy Island (Ellis-Evans 1981a), and in Victorialand (Andreoli *et al.* 1992; Roberts *et al.* 2004).

Current knowledge of polar lakes would place a majority (55–70% by Figure 9.1) in the ultra-oligotrophic category. Many large-scale surveys of Arctic lakes, based on single observations during the open water season, find averages below 1.5 mg Chl-am^{-3} (e.g. Levine and Whalen 2001; Jeppesen *et al.* 2003). However, ultra-oligotrophic

Table 9.3 Chlorophyll concentrations reported for plankton of Antarctic lakes.

Region	Lake	Chlorophyll concentration (mgChl-am^{-3})	Source
Peninsula and Scotia Arc (30–90°W)	Ponds	35–112	Parker (1972)
	9 lakes	0–550	Izaguirre et al. (1998)
	Boeckella	0–9.4	Allende and Izaguirre (2003)
	Chico	0–2.45	Allende and Izaguirre (2003)
	Heywood	0–40	Ellis-Evans (1981b); Hawes (1985)
	Kitiesh	0.74–1.67	Montecino et al. (1991)
	Moss	0.7–8	Ellis-Evans (1981a)
	Otero	0–202	Izaguirre et al. (2001)
	Sombre	0–9.5	Hawes (1985)
	Tranquil	0–4.2	Butler et al. (2000)
	Tres Hermanos	1.6–8.3	Unrein and Vinocur (1999)
Queen Maud Land (30°W–30°E)	3 lakes	0.05–0.8	Kaup (1994)
East Antarctica (30–90°E)	9 lakes	0.36–1.57	Tominaga (1977)
	12 lakes	0.1–2.69	Laybourn-Parry and Marchant (1992)
	Ace	0.7–5.7	Laybourn-Parry et al. (2005)
	Beaver	0.39–4.38	Laybourn-Parry et al. (2001)
	Crooked	0–1.05	Henshaw and Laybourn-Parry (2002)
	Druzhby	0–1.9	Henshaw and Laybourn-Parry (2002)
	Highway	0.32–12.66	Madan et al. (2005); Laybourn-Parry et al. (2005)
	Pendant	0.79–21.6	Madan et al. (2005)
	Watts	0.1–10	Heath (1988)
Victoria Land (150–180°E)	11 lakes	0.62–13.68	Andreoli et al. (1992)
	Bird	194–4,000	Spurr (1975)
	Bonney-East	0.03–3.8	Lizotte et al. (1996); Priscu et al. (1999)
	Bonney-West	0–7.4	Priscu (1995); Roberts et al. (2004)
	Fryxell	0.03–21.8	Priscu et al (1987)
	Hoare	0.1–6	Roberts et al. (2004)
	Joyce	0.3–6	Parker et al. (1982); Roberts et al. (2004)
	Miers	0.3–3.5	Parker et al. (1982); Roberts et al. (2004)
	Vanda	0.03–1.2	Parker et al. (1982); Priscu et al (1987)

status has been confirmed by multiple sampling in only a few cases, such as the Arctic Char Lake (Kalff and Welch 1974), and the Antarctic lakes of the Schirmacher Oasis (Kaup 1994), Crooked and Druzhby Lakes of the Vestfold Hills (Henshaw and Laybourn-Parry 2002), and Lake Vanda of the Dry Valleys (Vincent and Vincent 1982a). Given that the frequency distributions in Figure 9.1 include lakes for which we have only single observations rather than a time series from which to estimate the annual maximum Chl-a levels, it is possible that there are more polar lakes in the oligotrophic category than currently known. Seasonal studies

of Arctic tundra ponds and lakes Chl-a levels that even at the annual maximum indicate an oligotrophic status (Miller et al. 1978; Welch et al. 1980; Whalen and Alexander 1986b), as do several lakes in the McMurdo Dry Valleys (Priscu 1995, Roberts et al. 2004).

The standing stock of phytoplankton per unit area (for a water column profile) ranged from 1 to 165 mgChl-am^{-2} (Table 9.5). Standing stock provides a different perspective on the biological productivity of a lake because deeper oligotrophic lakes can accumulate substantial biomass rivaling levels observed in shallow eutrophic lakes.

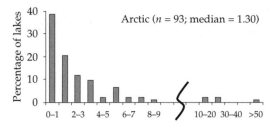

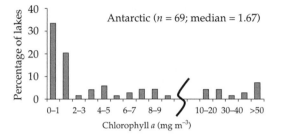

Figure 9.1 Frequency distribution of maximum observed chlorophyll concentrations in Arctic (top panel) and Antarctic (bottom panel) lakes. Note the differing scales on the x axes.

For example, the standing stocks in Lake Bonney, with a productive water column of over 15 m, can exceed that of the 6–9-m productive zones in GTH 112, GTH 114, Toolik Lake and Lake Kitiesh. It is likely that only eutrophic, shallow lakes (e.g. Lake Meretta, Lake Heywood) and hypertrophic lakes and ponds (e.g. Pony Lake, Lago Refugio) reach standing stocks of phytoplankton that might cause significant self-shading (impeding light quantity or quality to deeper populations) or shading out of benthic algae or plants.

There have been a few efforts to estimate the biomass of anoxygenic photosynthetic bacteria using bacteriochlorophyll (Bchl) concentrations. Lakes in the Vestfold Hills of Antarctica had concentrations of Bchl-c (green sulphur bacteria) as high as 490 mg m^{-3} (Burke and Burton 1988). Lake Mogilnoe had concentrations of Bchl-e (green sulphur bacteria) as high as 4600 mg m^{-3} (Lunina *et al.* 2005; see Figure 9.5 below).

Table 9.4 Trophic categories based on phytoplankton biomass (as reviewed by Padisak 2004) and example polar lakes.

	Chl-a (mg m^{-3})		Arctic lakes	Antarctic lakes
	Annual maximum	Annual mean		
Ultra-oligotrophic	<2.5	<1	A (Ellesmere), Char	Crooked, Druzhby, Vanda, Verkhneye
Oligotrophic	2.5–8	1–2.5	Methane, Oil, Toolik	Ace, Beaver, Bonney, Chico, Hoare
Mesotrophic	8–25	2.5–8	N2 (Alaska)	Boeckella, Fryxell, Highway, Moss, Pendant, Sombre, Watts
Eutrophic	25–75	8–25	Meretta	Heywood, Rookery
Hypertrophic	>75	>25		Otero

Table 9.5 Chlorophyll concentrations as standing stock reported for the plankton of polar lakes.

Region	Lake	Chlorophyll concentration, standing stock (mgChl-a m^{-2})	Source
Arctic	GTH 112	4.0–19.4	Whalen *et al.* (2006)
	GTH 114	2.6–10.2	Whalen *et al.* (2006)
	Toolik	2–12	Whalen and Alexander (1986b)
	Char	1–6	Kalff and Welch (1974)
	Meretta	1–40	Kalff and Welch (1974)
Antarctic	Heywood	15–165	Light *et al.* (1981)
	Kitiesh	6.7–12.6	Montecino *et al.* (1991)
	Bonney-East	1–32.4	Lizotte *et al.* (1996); Priscu *et al.* (1999)

9.4 Primary production

Phytoplankton photosynthesis results in the production of organic material, specifically increased cellular biomass and the associated release of extracellular materials. The latter include polymers, colonial matrices and structures, and dissolved excretions. A change in biomass over time, for example as cells, dry weight, carbon, or pigments, can be used to estimate net production, but this will underestimate the gross production due to losses from consumption by grazers, lysis (e.g. by viruses), and difficulty of clearly identifying the origin of many dissolved and particulate organic materials. Thus, researchers have most often estimated phytoplankton productivity based on either the production of the photosynthetic byproduct oxygen or the incorporation of inorganic carbon. Nearly all published estimates of polar lake phytoplankton production were made in the last 40 years, during which the use of radiocarbon tracers (^{14}C-labelled CO_2) have dominated the field. Radioisotope-based approaches have almost certainly been popular due to the higher sensitivity over biomass- or oxygen-based methods, a critical consideration when biomass is low and physiological rates are suppressed by low temperatures. The strengths of ^{14}C-based estimates of primary production include the ability to: trace all products (particulate and dissolved, cellular and excreted, even consumed); differentiate cellular/colonial production in size classes (important to food web studies); and biochemically separate different products (for understanding physiological acclimation, nutritional content, etc.). The major drawback to the ^{14}C methods are that they still underestimate gross production due to the potential for cells to respire some of the radiolabelled product. Thus ^{14}C-based estimates are best interpreted as between gross and net production rates.

Phytoplankton production is one of the sources of primary (autotrophic) production in lakes, in addition to aquatic plants, benthic macroalgae, attached photosynthetic microbes, and chemosynthetic bacteria. In many polar lakes, phytoplankton are the most frequently measured source of primary production; in many lakes, they are the only source measured. Phytoplankton primary production rates have been reported on hourly, daily, monthly, and yearly timescales; extrapolation between these estimates is difficult because of the need to factor in temporal changes in irradiance, nutrient supplies, competition, grazing losses, and other factors on appropriate timescales. Production rates are also reported on either a volumetric basis (e.g. per liter or cubic meter) or an areal basis (e.g. per square meter). The former relates best to the three-dimensional world of the phytoplankton cells, but the latter is relevant to understanding a system that can be limited by flux across a horizontal plane, from above (e.g. light; atmospheric deposition) or below (e.g. diffusion of recycled or reduced chemicals; anaerobic waters; vertically migrating grazers). The most frequently reported estimates for polar lakes have been for daily production per unit volume, and for annual production per unit area.

Daily phytoplankton production rates as high as 1500 mg C m^{-3} day^{-1} have been reported in polar lakes, but most observations are in the range of 1–100 mg C m^{-3} (Tables 9.6 and 9.7). As with the research on phytoplankton biomass, the far larger number of lakes in the Arctic than the Antarctic has led to more multilake surveys, particularly in Alaska and Svalbard. In the Antarctic, nearly all primary production estimates are from the Vestfold Hills (East Antarctica), the McMurdo Dry Valleys (Victoria Land), and Signy Island (Scotia Arc).

As described above for biomass, a data-set derived from the sources in Tables 9.6 and 9.7 was used to derive frequency distributions for polar lakes in each hemisphere (using only the highest value reported for a given lake). The results (Figure 9.2) showed that the Antarctic data were more skewed (52%) to the lowest bin of less than 10 mg C m^{-3} day^{-1}, while the Arctic data-set included more lakes (43%) with high values of more than 100 mg C m^{-3} day^{-1}. These distributions are reversed from those seen in biomass (Figure 9.1). Whereas the data-sets for biomass and production include studies from different lakes, and thus not directly comparable, the differences in the frequency distributions could be useful for generating hypotheses based on physical, chemical, physiological, and ecological differences between the Antarctic and the Arctic as settings for lakes.

Table 9.6 Photosynthetic rates reported for the plankton of Arctic lakes.

Region	Lake	Photosynthetic rate (mgC m⁻³)		Source
		Hourly	Daily	
Alaska	2 ponds		0–260	Kalff (1967)
	6 lakes		0–792	Howard and Prescott (1971)
	4 ponds	0–17	Up to 200	Alexander *et al.* (1980)
	10 lakes		1.9–230	Kling *et al.* (2000)
	39 lakes		6.5–101	Levine and Whalen (2001)
	2 lakes		41–220	Whalen *et al.* (2006)
	N2		5–20	O'Brien *et al.* (2005)
	Oil Lake		0–82	Miller *et al.* (1978)
	Pond C		0–130	Miller *et al.* (1978)
Canada	4 ponds	0.67–16.7		Wallen and Allen (1982)
Greenland	Lake 95		0–14	Maehl (1982)
Europe	2 lakes		3–1000	Laybourn-Parry and Marshall (2003)
	14 lakes		0–1500	Ellis-Evans *et al.* (2001)
	Mogilnoe		0.7–812	Ivanov *et al.* (2001)

Table 9.7 Photosynthetic rates reported for the plankton of Antarctic lakes.

Region	Lake	Photosynthetic rate (mgC m⁻³)		Source
		Hourly	Daily	
Peninsula and Scotia Arc (40–90°W)	Ponds	0.8–72		Parker (1972)
	Heywood		0–720	Ellis-Evans (1981b)
	Moss		0–28	Ellis-Evans (1981a)
	Sombre		0–40	Hawes (1983)
	Tranquil	0–4.5		Butler *et al.* (2000)
East Antarctica (30–90°E)	9 lakes		3.5–18	Tominaga (1977)
	5 lakes	0.15–240		Laybourn-Parry *et al.* (2002)
	Ace	0–32.1		Laybourn-Parry *et al.* (2005)
	Beaver		19.7–25.5	Laybourn-Parry *et al.* (2001)
	Crooked		0–70	Henshaw and Laybourn-Parry (2002)
	Druzhby		0–52	Henshaw and Laybourn-Parry (2002)
	Highway	0.56–45		Laybourn-Parry *et al.* (2005)
	Watts	0–6.4		Heath (1988)
Victoria Land (150–180°E)	Bonney-East		0–7.5	Parker *et al.* (1982)
	Bonney-West	Up to 0.03	0–9.5	Parker *et al.* (1982); Priscu (1995)
	Fryxell	0–3	0–73	Vincent (1981); Priscu *et al.* (1987)
	Hoare		0.4–5.0	Priscu (1995)
	Joyce	Up to 0.036	0–2	Parker *et al.* (1982)
	Vanda	0–0.17	0.1–1.6	Vincent and Vincent (1982a); Priscu (1995)

First, phytoplankton acclimate to lower light intensities by increasing pigment content, thus increasing light-harvesting capacity to maintain photosynthetic and growth rates. Higher Chl-*a* accumulation with lower productivity in Antarctica might be a symptom of lower light availability relative to Arctic lakes. While the range of latitudes are similar in both data-sets (61–79°), most of the Arctic

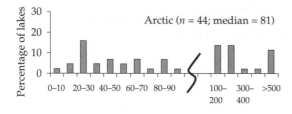

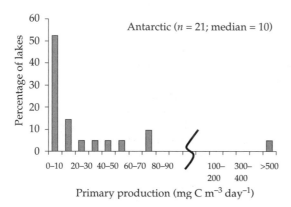

Figure 9.2 Frequency distribution of maximum observed primary production rates in Arctic (top panel) and Antarctic (bottom panel) lakes. Note the differing scales on the x axes.

lakes where primary production has been studied have ice-free seasons, whereas most of the Antarctic lakes listed (especially Victoria Land and East Antarctica) have perennial ice covers. At the substantially lower light levels beneath ice, it is likely that all primary production occurs at light-limited rates of primary production. The factors regulating the light environment in polar lakes include characteristics of the ice cover such as thickness, opacity, and accumulated snow, and optical constituents of the water such as cells, abiotic particles, and dissolved materials (see Chapter 4). Since photosynthetic organisms require a minimum irradiance to support cellular metabolism, the length of the growth season depends not only on the annual solar cycle, but varies greatly depending on ice, snow, and hydrology conditions.

Second, phytoplankton photosynthetic rates can be limited by nutrient availability. Some nutrients (e.g. nitrogen) are also necessary for pigment synthesis; thus nutrient-limited phytoplankton may also be diminished in their capacity to acclimate to low light. Studies of nutrient limitation of polar lake phytoplankton (reviewed below) and the predominance of the ultra-oligotrophic trophic classification described above suggest that nutrient availability is a common regulator of biomass accumulation and daily primary production rates. For example, higher Chl-*a* concentrations with lower photosynthetic rates in some Antarctic lakes may indicate that phytoplankton are present, but less able to access nutrients for growth. Arctic lakes can have much larger watersheds and links with substantial terrestrial ecosystems which may provide more or timely deliveries of nutrients (e.g. Kling *et al.* 2000).

Third, some members of the phytoplankton can gain nutrition from processes other than photosynthesis, which may affect the relationship between biomass and primary production. Mixotrophic species (e.g. many of the phytoflagellates such as cryptophytes, chrysophytes, dinoflagellates, and the ciliates) can consume bacteria and other microbes to gain nutrients and energy. Mixotrophy is an important strategy for survival, such as over the long, dark polar winter, and it may help organisms through acute or chronic shortages in light or nutrients during the growth season (Laybourn-Parry 2002). Mixotrophic cells usually retain pigments even while photosynthetic rates are low; thus we might hypothesize that the lower productivity of Antarctic lakes might be due to a greater prevalence of mixotrophic organisms in the phytoplankton.

Fourth, many of the Antarctic lakes are lacking or have poorly developed macrozooplankton and benthic invertebrate communities compared to Arctic lakes. This would allow phytoplankton in Antarctic lakes to accumulate to higher biomass at lower primary production rates. Conversely, increased grazing pressure in Arctic lakes would suppress phytoplankton biomass, but could also increase nutrient recycling to fuel higher daily primary production rates (and decrease the need for mixotrophic acquisition of nutrients).

It is important to recognize the limitations of frequency distributions and summary statistics. Another valuable approach is to compare systems, such as case studies of individual lakes that have helped us understand the relative roles of light and nutrients (see also Chapter 1).

Table 9.8 Annual primary production rates reported for plankton of polar lakes.

Region	Lake	Production rate (gCm^{-2})	Source
Arctic			
Alaska	2 ponds	0.38–0.85	Kalff (1967)
	4 ponds	0.3–0.7	Alexander *et al.* (1980)
	GTH 112	7.7	Whalen *et al.* (2006)
	GTH 114	10	Whalen *et al.* (2006)
	Peters	0.9	Hobbie (1964)
	Schrader	6.5–7.5	Hobbie (1964)
	Toolik	1.6–10.6	O'Brien *et al.* (1997)
Canada	Char	3.6–4.7	Kalff and Welch (1974)
	Far	9.7	Welch and Bergmann (1985)
	Maretta	9.4–12.5	Kalff and Welch (1974)
	P and N	15.4	Welch and Bergmann (1985)
	Spring	15.9	Welch and Bergmann (1985)
Greenland	Lake 95	4.7	Maehl (1982)
Antarctic			
Scotia Arc (30–90°W)	Heywood	100–270	Light *et al.* (1981)
Queen Maud Land (30°W–30°E)	Moss	14	Ellis-Evans (1981a)
	Verkhneye	0.58	Kaup (1994)
East Antarctica (30–90°E)	Watts	10.1	Heath (1988)
Victoria Land (150–180°E)	Alga	5.9–7.3	Goldman *et al.* (1972)
	Bonney-East	1.26–2.34	Priscu *et al.* (1999)
	Skua	45–133	Goldman *et al.* (1972)

- *Arctic Lake A and Char Lake*: Lake A receives less light because it is 1000 km north of Char Lake. Lake A also accumulates about 0.5 m of snow in spring, while Char Lake is often windswept. Yet Lake A has higher algal biomass than ultra-oligotrophic Char Lake (Van Hove *et al.* 2006).
- *Antarctic Lakes Fryxell and Vanda*: located in the Dry Valleys, Lake Vanda is renowned for its clean, clear, ice with 10–15% penetration by sunlight, while Lake Fryxell has a rough, milky ice cover passing about 1% of sunlight. Lake Fryxell has a much higher biomass and production of phytoplankton (Vincent 1981).
- *Arctic Char Lake and Meretta Lake*. These adjacent lakes provide a classic example of eutrophication. Similar in ice, snow, and incident irradiance conditions, they differ in primary production because of accelerated eutrophication of the latter through sewage discharge (Schindler *et al.* 1974).
- *Arctic and Antarctic open-water ponds*: most polar open-water ponds have clear waters with low chlorophyll values. The highest primary production is in the phytobenthos, where light is limiting but nutrients are abundant. However, the addition of nutrients from marine birds and mammals can increase phytoplankton to spectacular levels.

There have been few estimates of annual primary production in polar lakes (Table 9.8). The range of values reported is 0.3–270 gCm^{-2}year^{-1}, compared with ranges of 0.6–600 gCm^{-2}year^{-1} reported in global lake reviews (e.g. Schindler 1978). However, the most common estimates are in the order of magnitude of 1–10 gCm^{-2}year^{-1}, within the range of oligotrophic lakes worldwide. Schindler (1978) hypothesized that low annual production in polar lakes compared with more equatorial lakes corresponds to differences in light availability of approximately 50 times (due to seasonal irradiance and snow and ice cover), and differences of natural nutrient delivery of approximately 100 times (for example, due to precipitation rates, freezing, and soil microbe activity). The only polar lakes reported to achieve mesotrophic annual production levels exceeding $100\,g\,C\,m^{-2}\,year^{-1}$ are Antarctic lakes Heywood and Skua, which received large inputs of animal wastes from seabird and marine mammal colonies.

Estimating annual primary production is problematic because very few studies have recorded a full annual growth season. The seasonal progression of primary production in polar lakes is strongly tied to spring increases in solar irradiance, ice-cover melting, the summer solstice, and ice-cover freezing (Figures 9.3 and 9.4). The Antarctic lakes with perennial ice cover tend to peak months before the summer solstice at the end of December, suggesting nutrient limitation. Polar lakes that lose

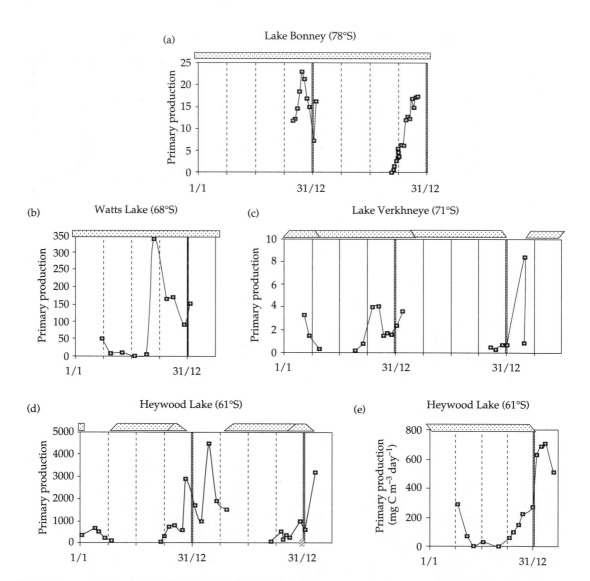

Figure 9.3 Seasonal to multiyear observations of areal primary production rates ($mgC\,m^{-2}\,day^{-1}$) in Antarctic lakes. Graph (e) has units of production per volume of water ($mgCm^{-3}\,day^{-1}$). Source data are from (a) Lizotte *et al.* (1996), (b) Heath (1988), (c) Kaup (1994), (d) Light *et al.* (1981), and (e) Ellis-Evans (1981).

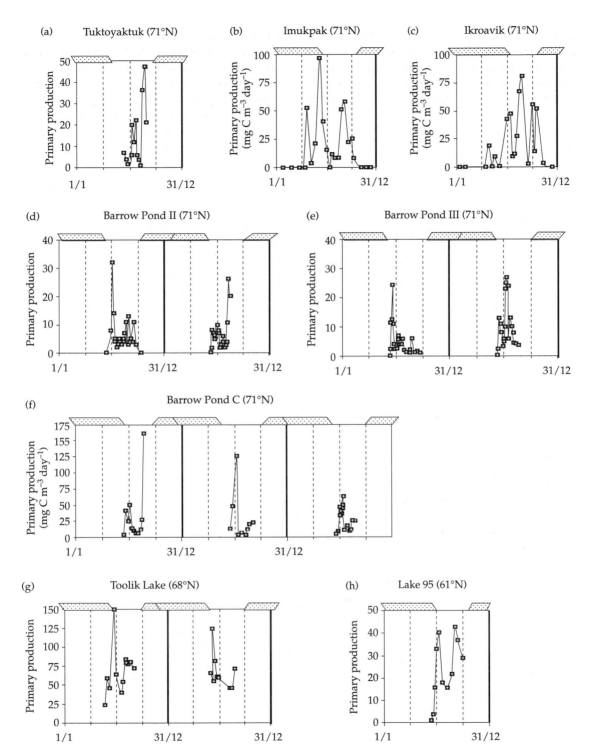

Figure 9.4 Seasonal to multiyear observations of areal primary production rates (mgCm^{-2}day^{-1}) in Arctic lakes. Graphs enclosed in boxes have units of production per volume of water (mgCm^{-3}day^{-1}). Source data are from (a) Sheath and Hellebust (1978), (b, c) Howard and Prescott (1971), (d, e) Kalff (1967), (f) Miller *et al.* (1978), (g) Whalen and Alexander (1986b), and (h) Maehl (1982).

their ice cover during the summer often show a peak within weeks of ice-out, and often another peak approximately 1–2 months later. Although increased light availability will increase primary production, ice-out is probably strongly correlated with increased runoff delivering nutrients to the lakes, and increased wind-induced entrainment of deeper, nutrient-rich waters into the euphotic zone. Later peaks in primary production may also be related to zooplankton population dynamics and life cycles, particularly for Arctic lakes.

9.5 Environmental stressors

Phytoplankton **primary production** can be limited by factors that reduce populations, limit physiological activity, or both. For example, population losses can occur due to grazing, disease outbreaks, or high flushing rates. Common factors limiting physiological activity include sub-optimal temperature, insufficient light, and nutrient shortages.

In polar lakes, the potential for grazing on phytoplankton can vary a great deal depending on food-web complexity. Many Arctic lakes have food webs as complex (in terms of trophic levels, if not species diversity) as found at lower latitudes, with microbial, invertebrate, and vertebrate members (e.g. O'Brien *et al.* 1997). However, shallower lakes can occasionally experience a winter ice-cover thickening that can kill off many overwintering macrofauna by concentrating organisms in a restricted deep part of the lake (with subsequent exhaustion of oxygen) or by freezing the entire water column. In these cases, and for geographically isolated lakes of Antarctica and higher latitudes of the Arctic, the grazers may be limited to protists and few small invertebrates (e.g. rotifers), and be lacking in vertebrates, crustaceans, and mollusks.

Diseases from viral or bacterial infection, or from fungal parasites, are poorly understood for phytoplankton. Sheath (1986) reported that chlorophytes suffered a hyphomycete (fungal) infection that led to a rapid population decrease in an Arctic lake. Virus-like particle counts in Antarctic lakes have not shown strong relationships with phytoplankton (e.g. Madan *et al.* 2005), although cyanophage may exert a control on the picocyanobacteria that are common in many polar lakes.

Temperature limitation of polar phytoplankton is nearly always due to temperatures less than the optimal range for physiological processes. The range of temperatures in polar lakes during the primary production season range from near freezing to more than 15°C. The latter may approach or exceed the optimum temperature for psychrophilic organisms. Species can be adapted to lower optimal temperatures or have more dynamic responses to temperature change (e.g. Q_{10} values much greater than the canonical 2 for a physiological rate). In a cross-system review, Markager *et al.* (1999) observed that the photosynthetic performance of polar-lake phytoplankton is often much lower than for algae in other cold environments (polar seas, sea ice, and in culture), and suggested that this may be a stress response resulting more from nutrient limitation than temperature effects.

Nutrient availability in polar lakes is highly variable geographically. High concentrations of nitrate (>100 µM) and phosphate (>5 µM) have been widely reported in Antarctic lakes on the Antarctic Peninsula (e.g. Izaguirre *et al.* 1998, 2001), Signy Island (Butler 1999), Ross Island (Vincent and Vincent 1982b), Vestfold Hills (e.g. Laybourn-Parry *et al.* 2002), and the Dry Valleys (e.g. Vincent 1981; Priscu 1995; Roberts *et al.* 2004). High nutrient conditions were found in lakes and ponds enriched by marine animals and in perennially ice-covered lakes that accumulate nutrients in deep monomolimnia. In the Arctic, nitrate levels of more than 10 µM have only rarely been recorded, including in lakes near marine animal colonies on Spitzbergen (Ellis-Evans *et al.* 2001; Laybourn-Parry and Marshall 2003), a Russian lake (Moiseenko *et al.* 2001), and in perennially ice-covered lakes in Canada (Van Hove *et al.* 2006); only the latter had high phosphate levels (>5 µM), in its deep anoxic zone. Maximum values of nitrate less than 4 µM and phosphate of less than 0.2 µM have been repeatedly reported in the Arctic for Alaskan lakes (e.g. Whalen and Alexander 1986a; Kling *et al.* 2000; Levine and Whalen 2001; Whalen *et al.* 2006), Canada (e.g. Pienitz *et al.* 1997; Vezina and Vincent 1997), and in Greenland (e.g. Maehl 1982). Such oligotrophic conditions have been observed in some remote Antarctic lakes (e.g. Tominaga 1977; Andreoli *et al.* 1992; Kaup 1994), but can also occur seasonally in the surface waters of perennially ice-covered lakes cited above.

Nutrient limitation of primary production has been frequently observed in polar lakes. Bioassays have been the most common method employed, where algae are monitored after the addition of one or more potentially limiting nutrients (usually phosphate, nitrate, and ammonium). This method has most often demonstrated phosphorus deficiency in the Antarctic (e.g. Vincent and Vincent 1982a; Kaup 1994; Priscu 1995; Dore and Priscu 2001) and Greenland (Brutemark *et al.* 2005). Phosphate deficiency in these lakes has also been supported by N:P ratios that are anomalously low in the dissolved nutrient pools and high in algal cells, and by measurements of phosphatase activity (Dore and Priscu 2001; Brutemark *et al.* 2005). Simultaneous additions of nitrogen and phosphorus occasionally show significant responses in Antarctic lakes (e.g. Priscu 1995; Dore and Priscu 2001) but are very common in Alaskan lakes (e.g. Whalen and Alexander 1986a; Levine and Whalen 2001). A strong nutrient response was observed in phytoplankton bioassays at Ward Hunt Lake, the northernmost lake in the Canadian High Arctic, while the phytobenthos showed no improved growth in response to enrichment (Bonilla *et al.* 2005). On larger and longer scales, Alaskan lakes have been studied through additions of nitrogen and phosphorus to mesocosms and whole lakes (O'Brien *et al.* 2005), allowing for a more sophisticated whole-ecosystem evaluation of enrichment. Scientists have also studied nutrient effects from eutrophication events caused by natural increases in animal colonies (Butler 1999) and human pollution (Kalff *et al.* 1975).

Phytoplankton acclimation to the low-light environment of polar lakes may often be diminished due to other stress factors. Photosynthetic rates, normalized to biomass (chlorophyll concentrations), rarely exceed 1 mg C (mg Chl)$^{-1}$ h^{-1} in the Arctic (Markager *et al.* 1999) or Antarctic (e.g. Tominaga 1977; Vincent 1981; Heath 1988; Kaup 1994). Exceptions include seasonally higher rates reported for nutrient-rich lakes on Signy Island (e.g. Hawes 1985) and the Antarctic Peninsula (Montecino *et al.* 1991). The efficiency of light harvesting, as measured by the maximal quantum yield (moles of carbon fixed per mole of photons absorbed), has also been found to be very low, approximately an order of magnitude lower than theoretical limits (Lizotte and Priscu 1994; Markager *et al.* 1999). Quantum efficiency is not a temperature-sensitive parameter, thus low values are often interpreted as a less ambiguous indicator of nutrient limitation reducing the efficiency of algal photosynthesis. Light limitation may also affect nutrient uptake and utilization directly (Priscu 1989). Natural events that deliver stream flow to polar lakes (e.g. from glacial melting) have an overall negative impact on primary production because the negative effects of light attenuation from sediments may reduce the ability of algae to utilize the increased nutrient load (Foreman *et al.* 2004; Whalen *et al.* 2006).

Many of the dimensions of phytoplankton dynamics in polar lakes come together in studies of the deep chlorophyll maxima observed in many polar lakes (Figure 9.5). The deep chlorophyll maxima examples in this figure are dominated by different organisms (cyanobacteria, phytoflagellates, non-flagellated algae, photosynthetic bacteria), and live in lakes of differing chemistry and physical conditions. Multiple deep chlorophyll maxima in a column are possible; for example, an oxygenic population above an anoxygenic bacterial layer. Shared characteristics include ice cover and very resilient pycnocline features (see Chapter 4). Deep chlorophyll maxima can be very stable features for repeated or long-term study in polar lakes. The main conceptual ideas about why deep chlorophyll maxima form include locating resources (such as a limiting nutrients, or prey for mixotrophs), locating optimal environmental conditions (e.g. temperature or oxygen levels), avoiding predation, avoiding harmful ultraviolet radiation, reducing energy needs (e.g. for swimming) by settling to a neutral buoyancy depth, accumulating more pigment per cell at depth to acclimate to the lower light levels, domination by species with different pigmentation, accumulation of dead or dying cells on the pycnocline, and combinations of the above. Burnett *et al.* (2006) recently reviewed the literature to test some of the pycnocline-related hypotheses. Their analyses suggest that for these lakes, deep chlorophyll maxima are primarily explained by resource availability (light from

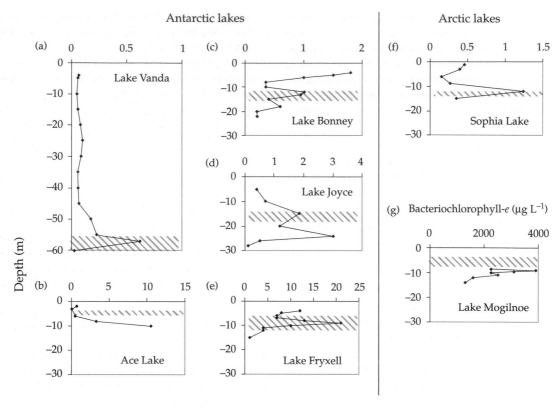

Figure 9.5 Examples of deep chlorophyll maxima in polar lakes. Note the differences in horizontal scale for Chl-*a* concentration (a–f; $mg\,m^{-3}$), and panel (g), which is in units of bacteriochlorophyll-*e* ($\mu g\,L^{-1}$). The hatched bar is the approximate location of the **pycnocline**. Data sources: (a) Vincent and Vincent (1982a); (b) Burch (1988); (c–e) Roberts *et al.* (2004); (f) Markager *et al.* (1999); and (g) Lunina *et al.* (2005).

above and nutrients from below; see Plate 14) and the presence of strong density gradients.

9.6 Conclusions

Our understanding of photosynthetic plankton and their productivity in polar lakes has advanced rapidly over the last few decades. These studies have shown that polar aquatic ecosystems are largely of ultra-oligotrophic or oligotrophic trophic status, and that they respond strongly to nutrient enrichment. However, the observations to date have been constrained by the sampling and analysis limitations common to all limnologists, and by the special logistical limitations of polar research. In the future, the diversity of microbes involved in lake autotrophy will be greatly expanded by the development of molecular approaches for detecting

and distinguishing rare or less conspicuous organisms. Second, autonomous samplers and instrumentation will greatly increase the capacity to conduct year-round studies in remote lakes. Third, increased sensitivity in chemical analyses (particularly mass spectrometry) are rapidly developing new methods for tracing for biological activity from gene products through food webs.

Polar limnologists can also address several issues that are emerging from past studies. The occurrence and role of nonoxygenic photosynthetic bacteria should be more widely explored in stratified lake systems, and it is possible that aerobic anoxygenic photoheterotrophs have gone undetected in oxygenated waters. While the challenge of detecting the wide variety of photosynthetic bacteria is making significant progress, studies of their activity remain rare and difficult. Estimating their contribution to primary production may require sampling and measurements on microlayers and fine-scale vertical gradients in oxygen, hydrogen sulphide, and nutrients, combined with detailed hydrodynamic and optical measurements. Such work will need to be complemented by experimental studies of their physiology under controlled laboratory conditions. Several studies have noted that very small cells, picoplankton, can make up a large component of the primary producers in oligotrophic polar lakes. Ongoing studies on this size class will be critical for understanding the food-web dynamics of polar lakes, which are often dominated by the microbial production and loss processes.

Acknowledgments

I wish to thank the editors and Stephen Whalen for constructive comments on the manuscript. This material is based upon work supported by the National Science Foundation under grant no. 0631888.

References

Adl, M.S. *et al.* (2005). The new classification of eukaryotes with emphasis on the taxonomy of protists. *Journal of Eukaryotic Microbiology* **52**, 399–451.

Alexander, V., Stanley, D.W., Daley, R.J., and McRoy, C.P. (1980). Primary producers. In Hobbie, J.E. (ed.), *Limnology of Tundra Ponds, Barrow, Alaska*, pp. 179–250. US/IBP Synthesis Series, vol. 13. Dowden, Hutchinson and Ross, Stroudsburg, PA.

Allende, L. and Izaguirre, I. (2003). The role of physical stability on the establishment of steady states on the phytoplankton community of two maritime Antarctic lakes. *Hydrobiologia* **502**, 211–224.

Andreoli, C., Scarabel, L., Sini, S., and Grassi, C. (1992). The picoplankton in Antarctic lakes of northern Victoria Land during summer 1989–1990. *Polar Biology* **11**, 575–582.

Bonilla, S., Villeneuve, V., and Vincent, W.F. (2005). Benthic and planktonic algal communities in a high arctic lake: pigment structure and contrasting responses to nutrient enrichment. *Journal of Phycology* **41**, 1120–1130.

Brutemark, A., Rengefors, K., and Anderson, N.J. (2005). An experimental investigation of phytoplankton nutrient limitation in two contrasting low arctic lakes. *Polar Biology* **29**, 487–494.

Burch, M.D. (1988). Annual cycle of phytoplankton in Ace Lake, an ice covered, saline meromictic lake. *Hydrobiologia* **165**, 59–75.

Burke, C.M. and Burton, H.R. (1988). The ecology of photosynthetic bacteria in Burton Lake, Vestfold Hills, Antarctica. *Hydrobiologia* **165**, 1–11.

Burnett, L., Moorhead, D., Hawes, I., and Howard-Williams, C. (2006). Environmental factors associated with deep chlorophyll maxima in Dry Valley lakes, South Victoria Land, Antarctica. *Arctic, Antarctic, and Alpine Research* **38**, 179–189.

Butler, H.G. (1999). Seasonal dynamics of the phytoplanktonic microbial community in a maritime Antarctic lake undergoing eutrophication. *Journal of Plankton Research* **21**, 2393–2419.

Butler, H.G., Edworthy, M.G., and Ellis-Evans, J.C. (2000). Temporal plankton dynamics in an oligotrophic maritime Antarctic lake. *Freshwater Biology* **43**, 215–230.

Coesel, P.F.M. (1996). Biogeography of desmids. *Hydrobiologia* **336**, 41–53.

Cremer, H. and Wagner, B. (2004). The diatom flora in the ultra-oligotrophic Lake El'gygytgyn, Chukotka. *Polar Biology* **26**, 105–114.

Dore, J.E. and Priscu, J.C. (2001). Phytoplankton phosphorus deficiency and alkaline phosphatase activity in the McMurdo Dry Valley lakes, Antarctica. *Limnology and Oceanography* **46**, 1331–1346.

Ellis-Evans, J.C. (1981a). Freshwater microbiology in the Antarctic: I. Microbial numbers and activity in oligotrophic Moss Lake, Signy Island. *British Antarctic Survey Bulletin* **54**, 85–104.

Ellis-Evans, J.C. (1981b). Freshwater microbiology in the Antarctic: II. Microbial numbers and activity in nutrient-enriched Heywood Lake, Signy Island. *British Antarctic Survey Bulletin* **54**, 105–121.

Ellis-Evans, J.C. (1996). Microbial diversity and function in Antarctic freshwater ecosystems. *Biodiversity and Conservation* **5**, 1395–1431.

Ellis-Evans, J.C., Galchenko, V., Laybourn-Parry, J., Mylnikov, A.P., and Petz, W. (2001). Environmental characteristics and microbial plankton activity of freshwater environments at Kongsfjorden, Spitsbergen (Svalbard). *Archiv für Hydrobiologie* **152**, 609–632.

Foreman, C.M., Wolf, C.F., and Priscu, J.C. (2004). Impact of episodic warming events on the physical, chemical and biological relationships of lakes in the McMurdo dry valleys, Antarctica. *Aquatic Geochemisty* **10**, 239–268.

Goldman, C.R., Mason, D.T., and Wood, B.J.B. (1972). Comparative study of the limnology of two small lakes on Ross Island, Antarctica. In Llano, G.A. (ed.), *Antarctic Terrestrial Biology*, pp. 1–50. Antarctic Research Series, vol. 20. American Geophysical Union, Washington DC.

Hawes, I. (1983). Turbulence and its consequences for phytoplankton development in ice covered Antarctic lakes. *British Antarctic Survey Bulletin* **60**, 69–82.

Hawes, I. (1985). Light climate and phytoplankton photosynthesis in maritime Antarctic lakes. *Hydrobiologia* **123**, 69–79.

Heath, C.W. (1988). Annual primary productivity of an annual continental lake: phytoplankton and benthic algal mat production strategies. *Hydrobiologia* **165**, 77–87.

Henshaw, T. and Laybourn-Parry, J. (2002). The annual patterns of photosynthesis in two large, freshwater, ultra-oligotrophic Antarctic Lakes. *Polar Biology* **25**, 744–752.

Hobbie, J.E. (1964). Carbon-14 measurements of primary production in two Alaskan lakes. *Verhandlungen Internationale Vereinigung fuer Theoretische und Angewandte Limnologie* **15**, 360–364.

Howard, H.H. and Prescott, G.W. (1971). Primary production in Alaskan tundra lakes. *American Midland Naturalist* **85**, 108–123.

Ivanov, M.V. *et al.* (2001). Microbial processes of the carbon and sulphur cycles in Lake Mogil'noe. *Microbiology* **70**, 583–593 [in Russian].

Izaguirre, I., Vinocur, A., Mataloni, G., and Pose, M. (1998). Phytoplankton communities in relation to trophic status in lakes from Hope Bay (Antarctic Peninsula). *Hydrobiologia* **369/370**, 73–87.

Izaguirre, I., Mataloni, G., Allende, L., and Vinocur, A. (2001). Summer fluctuations of microbial planktonic communities in a eutrophic lake – Cierva Point, Antarctica. *Journal of Plankton Research* **23**, 1095–1109.

Jeppesen, E. *et al.* (2003). The impact of nutrient state and lake depth on top-down control in the pelagic zone of lakes: a study of 466 lakes from the temperate zone to the Arctic. *Ecosystems* **6**, 313–325.

Jung, D.O., Achenbach, L.A., Karr, E.A., Takaichi, S., and Madigan, M.T. (2004). A gas vesiculate planktonic strain of the purple non-sulphur bacterium *Rhodoferax antarcticus* isolated from Lake Fryxell, Dry Valleys, Antarctica. *Archives of Microbiology* **182**, 236–243.

Kalff, J. (1967). Phytoplankton abundance and primary production rates in two Arctic ponds. *Ecology* **48**, 558–565.

Kalff, J. and Welch, H.E. (1974). Phytoplankton production in Char Lake, a natural polar lake, and in Meretta Lake, a polluted polar lake, Cornwallis Island, Northwest Territories. *Journal of the Fisheries Research Board of Canada* **31**, 621–636.

Kalff, J., Kling, H.J., Holmgren, S.H., and Welch, H.E. (1975). Phytoplankton, phytoplankton growth and biomass cycles in an unpolluted and in a polluted Polar lake. *Verhandlungen – Internationale Vereinigung fuer Theoretische und Angewandte Limnologie* **19**, 487–495.

Karr, E.A., Sattley, W.M., Jung, D.O., Madigan, M.T., and Achenbach, L.A. (2003). Remarkable diversity of phototrophic purple bacteria in a permanently frozen Antarctic lake. *Applied and Environmental Microbiology* **69**, 4910–4914.

Kaup, E. (1994). Annual primary production of phytoplankton in Lake Verkhneye, Schirmacher Oasis, Antarctica. *Polar Biology* **14**, 433–439.

Kling, G.W., O'Brien, W.J., Miller, M.C., and Hershey, A.E. (1992). The biogeochemistry and zoogeography of lakes and rivers in arctic Alaska. *Hydrobiologia* **240**, 1–14.

Kling, G.W., Kipphut, G.W., Miller, M.M., and O'Brien, W.J. (2000). Integration of lakes and streams in a landscape perspective: the importance of material processing on spatial patterns and temporal coherence. *Freshwater Biology* **43**, 477–497.

Laybourn-Parry, J. (2002). Survival mechanisms in Antarctic lakes. *Philosophical Transactions of the Royal Society of London Series B Biological Sciences* **357**, 863–869.

Laybourn-Parry, J. and Marchant, H.J. (1992). The microbial plankton of freshwater lakes in the Vestfold Hills, Antarctica. *Polar Biology* **12**, 405–410.

Laybourn-Parry, J. and Marshall, W.A. (2003). Photosynthesis, mixotrophy, and microbial plankton dynamics in two high Arctic lakes during summer. *Polar Biology* **26**, 517–524.

Laybourn-Parry, J., Quayle, W.C., Henshaw, T., Ruddell, A., and Marchant, H.J. (2001). Life on the edge: the plankton and chemistry of Beaver Lake, an ultra-oligotrophic epishelf lake, Antarctica. *Freshwater Biology* **46**, 1205–1217.

Laybourn-Parry, J., Quayle, W., and Henshaw, T. (2002). The biology and evolution of Antarctic saline lakes in relation to salinity and trophy. *Polar Biology* **25**, 542–552.

Laybourn-Parry, J., Marshall, W.A., and Marchant, H.J. (2005). Flagellate nutritional versatility as a key to survival in two contrasting Antarctic saline lakes. *Freshwater Biology* **50**, 830–838.

Levine, M.A. and Whalen, S.C. (2001). Nutrient limitation of phytoplankton production in Alaskan Arctic foothill lakes. *Hydrobiologia* **455**, 189–201.

Light, J.J., Ellis-Evans, J.C., and Priddle, J. (1981). Phytoplankton ecology in an Antarctic lake. *Freshwater Biology* **11**, 11–26.

Lizotte, M.P. and Priscu, J.C. (1994). Natural fluorescence and quantum yields in vertically stationary phytoplankton from perennially ice-covered lakes. *Limnology and Oceanography* **39**, 1399–1410.

Lizotte, M.P. and Priscu, J.C. (1998). Distribution, succession and fate of phytoplankton in the dry valley lakes of Antarctica, based on pigment analysis. In Priscu, J.C. (ed.), *Ecosystem Dynamics in a Polar Desert: The McMurdo Dry Valleys*, pp. 229–239. Antarctic Research Series, vol. 72. American Geophysical Union, Washington DC.

Lizotte, M.P., Sharp, T.R., and Priscu, J.C. (1996). Phytoplankton dynamics in the stratified water column of Lake Bonney, Antarctica: I. Biomass and productivity during the winter-spring transition. *Polar Biology* **16**, 155–162.

Lunina, O.N. *et al.* (2005). Seasonal changes in the structure of the anoxygenic phototrophic bacterial community in Lake Mogilnoe, a relict lake on Kil'din Island in the Barents Sea. *Microbiology* **74**, 588–596 [in Russian].

Madan, N.J., Marshall, W.A., and Laybourn-Parry, J. (2005). Virus and microbial loop dynamics over an annual cycle in three contrasting Antarctic lakes. *Freshwater Biology* **50**, 1291–1300.

Madigan, M.T. (2003). Anoxygenic phototrophic bacteria from extreme environments. *Photosynthesis Research* **76**, 157–171.

Maehl, P. (1982). Phytoplankton production in relation to physico-chemical conditions in a small, oligotrophic subarctic lake in South Greenland. *Holarctic Ecology* **5**, 420–427.

Markager, S., Vincent, W.F., and Tang, E.P.Y. (1999). Carbon fixation in high Arctic lakes: Implications of low temperature for photosynthesis. *Limnology and Oceanography* **44**, 597–607.

Miller, M.C., Alexander, V., and Barsdate, R.J. (1978). The effects of oil spills on phytoplankton in an Arctic lake and ponds. *Arctic* **31**, 192–218.

Moiseenko, T.I., Sanfimirov, S.S., and Kudryavtseva, L.P. (2001). Eutrophication of surface water in the Arctic region. *Water Resources* **28**, 307–316 [in Russian].

Montecino, V., Pizarro, G., Cabrera, S., and Contreras, M. (1991). Spatial and temporal photosynthetic compartments during summer in Antarctic Lake Kitiesh. *Polar Biology* **11**, 371–377.

O'Brien, W.J. *et al.* (1997). The limnology of Toolik Lake. In Milner, A.M. and Oswood, M.W. (eds), *Freshwaters of Alaska: Ecological Synthesis*, pp. 61–106. Springer, Dordrecht.

O'Brien, W.J. *et al.* (2005). Long-term response and recovery to nutrient addition of a partitioned arctic lake. *Freshwater Biology* **50**, 731–741.

Padisak, J. (2004). Phytoplankton. In O'Sullivan, P.E. and Reynolds, C.S. (eds), *The Lakes Handbook, Volume 1: Limnology and Limnetic Ecology*, pp. 251–308. Blackwell Publishing, Oxford.

Parker, B.C. (1972). Conservation of freshwater habitats on the Antarctic Peninsula. In Parker, B.C. (ed.), *Conservation Problems in Antarctica*, pp. 143–162. Allen Press, Lawrence, KA.

Parker, B.C., Simmons, Jr, G.M., Seaburg, K.G., Cathay, D.D., and Allnutt, F.C.T. (1982). Comparative ecology of plankton communities in seven Antarctic oasis lakes. *Journal of Plankton Research* **4**, 271–286.

Pienitz, R., Smol, J.P., and Lean, D.R.S. (1997). Physical and chemical limnology of 59 lakes located between the southern Yukon and the Tuktoyaktuk Peninsula, Northwest Territories (Canada). *Canadian Journal of Fisheries and Aquatic Sciences* **54**, 330–346.

Priscu, J.C. (1989). Photon dependence of inorganic nitrogen transport by phytoplankton in perennially ice-covered antarctic lakes. *Hydrobiologia* **172**, 173–182.

Priscu, J.C. (1995). Phytoplankton nutrient deficiency in lakes of the McMurdo Dry Valleys, Antarctica. *Freshwater Biology* **34**, 215–227.

Priscu, J.C., Priscu, L.R., Vincent, W.F., and Howard-Williams, C. (1987). Photosynthate distribution by microplankton in permanently ice-covered Antarctic desert lakes. *Limnology and Oceanography* **32**, 260–270.

Priscu, J.C. *et al.* (1999). Carbon transformations in a perennially ice-covered Antarctic Lake. *Bioscience* **49**, 997–1008.

Roberts, E.C., Priscu, J.C., Wolf, C., Lyons, W.B., and Laybourn-Parry, J. (2004). The distribution of microplankton in the McMurdo Dry Valley lakes, Antarctica: response to ecosystem legacy or present-day climatic controls? *Polar Biology* **27**, 238–249.

Schindler, D.W. (1978). Factors regulating phytoplankton production and standing crop in the world's freshwaters. *Limnology and Oceanography* **23**, 478–486.

Schindler, D.W. *et al.* (1974). Eutrophication in the high arctic: Meretta Lake, Cornwallis Island (75°N lat). *Journal of the Fisheries Research Board of Canada* **31**, 647–662.

Sheath, R.G. (1986). Seasonality of phytoplankton in northern tundra ponds. *Hydrobiologia* **138**, 75–83.

Sheath, R.G. and Hellebust, J.A. (1978). Comparison of algae in the euplankton, tychoplankton and periphyton of a tundra pond. *Canadian Journal of Botany* **56**, 1472–1483.

Spaulding, S.A., McKnight, D.M., Smith, R.L., and Dufford, R. (1994). Phytoplankton population dynamics in perennially ice-covered Lake Fryxell, Antarctica. *Journal of Plankton Research* **16**, 527–541.

Spurr, B. (1975). Limnology of Bird Pond, Ross Island, Antarctica. *New Zealand Journal of Marine and Freshwater Research* **9**, 547–562.

Tominaga, H. (1977). Photosynthetic nature and primary productivity of Antarctic freshwater phytoplankton. *Japanese Journal of Limnology* **38**, 122–130.

Toro, M. *et al.* (2007). Limnological characteristics of the freshwater ecosystems of Byers Peninsula, Livingston Island, in maritime Antarctica. *Polar Biology* **30**, 635–649.

Unrein, F. and Vinocur, A. (1999). Phytoplankton structure and dynamics in a turbid Antarctic lake (Potter Peninsula, King George Island). *Polar Biology* **22**, 93–101.

Van Hove, P., Belzile, C., Gibson, J.A.E., and Vincent, W.F. (2006). Coupled landscape-lake evolution in High Arctic Canada. *Canadian Journal of Earth Science* **43**, 533–546.

Vezina, S. and Vincent, W.F. (1997). Arctic cyanobacteria and limnological properties of their environment: Bylot Island, Northwest Territories, Canada (73°N, 80°W). *Polar Biology* **17**, 523–534.

Vincent, W.F. (1981). Production strategies in Antarctic inland waters: phytoplankton eco-physiology in a permanently ice-covered lake. *Ecology* **62**, 1215–1224.

Vincent, W.F. (2000a). Evolutionary origins of Antarctic microbiota: invasion, selection and endemism. *Antarctic Science* **12**, 374–385.

Vincent, W.F. (2000b). Cyanobacterial dominance in polar regions. In Whitton, B.A. and Potts, M. (eds), *The Ecology of Cyanobacteria*, pp. 321–340. Kluwer Academic, Dordrecht.

Vincent, W.F. and Vincent, C.L. (1982a). Factors controlling phytoplankton production in Lake Vanda (77°S). *Canadian Journal of Fisheries and Aquatic Sciences* **39**, 1602–1609.

Vincent, W.F. and Vincent, C.L. (1982b). Nutritional state of the plankton in Antarctic coastal lakes and the inshore Ross Sea. *Polar Biology* **1**, 159–165.

Wallen, D.G. and Allen, R. (1982). Variations in phytoplankton communities in Canadian Arctic ponds. *Naturaliste Canadien* **109**, 213–221.

Welch, H.E. and Bergmann, M.A. (1985). Winter respiration of lakes at Saqvaqjuac, N.W.T. *Canadian Journal of Fisheries and Aquatic Sciences* **42**, 521–528.

Welch, H.E. and Legault, J.A. (1986). Precipitation chemistry and chemical limnology of fertilized and natural lakes at Saqvaqjuac, N.W.T. *Canadian Journal of Fisheries and Aquatic Sciences* **43**, 1104–1134.

Welch, H.E., Rudd, J.W.M., and Schindler, D.W. (1980). Methane addition to an arctic lake in winter. *Limnology and Oceanography* **25**, 100–113.

Whalen, S.C. and Alexander, V. (1986a). Chemical influences on ^{14}C and ^{15}N primary production in an arctic lake. *Polar Biology* **5**, 211–219.

Whalen, S.C. and Alexander, V. (1986b). Seasonal inorganic carbon and nitrogen transport by phytoplankton in an Arctic lake. *Canadian Journal of Fisheries and Aquatic Sciences* **43**, 1177–1186.

Whalen, S.C., Chalfant, B.A., Fischer, E.N., Fortino, K.A., and Hershey, A.E. (2006). Comparative influence of resuspended glacial sediment on physicochemical characteristics and primary production in two arctic lakes. *Aquatic Sciences* **68**, 65–77.

CHAPTER 10

Benthic primary production in polar lakes and rivers

Antonio Quesada, Eduardo Fernández-Valiente, Ian Hawes, and Clive Howard-Williams

Outline

In polar ecosystems, liquid water is one of the most limiting factors for life, thus organisms are mainly distributed in the places and times in which liquid water is present. Freshwater ecosystems are typically considered as oases for life in the polar desert and tundra biomes. The benthic component of these lakes, ponds, rivers, and streams is often rich in biodiversity as well as biomass. The main communities in the benthic habitat are microbial mats, aquatic mosses, and green algal felts that inhabit both running and lentic waters. The substantial primary production in these benthic autotrophic communities, together with their reduced losses of assimilated carbon because of the low temperatures, low degradation rates, and minimal grazing pressure, can result in luxuriant growth and accumulation of these photosynthetic elements. In this chapter, we first describe the types of communities typically found in non-marine benthic habitats. We then review the primary production rates that ha ve been measured under different circumstances, drawing on information from the literature as well as unpublished results. We conclude the chapter by exploring why the benthic component (phytobenthos) often dominates over planktonic communities (phytoplankton) in polar aquatic ecosystems.

10.1 Introduction

In the polar non-marine aquatic ecosystems the benthic habitat is typically richer in diversity and biomass than the planktonic habitat (e.g. Bonilla *et al.* 2005; Moorhead *et al.* 2005). Benthic communities in polar freshwater systems are often dominated by microbes, with more complex organisms either absent or scarce (Vincent 1988). The reduced number or absence of multicellular invertebrates removes the top–down stresses of bioturbation and grazing and thus makes these systems stable and able to accumulate biomass at the lower trophic level over many years. The principle of accumulation over time applies to photosynthetically produced biomass and also to nutrients. This allows luxuriant growths to develop, despite low temperatures, a short growing season, and low fluxes of nutrients from barren catchments, making the benthic environment diverse and productive oases in polar desert and tundra landscapes.

Benthic environments in lakes, ponds, and streams are dominated by microbial mats in which cyanobacteria are typically the dominant element of the biota (Plate 10), although they are not always well developed, as in the lakes of the Vestfold Hills (Antarctica). Microbial mats are multilayered microbial consortia commonly found in polar aquatic or semi-aerophytic ecosystems in which they cover large expanses as soon as liquid water is available. They are thus widespread in lakes, ponds, streams, glaciers, and ice shelves, and they often dominate the total biomass and biological productivity in the polar environments (Vincent

2000). These mats typically consist of a matrix of mucilage in which cyanobacterial trichomes and other algal cells are embedded together with other heterotrophic and chemoautotrophic microorganisms and with sand grains and other inorganic materials (De los Ríos *et al.* 2004). In that way, they can be considered as complete ecosystems, with all trophic levels compressed into a layer only a few millimeters in thickness. The stratification of these communities can often be recognized by some degree of vertical color zonation due to the different pigmentation of phototrophic microorganisms (Vincent *et al.* 1993b), or associated with annual growth cycles by layers usually formed by seasonal sediment cycles rather than fast and slow growth patterns (Hawes *et al.* 2001). Advanced electron-microscopy techniques have shown that instead of a dense tissue, typical microbial mats are sponge-like structures with numerous holes allowing gliding filamentous cyanobacteria to adjust their position, vertically and horizontally, depending upon the environmental characteristics (De los Ríos *et al.* 2004). The ubiquitous presence of cyanobacterial mats and their taxonomic composition and physiological activities in streams, ponds, lakes, and meltwaters of different places in Antarctica are well documented (Howard-Williams and Vincent 1989; Howard-Williams *et al.* 1989a; Vincent *et al.* 1993a, 1993b; Ellis-Evans 1996; Fernández Valiente *et al.* 2001; Sabbe *et al.* 2004). Cyanobacterial mats are also common in shallow Arctic aquatic ecosystems (Villeneuve *et al.* 2001), but they have not yet been reported in the deep waters of Arctic lakes. On the contrary, benthic chlorophytes may only occasionally dominate in stream habitats, particularly in nutrient-rich conditions. For example, dense filamentous growths of *Mougeotia* and *Klebsormidium* were recorded in the Antarctic Peninsula area by Hawes (1989) and filamentous *Prasiola calophylla*, and *Binuclearia* and thalloid *Prasiola crispa* were common in some streams in the McMurdo Sound area (Broady 1989). Aquatic moss communities are found commonly in lakes of both polar regions (Imura *et al.* 1999; Sand-Jensen *et al.* 1999; Toro *et al.* 2007) and their standing stocks may be the dominant biomass component of some high-latitude aquatic ecosystems.

10.2 Types of benthic community

10.2.1 Microbial mats

10.2.1.1 Benthic communities in perennially ice-covered lakes

Perennially ice-covered lakes occur in many parts of the Antarctic and High Arctic. Doran *et al.* (2004) list five Arctic and 11 Antarctic lakes or lake regions that, at the moment, are perennially ice-covered. Perennial ice cover imparts quite different characteristics than seasonal ice cover. These effects include isolation from the atmosphere, which influences gas exchange and transfer of kinetic energy from wind to water, and the reduction in irradiance reaching the water column (see Chapter 4 in this volume).

A subset of lakes found in the McMurdo Sound, Dry Valley region of Antarctica (see Chapter 1) is the best-studied group of perennially ice-covered lakes. These are desert lakes, receiving low but variable water loadings each year, and most are endorheic. As the annual influx of water is highly variable over timescales of years to millennia, most have undergone sequential periods of positive and negative water balance, resulting in dry-down (concentrating salts) and refilling (dilute water influx) events. This, coupled with the lack of wind-induced mixing, has allowed the evolution and persistence of strong salinity gradients (see Chapter 8). The result can be stratification or stable gradients within the water column, through which the transfer of materials is dependent on diffusion, active movement of biota, and sedimentation. Diffusion is an effective process for the movement of solutes over short distances, up to millimeter spatial scales, but thick layers of stable water within density gradients severely restrict the flux of solutes and fine p articles (Spigel and Priscu 1998).

A second important feature of the McMurdo Dry Valley ice-covered lakes is that they are free of Metazoa large enough to cause substantial bioturbation of sediments or to graze substantially on benthic phototrophs. This is not the case in at least one Arctic perennially ice-covered lake, Ward Hunt Lake at 83°N on Canada's Ellesmere Island (Bonilla *et al.* 2005). Here, the sediments are densely populated by an unidentified chironomid larva, which enter and leave the lake via the ice-free margins during summer. The under-ice benthic vegetation in Ward Hunt Lake is sparse and dominated by diatoms and aquatic mosses, with cyanobacterial mats restricted to the lake margins.

10.2.1.2 The benthic flora of McMurdo Dry Valley lakes

Studies on the benthic communities of the McMurdo Dry Valley lakes consistently show that filamentous cyanobacteria, primarily species of *Lyngbya*, *Leptolyngbya*, *Oscillatoria*, and *Phormidium*, dominate the matrix of these microbial mats, with a range of pennate diatoms also present, increasing in relative abundance with increasing depth (Wharton *et al.* 1983; Wharton 1994; Hawes and Schwarz 1999). As with the cyanobacteria, all of the dominant diatoms are motile, with species of *Diadesmis*, *Achnanthes*, *Navicula*, *Muelleria*, *Luticola*, and *Stauroneis* commonly encountered. Mats form a variety of macroscopic structures on the lake floor, variously categorized by their gross morphology as pinnacle mats, lift-off mats, or columnar or prostrate mats (Wharton *et al.* 1983; Wharton 1994). In many situations, precipitation of calcite occurs within the mat, giving them further structural complexity.

Recent research has shown that fine-scale laminations (<1–3 mm) of mats in McMurdo Dry Valley lakes can be attributed to annual growth layering (Hawes *et al.* 2001), whereas coarser laminations are due to episodic events where pockets of sediment find their way through the ice and fall to the lake floor (Squyres *et al.* 1991). Fine-scale laminations become compressed as they are buried beneath new growth and the water that forms much of the mat matrix is excluded. The laminated, calcite-rich structures have been investigated as modern analogs of ancient stromatolites, currently the earliest complex communities for which we have any fossil record (Parker *et al.* 1981) and also as the possibility that they may be the nearest earthly analog of putative ice-covered aquatic communities that may have formed and perished on Mars during the early stages of its evolution (e.g. Doran *et al.* 1998).

Although the existence of these benthic microbial mats and the complex structures that they form have been known about for many years, their activity and contributions to lake carbon flow and nutrient flux have only recently begun to be fully appreciated. It is clear that, where light permits, benthic microbial mats are sinks for carbon in these lakes; they are net photosynthetic systems and sequester carbon, allowing the gradual accumulation of organic layers on the floor of the lakes (Hawes *et al.* 2001). Indeed, one of the remarkable features of these mats is the very high biomass that accumulates (Figure 10.1). Data collected from multiple sites and multiple years from Lake Hoare show a slight trend of a gradual decline in benthic chlorophyll *a* (Chl-*a*) concentration with depth, with a great degree of variability at any given depth and site (Figure 10.1).

Undoubtedly the absence of benthic invertebrates or strong physical mixing allows slow accumulation of benthic biomass to occur without disruption, but it seems that it is highly efficient photosynthetic processes that allow this live biomass to accumulate enough carbon in summer to persist year-round. The high efficiency of summer photosynthesis revolves around efficient capture and utilization of incoming solar radiation. The light that penetrates ice cover has a spectrum weighted to the blue-green wavelengths, peaking at 450–500 nm with very little light at wavelengths longer than 600 nm (Howard-Williams *et al.* 1998; Hawes and Schwarz 2001). The cyanobacteria that dominate mat under ice have a distinctive pink appearance (see Plate 10), which is due to a high content of phycoerythrin, a light-harvesting pigment that forms part of cyanobacterial phycobilisomes and which absorbs strongly in the blue-green region, at 500–550 nm (Hawes and Schwarz 2001). Phycoerythrin concentrations greatly exceed those of phycocyanin and even Chl-*a* in mats from under perennial ice cover, in contrast to mats from the moat regions of these lakes that thaw out annually, and where mats must cope with seasonally high irradiances (Hawes and Schwarz 1999).

Measurements show consistently that the pigment systems of benthic communities absorb over 50% of irradiance incident to a mat surface, and that this is utilized at quantum yields (carbon fixed per photon absorbed) which approach theoretical maxima of 12 photons per carbon atom fixed (Hawes and Schwarz, 2001; Vopel and Hawes 2006). In Lake Hoare, in the McMurdo Dry Valleys,

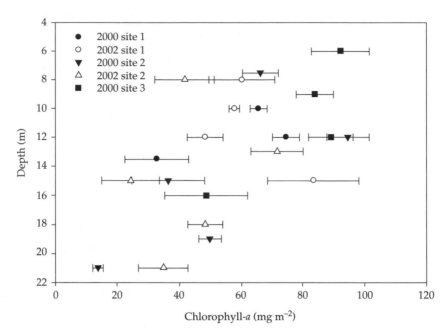

Figure 10.1 Biomass of benthic microbial mat as chlorophyll-*a* at three sites in Lake Hoare, McMurdo Dry Valleys, sampled in 2000 and 2002. Depths have been normalized to the nearest meter, based on lake depth in 2000. Methods used as described in Hawes and Schwarz (1999), unpublished author's data. Mean ± standard error, *n* = 5.

actively photosynthetic communities are found to at least 22 m depth, where *in situ* irradiance is not only restricted to a narrow waveband but also to extremely low values of 0.2–2 μmol photons m^{-2}s^{-1} (Hawes and Schwarz 1999). This barely exceeds 0.01–0.1% of midday summer irradiance in temperate regions. The low but persistent irradiance experienced during summer under perennial ice cover allows communities to acclimate to a narrow irradiance range in terms of total energy as well as spectral composition, and the light-utilization efficiency for these communities is very high.

Only recently have *in situ* observations confirmed that microbial mats are photosynthetically active under ambient conditions and that the rates of activity measured in laboratory incubations are indeed realistic estimates of *in situ* activity. Vopel and Hawes (2006) measured fine-scale gradients of oxygen concentration across the mat–water interface of Lake Hoare and confirmed that they are net sources of oxygen. Rates of net oxygen evolution measured were 100–500 μmol oxygen m^{-2}h^{-1} at an ambient irradiance of 1–5 μmol photons m^{-2}s^{-1}.

Vopel and Hawes (2006) were also able to confirm, using pulse amplitude-modulated fluorometric techniques, that layers remain photosynthetically active, albeit at much slower rates than surface layers, for many years after they have been formed, with active pigments persisting for up to seven or eight layers in mats from Lake Hoare. The number of active laminae decreases with depth.

The contribution of benthic communities to whole-lake production is strongly affected, as in all lakes, by the ratio of littoral to pelagic zones. However, as appears to be the case in oligotrophic lakes worldwide, within the littoral zone benthic production is considerably greater than that of the plankton in those ice-covered lakes for which comparisons can be made. Moorhead *et al.* (2005) estimated that in Lake Hoare benthic photosynthesis substantially exceeded that of phytoplankton on a whole-lake basis, a conclusion supported by Lawson *et al.* (2004) on the basis of stable isotopic ratios.

Several factors combine to allow benthic microbial mats to dominate biomass and activity in the

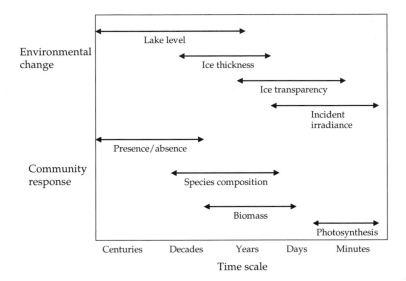

Figure 10.2 Schematic diagram of the correspondence between environmental change and community response in **benthic** microbial communities. At present our understanding is largely restricted to the right-hand end of this figure.

littoral zones of perennially ice-covered lakes. First, the absence of significant bioturbation or wave stress allows them to accumulate at millimeter rates to form thick, laminated sediments. Second, the unusual irradiance regime, with a relatively stable photon flux over the summer period allows them to photo-acclimate such that they are highly efficient over the range of irradiance that they experience. Finally the accumulation and recycling of nutrients within and beneath the mat complex appears to create a nutrient-rich microzone within an otherwise highly oligotrophic environment.

Whereas these observations conjure the image of a stable community in a stable environment, this is not necessarily the case. The physical environment within the lakes of the McMurdo Dry Valleys is not constant, but subject to climate-driven changes in both ice thickness and water level, over short and long time frames (Doran *et al.* 2002). The thickness and transparency of ice cover varies over time-scales of years to decades; for example, in our work at Lake Hoare we have measured transmission of photosynthetically available irradiance through the ice cover to vary from 0.1 to 2.5%. Snow cover is a key variable here. Such changes in transparency, resulting in a 25-fold range in irradiance reaching

the microbial mats, must affect their growth dynamics (Moorhead *et al.* 2005), although as yet this has to be effectively measured. A rise in lake level will both expose new substrates for microbial mat growth, and reduce the amount of light that reaches those already under the ice. Substantial lake-level reductions will expose mats to the air, and it has been suggested that this process, followed by drying and wind abrasion, is a significant in supplying fixed carbon to the soil ecosystem (Moorhead *et al.* 1999).

At any one time the microbial mats under the perennial ice cover are at best in a quasi-equilibrium state, as they adjust to daily, seasonal, decadal, and longer-term changes in their environment. The correspondence between time scales of habitat change and microbial responses is indicated in Figure 10.2. While we are beginning to understand the short-term responses, this understanding is still flawed since we are unable to place our experiments in the longer-term context of change and response. Paleolimnological studies offer some insights into long-term change (e.g. Spaulding *et al.* 1997). One of the key challenges facing research on these fascinating communities now is to understand the link between longer-term changes to their environment and community response. Approaching this problem, through a variety of approaches including physiological, long-term observation and paleolimnological methods, will allow us to much better understand the dynamics and ecosystem role of benthic microbial mats.

10.2.1.3 Mats in seasonally ice-covered freshwater ecosystems

Cyanobacterial mats are also widespread in shallow polar ecosystems that typically thaw in summer and may represent the most luxuriant phototrophic communities in non-marine ecosystems of the polar regions (Rautio and Vincent 2006). They are especially abundant in lakes and ponds, including thermokarst lakes, in High Arctic regions (Vézina and Vincent 1997), but also in ice-shelf ponds (see Chapter 6), and in ice-retreat ponds (Fernández-Valiente *et al.* 2007) in both polar regions. The physical structure and community composition of microbial mats from shallow ecosystem is apparently similar to that found in deeper water bodies. However, their environmental constraints are very different. Whereas in permanently ice-covered lakes one of the limitations for photosynthesis is the lack of light, in shallow lakes organisms are exposed to bright light, including high UV-A and UV-B radiation, and for a full 24-h day at high latitudes in summer. Under this situation benthic communities in shallow ecosystems need a number of strategies to minimize the pernicious effects of high radiation. First, the structure of the mat provides shade, in that way the surface layers are usually composed by material rich in carotenoids and sheath pigments (Quesada *et al.* 1999), characterized by very low oxygen evolution (e.g. Fernández-Valiente *et al.* 2007), indicating low physiological activity. On the contrary, active organisms are found in deeper layers, forming a bright blue-green band that is rich in photosynthetic pigments (Chl-*a* and phycobiliproteins; Quesada *et al.* 1999). Second, the capability of some cyanobacteria to migrate within the depth profile allows sensitive organisms to find the position with the most appropriate light climate (Quesada and Vincent 1997; Plate 10). Third, the capability of cyanobacteria to efficiently quench highly toxic, reactive oxygen species produced by exposure to high light by means of superoxide dismutase and quenching pigments allows them to alleviate the effects of high radiation (Quesada and Vincent 1997).

In shallow water bodies another important environmental constraint is temperature variation, which, unlike in perennially ice covered lakes, can change rapidly; for example from subzero freezing temperatures to +8°C within 24 h in the McMurdo Ice Shelf region (Figure 10.3). Photosynthetic rates in microbial mats increase with temperature, as has been shown by Velázquez *et al.* (2006) in Antarctic microbial mats, and similar results have been found in the High Arctic by Mueller *et al.* (2005). This indicates that many polar photosynthetic organisms are psychrotolerant rather than psychrophilic and are able to photosynthesize in a wider range of temperatures than usually encountered in their habitats.

In this way, the cyanobacterial mats from the two different ecosystems (permanently and seasonally ice-covered lakes) are subject to different ecological constraints and either are formed by different organisms, or show differences in the physical structure or comprise environmentally flexible organisms that can acclimate rapidly to highly variable conditions. In fact, recent studies using polyphasic methods suggest that the communities found in microbial mats from deep polar lakes may be different to the cyanobacterial consortia found in shallow ecosystems (see Chapter 12), although even the most refined genetic tools used to date have not identified particular phylotypes found exclusively in shallow or deep ecosystems.

Another important difference between both deep and shallow water ecosystems is the marginal zone. Small seasonally ice-covered fresh water ecosystems typically thaw first around the perimeter because of the direct effect of solar radiation over the edges of the water mass. Commonly in less extreme polar regions the liquid moat allows some movement to the still frozen ice plate, and the scouring of this ice on the littoral sediment and communities is one of the main constraints for the development of benthic communities, which are subsequently typically limited to the lake bottom at depths greater than the ice thickness. As a result, the littoral zone of the seasonally ice-covered lakes is little colonized by benthic communities and only

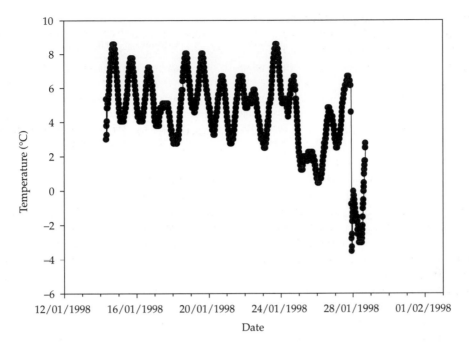

Figure 10.3 Temperature variations at the microbial mat surface in P70 pond in the McMurdo Ice Shelf in January 1998. E. Fernandez-Valiente and A. Quesada, unpublished results.

opportunistic organisms can inhabit these areas after the lake ice has melted. On the contrary perennially ice-covered lakes have a very rich littoral zone that receives high irradiance allowing higher growth rates, but without the physical disturbance of the ice moving and eroding the lake bottom. Even in those perennially ice-covered ecosystems, the littoral zone can melt but ice movement is rare. In these ecosystems ice melts at mid depths first, as the air is still cold while solar radiation heats up the water. Cold air keeps ice on the surface while the ice at 30–50 cm has already melted.

10.2.2 Aquatic mosses

Mosses are the commonest form of macrophytic vegetation in polar lakes, since at least in oligotrophic temperate lakes they tolerate extreme low irradiance. Typically they take the form of moss carpets (Priddle and Dartnall 1978), although 'moss pillars' – an assemblage of mosses and cyanobacteria forming a characteristic pillar structure – have also been described in some Antarctic lakes (Imura

et al. 1999). Mosses can grow at extraordinary depths in clear polar lakes, with a maximum depth reported so far of 81 m in Radok Lake, Amery Oasis, Antarctica. Mosses are quantitatively important components of the polar aquatic communities, often contributing to high biomass (Sand-Jensen *et al.* 1999) and to lake production (Welch and Kalff 1974; Priddle 1980a; Montecino *et al.* 1991; Ramlal *et al.* 1994).

Only a small number of moss genera are represented in the polar lake flora (Table 10.1), with a much larger diversity in high latitude wetlands. Within regions, similar taxa appear to dominate the flora and there are surprisingly similar genera in lakes of the northern and southern polar regions. Most aquatic species appear to be forms of the local terrestrial and semi-aquatic flora. Exceptions to this do occur, notably the genus *Leptobryum* that forms moss pillars in lakes near Syowa station (Plate 10) but that is apparently absent from the local terrestrial flora (Imura *et al.* 1999). Studies of these moss pillars provide a striking example of the rarity of aquatic mosses from the marginal zones of polar

Table 10.1 Mosses reported from various lakes in the Antarctic and High Arctic.

Mosses	Location	Latitude	Reference
Amblystegium sp., *Calliergon sarmentosum*, *Drepanocladus* sp.	Signy Island	60°S	Priddle (1980a)
Drepanocladus longifolius	Livingston Island	63°S	Toro *et al.* (2007)
Campyliadelphus polygamus	King George Island	64°S	Montecino *et al.* (1991)
Bryum pseudotriquetum	Bunger Hills	66°S	Savich-Lyubitskaya and Smirnova (1959)
Leptobryum pyriforme, Bryum pseudotriquetum	Syowa	69°S	Imura *et al.* (1992, 1999)
Plagiothecium simonovii	Schirmacher oasis	71°S	Savich-Lyubitskaya and Smirnova (1964)
Bryum pseudotriquetum	Amery Oasis	71°S	Wagner and Seppelt (2006)
Bryum cf *algens*	McMurdo Dry Valleys	78°S	Love *et al.* (1997)
Drepanocladus revolvens, Calliergon giganteum	Cornwallis Island	75°N	Sand-Jensen *et al.* (1999)
Drepanocladus revolvens, Calliergon giganteum	Axel Heiburg Island	79°N	Hawes *et al.* (2002)
Drepanocladus exannulatus, Scorpidium scorpioides	Greenland	81°N	Fredskild (1995)

lakes. Moss pillars occur throughout most ponds in this region, but never in the upper 2 m, which are ice-bound during winter (Imura *et al.* 1999). We have also noted that mosses are rare in the marginal zones of Arctic lakes, and that while moss propagules are present in shallow Antarctic ponds, such as those described above at Bratina Island (i.e. mosses grow out of axenic enrichment cultures of microbial mats), they are never seen in the ponds themselves.

The reason for the abundance of bryophytes in polar systems is often attributed to their abilities to grow and accumulate slowly, while losing little fixed carbon through respiration, decomposition, or grazing. This appears to be an inherent feature of aquatic mosses; they contribute much more to biomass and primary production than they do to fuelling secondary production (Heckey and Hesslein 1995). Low water temperatures in polar lakes appears to amplify this attribute through further suppression of respiratory losses (Priddle 1980b; Sand-Jensen *et al.* 1999), although most of the summer surplus production is still consumed under ice during winter (Welch and Kalff 1974). Welch and Kalff (1974) provided evidence that the lower depth limits of mosses in Char Lake, Canadian High Arctic, was set by the annual compensation point (where annual respiration equals annual photosynthesis). Because of the very fine balance between photosynthesis and respiration, Arctic mosses appear to grow just a few centimeters each

year, but can retain actively photosynthetic leaves on their stems for more than 10 years (Sand-Jensen *et al.* 1999; Hawes *et al.* 2002).

Priddle (1980a) estimated moss to cover 40% of the bottom of a small lake in the maritime Antarctic (Signy Island), and estimated total biomass, through a cleared-quadrat method, at 90 g m^{-2} as ash-free dry weight. Ramal *et al.* (1994) gave broadly similar estimates of biomass from a small lake in the Canadian Arctic (12–151 g dry weight m^{-2} for green tissue).

Comparison of photosynthesis in aquatic polar mosses is confused by the variety of methods and temporal coverage associated with different studies. The most complete, year-round or near-year-round studies are those of Priddle (1980a) at Signy Island in Antarctica and Welch and Kalff (1974) in the Canadian High Arctic. These authors carried out temporally extensive investigations and reported respective annual production of 400 and 1600 µg C cm^{-2} year^{-1} (converted from oxygen by Priddle 1980a). Shorter-term measurements have been made, but these offer limited insight into the production dynamics of these long-lived, slow-growing, and persistent moss-based autotrophic benthic communities that can dominate primary production in oligotrophic systems (e.g. Montecino *et al.* 1991; Ramlal *et al.* 1994).

Perhaps the best approach to assessing the performance of polar mosses is a reconstruction technique developed by Riis and Sand-Jensen (1997) and applied in three Arctic lakes (Sand-Jensen *et al.* 1999; Hawes *et al.* 2002). This exploits the annual banding of the mosses, evident as longitudinal waves of leaf length and spacing, which result from the extreme seasonality of irradiance at high latitude and with thick ice cover. During summer, large, closely spaced leaves are produced, with small, spaced-out leaves in spring and autumn. This growth pattern results in annual segments, which can be used to determine the annual growth. Sand-Jensen *et al.* (1999) recorded a remarkably constant year-to-year biomass accumulation in Char Lake, in the Canadian High Arctic, and Hawes *et al.* (2002) discussed the use of reconstruction to assess historical changes in growth conditions. We know of no applications of this technique to date in Antarctic waters.

As in lakes worldwide, the complex, three-dimensional matrix of benthic macrophytes creates habitats for a wide range of epiphytic flora and fauna. Turnover of carbon within the epiflora is likely to be much faster than that of the moss itself. Priddle and Dartnall (1978) have described a complex mixture of algae and invertebrates living within a moss carpet, with the diet of most invertebrates being algae rather than mosses. In Byers Peninsula (Livingston Island in maritime Antarctica), Toro *et al.* (2007) found that the populations of important limnetic invertebrates such as chironomid larvae (*Parochlus steinenii*) and the cladoceran *Macrothrix ciliate* were associated with the presence of the aquatic moss *Drepanocladus longifolius*. Other than these studies, experimental investigation of the ecological role of benthic moss carpets in polar lakes is rather sparse. The ecological reasons leading to the dominance of mosses and/or cyanobacteria are unknown.

10.2.3 Benthic communities in running waters

High-latitude polar streams whose source waters are from snow or glacier melt are highly seasonal, usually short (first- or second-order) and exposed with no riparian vegetation. In the Arctic, particularly in Canada and Russia, there are large rivers that flow into high latitudes but whose source waters arise from much lower latitudes. These are not dealt with here (see Chapter 5).

The hydrology of high-latitude melt streams, and hence the ecosystems they support, is determined from seasonal and daily heat balance on snow and ice surfaces rather than from precipitation. The resulting strong diurnal signals in flow superimposed on the seasonal melt and freeze cycles constitute a harsh environment for ecosystem function in both Antarctic and Arctic streams (Vincent *et al.* 1993c; Howard-Williams and Hawes 2007 and references therein). Relatively high ultraviolet radiation, desiccation, and daily freeze–thaw cycles impose constraints on ecosystem components but microbial mat communities dominated by cyanobacteria appear to be able to cope well with these. This community type is the most common in polar streams (e.g. as in Svalbard, 79°N; Kubečkova *et al.* 2001), although in some streams Chlorophyceae and Bacillariophyceae may dominate at some stages in the development of a stream benthic flora. Chlorophyceae are most common in nutrient-rich conditions (Moore 1974a, 1974b; Hawes 1989; Hawes and Brazier 1991). Cohesive mats in stream channels are especially common in Arctic and Antarctic streams where stable substrates (cobbles, boulders, or gravels) are found whereas epipsammic diatoms (*Navicula*, *Hantzchia*, and *Stauroneis*) may predominate on sandy substrates that may be disturbed by variable stream flows (Hawes and Howard-Williams 1998), and epipelic diatoms (*Achnanthes* and *Tabellaria* spp.) were the dominant benthic producers in fine sediment-based Arctic streams (68°N) on Baffin Island (Moore 1974a). Epilithic datioms dominate the flora in the Kuparuk River, North Slope, Alaska, with the genera *Hannaea*, *Cymbella*, *Achnanthes*, *Synedra*, and *Tabellaria* being most common. Short stream lengths and daily hydrological changes due to freeze–thaw cycles in summer mean that planktonic production in high-latitude streams is generally negligible.

In Antarctic streams dominated by benthic Chlorophyceae, the summer biomass is destroyed by winter desiccation and freezing and new biomass is formed each summer. For instance, the viability of the dense filamentous *Zygnema* community in streams at Signy Island depends on a few highly resistant cells that survive winter. In contrast, Antarctic stream-bed cyanobacterial mats may survive winter almost intact and in a 'freeze-dried' state, and rapidly recover photosynthetic capacity when meltwater floods the stream channels the next summer (Vincent and Howard-Williams 1986).

Vincent *et al* (1993a) summarized stream biomass values in Antarctica. The highest recorded mean values were those on Signy Island, ranging from 69 mg Chl-a m^{-2} for the chlorophytes *Mougeotia* and *Zygnema* to 198 mg Chl-a m^{-2} for cyanobacteria-dominated communities. Maximum biomass values were also those of cyanobacteria, particularly *Nostoc* (up to 226 mg Chl-a m^{-2}) and *Phormidium* (up to 400 mg Chl-a m^{-2}). These high biomass values were associated with high stream-bed cover by the autotrophic benthic communities of 50–100%. In sandy sections of the Onyx River dominated by benthic diatoms, a Chl-a biomass value of 8 mg m^{-2} was recorded (Vincent *et al*. 1993c). Diatom-dominated Chl-a biomass on the Kuparuk River, Alaska, varied considerably from not detectable to 21.9 mg m^{-2} depending on time of season and between low-flow (high-biomass) and high-flow (low-biomass) years. The average over seasons and sites for 2 years was 3.36 mg m^{-2} (Peterson *et al*. 1986; Slavik *et al*. 2004).

The major difference between High Arctic and Antarctic stream communities lies in the limited occurrence or even absence of grazers in the Antarctic. Grazing invertebrate larvae (e.g. the blackfly *Prosimulium*, the midge *Orthockadius*, and the stone fly *Nemoura*) are commonly found in Arctic streams and may play an important role in biomass control of benthic producers (Peterson *et al*. 1985; Miller and Stout 1989; Kubečkova *et al*. 2001). Moore (1974a) found that light and temperature were the primary controls on epipelic algal numbers with grazing and nutrient supply having little effect. In the Alaskan Arctic when long-term phosphorus fertilization was applied to the Kuparuk River, maximum photosynthesis (P_{max}) values increased from 2.5 to 8.9 μg C cm^{-2} h^{-1}, indicating nutrient limitation in this High Arctic river system (Peterson *et al*. 1993). Combined nitrogen and phosphorus enrichment of the river resulted in an eventual and dramatic shift from benthic algae to mosses (Plate 7; see Bowden *et al*. 1994). The enrichment effect of ntriuents is also seen in the sewage-influenced stream discharging into Meretta Lake, at Resolute Bay in the Canadian

High Arctic. In contrast with flowing waters nearby, this stream has prolific moss communities. In Antarctic streams, temperature was found to be the primary control on benthic biomass accumulation and photosynthesis (Vincent and Howard-Williams 1989) with nutrients determined to be not limiting, at least over timescales of days to weeks. Experimentally derived temperature optima for these communities are well above ambient (Hawes and Howard-Williams 1998).

High light and 24-hour days, combined with adequate nutrients from glacial or snowmelt and nitrogen fixation in some cases meant that photosynthetic rates of stream benthos in midsummer were similar to those in some temperate streams. Nitrogen fixation was found to contribute 10% to the total nitrogen uptake of the mats in Fryxell Stream, now officially named Canada Stream (details in Howard-Williams *et al*. 1989b). The growing season at latitudes above 70°S in the Antarctic is often very short due to a restricted period of available melt water (2–8 weeks) rather than solar radiation. The high biomass of benthic mats and filaments there can only be accounted for by a significant over-wintering stock that remains viable, combined with a lack of grazing.

10.3 Benthic primary production

The photosynthetic rate of the **benthic** component in different habitats and locations of polar regions is quite similar, with typical values within the range of 1–10 μg C cm^{-2} h^{-1} (Table 10.2). It is remarkable that photosynthetic rates are quite similar in mats from different latitudes within polar regions and from different types of habitats, indicating an ecological maximum for these communities in high-latitude regions. In mats from less extreme latitudes, even in tropical areas, the photosynthetic rates vary by three orders of magnitude, although most studies are within the same range as the polar communities (e.g. Paerl *et al*. 2000). Castenholz *et al*. (1992) attributed the large variance found to methodological differences.

In addition to photosynthesis, chemoautotrophic carbon assimilation can also contribute to primary production in autotrophic benthic communities. Studies performed in benthic mats from five ponds

Table 10.2 Primary productivity in polar regions.

Location	Habitat	Photosynthetic rate ($\mu g\ C\,cm^{-2}\,h^{-1}$)	Reference
Byers Peninsula (Livingston Island, Antarctica)	Ponds	1.5–6	Fernández- Valiente *et al.* (2007)
Lake Bonney (Dry Valleys, Antarctica)	Moat	0–18.8[a]	Parker and Wharton (1985)
McMurdo Ice Shelf (Antarctica)	Ponds	1–4	Vincent *et al.* (1993a)
Ward Hunt Ice Shelf (High Arctic)	Ponds	2.7–10	Mueller *et al.* (2005)
Lake Hoare (Dry Valleys)	Moat	2.8[a]	Hawes and Schwarz (1999)
Lake Hoare (Dry Valleys)	Deep (under perennial ice cover)	2[a]	Hawes and Schwarz (1999)

[a] Calculated from a daily rate (divided by five as in Castenholz *et al.* 1992).

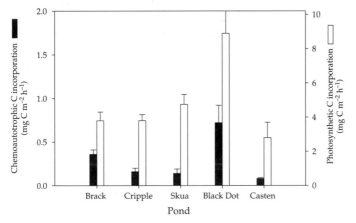

Figure 10.4 Comparison between photosynthetic and chemosynthetic carbon assimilation in five ponds of Bratina Island (McMurdo Ice Shelf). E. Fernández-Valiente, M. Sánchez-Contreras, and A. Quesada, unpublished results.

in Bratina Island, McMurdo Ice Shelf, showed consistent rates of chemoautotrophic carbon assimilation that were 3–10% of those of photosynthetic carbon assimilation (E. Fernández-Valiente and A. Quesada, unpublished results; Figure 10.4). However, this activity was not observed in mats from Byers Peninsula, Livingston Island (Fernández-Valiente *et al.* 2007).

In perennially ice-covered lakes, the photosynthetic rates of benthic elements may be less than that of the plankton, due probably to light limitation. Areal benthic Chl-*a* concentrations within mat communities can be very high in comparison with integrated areal concentrations of Chl-*a* for planktonic communities. For example, at 30 m depth in Lake Vanda, close to the mid-point of the photic zone, benthic undegraded Chl-*a* was measured at

151 mg m^{-2} in December 1999 (Hawes and Schwarz 2001). Integrating planktonic Chl-*a* from 5 to 65 m (that is, the planktonic photic zone), for the same time period, gives a total of only 5.3 mg m^{-2} (I. Hawes, unpublished results). The differences are much more remarkable in ponds or shallow lake ecosystems, seasonally ice-covered, in which the benthic component of the ecosystem shows standing stocks and primary production to be several orders of magnitude larger than the planktonic component. For instance in Limnopolar lake (Byers Peninsula, Livingston Island, Antarctica) the integrated planktonic standing stock of photosynthetic organisms ranges between 2.6 and 3.6 µg C cm^{-2}. However, the estimated standing stock of the lake bottom – which is covered by a very rich community of subaquatic mosses (*Drepanocladus longifolius*;

Toro *et al.* 2007), and with cyanobacteria (*Nostoc* sp. colonies), green algae, and diatoms as epiphytes – is about three orders of magnitude larger than the planktonic standing stock (E. Fernández-Valiente, A. Justel, C. Rochera and A. Quesada, unpublished results). Nevertheless, the photosynthetic rate is 2-fold greater in the planktonic than in the benthic habitat in Limnopolar Lake, in agreement with ponds carpeted by microbial mats (see above). The differences in standing stock and production of the communities from both habitats can be explained by different temporal strategies. The autotrophic benthic communities follow the 'lichen strategy': they grow slowly, but are capable of remaining under very adverse conditions and as soon as the environment allows them to grow, they increase their biomass with lower grazing pressure. On the other hand, the planktonic community is more opportunistic, showing high growth rates that are necessary to compensate the loss processes, including higher grazing pressure and wash-out events.

We hypothesize that the high benthic production in ice-covered lakes is due to the high biomass, which accumulates over long time periods, decades, or even longer. This accumulation is due to slow loss processes and may be further supported by cycling of nutrients within a relatively closed sediment-mat system. Earlier we suggested that the fluxes of solutes in stratified lakes were restricted to diffusion, an ineffective process at large spatial scales. However, microbial mats are structured on scales where diffusion can be effective (Paerl *et al.* 2000). Close alignment of microbes that mediate cascade-like nutrient transformations is possible within steep gradients of metabolite concentration, for example those responsible for nitrogen cycling, means that diffusive flux is an effective means of transport. We hypothesize that when the active surface layer of mat has a sufficiently high affinity for nutrients to prevent these being returned to the water column, it allows the mat structure to recycle and accumulate nutrients delivered from the water column, acting as a one-way trap. In this way the benthic autotrophic community will act as a long-term sink for nutrients entering the lake and the constant supply of recycled nutrients from underlying layers fuels growth in those upper layers that receive sufficient irradiance.

In support of this hypothesis, data show that benthic communities are sinks for carbon, nitrogen and phosphorus, as the accumulating layers that are slowly buried under new growth contain these elements (Figure 10.5). Furthermore, the limited data available on concentrations of nitrogen and phosphorus within the mat pore water (i.e. within the microbial mat matrix) are higher than in the water column, with a diffusion gradient from deep in the mat to the active layer (Table 10.3). Similar nutrient distribution has been described in mats from Arctic shallow ecosystems (Bonilla *et al.* 2005). In some polar microbial mats it is common to find thick layers of dead but undegraded biological material at the bottom of the mat, indicating slow decomposition processes, and thus nutrient recycling, in these extreme environments. However, nutrient concentrations within mat interstitial waters (Table 10.3) are typically very high, indicating that growth of the mat primary producers should not be nutrient-limited, as shown by Bonilla *et al.* (2005) in enrichment experiments with Arctic microbial mats.

In spite of the apparent nutrient sufficiency in complete microbial mats, nitrogen-fixing microorganisms are abundant in polar cyanobacterial mats and nitrogen fixation, determined *in situ*, has been shown to be a considerable fraction of the nitrogen input into the ecosystem. It was found to contribute at least one-third of the nitrogen requirements of microbial mats in the McMurdo Ice Shelf communities (Fernández-Valiente *et al.* 2001). In mats from Byers Peninsula (maritime Antarctica) nitrogen fixation constituted up to 23% of the assimilated nitrogen (Fernández-Valiente *et al.* 2007). Many of the nitrogen fixers, notably species of *Nostoc*, are located at the mat surface, or in some ponds at the water surface, often in a foliose form spatially separated from the underlying oscillatorian matrix, where interstitial nitrogen concentration may be high. However, other attributes such as the ability to tolerate desiccation and high ultraviolet irradiance, may explain their presence within mats.

Photosynthetic parameters for Antarctic stream benthic mats are summarized in Table 10.4. Polar stream benthic producers are usually exposed to full sunlight in shallow streams and most are characterized as 'high light' communities in terms

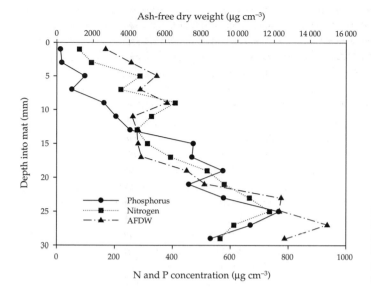

Figure 10.5 Accumulation of ash-free dry weight (AFDW), nitrogen, and phosphorus with depth into a core of microbial mat from 12.6 m depth in Lake Hoare, McMurdo Dry Valleys. The data are means of two replicate cores which were frozen and then sectioned into 2 mm segments.

Table 10.3 Concentrations of nitrogen and phosphorus species in interstitial waters within three layers of a microbial mat from 12.6 m and in overlying water in Lake Hoare, McMurdo Dry Valleys, in December 2000. Overlying water was collected into acid-washed syringes, while mat water was extracted by gently squeezing mat layers in an acid-washed, 60-ml syringe that had been pre-rinsed by making a preliminary extraction. Upper, mid, and lower mat layers refer to the cohesive and pigmented, photosynthetic layer approximately 10 mm thick, a cohesive but unpigmented layer immediately below, also roughly 10 mm thick, and finally an uncohesive layer below this cut to 1 cm thick, respectively. Overlying and expressed mat water was filtered through a glass fibre filter (type GF/F), frozen, and analysed in New Zealand using standard methods. Values are means of three replicates. FRP, filterable reactive phosphorus; FOP and FON, filterable organic phosphorus and nitrogen (I Hawes, unpublished results).

	Concentration (mg m^{-3})				
	FRP	NH$_4$'$-$N	NO$_3^-$$-$N	FOP	FON
Water	2.1	16.9	58.8	0.7	236
Upper mat extract	7.8	14.1	42.9	43.8	778
Middle mat layer	15.5	52.15	72.7	37.2	838.5
Lower mat extract	33.2	101.3	147.5	80.25	2068.5

of their photosynthetic parameters (Howard-Williams and Vincent 1989). These show high levels of maximum photosynthesis (P_{max}), and a high light saturation value (E_k) of approximately 100 µmol m^{-2}s^{-1} or more (Table 10.4). Maximum rates of gross photosynthesis normalized to unit area were particularly high in Antarctica for dense *Nostoc*-dominated communities. Recorded P_{max} values ranged from 0.4 and 12 µg C cm^{-2}h^{-1} in

the Antarctic streams and 0.8–2.5 µg C cm^{-2}h^{-1} in the Arctic, with the highest values recorded in the Nostocales, largely because of significant accumulations of *Nostoc* biomass in some streams often as high as 60 µg Chl-*a* cm^{-2} (Nigoyi *et al.* 1997; McKnight *et al.* 1998). These values of P_{max} are comparable with those in temperate streams. The exceptions were *Gloeocapsa* and *Prasiola calophylla* that occurred in crevices between rocks or in shade under boulders

Table 10.4 Summary of ranges of photosynthesis–irradiance relationships for Antarctic stream communities (adapted from Hawes and Howard-Williams 1998). P_{max}, maximum rate of gross photosynthesis normalized to unit area; E_k, light saturation onset parameter; α, slope of photosynthesis–irradiance curve at low irradiance; R, dark respiration; ND, no data. References: 1, Hawes (1993); 2, Howard-Williams and Vincent (1989); 3, Vincent and Howard-Williams (1986); 4, Hawes and Howard-Williams (1998); 5, J.H. Duff and C.M. Tate, unpublished results.

Community	P_{max} (μg C cm^{-2} h^{-1})	E_k (μmol m^{-2} s^{-1})	α (P_{max}/E_k)	R (μg C cm^{-2} h^{-1})	References
Oscillatoriaceae	0.4–6.5	75–300	0.01–0.06	0.07–5.7	1–5
Nostocaceae	0.2–8.0	60–150	0.02–0.15	0.24–3.2	2, 3, 5
Gloeocapsa	0.4	40	0	ND	4
Chlorophyceae, *Prasiola calophylla*	3.5	20	0.17	ND	2
Chlorophyceae, *Binuclearia*	2.8	70	0.04	ND	2

(Hawes and Howard-Williams 1998). Hence these were better adapted to moderate or low-light conditions. Assimilation numbers (photosynthetic rates per unit of Chl-*a* biomass) were very low, often less than 0.1 μg C (μg Chl-*a*)$^{-1}$ h^{-1} (Vincent and Howard-Williams 1986) and generally not comparable to temperate stream communities. Low assimilation numbers are due to the high accumulations of Chl-*a* that degrades very slowly in these cold environments (Vincent and Howard-Williams 1989). Consequently much of the recorded biomass may be non-active Chl-*a* that accumulates in the benthic mats, skewing the calculation of assimilation numbers. Further discussion on low assimilation numbers in Antarctic microbial mats was provided in Howard-Williams *et al.* (1989a) who showed that Chl-*a* when determined spectrophotometrically over-estimated high-performance liquid chromatography (HPLC)-measured Chl-*a* by 21–45% as it also included the non-active epimer Chl-*a'* and some of the degradation product chlorophyllide.

10.4 Conclusions

Benthic autotrophic communities are extremely important in the polar regions where they frequently dominate the biomass of inland water ecosystems. This appears to be due to the long-term strategy of slow growth and accumulation each year, coupled with multiyear persistence and activity. Recognizable annual features in the growth of mosses in seasonally ice covered Arctic lakes and annual layering in cyanobacterial mats in perennially ice-covered Antarctic lakes are clearly demonstrated: photosynthetically active elements can be found in systems many years old, and the annual growth is a minute fraction of the standing biomass. The marked persistence of such communities in waters that do not freeze each year (deep, ice-capped lakes) requires that net photosynthesis in summer offsets winter respiration, and this appears to be a fine balance in this type of community (Welch and Kalff 1974; Hawes *et al.* 2001).

Benthic microbial mats in shallow water environments also reflect a perennial, *K*-type strategy, although their stresses are different to those discussed above for deep-water communities. It is notable that the respiration/photosynthesis ratio in mature shallow-water communities appears to be close to a balance during the summer growth period. This would appear to offer little or no scope for utilization of summer-photosynthate during winter darkness. We suggest that this simply reflects the fact that these communities are frozen for most of the dark period, thus metabolic requirements are exceedingly small. We suggest that communities from annually frozen and unfrozen areas are thus fundamentally different; one is structured around the stresses of low annual irradiance and prolonged darkness, the other around stresses of high summer irradiance and winter freezing. Lotic meltstream habitats present a special case with the autotrophic benthic communities needing to survive winter dessication, summer freeze–thaw cycles, and high irradiance.

The ability to maintain biomass in an over-wintering state that allows rapid re-establishment of metabolic processes the following summer is a key adaptation to the polar environment. Benthic cyanobacteria appear to be especially successful under all of these conditions.

Acknowledgments

Part of this research was funded by the Spanish Polar Programme grants REN2000-0435/ANT and CGL2005-06549-CO2-01. We thank Dr de los Ríos and Dr Ascaso for the electron microscopy, Dr Imura for the moss pillar, and Mr Toro for the stream pictures. We also thank the New Zealand Foundation for Research Science and Technology for continuing support grants since 1992. We are also grateful for the comments of Dr W.F. Vincent and external reviewers.

References

Bonilla, S., Villeneuve, V., and Vincent, W.F. (2005). Benthic and planktonic algal communities in a high Arctic lake: pigment structure and contrasting responses to nutrient enrichment. *Journal of Phycology* **41**, 1120–1130.

Bowden, W.B., Finlay, J.C., and Maloney, P.E. (1994). Long-term effects of PO_4 fertilization on the distribution of bryophytes in an arctic river. *Freshwater Biology* **32**, 445–454.

Broady, P.A. (1989). Broadscale patterns in the distribution of aquatic and terrestrial vegetation at 3 ice-free regions on Ross Island, Antarctica. *Hydrobiologia* **172**, 77–95.

Castenholz, R.W., Bauld, J., and Pierson, B.K. (1992). Photosynthetic activity in modern microbial mat-building communities. In Schopf, J.W. and Klein, C. (eds), *The Proterozoic Biosphere. A Multidisciplinary Study*, pp.279–286. Cambridge University Press, Cambridge.

De los Ríos, A., Ascaso, C., Wierzchos, J., Fernández-Valiente, E., and Quesada, A. (2004). Microstructural characterization of cyanobacterial mats from the McMurdo Ice Shelf, Antarctica. *Applied Environmental Microbiology* **70**, 569–580.

Doran, P.T., Wharton, Jr, R.A., Marais, D.J.D., and McKay, C.P. (1998). Antarctic paleolake sediments and the search for extinct life on Mars. *Journal of Geophysical Research* **103**, 28481–28493.

Doran, P.T. *et al.* (2002). Antarctic climate cooling and terrestrial ecosystem response. *Nature* **415**, 517–520.

Doran, P.T. *et al.* (2004). Palaeolimnology of extreme cold terrestrial and extraterrestrial environments. In Pienitz, R., Douglas, M.S.V., and Smol, J.P. (eds), *Long term Environmental Change in Arctic and Antarctica Lakes*, pp. 475–507. Springer, Netherlands.

Ellis-Evans, J.C. (1996) Microbial diversity and function in Antarctic freshwater ecosystems. *Biodiversity Conservation* **5**, 1395–1431.

Fernández-Valiente, E., Quesada, A., Howard-Williams, C., and Hawes, I. (2001). N_2-fixation in cyanobacterial mats from ponds on the McMurdo Ice Shelf, Antarctica *Microbial Ecology* **42**, 338–349.

Fernández-Valiente, E. *et al.* (2007). Community structure and physiological characterization of microbial mats in Byers Peninsula, Livingston Island (South Shetland Islands, Antarctica). *FEMS Microbiology Ecology* **59**, 377–385.

Fredskild, B. (1995). Palymnology and sediment slumping in a High Arctic Greenland lake. *Boreas* **24**, 345–354.

Hawes, I. (1989). Filamentous green-algae in fresh-water streams on Signy Island, Antarctica. *Hydrobiologia* **172**, 1–18.

Hawes, I. and Brazier, P. (1991). Fresh-water stream ecosystems of James-Ross Island, Antarctica. *Antarctic Science* **3**, 265–271.

Hawes, I. (1993). Photosynthesis in thick cyanobacterial films; a comparison of annual and perennial Antarctic mat communities. *Hydrobiologia* **252**, 203–209.

Hawes, I. and Howard-Williams, C. (1998). Primary production processes in streams of the McMurdo Dry Valleys, Antarctica. In Priscu, J.C. (ed.), *Ecosystem Dynamics in a Polar Desert: The McMurdo Dry Valleys, Antarctica*, pp. 129–140. Antarctic Research Series, vol. 72. American Geophysical Union, Washington DC.

Hawes, I. and Schwarz, A.-M. (1999). Photosynthesis in an extreme shade environment: benthic microbial mats from Lake Hoare, a permanently ice-covered Antarctic lake. *Journal of Phycology* **35**, 448–459.

Hawes, I. and Schwarz A.-M. (2001). Absorption and utilization of low irradiance by cyanobacterial mats in two ice-covered Antarctic lakes. *Journal of Phycology* **37**, 5–15.

Hawes, I., Moorhead, D., Sutherland, D., Schmeling, J., and Schwarz, A.-M. (2001). Benthic primary production in two perennially ice-covered Antarctic Lakes: patterns of biomass accumulation with a model of community metabolism. *Antarctic Science* **13**, 18–27.

Hawes, I., Andersen, D., Pollard, T., and Wayne, H. (2002). Submerged aquatic bryophytes in Colour Lake, a naturally acidic polar lake with occasional year-round ice-cover. *Arctic* **55**, 380–388.

Heckey, R.E. and Hesslein, R.H. (1995). Contributions of benthic algae to lake food webs as revealed by stable isotope analysis. *Journal of the North American Benthological Society* **14**, 631–653.

Howard-Williams, C. and Vincent, W.F. (1989). Microbial communities in southern Victoria Land streams (Antarctica) I: Photosynthesis. *Hydrobiologia* **172**, 27–38.

Howard-Williams, C. and Hawes, I. (2007). Ecological processes in Antarctic inland waters: interactions between physical processes and the nitrogen cycle. *Antarctic Science* **19**, 205–217.

Howard-Williams, C., Pridmore, R., Downes, M.T., and Vincent, W.F. (1989a). Microbial biomass, photosynthesis and chlorophyll *a* related pigments in the ponds of the McMurdo Ice Shelf, Antarctica. *Antarctic Science* **1**, 125–131.

Howard-Williams, C., Priscu, J.C., and Vincent, W.F. (1989b). Nitrogen dynamics in two Antarctic streams. *Hydrobiologia* **172**, 51–61.

Howard-Williams, C., Schwarz, A.-M., Hawes, I., and Priscu, J.C. (1998). Optical properties of the McMurdo Dry Valley lakes, Antarctica. In Priscu, J.C. (ed.), *Ecosystem Dynamics in a Polar Desert: The McMurdo Dry Valleys, Antarctica*, pp. 189–203. Antarctic Research Series, vol. 72. American Geophysical Union, Washington DC.

Imura, S., Higuchi, M., Kanda, H., and Iwatski, Z. (1992). Culture of rhizoidal tubers on an aquatic moss in the lakes near Syowa Station area, Antarctica. *Proceedings NIPR Symposium on Polar Biology* **5**, 114–117.

Imura, S., Bando, T., Saito, S., Seto, K., and Kanda, H. (1999). Benthic moss pillars in Antarctic lakes. *Polar Biology* **22**, 137–140.

Kubečkova, K., Elster, J., and Kanda, H. (2001). Periphyton ecology of glacial and snowmelt streams, Ny-Ålesund, Svalbard: presence of mineral particles in water and their erosive activity. *Nova Hedwigia* **123**, 141–172.

Lawson, J., Doran, P.T., Kenig, F., des Marias, D.J., and Priscu, J.C. (2004). Stable carbon and nitrogen isotopic composition of benthic and pelagic organic matter in lakes of the McMurdo Dry Valleys, Antarctica. *Aquatic Geochemistry* **10**, 269–301.

Love, F.G., Simmons, Jr, G.M., Parker, B.C., Wharton, Jr, R.A., and Seaburg, K.G. (1997). Modern conophyton-like microbial mats discovered in Lake Vanda, Antarctica. *Geomicrobiology Journal* **31**, 33–48.

McKnight, D.M., Alger, A., Tate, C.M., Shuoe, G., and Spaulding, S. (1998). Longitudinal patterns in algal abundance and species distribution in meltwater streams in Taylor Valley, Southern Victoria Land. Antarctica. In Priscu, J.C. (ed.), *Ecosystem Dynamics in a Polar Desert: The McMurdo Dry Valleys, Antarctica*, pp. 109–127. Antarctic Research Series, vol. 72. American Geophysical Union, Washington DC.

Miller, M.C. and Stout, J.R. (1989).Variability of macroinvertebrate community composition in an Arctic and subarctic stream. *Hydrobiología* **172**, 111–127.

Montecino, V., Pizarro, G., Cabrera, S., and Contreras, M. (1991). Spatial and temporal photosynthetic compartments during summer in Antarctic Lake Kitiesh. *Polar Biology* **11**, 317–377.

Moore, J.W. (1974a). The benthic algae of Southern Baffin Island. I. Epipelic communities in rivers. *Journal of Phycology* **10**, 50–57.

Moore, J.W. (1974b). The benthic algae of Southern Baffin Island. II. Epilithic and epiphytic communities. *Journal of Phycology* **10**, 456–462.

Moorhead, D., Schmeling, J., and Hawes, I. (2005). Contributions of Benthic Microbial Mats to Net Primary Production in Lake Hoare, Antarctica. *Antarctic Science* **17**, 33–45.

Moorhead, D.L. *et al.* (1999). Ecological legacies: impacts on ecosystems of the McMurdo Dry Valleys. *BioScience* **49**, 1009–1019.

Mueller, D.R., Vincent, W.F., Bonilla, S., and Laurion, I. (2005). Extremotrophs, extremophiles and broadband pigmentation strategies in a high arctic ice shelf ecosystem. *FEMS Microbiology Ecology* **53**, 73–78.

Nigoyi, D.K., Tate, C.M., McKnight, D.M., Duff, J.H., and Alger, A.S. (1997). Species composition and primary production of algal communities in Dry Valley streams in Antarctica: examination of the functional role of biodiversity. In Lyons, W.B., Howard-Williams, C., and Hawes, I. (eds), *Ecosystem Processes in Antarctic Ice-free Landscapes*, pp. 171–180. Balkema, Rotterdam.

Paerl, H., Pinckney, J.L., and Steppe, T.F. (2000). Cyanobacterial-bacterial mat consortia: examining the functional unit of microbial survival and growth in extreme environments. *Environmental Microbiology* **2**, 11–26.

Parker, B.C., Simmons, Jr, G.M., Love, F.G., Wharton, Jr, R.A., and Seaburg, K.G. (1981). Modern stromatolites in Antarctic Dry Valley lakes. *BioScience* **31**, 656–661.

Parker, B.C. and Wharton, R.A. (1985). Physiological ecology of blue-green algal mats (modern stromatolites) in Antarctic oasis lakes. *Algological Studies* **38/39**, 331–348.

Peterson, B.J., Hobbie, J.E., and Corliss, T.L. (1986). Carbon flow in a tundra stream ecosystem. *Canadian Journal of Fisheries and Aquatic Sciences* **43**, 1259–1270.

Peterson, B.J. *et al.* (1985). Transformation of a tundra river from heterotrophy to autotrophy by addition of phosphorus. *Science* **229**, 1383–1386.

Peterson, B.J. *et al.* (1993). Biological responses of a tundra river to fertilisation. *Ecology* **74**, 653–672.

Priddle, J. (1980a). The production ecology of benthic plants in some Antarctic lakes. 1. In situ production studies. *Journal of Ecology* **68**, 141–153.

Priddle, J. (1980b). The production ecology of benthic plants in some Antarctic lakes. 2. Laboratory physiology studies. *Journal of Ecology* **68**, 155–166.

Priddle, J. and Dartnall, H.J.G. (1978). Biology of an Antarctic aquatic moss community. *Freshwater Biology* **8**, 469–480.

Quesada, A. and Vincent, W.F. (1997). Strategies of adaptation by Antarctic cyanobacteria to ultraviolet radiation. *European Journal of Phycology* **32**, 335–342.

Quesada, A., Vincent, W.F., and Lean, D.R.S. (1999). Community and pigment structure of Arctic cyanobacterial assemblages: the occurrence and distribution of UV-absorbing compounds. *FEMS Microbiology Ecology* **28**, 315–323.

Ramlal, P.S. *et al.* (1994). The organic carbon budget of a shallow arctic tundra lake on the Tuktoyaktuk Peninsula, NWT, Canada. *Biogeochemistry* **24**, 145–172.

Rautio, M. and Vincent, W.F. (2006). Benthic and pelagic food resources for zooplankton in shallow high-latitude lakes and ponds, *Freshwater Biology* **51**, 1038–1052.

Riis, T. and Sand-Jensen, K. (1997). Growth reconstruction and photosynthesis of aquatic mosses: Influence of light, temperature and carbon dioxide at depth. *Journal of Ecology* **85**, 359–372.

Sabbe, K. *et al.* (2004). Salinity, depth and the structure and composition of microbial mats in continental Antarctic lakes. *Freshwater Biology* **49**, 296–319.

Sand-Jensen, K., Riis, T., Markager, S., and Vincent, W.F. (1999). Slow growth and decomposition of mosses in Arctic Lakes. *Canadian Journal of Fisheries and Aquatic Sciences* **56**, 388–393.

Savich-Lyubitskaya, L.I. and Smirnova, Z.N. (1959). A new species of the genus *Bryum* Hedw. from the Bunger's Oasis. *Sovetskaia Antarkticheskaia Ekspeditsiia, Informatsionnyi Biulleten* **7**, 34–39 [in Russian].

Savich-Lyubitskaya, L.I. and Smirnova, Z.N. (1964). A deepwater member of the genus *Plagiothecium* Br. and Sch. from Antarctica. *Sovetskaia Antarkticheskaia Ekspeditsiia, Informatsionnyi Biulleten* **49**, 33–39 [in Russian].

Slavik, K. *et al.* (2004). Long-term responses of the Kuparuk River ecosystem to phosphorus fertilisation. *Ecology* **85**, 939–954.

Spaulding, S.A., McKnight, D.M., Stoermer, E.F., and Doran, P.T. (1997). Diatoms in sediments of perennially ice-covered Lake Hoare, and implications for interpreting lake history in the McMurdo Dry Valleys of Antarctica. *Journal of Paleolimnology* **17**, 403–420.

Spigel, R.H. and Priscu, J.C. (1998). Physical limnology of McMurdo Dry Valley lakes. In Priscu, J.C. (ed.), *Ecosystem Dynamics in a Polar Desert: The McMurdo Dry Valleys, Antarctica*, Antarctic Research Series, vol. 72, pp. 153–187. American Geophysical Union, Washington DC.

Squyres, S.W., Andersen, D.W., Nedell, S.S., and Wharton, Jr, R.A. (1991). Lake Hoare, Antarctica: sedimentation through a thick perennial ice cover. *Sedimentology* **38**, 363–379.

Toro, M. *et al.* (2007). Limnology of freshwater ecosystems of Byers Peninsula (Livingston Island, South Shetland Islands, Antarctica. *Polar Biology* **30**, 635–649.

Velázquez, D., Rochera, C., Justel, A., Toro, M., and Quesada, A. (2006). Temperature dependency of photosynthesis and N_2-fixation in microbial mats from Byers Peninsula (Livingston Island, South Shetland Islands), Antarctica. *International Conference on Alpine and Polar Microbiology*, p. 86. Innsbruck, Austria

Vézina, S. and Vincent, W.F. (1997). Arctic cyanobacteria and limnological properties of their environment: Bylot Island, Northwest Territories, Canada (73 degrees, N., 80 degrees W). *Polar Biology* **17**, 523–534.

Villeneuve, V., Vincent, W.F., and Komárek, J. (2001). Community structure and microhabitat characteristics of cyanobacterial mats in an extreme high artic environment: Ward Hunt Lake. *Nova Hedwigia* **123**, 199–224.

Vincent, W.F. (1988). *Microbial Ecosystems of Antarctica*. Cambridge University Press, Cambridge.

Vincent, W.F. (2000). Cyanobacterial dominance in the polar regions, In Whitton, B.A. and Potts, M. (eds), *Ecology of the Cyanobacteria: their Diversity in Space and Time*, pp. 321–340. Kluwer Academic Press, Dordrecht.

Vincent, W.F. and Howard-Williams, C. (1986). Antarctic stream ecosystems: physiological ecology of a blue-green algal epilithon. *Freshwater Biology* **16**, 219–233.

Vincent, W.F. and Howard-Williams, C. (1989). Microbial communities in southern Victoria Land streams (Antarctica) II: The effects of low temperature. *Hydrobiologia* **172**, 39–49.

Vincent, W.F., Castenholz, R.W., Downes, M.T., and Howard-Williams, C. (1993a). Antarctic cyanobacteria: light, nutrients, and photosynthesis in the microbial mat environment. *Journal of Phycology* **29**, 745–755.

Vincent, W.F., Downes, M.T., Castenholz, R.W., and Howard-Williams, C. (1993b). Community structure and pigment organisation of cyanobacteria-dominated microbial mats in Antarctica. *European Journal of Phycology* **28**, 213–221.

Vincent, W.F., Howard-Williams, C., and Broady, P.A. (1993c). Microbial communities and processes in Antarctic flowing waters. In Friedman, E.I. (ed.), *Antarctic Microbiology*, pp. 543–569. Wiley-Liss, New York.

Vopel, K. and Hawes, I. (2006). Photosynthetic performance of benthic microbial mats in Lake Hoare, Antarctica. *Limnology and Oceanography* **51**, 1801–1812.

Wagner, B. and Seppelt, R. (2006). Deep water occurrence of the moss *Bryum pseudotriquetrum* in Radok Lake, Amery Oasis, East Antarctica. *Polar Biology* **29**, 791–795.

Welch, H.E. and Kalff, J. (1974). Benthic photosynthesis and respiration in Char Lake. *Canadian Journal of Fisheries and Aquatic Sciences* **31**, 609–620.

Wharton, R.A. (1994). Stromatolitic mats in Antarctic lakes. In Bertrand-Sarfati, J. and Monty, C. (eds), *Phanerozoic Stromatolites II*, pp. 53–70. Kluwer Academic Publishers, Dordrecht.

Wharton, R.A., Parker, B.C., and Simmons, G.M. (1983). Distribution, species composition, and morphology of algal mats in Antarctic Dry Valley lakes. *Phycologia* **22**, 355–365.

CHAPTER 11

Heterotrophic microbial processes in polar lakes

John E. Hobbie and Johanna Laybourn-Parry

Outline

This review is limited to lakes on the Antarctic continent and Arctic lakes and ponds of Europe, Asia, North America, and Greenland. Our emphasis is on heterotrophy of microbes (bacteria, flagellates, and ciliated protozoans), the food web that consumes almost all of the organic carbon produced in these inland waters (autochthonous carbon) or produced on land and transported into these waters (allochthonous carbon). Primary production and the factors that control it are treated in Chapter 9, and secondary production for non-microbial organisms is covered in Chapters 13 and 15. The linkages between all these elements in polar aquatic food webs are detailed in Chapter 1. Among the photosynthetic flagellates there are elements that sit on the boundary between heterotrophy and phototrophy (mixotrophs) that are dealt with here. In some instances these mixotrophic microbes may be seasonally important as consumers of bacteria.

11.1 Introduction

A major difference between Arctic and Antarctic inland waters is that most Arctic lakes and ponds receive large amounts of organic matter from their catchments and immediate surroundings (allochthonous organic matter) while the organic matter in Antarctic waters is almost all autochthonous, produced from primary production within the pond or lake. Arctic lakes and ponds form a continuum from those close to the treeline in the south, surrounded by tundra and very like Sub-Arctic and temperate lakes, to those in vegetation-free rock basins in the far north where lakes may have year-round ice cover. Even the northernmost Antarctic lakes resemble the most extreme Arctic lakes with no plants in their drainage basin and some with year-round ice covers. Other differences are the consequence of a more severe climate in the Antarctic than in most of the Arctic. The Arctic is largely composed of the northern regions of continents (North America, Asia, and Europe),

while Antarctica is an isolated continent. This fact has major implications for colonization, which is explored in Chapter 12 in this volume.

As outlined in Chapter 1 the polar regions contain a wide variety of lakes ranging from small shallow ponds to large deep lakes and epishelf lakes that are unique to these regions. Freshwater, brackish, saline, and hypersaline lakes occur. Apart from the more severe climate of southern polar regions, another major difference between the Arctic and Antarctic is the presence of vegetation; e.g. grasses, sedges, willow, and birch, in the catchment of many Arctic lakes, particularly in the tundra zones. Most of Antarctica is covered by an ice cap up to 4 km thick. The ice-free 2% of the continent does have lakes set amidst bare rock where one may find sparse colonies of mosses and lichens. As a result many Arctic lakes receive allochthonous carbon and other nutrients derived from soils and vegetation but Antarctic lakes lack such inputs unless they are adjacent to a penguin rookeries or seal wallows. It should be noted that almost all (85–90%) of the allochthonous material is dissolved organic carbon (DOC), dissolved organic nitrogen (DON), and dissolved organic phosphorus (DOP). The quantities of DOC entering lakes can be many times higher than the algal primary production; in Toolik Lake, Alaska, Whalen and Cornwell (1985) measured the annual loading as $98\,g\,C\,m^{-2}$ compared to an algal productivity of $14\,g\,C\,m^{-2}\,year^{-1}$. It is important to note that microbes in Toolik Lake make use of less than 10% of the allochthonous DOC, most of which $(86\,g\,C\,m^{-2})$ flows out of the lake.

The limnology of the Arctic must be regarded as simply an extension of temperate limnology. In the low Arctic, chemical, physical, and biological processes and species found in ponds and lakes are very similar to those of oligotrophic temperate ponds and lakes. The more northerly an Arctic water body, the more oligotrophic is the food web and the lower the biodiversity. Entire trophic levels drop out until only algae, bacteria, ciliates, and heterotrophic and mixotrophic flagellates persist. The limnology of the Antarctic is similar to the more northerly Arctic lakes in that there are no fish and few metazoans. Some types of metazoans may be missing because of biogeography; that is, the physical isolation of Antarctica. However, the physics and chemistry of a number of Antarctic lakes are very different from nearly all Arctic lakes. Many lakes are saline or hypersaline because they are derived from either seawater or the evaporation of fresh water. The combination of ice cover, salinity, and penetration of solar radiation into the clear waters of lakes creates unique warm-water columns. There are even a number of lakes beneath the 4-km-thick ice sheet, although little is yet known about their biology (see Chapter 7).

11.2 Food webs

The structure of polar lake food webs outlined in Chapter 1 shows that Antarctic lakes are species-poor and possess simplified, truncated food webs dominated by the microbial loop (Figure 1.6). In contrast, Arctic lakes possess a variety of food webs ranging from those with well-developed zooplankton and fish communities (Figure 1.5) to depauperate food webs with no planktonic metazoans (Figure 1.6).

Whereas Antarctic lake food webs are simple, and none contain fish (Chapter 1, Figure 1.6), there are variations across the continent. Lakes on coastal ice-free areas, like the Vestfold Hills lying between 68°25′S 77°50′E and 68°40′S 78°35′E usually have a single planktonic crustacean. In the freshwater and slightly brackish lakes (e.g. Highway Lake) sparse populations of the endemic Antarctic cladoceran *Daphniopsis studeri* (Plate 12) occur, while in the marine-derived saline lakes with salinities up to that of seawater, the marine copepod *Paralabidocera antarctica* occurs. Both species remain active throughout the year. The McMurdo Dry Valleys lie much further south, between 76°S and 78°S in western Antarctica. Here planktonic Crustacea are lacking, though a few copepod nauplii were found on the chemocline of Lake Joyce (upper Taylor Valley) (Laybourn-Parry 2008). *Philodina gregaria* (Plate 3), a common component of Antarctic algal mats in lakes and ponds on ice shelves (Suren 1990), has also been found in the plankton of the dry valley lakes.

In conclusion, there is a great deal of similarity in the food webs of Arctic and Antarctic lakes and ponds at the level of bacteria, nanoflagellates, and ciliates. For ponds and lakes at the same trophic status across the poles, the genera are similar, as are the numbers of organisms per litre. The similarity does not hold, however, for higher trophic levels (see Chapter 15). There are only a few genera of rotifers and crustacean in Antarctic lakes and ponds while rotifers and crustaceans are relatively abundant in Arctic waters.

11.3 Bacteria

The concentration of heterotrophic bacteria in polar lakes is a function of the concentration of organic carbon, inorganic and organic nitrogen and phosphorus, and the abundance of the heterotrophic and mixotrophic nanoflagellates, the principal grazer on bacteria. When all the organic matter comes from the plankton, bacterial numbers are low, reflecting the oligotrophic nature of the waters, as is the case in Antarctic and High Arctic lakes. In contrast, tundra lakes and ponds in the Arctic receive relatively high amounts of organic matter from the surrounding watershed and the bacteria numbers are correspondingly higher (Table 11.1).

Table 11.1 Bacterial, heterotrophic nanoflagellate (HNAN), and autotrophic nanoflagellate (PNAN) abundances in Arctic and Antarctic lakes.

Lake	Bacteria ($\times 10^9$ L^{-1})	HNAN ($\times 10^5$ L^{-1})	PNAN ($\times 10^6$ L^{-1})	Reference
Antarctica				
Crooked Lake, Vestfold Hills	0.15–0.48	0.04–5.09	0.81–0.68	Henshaw and Laybourn-Parry (2002), Laybourn-Parry et al. (2004)
Lake Nottingham, Vestfold Hills	0.094–1.16	0.91–6.44	0.19–1.38	J. Laybourn-Parry, T. Henshaw, and W. Quayle, unpublished results
Ace Lake, Vestfold Hills	0.13–7.28	9.2–402	1.52–215.0	Bell and Laybourn-Parry (1999)
Beaver Lake, MacRobertson Land	0.093–0.14	2.26–13.38	0.35–1.31	Laybourn-Parry et al. (2006)
Pendant Lake, Vestfold Hills	0.20–7.77	2.76–10.12	7.6–424.1	Laybourn-Parry et al. (2002)
Lake Fryxell, Dry Valleys	0.4–3.4	6.9 max.	14.5 max.	Roberts et al. (2000)
Lake Hoare, Dry Valleys	1.18 max.	3.4 max.	2.1 max.	Roberts et al. (2004)
Arctic				
Tvillingvatnet, Svalbard	0.6–2.98	0.5–3.2	0.35–90.6	Laybourn-Parry and Marshall (2003)
Toolik Lake, Alaska	0.2–3.0	1–6	≈2	O'Brien et al. (1997), Hobbie et al. (2000)
Pond C, Barrow, Alaska	2–6		1–6	Alexander et al. (1980), Hobbie et al. (1980)
Meretta Lake, Canada	0.2–8.3			Morgan and Kalff (1972)
Char Lake, Canada	0.01–0.02			Morgan and Kalff (1972)
Franz Josef Land	0.9–1.7	0.7–1.7		Panzenbock et al. (2000)

Crooked Lake in the Antarctic (68°S), one of the few polar lakes sampled year-round, has very low numbers of bacteria – $(0.15–0.5) \times 10^9$ cells L^{-1} – which is exactly the concentration found in the upper oxygenated water column of meromictic Lake Fryxell. In the High Arctic, a lake at 81°N in the Franz Josef archipelago had summer numbers close to 1.5×10^9 cells L^{-1}.

Molecular biology has allowed the development of a growing database on polar bacterial biodiversity. Although there is some evidence of endemicity in the Antarctic, there are as yet insufficient data to substantiate a consensus (Laybourn-Parry and Pearce 2007); nonetheless, there is now an expanding database on the molecular diversity of Antarctic bacteria (see Chapter 12). There are only a few studies of Arctic lakes, mostly of Toolik Lake where the bacterial community has limited diversity compared with that of temperate lakes. The bacteria in this oligotrophic lake were identified as members of globally distributed freshwater phylogenetic clusters within the *Proteobacteria*, the *Cytophaga-Flavobacteria-Bacteroides* group, and the *Actinobacteria*. Toolik studies also offer some insight into community changes over the year and over a region. In a seasonal study, Crump *et al.*

(2003) found that bacterioplankton communities were composed of persistent populations present throughout the year and transient populations that appeared and disappeared. Most of the transient populations were either the populations that were advected into the lake with terrestrial dissolved organic matter during the spring runoff or those that grew up from low concentrations during the development of the phytoplankton community in early summer. A spatial study over lakes and streams in some 50 km^2 of the Toolik Lake drainage basin found that communities of bacteria in lakes and streams were quite different. Nine hydraulically connected lakes were all quite similar yet the communities in these lakes were somewhat different from the community in one nearby but unconnected lake. Thus across a landscape, geographic distance and connectivity greatly influence the distribution of bacterial communities.

11.4 Photosynthetic plankton: autotrophs and mixotrophs

The photosynthetic picoplankton, 0.2–2.0-μm-sized cyanobacteria, can be abundant in polar lakes and sometimes may even dominate the biomass of the phytoplankton community. *Synechocystis* dominated the deep chlorophyll maximum in Lake Vanda in the McMurdo Dry Valleys and in Ace Lake (Vestfold Hills) *Synechococcus* was found at <$10^7 L^{-1}$ during the winter and at remarkably high peak concentrations of $8×10^9 L^{-1}$ during the summer (Vincent 2000). Picocyanobacteria and eukaryotic photosynthetic picoplankton have also been identified in ponds near Terra Nova Bay (Antarctica). In several cases more than 50% of the planktonic chlorophyll was in the photosynthetic picoplankton (Andreoli *et al.* 1992). However, in many Antarctic lakes they are not significant. There are picocyanobacteria in oligotrophic Arctic lakes (Vincent 2000). In Toolik Lake, Alaska, picocyanobacteria abundance was $(1.5–64)×10^6$ cells L^{-1} (June-August), and at Bylot Island, Canada (73°N), these were also among the dominant planktonic organisms (Vézina and Vincent 1997). *Synechococcus* is also important in lakes along the northern coastline of Ellesmere Island (P. Van Hove *et al.*, unpublished results). In shallow lakes in both polar regions, benthic cyanobacteria are often abundant and may dominate biomass within the ecosystem.

The phytoplankton of Antarctic lakes is often dominated by a few species among which cryptophytes are common (Plate 11). These photosynthetic flagellates are often mixotrophic ingesting bacteria and in some cases possibly taking up dissolved organic matter to supplement their carbon budgets and/or gain nutrients for photosynthesis (Laybourn-Parry *et al.* 2005). *Pyramimonas* sp. (Plate 11) is common and in both the dry valleys and the Vestfold Hills lakes is also mixotrophic, but has not been reported so at lower latitudes. In Arctic lakes phytoplankton are almost always dominated by small flagellates, less than 6 μm in diameter. In Toolik Lake, flagellates belonging to the Chrysophyceae and Cryptophyceae dominate, although small planktonic centric diatoms, *Cyclotella compta* and *Cyclotella bodanica*, are also present. The larger flagellates *Dinobryon* (Plate 11) and *Cryptomonas* are well-known mixotrophs and were found by Laybourn-Parry and Marshall (2003) in an Arctic lake at 78°N, and are also common in Char Lake and Ward Hunt Lake in the Canadian High Arctic (Bonilla *et al.* 2005). The mixotrophs appear to become abundant earlier in the year than

the strict phototrophs, perhaps because they can gain nutrients from bacteria. In Langemandssø, a lake in Greenland (74°N), there was a remarkable and unexplained shift from dominance by dinophytes in a short, cold summer to chrysophytes in a long, warm summer (Wrona *et al.* 2006).

11.5 Heterotrophic microplankton: flagellates, ciliates, and rotifers

Our information on the biodiversity of heterotrophic nanoflagellates, major grazers of heterotrophic bacterioplankton and photosynthetic picoplankton, is so far restricted to classic morphologically based identifications. In the Antarctic, *Paraphysomonas* sp., *Heteromita* sp., and *Monosiga consociata* have been identified from Vestfold Hills freshwater lakes (Tong *et al.* 1997). Some of the marine-derived saline lakes are dominated by marine choanoflagellates, for example *Diaphanoeca grandis* (Plate 11) in brackish Highway Lake (5‰), and in hypersaline Organic Lake (176‰) a species resembling *Acanthocorbis unguiculata* that is found in the sea off the Vestfold Hills occurs (Van den Hoff and Franzmann 1986). In the Arctic, Chrysophyceae are dominant in Toolik Lake and the genera *Chromulina*, *Ochromonas*, *Spiniferomonas*, *Pseudopedinella*, *Pseudokephyrion*, *Paraphysomonas*, and *Kephyrion* are present. Some of these have colorless forms or are mixotrophic (Hobbie *et al.* 2000). In general the abundance of the heterotrophic nanoflagellates depends on the density of their bacterial food source (Table 11.1).

The ciliate communities of the Antarctic continent display distinct differences not only across freshwater to saline habitats, but also across the continent. In the Vestfold Hills the freshwater lakes have low species diversity that includes *Askenasia* sp., *Strombidium viride*, and scuticociliates. Moreover, this simple community is found in all the freshwater lakes in this region, including very young lakes that have formed adjacent to the ice cap as it is retreating. Concentrations are usually low, less than $100 L^{-1}$. The much older meromictic lakes of the Dry Valleys have 11 genera. These include *Plagiocampa*, *Askenasia*, *Cyclidium*, *Monodinium*, *Euplotes*, *Vorticella*, and the predatory suctorian ciliate *Sphaerophrya* that preys on other ciliates (Kepner *et al.* 1999). In these perennially ice-covered lakes the ciliates occupy distinct strata in the stable water columns (Roberts *et al.* 2000, 2004). The same is true of the only epishelf lake that has been studied in detail, Beaver Lake in MacRobertson Land (70°48′S, 68°15′E). This lake, which is believed to predate the last major glaciation, has species from genera found in both the McMurdo Dry Valleys and Vestfold Hills: *Askenasia*, *Monodinium*, *Strombidium*, and *Mesodinium* spp. as well as a range of scuticociliate species. Beaver Lake is ultra-oligotrophic and supports only low numbers of ciliates, usually around 50 L^{-1} or less (Laybourn-Parry *et al.* 2006).

The dominant ciliate in the marine-derived brackish and saline lakes of the Vestfold Hills is the remarkable marine ciliate *Mesodinium rubrum* (Plate 11) that occurs worldwide in seas and estuaries where it may form red tides. It is photosynthetic and was thought to contain an endosymbiotic cryptophycean, but there is some evidence to suggest that some strains may be sequestering plastids from ingested cryptophytes. It lacks a cell mouth or cytostome. This protozoan also has the potential to take up DOC but it is not known whether it gains significant energy by this mode. *Mesodinium rubrum* can reach concentrations of around 100 000 L^{-1}, and as in the dry valley lakes there is considerable inter-annual variation. The population numbers drop in winter, though active cells remain. Many of the individuals encyst and in spring the actively growing population is augmented by excysting cells (Bell and Laybourn-Parry 1999).

In the Arctic, the structure of the ciliate/rotifer/copepod (microplankton) community (20–200-μm-size fraction) is dependent upon the trophic state of the lake or pond. Biogeography seems to play little role. When oligotrophic Barrow ponds in the Arctic were compared with similar oligotrophic temperate ponds, the diversity of ciliates in the benthos was very similar. Even the species were exactly the same (Hobbie *et al.* 1980). In Toolik Lake, the genera of protozoans found are similar to those found in oligotrophic lakes in the Antarctic, mainly the oligotrichs *Strobilidium* sp. and *Strombidium* (Plate 11), *Urotricha* sp., and *Askenasia* sp. as well as small, medium, and large unidentified ciliate species. The number of ciliates ranged from 40 to 1000 L^{-1}

(Arctic LTER database, http://ecosystems.mbl.edu/ARC). A fertilization experiment on an oligotrophic lake near Toolik Lake demonstrated that planktonic microplankters are also controlled by trophic state. When the lake was fertilized with nitrogen and phosphorus for 5 years at two to three times the annual loading of lakes in the area both the chlorophyll and primary productivity increased 10-fold. Prior to fertilization the average biomass of the protozoans was 0.5 μg C L^{-1}, with values never exceeding 1.4 μg C L^{-1}. As a result of fertilization the structure of the microplankton community changed from a nutrient-limited system, dominated by oligotrich protozoans (*Strombidium*, *Strobilidium*, and *Halteria* spp.) and small-particle-feeding rotifers (*Conochilus unicornis* and *Keratella cochlearis*), to a system dominated by a succession of peritrich protozoans and predatory rotifers (Bettez *et al.* 2002). The average biomass of the rotifers and protozoans was more than seven and a half times larger by the end of fertilization than it was initially. Bacterivore peritrichs, which are common in lower-latitude mesotrophic and eutrophic lakes, responded to increased bacteria numbers. The rotifers that responded, *Synchaeta* and *Polyarthra*, both feed on flagellated algae. The predatory rotifers (Trichodercidea) increased strongly during the fourth year of fertilization. Thus, this artificial increase in the trophic state of this lake allowed new functional groups, the bacterivores and predators, to become a prominent part of the microplankton community structure.

11.6 Viruses

Viruses can cause lysis of bacterial and phytoplankton cells, thereby short-circuiting the microbial loop and returning dissolved and particulate carbon to the carbon pool. They also play a role in transferring genetic material between bacteria by transduction. There is now a growing database on viruses in polar lakes, where their numbers are high, for example in saline lakes of the Vestfold Hills (0.89–12.02)×10^{10} L^{-1} over an annual cycle; Madan *et al.* 2005), in the Dry Valley lakes (0.023–5.56)×10^{10} L^{-1}; Lisle and Priscu 2004), and in ultra-oligotrophic freshwater lakes (0.016–0.16)×10^{10} L^{-1}; Säwström *et al.* 2007). The values for saline lakes overlap and exceed reported values for marine systems, while those in freshwater lakes are lower than values reported for temperate lakes. Whereas virus to bacteria ratios were high in the saline lakes (18.6–126.7) they were low in the freshwater lakes (1.2–8.4); however, analysis of visibly infected cells showed a high level of infection compared with lower-latitude lakes with 11.4–66.7% of bacterial cells containing phage. Although burst sizes of infected bacteria were low (around 2–15) compared with lower-latitude systems, it is clear from the accumulating information that viruses are playing a role in recycling carbon in polar lakes, particularly those lakes dominated by the microbial loop.

11.7 Microbial heterotrophic processes and controls

The base of the heterotrophic food web is DOC derived from algal photosynthesis or terrestrial plant photosynthesis (allochthonous inputs) that provides a substrate for bacterial growth. The DOC pool is chemically complex and is derived from exudation of photosynthate by the phytoplankton and benthic algae, sloppy feeding by zooplankton, and dissolution of fecal pellets as well as from allochthonous inputs from the catchment. Given the major differences in the catchments and food webs of Arctic and Antarctic lakes one would expect differences in the concentration of DOC and its chemical makeup. In general, DOC concentration is negatively correlated with latitude (Vincent and Hobbie 2000). Both Antarctic lakes and far-north Arctic lakes have very low concentrations of DOC (Figure 11.1). In the Arctic there is also a strong correlation with treeline; the farther north one moves from the treeline the lower the DOC concentration (Figure 11.1). The higher DOC is directly correlated with the coniferous vegetation.

Allochthonous inputs are negligible in Antarctic lakes, the bulk of DOC is derived from algal production. Where lakes are fertilized, for example Rookery Lake close to a penguin rookery in the Vestfold Hills, the levels of bacterial production are greatly increased (177–7848 µg C L^{-1} day^{-1}; see Table 11.2 for comparison); however, turbidity inhibits photosynthesis. Concentrations of DOC in

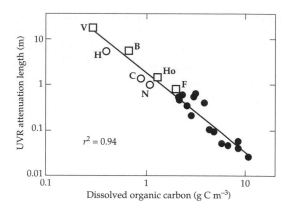

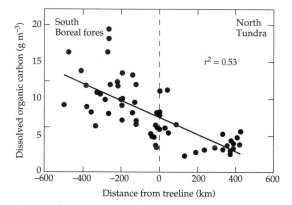

Figure 11.1 Dissolved organic carbon (DOC) in high-latitude lakes. Top: decreasing penetration (attenuation length) of ultraviolet radiation (UVR) with increasing DOC concentration in the upper water columns of Antarctic polar desert lakes (open squares; V, Lake Vanda; Ho, Lake Hoare; B, Lake Bonney; F, Lake Fryxell), Arctic polar desert lakes (open circles; H, Lake Hazen; C, Char Lake; N, North Lake), and forest-tundra lakes, Québec (closed circles). Bottom: DOC concentration as a function of distance from the treeline in Sub-Arctic Québec lakes. Adapted from Vincent (1997) and Vincent and Pienitz (1996).

freshwater Antarctic systems are usually low and close to the 1.3 mg DOC L^{-1} in Lake Fryxell (Roberts *et al.* 2000). Low DOC concentrations in the epishelf Beaver Lake (0.1–0.3 mg DOC L^{-1}) (Laybourn-Parry *et al.* 2006) and lakes abutting the ice cap, like Lake Nottingham (0.1–1.4 mg DOC L^{-1}), suggest that at times that the bacterial community may be organic-carbon-limited. In contrast, the saline lakes of Antarctica have higher DOC concentrations, particularly in the monimolimnion of meromictic lakes like Ace Lake (2.9–22.8 mg DOC L^{-1}).

Table 11.2 Primary and bacterial production in Arctic and Antarctic lakes.

Lake	Primary productivity (μg C L^{-1} day^{-1})	Bacterial productivity (μg C L^{-1} day^{-1})	References
Antarctica			
Crooked Lake, Vestfold Hills	0–13.4*	0–11.4*	Bayliss et al. (1997), Laybourn-Parry et al. (2004)
Lake Druzhby, Vestfold Hills		0–8.3*	Laybourn-Parry et al. (2004)
Beaver Lake, MacRobertson Land	2.2–13.8	0.05–0.29	Laybourn-Parry et al. (2006)
Ace Lake, Vestfold Hills	0–195.6	4.6–25.1	Laybourn-Parry et al. (2007)
Pendant Lake, Vestfold Hills	28.2–152.2	6.5–77.3	Laybourn-Parry et al. (2007)
Lake Fryxell, Dry Valleys	Up to 30	0–9.0	Spaulding et al. (1994), Takacs and Priscu (1998)
Lake Hoare, Dry Valleys	–	0–0.7	Takacs and Priscu (1998)
Lake Verkhneye, Schirmacher Oasis	0.1–10.1	–	Kaup (1994)
Arctic			
Tvillingvatnet, Svalbard	24.5–1000	–	Laybourn-Parry and Marshall (2003)
Cryoconites, Svalbard	8.2–3767	–	Säwström et al. (2002)
Toolik Lake (10-m column)	8.3–12.7	1.6–22.4	Miller et al. (1986), O'Brien et al. (1997)
Franz Joseph Land lake	22	1.2–3.9	Panzenböck et al. (2000)

*Over an annual cycle.

The tremendous range in concentrations of DOC in the Arctic is caused by a large variation in the amount of allochthonous carbon derived from the surrounding terrestrial ecosystem. Where vegetation cover of the drainage basin is very low the DOC is low; for example, 1 mg L^{-1} in Char Lake and in another High Arctic lake in Franz Joseph Land. When there is a continuous vegetation cover the concentrations that run off from land are high. For example, spring runoff for a stream entering Toolik Lake had a DOC loading of 12.1 mg DOC L^{-1} (Crump et al. 2003), which contributed to a summer average of 6 mg DOC L^{-1} within the lake (O'Brien et al. 1997). Other lakes and ponds in vegetated regions possess high DOC concentrations such as Barrow ponds (10–14 mg DOC L^{-1}), and lakes above treeline in Finnish Lapland with a mean of 18 mg DOC L^{-1} (Rautio and Korhola 2002).

Where chemical analysis of the DOC pool has been undertaken, bulk dissolved amino acids appear to predominate throughout the year in freshwater Crooked Lake and Lake Druzhby. Amino acids and dissolved carbohydrates accounted for up to 64% of total DOC in these lakes. The low fraction of dissolved carbohydrates suggests that this was the preferred fraction by the bacterial community. Both fractions can be used and in lower-latitude systems there is a preference for amino acids. In brackish Pendant, Ace, and Highway lakes and hypersaline Lake Williams dissolved free amino acids also predominated over carbohydrates. In late summer they contributed between 72 and 100% of the DOC pool, while in meromictic Ace Lake they contributed only 28%, indicating that other fractions predominated (Laybourn-Parry et al. 2002, 2004). Chromophoric dissolved organic matter composed of aromatic humic and fulvic acids derived from vegetation and leaf litter in the catchment provide protection from UV-B radiation. While many Arctic lakes do have high chromophoric dissolved organic matter, Antarctic and Arctic desert lakes do not (Vincent et al. 1998; Laurion et al. 1997). Thus in polar desert lakes there is the potential for inhibition of primary production by ultraviolet radiation.

Heterotrophic processes in Antarctic lakes are almost exclusively driven by carbon fixed during photosynthesis by the phytoplankton and benthic algal mats, because there is negligible alloch-thonous carbon input from the catchment. This contrasts with Arctic lakes where there may be significant allochthonous inputs. Rates of photo-synthesis in Antarctica are low (Table 11.2; see also Chapter 9). However, photosynthesis is not entirely confined to the short summer. Year-long studies have shown that both the freshwater and saline lakes of the Vestfold Hills have measurable photo-synthesis in late winter and early spring and this drives microbial heterotrophic processes all year around. Viral lysis of bacteria and the recycling of carbon probably also contributes to sustaining heterotrophic microbial processes over the winter. As yet we do not know what happens in winter in the Dry Valley lakes as they are only investigated during summer. However, an early season study in September showed that photosynthesis was occurring (Lizotte et al. 1996). The evidence sug-gests that these more southerly lakes operate in the same way as the coastal lakes in the Vestfold Hills where biological processes continue over winter allowing communities to enter the short austral summer with actively growing populations. This contrasts to many Arctic systems, where under snow-covered ice primary production is inhib-ited by lack of photosynthetically active radiation. Snow cover is rare on Antarctic lakes as it is blown off by katabatic winds.

Bacterial production occurs throughout the year in the lakes of the Vestfold Hills, but the rates are low (Table 11.2). The saline lakes have higher DOC concentrations and consequently support higher production. Interestingly, highest bacterial growth is often recorded in winter and tends to follow peaks in DOC. It is likely that viruses play an important role in the recycling of carbon to support bacterial growth, as they are present in relatively high num-bers with high virus to bacteria ratios (see above). In lower-latitude lakes aggregates or 'lake snow' of organic carbon support higher concentrations of bacteria and production than the surround-ing water. In Antarctic freshwater lakes (Crooked Lake and Lake Druzhby, Vestfold Hills) attached bacteria were larger than free-floating bacteria and production in aggregate-associated bacteria higher by 82–94% and 38–100% respectively. On only one occasion was production in free bacteria higher

(Laybourn-Parry et al. 2004). In the more extreme lakes of the Dry Valleys bacterial production was lower, with rates at the lower end of the spectrum reported from continental lakes (Table 11.2). This reflects lower bacteria concentrations and biomass (Table 11.1). However, here too the saline mero-mictic lake (Lake Fryxell) had higher production than freshwater Lake Hoare. There are less data on bacterial production in algal mats and they are dif-ficult to compare directly with planktonic rates as they are expressed in production per unit area. In the High Arctic ice-shelf mats had high bacterial production rates of 0.037–0.21 mgC m^{-2}h^{-1}, although these rates were three orders of magnitude lower than primary production (Mueller et al. 2005).

As indicated in Figure 1.6, the bacterioplank-ton form a crucial energy source for heterotrophic nanoflagellates, and also at times for mixotrophic nanoflagellates. It is possible to measure the grazing rates of flagellates and ciliates in several ways. The bacteria ingested can be counted micro-scopically or the number of bacteria consumed can be calculated from changes in numbers of bacteria. The uptake of fluorescently labelled bac-teria (FLB) is now the most widely used technique. These bacteria are heat-killed and stained with the fluorochrome 5-[(4,6-dichlorotriazine-2-yl)amino]-fluoroscein (DTAF). The protozoans take up these bacteria despite the fact that the cells walls of the food have been altered by death and a stain. The assumption has been that the protozoans cannot discriminate between live and dead cells or strains of bacteria. However, it is now possible to insert different colored fluorescent marker genes into bacteria, such as green fluorescent proteins (GFPs). The bacteria are strongly fluorescent and have the advantage that they are alive and their cell walls are unaffected by the marker gene. Moreover, grazing rates measured using cells with GFP are up to 45% higher than rates measured with the dead DTAF-stained cells. The flagellates evidently 'prefer' the live GFP-containing cells. Many of the grazing rates that will be discussed here were determined using FLB; thus the reader should be aware that these may represent underestimates of grazing rates and impact.

Grazing plays a role in controlling the numbers of bacteria in lakes but this is difficult to demonstrate in natural waters. In an experiment in a fertilized 92 m³ limnocorral in Toolik Lake (Figure 11.2a), the bacterial population at first grew rapidly and peaked at 8×10⁹ L⁻¹ in response to high rates of algal photosynthesis (O'Brien *et al.* 1992). The flagellates grew more slowly and did not reach high numbers until day 19 when they soon reduced the bacterial numbers. Note that there are three clear peaks of flagellate abundance (day 4, 22, and 27) that correlate with a decrease in bacterial abundance. Bacterial production was at its peak (both per cell and total) when the flagellates reached their maximum and the bacterial numbers were falling rapidly. It is extremely difficult to find examples of similar predator/prey cycles in natural waters but there is one example (Figure 11.2b) from Tvillingvatnet

(Svalbard) where Laybourn-Parry and Marshall (2003) found low numbers of bacteria for most of the season (≈0.6×10⁹ L⁻¹) followed by a sudden rise in late summer to approximately 3×10⁹ L⁻¹ at the same time as the heterotrophic nanoflagellates decreased. Differences in bacterial concentrations in Char Lake and Meretta Lake (Table 11.1), were caused by sewage nutrients entering Meretta Lake and stimulating primary productivity by 8-fold. Higher bacterial numbers in the summer months in Toolik Lake, between 1 and 2×10⁹ L⁻¹, are high for an oligotrophic lake but are a result of allochthonous organic matter input (the DOC in the lake is 6 mg C L⁻¹).

Heterotrophic nanoflagellates vary considerably in abundance across Antarctic lakes, reflecting the

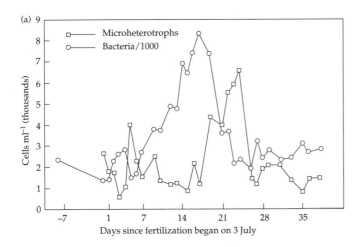

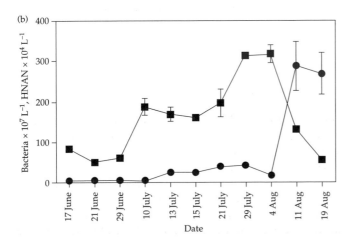

Figure 11.2 (a) Bacterial (10⁹ L⁻¹) and flagellate (10⁶ L⁻¹) abundance in corral B in Toolik Lake, Alaska. The corral volume of 92 m³ was fertilized daily with 2.77 g N as ammonium and 0.465 g P. Redrawn from data in Hobbie and Helfrich (1988) and O'Brien *et al.* (1992). (b) Bacterial (circles) and heterotrophic nanoflagellate (squares) abundance in Tvillingvatnet, Svalbard. Redrawn from data in Laybourn-Parry and Marshall (2003).

abundances of their prey (Table 11.1). Generally bacterial numbers are higher in the saline lakes (e.g. Pendant, Ace, Fryxell) compared with freshwater lakes, which tend to be ultra-oligotrophic. Heterotrophic nanoflagellate grazing impact also varies and this relates to their grazing rates. Choanoflagellates tend to have higher grazing rates than other heterotrophic nanoflagellates. In Lakes Fryxell and Hoare choanoflagellate species predominated, accounting for higher ingestion rates compared with heterotrophic nanoflagellates in the other Antarctic lakes listed in Table 11.3. In Highway Lake the predominant flagellate is *Diaphanoeca grandis* (Plate 11) a marine choanoflagellate that had relatively high clearance rates compared with those of the heterotrophic nanoflagellates in neighboring Ace Lake. The impact of heterotrophic nanoflagellate grazing can be

very significant at times, for example in Beaver Lake where at midsummer all of daily bacterial production is grazed, causing a decline in bacterioplankton, followed by a decline in the heterotrophic nanoflagellates. In Lakes Fryxell and Hoare flagellate grazing has been implicated in marked decreases in bacterial biomass in summer (Takacs and Priscu 1998). In other lakes the heterotrophic nanoflagellates have a negligible impact of bacterial production (Table 11.3).

Nutritional versatility appears to be an important survival strategy among Antarctic phytoflagellates. Some of the most successful phytoflagellates, especially cryptophytes (Plate 11), are mixotrophic, ingesting bacteria to supplement their carbon budgets or to acquire nitrogen and phosphorus (for grazing rates see Table 11.3). Cryptophytes dominate the plankton of several dry

Table 11.3 Protozoan grazing rates and grazing impact (% of bacterial production removed) in the plankton of Arctic and Antarctic lakes. HNAN, heterotrophic nanoflagellates; PNAN, mixotrophic phytoflagellates; HC, heterotrophic ciliates.

Lake	Temperature (°C)	Clearance rate (nL cell^{-1} h^{-1})	Grazing rate (bacterial cells h^{-1})	Bacterial biomass grazed (day^{-1})	References
Antarctic					
Crooked Lake HNAN, Vestfold Hills	2–4	–	0.2	0.1–9.7% of production	Laybourn-Parry et al. (1995)
Beaver Lake HNAN, MacRobertson Land	1	1.11	0.113–0.15	Up to 100% of production	Laybourn-Parry et al. (2006)
Highway Lake HNAN Vestfold Hills PNAN	2 / 2	0.02–1.80 / 0.04–1.05	0.02–2.4 / 0.002–1.56	0.51–5.8% / 1% max	Laybourn-Parry et al. (2005)
Ace Lake HNAN, Vestfold Hills PNAN	2 / 2	0.04–0.37 / 0.002–0.21	0.01–1.45 / 0.003–1.04	<1% / 1% max	Laybourn-Parry et al. (2005)
Lake Fryxell HNAN Dry Valleys PNAN	0.41 / 0.41	–	4.0–5.4 / 1.6–3.6	1–4% / 5.0–13%	Roberts and Laybourn-Parry (1999)
Lake Hoare HNAN Dry Valleys PNAN	2	–	6.2–6.7 / 0.2–1.0	1–2% / <1–3%	Roberts and Laybourn-Parry (1999)
Arctic					
Ossian Sars Lake HNAN Svalbard PNAN	4 / 4	2.1–7.2 / 0.78–4.2	0.79–16.9 / 0.28–1.12	Up to 20% / <1%	Laybourn-Parry and Marshall (2003)
N1 (near Toolik) *Vorticella* HC *Epistylis* HC		150–280 / 200–1300	150–560 / 200–260	6% / 3%	Bettez (1996)
Toolik fertilized corral HNAN+PNAN	11–13	2–20	3–160	50–200%	Hobbie and Helfrich (1988)
Barrow pond HNAN, sediment HC				20% of production	Hobbie et al. (1980)

valley lakes (Hoare and Fryxell) and in the brackish and saline lakes of the Vestfold Hills. In the McMurdo dry valleys they never become entirely photosynthetic, but the dependence on mixotrophy decreases towards midsummer and in meromictic Lake Fryxell, higher in the water column above the chemocline. The evidence suggests that they are heterotrophic in winter and become progressively more autotrophic towards summer. At times they remove more bacterial biomass than the heterotrophic nanoflagellates in Lake Fryxell, Table 11.3. In the less extreme lakes of the Vestfold Hills cryptophytes are also mixotrophic, but not all year around (Laybourn-Parry *et al.* 2005). Cryptophytes do practice mixotrophy in lower-latitude environments, but it is rare. *Pyramimonas* is not a genus reported as mixotrophic at lower latitudes, but in the lakes of the dry valleys and the Vestfold Hills it takes up bacteria and DOC. This nutritional versatility enables these organisms to remain active all winter. Contrary to what one might expect, the planktonic communities of Antarctic lakes remain functional in winter, including the few zooplankton species that tend to contain very obvious fat globules that probably provide endogenous energy reserves to supplement their diet.

Despite the bacteria out-competing the phytoplankton in head-to-head competition for DOC, when isotopically or fluorescently labelled organic compounds are added to lake samples some of the label does enter the phytoflagellate cells. For example, Laybourn-Parry *et al.* (2005) showed that dextran with a fluorescent tag entered the feeding vacuoles of cryptophytes and *Pyraminomas* in saline Antarctic lakes. Vincent and Goldman (1980) found that ^{14}C-acetate did enter algal cells collected from deep in Lake Tahoe. These authors, however, concluded that it is not sufficient to demonstrate a potential pathway. In addition, uptake rates must be combined with concentrations of substrate to prove that the algae can survive and grow as DOC heterotrophs. Vincent and Goldman (1980) argued that final proof of algal heterotrophy *in situ* must ultimately include the following: first, an analytical description of the ambient concentrations and rate of supply of utilizable substrates; second, *in situ* evidence that the algae have high-affinity transport systems which enable them to take up organic

compounds effectively at the ambient concentrations; third, evidence that the phytoplankton are physiologically equipped to use these substrates for growth; and last, an assessment of the relative contribution *in situ* of the dissolved organic substances compared with light and carbon dioxide as carbon and energy sources for phytoplankton growth. The take-home message is that when organic compounds are added to plankton they move across cell membranes or into feeding vacuoles but rate measurements are necessary to prove the importance of DOC for algal nutrition. Perhaps the picoplankton, close to heterotrophic bacteria in size, can better compete with bacteria for DOC but this has not been tested. In conclusion, phytoplankton survival by heterotrophy on DOC remains controversial.

The ciliate communities of the ultra-oligotrophic Antarctic lakes are sparse and probably play a small role in control of the heterotrophic food web. In Beaver Lake, Crooked Lake, and Lake Nottingham, for example, their numbers are usually below $100 L^{-1}$. In more productive Lake Hoare ciliates reached concentration up to $23 400 cells L^{-1}$, but there are very significant inter-annual variations. Here the dominant species was *Plagiocampa* that contributed up to 64% of biomass. In meromictic brackish Lake Fryxell this species also predominated accounting for more than 80% of ciliate biomass where maximum ciliate abundance on the chemocline reached $58 000 cells L^{-1}$. In Lake Fryxell *Plagiocampa* appears to depend very largely on cryptophytes as food. The cryptophytes remain undigested for several days and retain their autofluorescence. It is possible the *Plagiocampa* sequesters its prey for a period before digestion, allowing the cells to undertake photosynthesis in a nutrient-rich environment and benefiting from that production. In these Dry Valley lakes the ciliates probably do play some role in carbon cycling.

The sparse crustacean populations of Antarctic waters probably do not exert any significant top-down control. Modeling of carbon cycling has indicated that this is the case in Ace Lake where *P. antarctica* occurs. Beaver Lake supports a sparse population of a dwarf form of *Boeckella poppei* that has only been recorded in the lakes of MacRobertson Land in eastern Antarctica.

Normal-sized individuals form populations in lakes in western Antarctica and South America. The females carry few eggs in their egg sacs, suggesting low fecundity. They probably grow and reproduce very slowly in the cold unproductive waters of Beaver Lake. Their fecal pellets persist in the water column and are well colonized by bacteria and flagellates. It is likely that *Boeckella* practices coprophagy, thereby benefiting from microbial production.

The crustaceans are also sparse in Arctic lakes and O'Brien et al. (1997) concluded that their grazing rates on the phytoplankton of Toolik Lake (Alaska) are very low. The cladoceran filtering rate is approximately $50\,ml\,L^{-1}day^{-1}$ and the copepod rate is $100\,ml\,L^{-1}day^{-1}$. At its greatest, the zooplankton might turn over the epilimnetic water once a week in late June and early July and once a month in August. One reason the crustaceans are sparse is predation by fish such as the Arctic grayling and small lake trout. These, as well as whitefish, use planktonic and benthic sources of food and so their numbers are high enough that their grazing can be a control on zooplankton abundance. In one whole-lake experiment near Toolik Lake, most of the fish were removed by gill netting followed by an explosion of *Daphnia* numbers.

11.8 Carbon and microbial heterotrophy

The detailed flux of carbon in the heterotrophic food web has only been measured a few times in both polar regions. The pond studied at Barrow, Alaska (Figure 11.3), is very shallow and the carbon input is dominated by the emergent graminoids *Carex aquatilis* and *Arctophyla fulva*. Feeding rates of protozoans and micrometazoans (rotifers, nematodes, chaetonotids, and microcrustaceans) were determined by a variety of means. The bacteria actively break down the plant remains to DOC and their productivity adds some 5% to the bacterial biomass each day. Protozoans and micrometazoans graze only 1% of the bacterial biomass each day but this is some 20% of the bacterial production. Another food web is based on the production of benthic algae (mostly cyanobacteria and green algae). Their primary production adds 5% to the standing stock of microalgae; animal grazing removes the same amount (Hobbie *et al.* 1980).

The carbon dynamics of Lake 18, a typical Tuktoyaktuk Peninsula (Canada) freshwater lake, which has a mean depth of 1.45 m and a maximum depth of 5.2 m, indicated that 50% of the carbon was produced by the benthic community and 20% by plankton, and that 30% entered as allochthonous

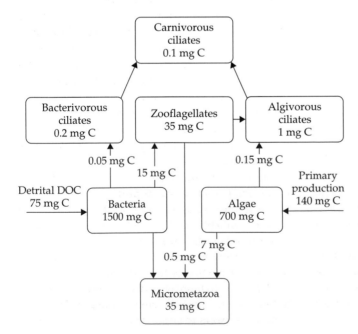

Figure 11.3 Carbon flow through the bacteria, algae, protozoans, and micrometazoa of the sediment of a tundra pond, Barrow, Alaska, on 12 July 1971. Units are mg C m⁻² for the standing stock (rectangles) and mg C m⁻² day⁻¹ for the rates (arrows). Based on information in Fenchel (1975) and Hobbie *et al.* (1980).

matter (Ramlal *et al.* 1994). While allochthonous carbon dominated in this lake, a lake in the far north (81°N) had only autochthonous carbon. In this lake in the Franz Joseph Land archipelago, Panzenböck *et al.* (2000) measured algal productivity, photosynthetic extracellular release, and bacterial productivity. Rotifers were the highest trophic level. Photosynthetic extracellular release made up 31–96% of total primary productivity. Bacterial growth rates (0.1–0.3 day⁻¹) were closely tied to the extracellular release rates indicating that algal exudates are the major carbon source for the bacteria. Bacterial carbon demand (assuming a 50% growth efficiency) amounted to 19% of total primary productivity and 31% of the exudate production.

Detailed studies on the plankton of simple microbially dominated Antarctic lakes allow models of carbon cycling to be developed. Simulations on the food web of Lake Fryxell over 2 years revealed little evidence of top-down control. Major fluctuations in total community biomass followed general patterns of seasonal and inter-annual photosynthetically active radiation levels, clearly indicating bottom-up control (McKenna *et al.* 2006).

11.9 Conclusions

The organisms and processes involved in heterotrophy and microbial food webs are very similar in the two polar regions. Differences are matters of degree. For example, the Antarctic is colder than the Arctic, the Arctic has more biotic diversity in the heterotrophs and more complex microbial food webs, allochthonous organic matter is important in the Arctic, and mixotrophy is more prevalent in the Antarctic. Differences between heterotrophic microbial processes are mostly linked to the absence of some aquatic metazoans.

The extreme climate of Antarctic and Arctic lakes and ponds affects the heterotrophic food webs in direct and indirect ways. The direct effect is that the cold temperatures and winter darkness cause slow growth rates and limit the length of the growing season for the phytoplankton. There is, of course, a difference in the climate of the two polar regions as the Antarctic continent is far colder than the Arctic region. In spite of this, there are similar

types of lakes in both polar regions: lakes with a perennial ice cover, lakes with a very short ice-free season, meromictic lakes, and saline lakes. There are, however, no Antarctic lakes similar to the low Arctic lakes situated in tundra-covered watersheds. If Arctic lakes and ponds can be thought of as a continuum from the climate of the Sub-Arctic and temperate regions to the High Arctic, then the Antarctic lakes cover only the colder half of this range.

Bottom-up control of food webs is important in the polar regions where virtually all the lakes and ponds are oligotrophic and some ultra-oligotrophic. In the polar regions, the low trophic status of the entire heterotrophic food web results in shortened or truncated food webs (Figures 1.5 and 1.6). In the extreme latitudes of the polar regions, the entire heterotrophic food web may consist of bacteria, and mixotrophic and heterotrophic small flagellates. As the trophic level of the waters increase towards lower latitudes, rotifers, and other microheterotrophs become relatively abundant. One aspect of the resource or bottom-up control is the increased importance of mixotrophy or the consumption of bacteria by photosynthetic flagellates. Mixotrophy, as exemplified by the chrysophyte flagellate *Dinobryon*, is found in oligotrophic lakes worldwide. There is good evidence that polar lakes, where the phytoplankton is usually dominated by small flagellates, exhibit a great deal of mixotrophy. This ability of the mixotrophs to consume bacteria is likely the reason that the small flagellates are active even under the ice and darkness of the Antarctic winter. There is also evidence that mixotrophy is more important in the Antarctic than in the Arctic. However, it is also true that the process has been more extensively studied in Antarctic than in Arctic waters.

As we have seen, Antarctic lakes lack subsidies of allochthonous carbon from their catchments, whereas many Arctic waters receive large amounts of dissolved organic carbon, nitrogen, and phosphorus when they are surrounded by tundra vegetation. In the spring runoff much of this allochthonous material is available for microbial breakdown but later in the season it is resistant to breakdown. Overall most of the allochthonous material appears to pass through the lakes and is discharged through the outlet. The material that remains is used by microbes. As a result, bacteria in those lakes with large allochthonous loadings have a much higher bacterial production than other lakes with a similar primary production. However, the added DOC will prevent the penetration of UV light into lakes, which in turn will reduce the rate of photosynthesis and increase the release of nutrients from complex organic compounds of DOC.

The final topic is the regulation of the planktonic heterotrophic food web by grazers, called top-down control. These relationships are difficult to study in any system but the oligotrophic state of polar lakes makes it especially difficult to tease apart the controls. In the most oligotrophic lakes the upper ranges of bacterial concentration are clearly controlled by the mixotrophic and heterotrophic flagellates. In Toolik Lake (Alaska) the crustaceans are so small and sparse that that they take weeks to a month to filter all of the water in the epilimnion. Although the zooplankters are too sparse to support the fish population of Toolik Lake, which depends in great part on benthic sources of food, the fish do control the abundance of zooplankton.

The microbial food web and the primary producers and grazers that feed it with carbon and energy and control the structure of the communities involved are an important part of the ecology of polar lakes. In some cases they are the whole of the ecosystem. While there is a surge in process measurements today, there are still only a few places in the polar regions where bacterial productivity, transfer of carbon and energy, and controls by grazing of flagellates, ciliates, and higher level animals, have been measured. These measurements are only the first steps in developing the understanding of the microbial food webs necessary to predict the effect of changes in climate and species.

References

Alexander, V., Stanley, D.W., Daley, R.J., and McRoy, C.P. (1980). Primary producers. In Hobbie, J.E. (ed.), *Limnology of Tundra Ponds, Barrow, Alaska*, pp. 179–250. Dowden, Hutchinson & Ross, Stroudsburg, PA.

Andreoli, C.L., Scarabel, L., Spini, S., and Grassi, C. (1992). The picoplankton in Antarctic lakes of northern Victoria Land during summer 1989–1990. *Polar Biology* **11**, 575–582.

Bayliss, P., Ellis-Evans, J.C., and Laybourn-Parry, J. (1997). Temporal patterns of primary production in a large ultra-oligotrophic Antarctic freshwater lake. *Polar Biology* **18**, 363–370.

Bell, E.M. and Laybourn-Parry, J. (1999). Annual plankton dynamics in an Antarctic saline lake. *Freshwater Biology* **41**, 507–519.

Bettez, N. (1996). *Changes in Abundance, Species Composition and Controls within the Microbial Loop of a Fertilized Arctic Lake*. MSc Thesis, University of North Carolina, Greensboro, NC.

Bettez, N., Rubless, P, O'Brien, W.J., and Miller, M.C. (2002). Changes in composition and controls within the plankton of fertilized Arctic lakes. *Freshwater Biology* **47**, 303–311.

Bonilla, S., Villeneuve, V., and Vincent, W.F. (2005). Benthic and planktonic algal communities in a high Arctic lake: pigment structure and contrasting responses to nutrient enrichment. *Journal of Phycology* **41**, 1120–1130.

Crump, B.C., Kling, G.W., Bahr, M., and Hobbie, J.E. (2003). Bacterioplankton community shifts in an arctic lake correlate with seasonal changes in organic matter source. *Applied and Environmental Microbiology* **69**, 2253–2268.

Fenchel, T. (1975). The quantitative importance of the benthic microfauna of an arctic tundra pond. *Hydrobiologia* **46**, 445–464.

Henshaw, T. and Laybourn-Parry, J. (2002). The annual patterns of photosynthesis in two large, freshwater, ultra-oligotrophic Antarctic lakes. *Polar Biology* **25**, 744–752.

Hobbie, J.E. and Helfrich, III, J.V.K. (1988). The effect of grazing by microprotozoans on production of bacteria. *Archiv für Hydrobiologie* **31**, 281–288.

Hobbie, J.E., Bahr, M., Bettez, N., Rublee, P.A. (2000). Microbial food webs in oligotrophic arctic lakes. *Microbial Biosystems: New Frontiers, Proceedings of the 8th International symposium on Microbial Ecology. Atlantic Canada society for Microbial Ecology*, pp. 293–298. Halifax, Canada.

Hobbie, J.E. *et al.* (1980). Decomposers, bacteria, and microbenthos. In Hobbie, J.E. (ed.), *Limnology of Tundra Ponds*, pp. 340–387. Dowden, Hutchinson, and Ross, Stroudsburg, PA.

Kaup, E. (1994). Annual primary production of phytoplankton in Lake Verkhneye, Schirmacher Oasis, Antarctica. *Polar Biology* **14**, 433–439.

Kepner, R.L., Wharton, R.A., and Coats, D.W. (1999). Ciliated protozoa of two Antarctic lakes: analysis by quantitative protargol staining and examination of artificial substrates. *Polar Biology* **21**, 285–294.

Laurion, I., Vincent, W.F., and Lean, D.R.S. (1997) Underwater ultraviolet radiation: Development of spectral models for northern high latitude lakes. *Photochemistry and Photobiology* **65**, 107–114.

Laybourn-Parry, J. (2008). Inland waters. In Thomas, D. et al., *The Biology of Polar Regions*, pp. 116–142. Oxford University Press, Oxford.

Laybourn-Parry, J. and Marshall, W.A. (2003). Photosynthesis, mixotrophy and microbial plankton dynamics in two high Arctic lakes during summer. *Polar Biology* **6**, 517–524.

Laybourn-Parry, J. and Pearce, D. (2007). Biodiversity and ecology of Antarctic lakes – models for evolution? *Philosophical Transactions of the Royal Society of London Series B Biological Sciences* **362**, 2273–2289.

Laybourn-Parry, J., Bayliss, P. and Ellis-Evans, J.C. (1995). The dynamics of heterotrophic nanoflagellates and bacterioplankton in a large ultra-oligotrophic Antarctic lake. *Journal of Plankton Research* **17**, 1835–1850.

Laybourn-Parry, J., Henshaw, T., and Quayle, W.C. (2002). The evolution and biology of Antarctic saline lakes in relation to salinity and trophy. *Polar Biology* **25**, 542–552.

Laybourn-Parry, J., Henshaw, T., Jones, D.J., and Quayle, W. (2004). Bacterioplankton production in freshwater Antarctic lakes. *Freshwater Biology* **49**, 735–744.

Laybourn-Parry, J., Marshall, W.A., and Marchant, H.J. (2005). Nutritional versatility as a key to survival in Antarctic phytoflagellates in two contrasting saline lakes. *Freshwater Biology* **50**, 830–838.

Laybourn-Parry, J., Madan, N.J., Marshall, W.A., Marchant, H.J., and Wright, S.W. (2006). Carbon dynamics in a large ultra-oligotrophic epishelf lake (Beaver Lake, Antarctica) during summer. *Freshwater Biology* **51**, 1119–1130.

Laybourn-Parry, J., Marshall, W.A., and Madan, N.J. (2007). Viral production and lysogeny in Antarctic saline lakes. *Polar Biology* **30**, 351–358.

Lisle, J.T. and Priscu, J.C. (2004). The occurrence of lysogenic bacteria and microbial aggregates in the lakes of the McMurdo Dry Valleys, Antarctica. *Microbial Ecology* **47**, 427–439.

Lizotte, M.P., Sharp, T.R., and Priscu, J.C. (1996). Phytoplankton dynamics in the stratified water column of Lake Bonney, Antarctica. 1. Biomass and productivity during the winter-spring transition. *Polar Biology* **16**, 155–162.

Madan, N.J., Marshall, W.A., and Laybourn-Parry, J. (2005). Virus and microbial loop dynamics over an annual cycle in three contrasting Antarctic lakes. *Freshwater Biology* **50**, 1291–1300.

McKenna, K.C., Moorhead, D.L., Roberts, E.C., and Laybourn-Parry, J. (2006). Simulated patterns of carbon flow in the pelagic food web of Lake Fryxell, Antarctica: little evidence of top-down control. *Ecological Modelling* **192**, 457–472.

Miller, M.C., Hater, G.R., Spatt, P., Westlake, P., and Yeakel, D.P. (1986). Primary production and its control in Toolik Lake, Alaska. *Archives Hydrobiologia. Supplement* **74**, 97–131.

Morgan, K.C. and Kalff, J. (1972). Bacterial dynamics in two high-arctic lakes. *Freshwater Biology* **2**, 217–228.

Mueller, D.R., Vincent, W.F., Bonilla, S., and Laurion, I. (2005) Extremotrophs, extremophiles and broadband pigmentation strategies in a high artic ice shelf ecosystem. *FEMS Microbial Ecology* **53**, 73–87.

O'Brien, W.J. et al. (1992). Control mechanisms of arctic lake ecosystems: a limnocorral experiment. *Hydrobiologia* **240**, 143–188.

O'Brien, W.J. et al. (1997). The limnology of Toolik Lake. In Milner, A.M. and Oswood, M.W. (eds), *Freshwaters of Alaska Ecological Synthesis*, pp. 62–106. Springer, New York.

Panzenböck, M., Möbes-Hansen, B., Albert, R., and Herndl, G.J. (2000). Dynamics of phyto- and bacterioplankton in a high Arctic lake on Franz Joseph Land archipelago. *Aquatic Microbial Ecology* **21**, 265–273.

Ramlal, P.S. et al. (1994). The organic carbon budget of a shallow arctic tundra lake on the Tuktoyaktuk Peninsula, NWT, Canada: Arctic lake carbon budget. *Biogeochemistry* **24**, 145–172.

Rautio, M. and Korhola, A. (2002) Effects of ultraviolet radiation and dissolved organic carbon on the survival of subarctic zooplankton. *Polar Biology* **25**, 460–468.

Roberts, E.C. and Laybourn-Parry, J. (1999). Mixotrophic cryptophytes and their predators in the Dry Valley lakes of Antarctica. *Freshwater Biology* **41**, 737–746.

Roberts, E.C., Laybourn-Parry, J., McKnight, D.M., and Novarino, G. (2000). Stratification and dynamics of microbial loop communities in Lake Fryxell, Antarctica. *Freshwater Biology* **44**, 649–662.

Roberts, E.C., Priscu, J.C., and Laybourn-Parry, J. (2004). Microplankton dynamics in a perennially ice-covered Antarctic lake – Lake Hoare. *Freshwater Biology* **49**, 853–869.

Säwström, C., Hodson, A., Marshall, W., and Laybourn-Parry, J. (2002). The microbial communities and primary productivity of cryoconite holes. *Polar Biology* **25**, 591–596.

Säwström, C., Granéli, W., Laybourn-Parry, J., and Anesio, A.M. (2007). High viral infection rates in Antarctic and Arctic bacterioplankton. *Environmental Microbiology* **9**, 250–255.

Spaulding, S.A., McKnight, D.M., Smith, R.L., and Dufford, R. (1994) Phytoplankton population dynamics in perennial ice-covered Lake Fryxell, Antarctica. *Journal of Plankton Research* **16**, 527–541.

Suren, A. (1990) Microfauna associated with algalmats in melt ponds of the Ross Ice Shelf. *Polar Biology* **10**, 329–335.

Takacs, C.D. and Priscu, J.C. (1998). Bacterioplankton dynamics in the McMurdo Dry Valley lakes, Antarctica: production and biomass loss over four seasons. *Microbial Ecology* **78**, 473–476.

Tong, S., Vørs, N., and Patterson, D.J. (1997). Heterotrophic flagellates, centrohelid heliozoa and filose amoebae from marine and freshwater sites in the Antarctic. *Polar Biology* **18**, 91–106.

Van den Hoff, J. and Franzmann, P.D. (1986). A choano-flagellate in a hypersaline Antarctic lake. *Polar Biology* **6**, 71–73.

Vézina, S. and Vincent, W.F. (1997). Arctic cyanobacteria and limnological properties of their environment. Bylot Island, Northwest Territories, Canada (73°N, 80°W). *Polar Biology* **17**, 523–534.

Vincent, W.F. (1997). Polar desert ecosystems in a changing climate: a North-South perspective. In Lyons, W.B., Howard-Williams, C., and Hawes, I. (eds), *Ecosystem Processes in Antarctic Ice-Free Landscapes*, pp. 3–14. A.A. Balkema Publishers, Rotterdam.

Vincent, W.F. (2000). Cyanobacterial dominance in the polar regions. In Whitton, B.A. and Potts, M. (eds), *The Ecology of Cyanobacteria*, pp. 3321–3340. Kluwer Academic Publishers, Dordrecht.

Vincent, W.F. and Goldman, C.R. (1980). Evidence for algal heterotrophy in Lake Tahoe, California-Nevada. *Limnology and Oceanography* **25**, 89–99.

Vincent, W.F. and Pienitz, R. (1996). Sensitivity of high latitude freshwater ecosystems to global change: temperature and solar ultraviolet radiation. *Geoscience Canada* **25**, 231–236.

Vincent, W.F. and Hobbie J.E. (2000). Ecology of arctic lakes and rivers. In Nuttal, M. and Callaghan, T.V. (eds), *The Arctic: Environment, People, Policy*, pp. 197–232. Harwood Academic, London.

Vincent, W.F., Rae, R., Laurion, I., Howard-Williams, C., and Priscu, J.C. (1998) Transparency of Antarctic ice-covered lakes to solar UV radiation. *Limnology and Oceanography* **43**, 618–624.

Whalen, S.C. and Cornwell, J.C. (1985). Nitrogen, phosphorus and organic carbon cycling in an arctic lake. *Canadian Journal of Fisheries and Aquatic Sciences* **42**, 797–808.

Wrona, F.J. *et al.* (2006). Climate change effects on aquatic biota, ecosystem structure and function. *Ambio* **35**, 359–369.

CHAPTER 12

Microbial biodiversity and biogeography

David A. Pearce and Pierre E. Galand

Outline

Microorganisms dominate high-latitude aquatic ecosystems, although much of the microbial biodiversity in the polar regions remains poorly described. Here, we summarize the major groups of microorganisms found in high-latitude aquatic ecosystems and the methodological approaches used to study their biodiversity. We consider the underlying mechanisms which influence this microbial biodiversity, including both the survival (selection pressure) and dispersal (colonization) of microorganisms in polar aquatic environments. Finally, we take a look at the biogeography of microorganisms (i.e. the distribution of their biodiversity) in polar aquatic ecosystems and present studies that reveal a degree of endemism. The polar regions, due to their remoteness and relatively extreme environmental conditions, can be regarded as discrete ecosystems with diverse aquatic chemistry. This leads to novel selection pressures that can be influenced by low temperature and slow metabolism. To date, studies have shown that among the microorganisms of polar aquatic habitats, morphological class hides a much greater genetic diversity, that the extent of the biodiversity uncovered is proportional to sample effort, that the vast majority of biodiversity is as yet uncultivated, that cosmopolitan groups are present, and that the key environmetal factors controling microbial distribution are temperature and biotic interactions.

12.1 Microbial biodiversity

Biodiversity is the variety of life within a given ecosystem. Biodiversity can have a profound effect on ecosystem function (Hooper *et al.* 2005), and the investigation of biodiversity in microbial systems has led to the detection of novel physiologies, the description of new biochemical pathways and the discovery of biochemicals such as enzymes that function in different ways. Microorganisms inhabit all available ecosystems on the planet and contribute the greatest genetic diversity in the biosphere (www.microbes.org/biodiversity.asp); however, according to the *Global Biodiversity Assessment*, even though an estimated 1 000 000 bacterial species exist on Earth, so far, fewer than 4500 have been described.

Microorganisms are fundamental members of the trophic chain in aquatic ecosystems, where they recycle nutrients and transform chemical components. They catalyze reactions involving molecules such as carbon and nitrogen, which are part of larger biogeochemical cycles; for example, the cyanobacteria and photosynthetic sulphur bacteria are dominant primary producers in these ecosystems. Thus aquatic microorganisms directly influence global processes. Until recently, however, the diversity and role of microorganisms in defined ecosystems were not acknowledged because of the lack of data, mainly due to methodological limitations. However, the recent development of molecular techniques targeting phylogenetic genes has dramatically improved our ability to study microorganisms in the environment (Table 12.1).

Table 12.1 A summary of the methods most commonly used to study the biodiversity of microorganisms in the environment.

Methods	Principles	Use
Traditional		
Culture-based approaches	Isolate microorganisms through cultivation	Identify and characterize the physiology
Microscopy	Direct observation of microorganisms through magnification	Identify and enumerate
Flow cytometry	Automated enumeration and size fractionation	Quantification and diversity estimates
Molecular		
Polymerase chain reaction (PCR)	Specifically amplify targeted genes	Essential step to further uncover the identity and function of microorganisms
Molecular fingerprinting	Separate from each other the different types of sequences amplified by PCR	Describe and compare microbial diversity in the environment
Molecular cloning	Separate different sequences using the cellular machinery within bacteria	Determine the identity of organisms present in the environment
Sequencing	Process by which the order of nucleotides forming a DNA strand is determined	Determine the identity of organisms and the composition of their genome
Quantitative PCR (qPCR)	Quantify the abundance of specific targeted genes	Monitor the abundance of microorganisms or their functional genes
Metagenomics	Sequence and analyse an ensemble of genes from an environment or community	Uncover microbial diversity and functional genes from an environmental sample or community
Microautoradiography	Enables direct detection of active bacteria in complex microbial systems at the single-cell level	Link key physiological features to the identity of the microorganism

The introduction by Carl Woese of the 16S rRNA gene as a phylogenetic marker (Woese *et al.* 1975) enabled the description of the three domains of life: Eukarya, Archaea, and Bacteria. Archaea and Bacteria are two purely microbial domains of life which include only prokaryotic organisms (i.e. without a cell nucleus). The third domain, Eukarya, comprises eukaryotic organisms and includes a large diversity of microorganisms such as fungi, protozoa, and algae. Recently more attention has been directed toward describing the diversity of yet another important group of microorganisms found in the environment, the viruses.

The total biodiversity of polar limnetic systems is generally regarded as low, as there are a relatively low number of described species and restricted biomass. However, the extent of the biodiversity uncovered is often determined by the sample effort, and the more these systems are investigated, the higher the apparent microbial biodiversity. With a relatively low diversity and low species richness, polar limnetic systems can provide excellent models in which to study the interaction of microorganisms

with their environment. For example, Antarctic lakes can exhibit periods when a maximum of only three species comprise more than 80% of the standing crop (Allende and Izaguirre 2003). Relatively simple microbial communities, or conspicuous species within them, might be used as indicators of microbial processes and responses to environmental change. However, to date, no polar limnetic ecosystem has been sampled to saturation.

The notion of species, which is widely used to describe the biodiversity of larger organisms, is not normally applicable to microorganisms, because they lack specific morphological characteristics and a sexual life cycle. Microbial diversity in the environment is therefore often characterized by comparing the level of similarity between microbial DNA sequences and is thus described as phylogenetic diversity or genetic diversity. In recent years, the 16S rRNA gene has been recognized as a universal marker to determine the phylogenetic position of microorganisms colonizing the environment, and increased attention has been focused on microbial biodiversity in polar ecosystems. In

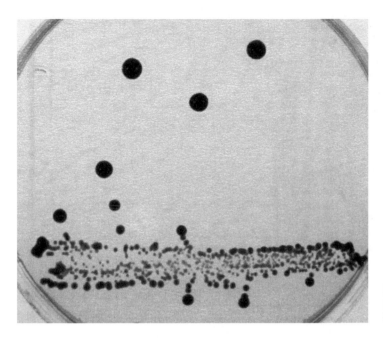

Figure 12.1 Highly pigmented *Chromobacter* isolated from Signy Island in the South Orkney Islands. Photograph: D. Wynn-Williams.

Antarctic lakes, microorganisms are confronted by continuous low temperatures as well as a poor light climate and nutrient limitation. Such extreme environments support truncated food webs with no fish, few metazoans, and a dominance of the microbial plankton (Tranvik and Hansson 1997; Laybourn-Parry 2002). Antarctic prokaryotes are considerably diverse and most evolutionary groups are represented, including representatives of both Archaea and Bacteria (Franzmann 1996; Tindall 2004). The application of molecular techniques has also allowed the first detailed descriptions of microbial consortia at high latitudes, and even though the number of studies remain limited, microbial communities have now also been described in a range of Arctic habitats.

12.2 Bacteria

The first descriptions of high-latitude aquatic microorganisms were intended primarily to demonstrate the presence of pathogens in water supplies. Bacteria were therefore the first microorganisms targeted, and early studies described the presence of coliforms, thermophilic bacteria (McBee and McBee 1956), and the bacteria responsible for

bacillary dysentery (Pauls, 1953). Boyd and Boyd (1963) broadened the description and reported a variety of species in the coastal lakes of northern Alaska. Isolated strains were non-pathogenic, such as species of *Chromobacterium* (Figure 12.1), *Flavobacterium*, and *Pseudomonas*, among which were some isolates capable of denitrification. Although the first descriptions of bacteria in Arctic lake ecosystems were published in the 1950s, the first molecular description of an Arctic freshwater system was not published until much later. In this first study, Bahr *et al.* (1996) reported a diversity of Bacteria in Toolik Lake (Alaska, USA) with isolated cultures grouping into 11 phenotypes including one Gram-positive species, one *Flavobacterium*, and several *Proteobacteria* of the *gamma* and *beta* subclasses. Sequences from *Betaproteobacteria* were dominant in their clone library giving a first indication of the importance of that group of bacteria in Arctic freshwater systems. They also described the presence of *Alphaproteobacteria* forming a freshwater SAR11 clade. SAR11 was first detected in marine waters (Giovannoni *et al.* 1990) and later described as representing the most common group of bacteria in the ocean surface (Morris *et al.* 2002). The freshwater clade of SAR11 first detected in Toolik Lake was later named LD12 by Zwart *et al.* (2002), and is similar to the alphaV cluster of Glöckner *et al.* (2000). In a more recent molecular survey of Toolik Lake, Crump *et al.* (2003) revealed the presence of members of globally distributed freshwater phylogenetic clusters including *Alpha-* and *Betaproteobacteria*, the *Cytophaga-Flavobacterium-Bacteroidetes* (CFB) group and *Actinobacteria*. Among the *Alphaproteobacteria*, members of the LD12 cluster were detected again, and among the *Betaproteobacteria* freshwater clusters such as BAL47 and *Polynucleobacter necessarius* were common.

In addition to polar limnetic ecosystems, there are a wide variety of other freshwater environments in high-latitude systems and these are each potential habitats for freshwater microorganisms. Such sources include ponds, rivers, streams, ice, and snow. Within these systems, microorganisms are often associated with other components of the ecosystem, for example, the plankton (particle-associated and free-living), benthos, microbial mats (Buffan-Dubau *et al.* 2001; Van Trappen *et al.* 2002; Spring *et al.* 2003; Stackebrandt *et al.* 2004; Taton *et al.* 2006a, 2006b), sediment (Battin *et al.* 2001; Purdy *et al.* 2003), glacial meltwater (Dietrich and Arndt 2004), subglacial meltwaters, polythermal glaciers, ice caps, Greenland glaciers, Antarctic ice sheets (Mikucki *et al.* 2005), subglacial lakes (Priscu *et al.* 1999; Price, 2000; Poglazova *et al.* 2001; Christner *et al.* 2006), and cryoconites (Christner *et al.* 2003).

Glacial meltwaters have recently been described as a habitat for microbial communities in the Arctic. A study combining culture and microscopical observation of microorganisms revealed the presence of a range of bacteria in this environment, where both the solute concentration and temperature are low (Skidmore *et al.* 2000). Microorganisms present included aerobic chemoheterotrophs, anaerobic nitrate reducers, sulphate reducers, and methanogens. Further description of subglacial communities revealed a wide range of different bacterial phylotypes (Skidmore *et al.* 2005). The authors studied meltwater samples from two glaciers, Bench Glacier situated in Alaska, and John Evans Glacier on Ellesmere Island, Canada. At both locations the bacterial communities were dominated by members of the *Betaproteobacteria*. *Betaproteobacteria*

have been described as a group commonly found in freshwater systems over a wide range of climates (Zwart *et al.* 2002). Dominant sequences from each site belonged or were closely related to typical freshwater groups previously named BAL 47 and GKS16 (Zwart *et al.* 2002). The second most abundant groups belonged to the CFB group at the John Evans Glacier and *Gamma-* and *Epsilonproteobacteria* at the Bench Glacier, also common in freshwater ecosystems. Samples also contained several other bacterial groups reflecting a relatively high diversity of Bacteria in that particular type of environment. These groups included representatives from the *Alpha-*, *Gamma-*, *Delta-*, and *Epsilonproteobacteria*, *Actinobacteria*, and *Verrucomicrobia*. A recent study of the John Evans glacier by Cheng and Foght (2007) reported that *Betaproteobacteria* sequences were the most abundant in clone libraries of glacial meltwater, thus confirming that the group was also well represented in that Arctic environment.

Very little attention has been directed toward describing microbial diversity in rivers generally and toward Arctic rivers especially. In the first study on an Arctic river, Garneau *et al.* (2006) evaluated the proportion of *Alpha-*, *Beta-*, and *Gammaproteobacteria* and CFB group in the Mackenzie River and adjacent Beaufort Sea. They described a microbial community largely dominated by members of the *Betaproteobacteria* in the Mackenzie River; the other targeted groups had an equally low abundance in the sample. The phylogenetic identity of the bacterial groups was later identified by Galand *et al.* (2008). *Betaproteobacteria* dominated the clone libraries followed by *Bacteroidetes*. *Betaproteobacteria* sequences belonged to typical freshwaters groups, mostly to the BAL47, GKS16, and *P. necessarius* clusters. The study also revealed that the river contained a high diversity of bacteria with the estimated presence of more than 73 phylotypes for a 3% distance level between sequences.

Antarctic lakes contain a wide variety of common freshwater bacterial genera (Table 12.2) and many as yet uncultivated groups. Bowman *et al.* (2000b) examined the biodiversity and community structure within anoxic sediments of several marine-type salinity meromictic lakes, and a coastal marine basin located in the Vestfold Hills

Table 12.2 Diverse range of bacteria identified from freshwater lakes on Signy Island, in the South Orkney Islands

Phylum	Class/order	Genus
Actinobacteria	Actinobacteria	*Acidothermus*
		Actinobacterium
		Actinomycetales
		Actinoplanes
		Arcanobacterium
		Cellulomonas
		Clavibacter
		Micrococcus
		Nocardioides
		Rhodococcus
		Tsukamurella
Proteobacteria	Alpha	*Nordella*
		Rhodobacter
		Sphingomonas
Proteobacteria	Beta	*Achromobacter*
		Acidovorax
		Alcaligenes
		Bordetella
		Burkholderia
		Comamonadaceae
		Neisseria
		Pelistega
		Polaromonas
		Polynucleobacter
		Rhodoferax
		Thauera
		Variovorax
Proteobacteria	Gamma	*Acinetobacter*
		Pseudomonas
		Stenotrophomonas
Bacteroidetes	Flavobacteria	*Flavobacterium*
	Sphingobacteria	*Cytophaga*
		Flectobacillus
		Flexibacter
		Haliscomenobacter
Firmicutes	Bacilli	*Paenibacillus*
		Staphylococcus
Verrucomicrobia	Verrucomicrobiae	*Verrucomicrobium*

area of eastern Antarctica. Importantly, they found that a number of phylotypes predominated, with a group of 10 phylotypes (31% of clones) forming a novel deep branch within the low G+C Gram-positive division. Other abundant phylotypes detected in several different clone libraries

grouped with *Prochlorococcus* cyanobacteria, diatom chloroplasts, delta proteobacteria (*Desulfosarcina* group, *Syntrophus*, the genera *Geobacter*, *Pelobacter*, and *Desulfuromonas*, and the orders Chlamydiales (Parachlamydiaceae) and Spirochaetales (wall-less Antarctic spirochaetes).

Van Trappen *et al.* (2002) isolated 746 bacterial strains from mats collected from 10 lakes in the Dry Valleys. The strains belonged to the *Alpha*, *Beta*, and *Gamma* subclasses of the *Proteobacteria*, the high and low percentage G+C Gram-positives, and to the *Cytophaga-Flavobacterium-Bacteroidetes* branch. They also found that the clusters some represented, belonged to taxa that have not been sequenced yet or represented as yet unnamed new taxa, related to *Alteromonas*, *Bacillus*, *Clavibacter*, *Cyclobacterium*, *Flavobacterium*, *Marinobacter*, *Mesorhizobium*, *Microbacterium*, *Pseudomonas*, *Saligentibacter*, *Sphingomonas*, and *Sulfitobacter*. From similar microbial mats, Taton *et al.* (2006b) isolated 59 cyanobacterial strains and characterized their morphological and genotypic diversity. They included a majority of Oscillatoriales assigned to seven morphospecies, a few Nostocales corresponding to the genera *Nostoc*, *Calothrix*, *Petalonema*, and *Coleodesmium*, and one unicellular *Chondrocystis*. However, the 16S rRNA gene analysis detected 21 operational taxanomic units (OTUs; groups of sequences with more than 97.5% similarity), showing that simple morphologies hide a larger genetic diversity. Nine new OTUs were detected at this time. Moosvi *et al.* (2005) described a diverse facultatively methylotrophic microbiota which exists in some Antarctic locations, including *Methylobacterium*, *Hyphomicrobium*, *Methylobacterium*, *Methylomonas*, *Afipia felis*, and a methylotrophic *Flavobacterium* strain.

A new and exciting environment in the Antarctic is that of the subglacial lakes. Lake Vostok, the largest subglacial lake in Antarctica, is separated from the surface by approximatly 4 km of glacial ice. It has been isolated from direct surface input for at least 420 000 years, and the possibility therefore exists of a novel environment, with distinct selection pressures. Lake Vostok water has not been sampled, but an ice core has been recovered that extends into the ice accreted below glacial ice by freezing of Lake Vostok water. *Brachybacteria*, *Methylobacterium*, *Paenibacillus*, and *Sphingomonas* lineages have all been identified from a sample of meltwater from this accretion ice that originated 3593 m below the surface. This meltwater also contained sequences from *Alpha-* and *Betaproteobacteria*, low- and high-G+C Gram-positive bacteria and a member of the CFB lineage (Christner *et al.* 2001). Elsewhere (in South Pole snow), *Deinococcus*, a radiation-resistant genus, has also been observed (Carpenter *et al.* 2000; see Chapter 7).

12.3 Cyanobacteria

For the Cyanobacteria, Oscillatoriacae and *Nostoc* sp. (Buffan-Dubau *et al.* 2001) are typical Antarctic taxa identified, with *Nostoc* sp. being dominant in two of three Signy Island lakes studied (Pearce 2003; Pearce *et al.* 2003, 2005). Clone libraries of five benthic microbial mats of four lakes on the eastern Antarctic coast showed the presence of 28 OTUs, of which 11 were novel and generally present in only one lake (Taton *et al.* 2006a; see also a review of Cyanobacteria in polar aquatic habitats by Zakhia *et al.* (2007). Picocyanobacteria have been analyzed in the Mackenzie River (Canadian Arctic) and the adjacent Beaufort Sea, and the results suggest that the marine assemblages are derived from the river (Waleron *et al.* 2007).

12.4 Archaea

The largest proportion and greatest diversity of Archaea exist in cold environments (Cavicchioli 2006). Most of the Earth's biosphere is cold, and Archaea therefore represent a significant fraction of the biomass present. The description of archaeal communities started with the development of molecular techniques, and Archaea were subsequently detected in a wide range of ecosystems. However, there are currently no descriptions of archaeal communities in Arctic lakes or glacial meltwater. The only Arctic limnic system from which Archaea have been described so far is the Mackenzie River. In this ecosystem, Archaea were found to represent a relatively large fraction of the prokaryotic community (Garneau *et al.* 2006) and the phylogenetic composition of the archaeal community was then described by Galand *et al.* (2006) in the first report of an Arctic limnic archaeal

community. Their sequencing effort on autumn samples revealed a remarkably high diversity of Archaea in the river. These grouped mainly in two environmental clusters of uncultured Euryarchaeota named RC-V (Rice Cluster-V) and LDS (Lake Dagow sediment) and were related to sequences previously retrieved from freshwater environments such as lake sediment and rice-field soil (Figure 12.2). A second study of samples from the Mackenzie River and the downstream stamukhi Lake Mackenzie confirmed the high diversity of Archaea (Galand *et al.* 2008). Euryarchaeaota from the RC-V and LDS clusters were still the dominant groups in that sample but the authors also detected a large number of Crenarchaeota from Group I.1a. The authors named the group Freshwater Group I.1a and defined it as crenarchaeotal Group I.1a containing freshwater representatives.

In the Antarctic, Bowman *et al.* (2000b) found that most archaeal clones detected (3.1% of clones) belonged to a highly diverged group of Euryarchaeota clustering with clones previously detected in rice soil, aquifer sediments, and hydrothermal vent material. Little similarity existed between the phylotypes detected in this study and other clone libraries based on marine sediment, suggesting that an enormous prokaryotic diversity occurs within marine and marine-derived sediments.

12.5 Eukaryotes

Small eukaryotes are bigger than prokaryotes and thus can be visualized with microscopy and morphological traits used to separate species. This characteristic has allowed early descriptions of eukaryotic communities with distinct morphological structure, such as the diatoms. The first descriptions date from as early as the 1880s in Greenland and over the years a number of species were described from various Arctic limnic systems. Diatom communities have been described through layers of sediments from a paleolimnological perspective (see Chapter 3) but a number of studies have also reported the planktonic or benthic diversity of living diatoms. A detailed review of the available literature on diatom flora in the Arctic can be found in Cremer *et al.* (2001) and

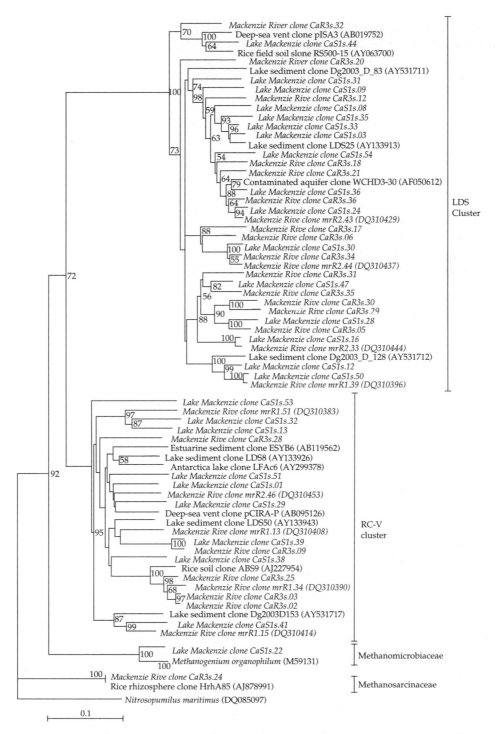

Figure 12.2 Phylogenetic tree illustrating the high diversity of archaeal sequences found in an Arctic lake and river. Sequences in italics were retrieved from the Mackenzie River and stamukhi Lake Mackenzie. Most of the sequences detected in that environment belonged to two clusters of uncultured Euryarchaeota (LDS and RC-V clusters). From Galand *et al.* (2008), by permission of the American Society of Limnology and Oceanography.

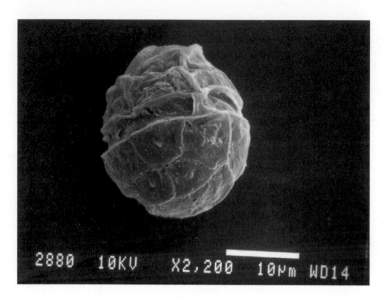

Figure 12.3 An Antarctic dinoflagellate. Photograph: G. Nash.

Antoniades *et al.* (2008). Recent publications have added additional taxonomical information on the species present in Arctic lakes, for example from northeast Greenland (Cremer and Wagner, 2004; Cremer *et al.* 2005), northeastern Siberia (Cremer and Wagner, 2003), and northern Canada (Stewart *et al.* 2005).

Chrysophytes also have traits which allow identification using microscopy. They have been described in Arctic lakes (e.g. the mixotrophic species *Dinobryon*, Plate 11) and dominate assemblages at various locations (Douglas and Smol, 2000; Laybourn-Parry and Marshall 2003; Rasch and Caning 2003; Bonilla *et al.* 2005). Other groups of small eukaryotes described in Arctic lakes include dinoflagellates (also found in Antarctic lakes; Figure 12.3) in Greenland and Spitsbergen (Laybourn-Parry and Marshall 2003; Rasch and Caning 2003), and ciliates in Spitsbergen (Petz 2003; Petz *et al.* 2007).

The identification and description of picoeukaryotes (between 0.2 and 3 µm in diameter) is more difficult with conventional microscopy due to the small cell size. The development of molecular techniques has allowed the discovery of a large diversity of picoeukaryotes in the oceans but their description in limnic systems remains limited to few temperate lakes. Galand *et al.*

(2008) reported the first molecular description of picoeukaryote community in the Arctic. They described sequences obtained from the Mackenzie River and reported the presence of a large diversity of organisms dominated by members of the alveolates (*Choreotrichia*), stramenopiles (*Spumella/ Ochromonas*), and *Cryptophyta*. The diversity of picoeukaryotes in other Arctic freshwater systems remains unknown but it is hoped that the use of molecular tools will, in future, uncover the true diversity of that group of microorganisms.

Some of the important microbial eukaryotes (Plate 11) that have been studied in Antarctic limnetic systems are the fungi (Figure 12.4), algae, heterotrophic flagellates, centrohelid Heliozoa, and filose amoebae. Copepods are known to play a crucial role in the regulation of microbial food webs in certain polar lakes (Tong *et al.* 1997; Bayly *et al.* 2003; see also Chapters 1 and 11). Allende and Izaguirre (2003) studied two Antarctic lakes near Hope Bay. In Lake Boeckella, *Chlamydomonas* sp. followed by *Ochromonas* sp. were the most frequently encountered taxa in the nanophytoplankton fraction; however, the latter species was dominant when the lake froze. In Lake Chico, the major contribution to this fraction was due to different genera of flagellated Chrysophyceae (*Ochromonas* sp., *Chromulina* sp. cf. *Chrysidalis*).

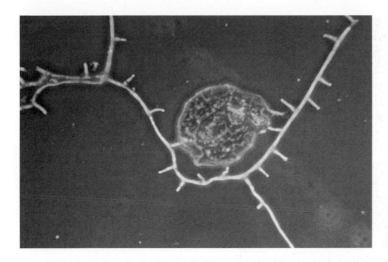

Figure 12.4 Antarctic rotifer trapped by an aquatic Antarctic fungus (McInnes 2003). Photograph: S.J. McInnes.

Unrein *et al.* (2005) studied the composition of planktonic eukaryotes in the size fraction 3–20 µm of 10 maritime Antarctic lakes using denaturing gradient-gel electrophoresis (DGGE). Microscopic observations were also carried out to compare the results obtained by this molecular fingerprinting technique with morphological data. Six lakes from Hope Bay (Antarctic Peninsula) and four from the Potter Peninsula (King George Island) were sampled. The most important organisms in these lakes were the Chrysophyceae, including uniflagellated and biflagellated naked organisms: sequences belonged to the Chrysosphaerales and to the Ochromonadales (unicellular biflagellates). Sequences related to Chlorophyceae, Bacillariophyceae, and probably Cercozoa were also retrieved. A Dictyochophyceae belonging to the order Pedinellales was reported for the first time in freshwater Antarctic ecosystems. Microscopic observations suggested that this phytoplanktonic organism most likely corresponded to *Pseudopedinella*.

12.6 Viruses

Viruses are now generally accepted to be important components of nearly all ecosystems, and polar limetic ecosystems are no exception (Pearce and Wilson 2003). The frequency of visibly phage-infected bacterial cells and the average number of

phages per cell have been determined in Antarctic and Arctic ultra-oligotrophic freshwater environments. Data showed the highest frequency of visibly phage-infected bacterial cells (average 26.1%, range 5.1–66.7%) and the lowest number of phages per cell (average 4, range 2–15) ever reported in the literature (Sawström *et al.* 2007). In the Arctic, tomato mosaic virus has been isolated from the Greenland ice core (Castello *et al.* 1999), raising speculation at the time that one day as the ice caps might melt, such pathogens might be relased back into the biosphere. Viruses are also potential regulators of microbial biodiversity by selectively lysing particularly successful or dominant groups and can be responsible for the transfer of genetic information from one organism to another.

12.7 Survival

One key determinant of biodiversity is survival, largely determined through the action of selection pressure. The often intense selection pressures found in polar aquatic ecosystems include competition for photosynthetically active radiation (photoperiod, wavelength, ice shadow), low temperature, high dilution rate, slow metabolic rate, slow growth rate, incidence of ultraviolet radiation, freeze–thaw cycling (and rate), low turbidity, salinity, and also their combined effects, for example, the availability of key environmental resources, such as water and nutrients, are likely to have a greater impact upon continental Antarctic ecosystems than the effects of fluctuations in temperature alone (Wasley *et al.* 2006).

Key selection pressures in high-latitude aquatic environments are nutrient availability, temperature, biotic interactions, dilution rate, and salinity. Nutrient availability can include the availability of key trace metals (Bratina *et al.* 1998) and high-latitude lake systems can have specific mechanisms of nutrient enrichment (Bell and Laybourn-Parry 1999a, 1999b; Pearce *et al.* 2005). However, Laybourn-Parry *et al.* (2006) suggested using from data from Beaver Lake that primary production and bacterial production were not limited by nutrient availability, but by other factors; for example, in the case of bacterial production by organic carbon concentrations and primary production by low temperatures. Crooked Lake and Lake Druzhby data indicated that bacterial production was limited by continuous low temperatures and the limited availability of suitable dissolved organic carbon substrate. However, the bacterioplankton functions year round, responding to factors other than temperature (Laybourn-Parry *et al.* 2004). In at least three other polar studies, the key factors controlling bacterioplankton community structure were found to be temperature and biotic interactions (Lindström 2000, 2001; Pearce 2005; Yannarell and Triplett 2005). In addition, Upton *et al.* (1990) identified at least four different groups of bacteria that occupied separate so-called temperature niches.

Biotic interactions come in a variety of different forms, which include grazing, competition, and viral lysis. Bell and Laybourn-Parry (2003) calculated that the grazing *Pyramimonas gelidicola* community could remove between 0.4% and approximately 16% of *in situ* bacterial biomass, equivalent to between 4% and more than 100% of *in situ* bacterial production. Elsewhere, it has been observed that there was a positive correlation between bacterial and total phytoplankton (Izaguirre *et al.* 2001). In Beaver Lake, an extreme lacustrine ecosystem where heterotrophic processes occasionally appear to be carbon-limited, highest rates coincided with times of highest primary production (Laybourn-Parry *et al.* 2006; see Chapter 11).

Dilution rates and storms can also affect microbial numbers, and material flowing into lakes can determine the composition of the bacterial community present. The extent to which this happens depends upon the volume of the lake and the flow rate of the inflow (Lindström, 1998). Battin *et al.* (2001), studied the abundance of sediment-associated Archaea, and found that they decreased downstream of the glacier snout; in addition, a major storm increased their relative abundance by a factor of 5.5–7.9. Bacteria of the CFB group were also 6–8-fold more abundant in the storm aftermath.

Salinity may also be important in survival. Bowman *et al.* (2000a) used 16S rRNA clone library analysis to examine the biodiversity and community structure within sediments of three hypersaline Antarctic lakes. Compared to sediment of low to moderate salinity Antarctic lakes the species richness of the hypersaline lake sediments was 2–20 times lower.

Finally, high-latitude aquatic ecosystems are currently subject to interest in relation to global change (Plate 16). In the Arctic, increased melting is potentially leading to more drainage and the increased influence of soils. In the Antarctic more freshwater lakes might appear in association with the temperature increase, especially along the Antarctic Peninsula, which is warming at a rate three times higher than the global average (Quayle *et al.* 2002; Vaughan *et al.* 2004). These increases in temperature will have consequences for microorganisms and could influence the effect of microorganisms in global climate processes (for example, in the production of greenhouse gas).

12.8 Dispersal

Biodiversity is also dependent upon dispersal. The dispersal of microorganisms present in high-latitude ecosystems is dependent upon a number of key factors. These include historical factors; that is, which organisms were present in the environment prior to lake formation and factors related to colonization, mechanisms of inflow, survival and input rates, local and global sources, and evolution rates and colonization rates.

The time of colonization (or recolonization) of Antarctic environments by individual species may have been very recent in evolutionary time scales (Franzmann 1996). Use of the 16S rRNA gene as a molecular clock would suggest that the majority of Antarctic prokaryotes diverged from their nearest known non-Antarctic relatives long before a stable ice sheet developed in Antarctica (Franzmann 1996). Judging from â-tubulin data, the species emerged 30–40 million years ago, when the Antarctic continent reached its current position (De Hoog *et al.* 2005). Most Antarctic environments, however, continue to receive microbial propagules from outside the region, as indicated by spore-trap data, the microflora found in Antarctic snow and ice, the colonizing taxa at geothermal sites, and the high frequency of apparently cosmopolitan species in most habitats.

Many freshwater microorganisms could originate from marine environment. Franzmann and Dobson (1993) sequenced the 16S rRNAs of nine species of prokaryotes, isolated from four lakes of the Vestfold Hills, and found that generally they were most closely related to organisms from marine environments. The sequence dissimilarity between the rRNA gene sequences of the Antarctic strains and their nearest known relatives suggest they diverged from each other much earlier than the establishment of their modern Antarctic habitat. In a study of freshwater Heywood Lake on Signy Island, clones with closest matches to clones of marine origin were also observed (Pearce *et al.* 2005).

Aerial transport is also a major source of microbial propagules into the Antarctic (Hughes *et al.* 2004). For example, in a study by Christner *et al.* (2003), biotic communities did not reflect the composition of the community in the immediately surrounding environments, suggesting the effects of eolian mixing and transport of sediments and biota across the valley. Gordon *et al.* (2000) investigated the phylogenetic diversity of bacteria and cyanobacteria colonizing sediment particles in the permanent ice cover of an Antarctic lake. The results demonstrated the presence of a diverse microbial community dominated by cyanobacteria in the lake ice, and also showed that the dominant members of the lake-ice microbial community were found in terrestrial mats elsewhere in the area. The lake ice microbial community appeared to be dominated

by organisms that are not uniquely adapted to the lake-ice ecosystem, but instead were species that originated elsewhere in the surrounding region and opportunistically colonized the unusual habitat provided by the sediments suspended in the lake ice. Conversely, when lake mats detach from the bottom because they are loaded with oxygen bubbles that are stuck in the mucilage, they enter the ice cover from below. Due to albedo, they can melt the ice locally, rising until they reach the surface. From the surface, they can be blown by the wind to the surrounding soil and rocks. So, the exchange of material goes in both directions.

Feet and feathers of birds are also excellent sources of inocula. The genus *Thelebolus* is known to combine psychrophily with growth on dung and guano, suggesting that strains in lake biomats have originated from bird vectors (De Hoog *et al.* 2005).

12.9 Biogeography

Biogeography is the study of biodiversity distribution over space and time. It aims to reveal where organisms live, at what abundance, and why (Martiny *et al.* 2006). These distribution patterns are often related to physical or biotic factors. Currently microorganisms are found in all ecosystems on Earth, including freshwater systems in the Arctic and Antarctic, and with the increased awareness of their important function in ecosystems, microbiologists are now directing their attention to finding out what species are present, and what role they might have in particular habitats. Increasingly, studies are now focusing on mapping the biodiversity of microorganisms within the environment.

The bipolar geographic separation of the Antarctic and Arctic regions represents the most extreme geographic separation possible on Earth (Petz 2003; Petz *et al.* 2007; Pearce *et al.* 2007), yet these two high-latitude environments share similar selection pressures, and, as such, offer the unique opportunity to test theories and investigate patterns of global distribution. There are also marked differences between these two types of environment. Big rivers are present in the Arctic but not in the Antarctic, the types of lakes differ, the influence of soil is present in the Arctic but not often in the Antarctic, and there are differences in temperature and seasonal cycling. More fundamentally, the Arctic Ocean is surrounded by land, and subject to terrestrial inputs, whereas the Antarctic is an icy continent surrounded by water and subject to marine inputs. This also acts as a major barrier to dispersal. Lakes and ponds on land may also serve as 'islands' that facilitate the selection and separation of unique species (Naganuma *et al.* 2005), particularly in view of the large distances involved in the geographical separation of psychrotrophic limnetic environments. Other psychrotrophic lake environments occur at altitude (for example, the alpine lake systems) and mountain ranges can in themselves act as islands, which reduce the potential for gene flow.

The Arctic, with its very strong seasonality and vast geographical span, is an important environment for the study of microbial biogeography. However, such studies are rare. Nevertheless, some reports have recently been published giving the first indications on the distribution and factors controlling Arctic limnic microbial populations. For instance, Skidmore *et al.* (2005) compared the composition of glacier meltwater microbial communities between the Bench Glacier in Alaska and the John Evans Glacier on Ellesmere Island, northern Canada, and revealed an interesting biogeography. The two glaciers included in their study are separated by thousands of kilometers but their microbial communities were dominated by the same phylogenetic group of *Betaproteobacteria*. The authors noticed, however, some clear differences in community composition indicative of distinct populations. The most interesting feature was the presence of a large number of CFB group bacterial sequences at the John Evans glacier and *Gamma-* and *Epsilonproteobacteria* at the Bench Glacier. The authors suggested that the difference in composition was due to the presence of iron- and sulphur-oxidizing organisms at the Bench Glacier and correlated the microbial populations with the observed aqueous geochemistry.

Regarding seasonal variations, Crump *et al.* (2003) described the diversity of bacterial communities in Toolik Lake, Alaska, from spring to summer, and correlated diversity with the quality of terrestrial dissolved organic matter (DOM). They demonstrated a seasonal shift in the composition of bacterial populations which they correlated to changes in the source and lability of DOM. The community shift corresponded first to the arrival of large influx of DOM associated with snow meltwater and then to the development of phytoplankton communities. They also showed that some bacteria were present all year long while others were transient populations which appeared and disappeared following the seasons. Some of the transient communities were advected to the lake while others grew when the phytoplankton developed. Similar observations have also been made for Signy Island lake communities in the Antarctic (Ellis-Evans and Wynn-Williams 1985).

There are no specific studies on the biogeography of Archaea in the Arctic to date, but the archaeal community of the Mackenzie River has been described for two different seasons: spring (June; Galand *et al.* 2008) and autumn (October; Galand *et al.* 2006). The composition of the spring archaeal community differed from the autumn community; the main difference was illustrated by a clear presence of members of *Crenarchaeota* Group I.1a sequences in spring and a quasi absence in autumn. We can speculate that the difference in community composition represents a seasonal variation of the community possibly linked to a change in water regime, as shown for bacteria in Toolik Lake (Crump *et al.* 2003). However, this observation relies on few samples and extended analyses would be required to demonstrate seasonal biogeography of Archaea in Arctic systems.

For eukaryotes, a recent study of diatoms in sediments from 62 lakes from the Canadian Arctic Archipelago revealed interesting patterns of spatial biogeography (Bouchard *et al.* 2004). The authors showed that although many taxa were found across the Arctic, there were clear regional differences in diatom assemblages. Variability in assemblages was explained mostly by lake water chemistry but also by other environmental factors such as water inflows or terrestrial vegetation. Their result differed from an early study on diatom assemblages in lakes from Arctic Lapland which indicated that spatial distribution of diatoms was markedly regulated by temperature (Weckström and Korhola 2001). For seasonal variation, Stewart *et al.* (2005) recently reported the temporal survey of diatom communities in the Lord Lindsay River, Nunavut, Canada. In the first study of its kind, the authors monitored possible changes in diatom assemblages throughout the melt season. They showed that the composition of diatom assemblages remained relatively constant throughout the season and that those assemblages only responded to dramatic hydrological changes.

In the Antarctic, there have been fewer studies, but the cyanobacterial distribution on the Antarctic continent has been studied in detail (Taton *et al.* 2006a). A synthesis of existing 16S rRNA gene sequences in databases showed the presence of 53 OTUs, of which 68% had not been recorded for non-Antarctic sites at the time, suggesting endemism (i.e. it is unique to that location). There was also a tendency for OTUs found outside Antarctica to be detected in more than one region, suggesting a larger tolerance to environmental factors and dissemination conditions. Each new sample also brought new diversity, suggesting that there is much more diversity left to uncover.

12.10 Endemism

Antarctic isolation makes it an ideal place to study the occurrance of microbial endemism (Vincent 2000). Common freshwater clusters of cosmopolitan distribution have already been suggested (Glöckner *et al.* 2000), and this might be expected given the colonization of lake systems by birds and mammals, some of which might act as common sources of inoculation. Zwart *et al.* (2002) compared 689 bacterial 16S rRNA gene sequences from available freshwater data-sets using BLAST. They found that the majority of the sequences (54%) were similar to other sequences found in freshwater habitats. However, in several studies from Lake Fryxell there is growing evidence that polar endemism can occur. Spring *et al.* (2003) conducted taxonomic studies on four strains of anaerobic, Gram-positive, spore-forming bacteria originally isolated from a mat sample. Phylogenetic analyses indicated that these strains were affiliated with cluster I Clostridia and formed a coherent group with *Clostridium estertheticum* and *Clostridium laramiense*. However, despite the close phylogenetic relationship, several distinguishing phenotypic

traits were found. Van Trappen *et al.* (2002) also isolated bacterial strains from mats collected from 10 lakes in the McMurdo Dry Valleys. Some of the clusters that they represented belonged to taxa that have not been sequenced yet or are as yet unnamed new taxa. Stackebrandt *et al.* (2004) reported a high degree of as-yet-uncultured prokaryotes, and many of the taxa could be regarded novel species as judged from the distance of their gene sequences to those of their nearest cultured phylogenetic neighbors.

Elsewhere, Taton *et al.* (2003) used molecular methods to study the diversity of cyanobacteria in a field microbial mat sample. They found that the 16S rRNA gene sequences were distributed in 11 phylogenetic lineages, three of which were exclusively Antarctic and two that appeared to be novel (more than 2.5% dissimilarity of the 16S rRNA gene sequences with the GenBank data at the time). They suggested that Antarctic endemic species could be more abundant than has been estimated on the basis of morphological features alone. Jung *et al.* (2004) studied strains of two strains of *Rhodoferax antarcticus*. They concluded that although the strains might indeed be the same species, their ecologies were quite different. In the Unrein *et al.* (2005) study of planktonic eukaryotes, most lakes were shown to share several common sequences, which suggests the existence of well-adapted nanoplanktonic species dispersed throughout the Antarctic lakes.

12.11 Conclusions

The very low number of studies of microbial communities in high-latitude aquatic ecosystems suggests that caution is necessary in drawing general conclusions about the biodiversity and biogeography of microorganisms in that environment. In common with other freshwater environments, for bacteria, the *Betaproteobacteria* appears as the dominant group in glacial meltwater, lakes, and rivers and thus is probably the most abundant group in all polar freshwater systems. Dominant freshwater lake bacterioplankton with a bipolar distribution do occur, and most of the bacterioplankton groups encountered in both Arctic and Antarctic freshwater lake systems might fit into emerging freshwater groups described elsewhere (Zwart et al. 1998, 2002; Glöckner et al. 2000; Hahn 2003). However, although some similarities exist, individual lake differences in microbial populations are evident (Lindström and Leskinen 2002; Pearce 2003; Pearce et al. 2003, 2005), and the presence of distinct sequences at each pole does suggest some degree of geographical isolation for freshwater bacterioplankton (Pearce et al. 2007). For Archaea, the first insight into their molecular diversity indicates that Arctic rivers harbour a very diverse community of *Euryarchaeota* and fewer representatives of Freshwater Group I.1 *Crenarchaeota*. However, the study of archaeal diversity in the Arctic is still in its infancy, with no reports on Archaea in Arctic lakes or glacier meltwaters. For eukaryotes, several studies have reported the diversity of small eukaryotes identifiable by microscopy but the molecular diversity of picoeukaryotes has only been described once. The first study targeting the Mackenzie River unveiled the presence of a diverse community with potentially important functions in the ecosystem. As for Archaea, very little is known about picoeukaryotes in the both the Arctic and the Antarctic, and further research could reveal their diversity and role.

Specific studies of bacteria in the Antarctic have shown that there are undoubtedly ubiquitous bacteria, and that unique Antarctic selection pressures do generate unique communitites. In comparisons of species diveristy and species richness, it becomes apparent that generic diversity is low, although species diversity can be very high (Pearce 2003; Pearce et al. 2005). Clearly there are also seasonal variations to contend with (Pearce 2005) and microbial biodiversity can be affected by patterns in the environment such as latitudinal, altitudinal or chemical gradients (Pearce and Butler 2002). The seed bank (or range of dormant propagules present) represented by the underexplored polar diversity could well offer potential for the discovery of new organisms with potential for bioremediation or provide a reservoir for recolonization following climate change.

In conclusion, a great deal of effort needs to be directed toward analyzing microbial biodiversity and biogeography in high-latitude aquatic ecosystems in order to get an insight into how these ecosystems function and how they might respond to environmental change. Further work will allow us to identify the specific role of these microorganisms in global biogeochemical cycles and to understand which factors control their distribution and activity.

Acknowledgments

We would like to thank the Natural Environment Research Council for funding much of the research outlined in the Antarctic; and the Natural Sciences and Engineering Research Council of Canada, the Canada Research Chair program, Fonds québécois de recherche sur la nature et les technologies, and Indian and Northern Affairs Canada for funding P.E. Galand's research in the Arctic. We would like to thank the reviewers for helpful comments and contributions to the manuscript, in particular Annick Wilmotte, Johanna Laybourn-Parry, Warwick F. Vincent, and Sandra J. McInnes.

References

Allende, L. and Izaguirre, I. (2003). The role of physical stability on the establishment of steady states in the phytoplankton community of two maritime Antarctic lakes. *Hydrobiologia* **502**, 211–224.

Antoniades, D., Hamilton, P.B., Douglas, M.S.V., and Smol, J.P. (2008). Diatoms of North America: the freshwater floras of Prince Patrick, Ellef Ringnes and northern Ellesmere Islands from the Canadian Arctic Archipelago. *Iconographia Diatomologica* **17**.

Bahr, M., Hobbie, J.E., and Sogin, M.L. (1996). Bacterial diversity in an Arctic lake: a freshwater SAR 11 cluster. *Aquatic Microbial Ecology* **11**, 271–277.

Battin, T.J., Wille, A., Sattler, B., and Psenner, R. (2001). Phylogenetic and functional heterogeneity of sediment biofilms along environmental gradients in a glacial stream. *Applied and Environmental Microbiology* **67**, 799–807.

Bayly, I.A.E., Gibson, J.A.E., Wagner, B., and Swadling, K.M. (2003). Taxonomy, ecology and zoogeography of two East Antarctic freshwater calanoid copepod species: *Boeckella poppei* and *Gladioferens antarcticus*. *Antarctic Science* **15**, 439–448.

Bell, E.M. and Laybourn-Parry, J. (1999a). Annual plankton dynamics in an Antarctic saline lake. *Freshwater Biology* **41**, 507–519.

Bell, E.M. and Laybourn-Parry, J. (1999b). The plankton community of a young, eutrophic, Antarctic saline lake. *Polar Biology* **22**, 248–253.

Bell, E.M. and Laybourn-Parry, J. (2003). Mixotrophy in the Antarctic phytoflagellate, *Pyramimonas gelidicola* (Chlorophyta : Prasinophyceae). *Journal of Phycology* **39**, 644–649.

Bonilla, S., Villeneuve, V., and Vincent W.F. (2005). Benthic and planktonic algal communities in a high arctic lake: pigment structure and contrasting responses to nutrient enrichment. *Journal of Phycology* **41**, 1120–1130.

Bouchard, G., Gajewski, K., and Hamilton, P.B. (2004). Freshwater diatom biogeography in the Canadian Arctic Archipelago. *Journal of Biogeography* **31**, 1955–1973.

Bowman, J.P., McCammon, S.A., Rea, S.M., and McMeekin, T.A. (2000a). The microbial composition of three limnologically disparate hypersaline Antarctic lakes. *FEMS Microbiology Letters* **183**, 81–88.

Bowman, J.P., Rea, S.M., McCammon, S.A., and McMeekin, T.A. (2000b). Diversity and community structure within anoxic sediment from marine salinity meromictic lakes and a coastal meromictic marine basin, Vestfold Hills, Eastern Antarctica. *Environmental Microbiology* **2**, 227–237.

Boyd, W.L. and Boyd, J.W. (1963). A bacteriological study of an Arctic coastal lake. *Ecology* **44**, 705–710.

Bratina, B.J., Stevenson, B.S., Green, W.J., and Schmidt, T.M. (1998). Manganese reduction by microbes from oxic regions of the Lake Vanda (Antarctica) water column. *Applied and Environmental Microbiology* **64**, 3791–3797.

Buffan-Dubau, E., Pringault, O., and de Wit, R. (2001). Artificial cold-adapted microbial mats cultured from Antarctic lake samples. 1. Formation and structure. *Aquatic Microbial Ecology* **26**, 115–125.

Carpenter, E.J., Lin, S.J., and Capone, D.G. (2000). Bacterial activity in South Pole snow. *Applied and Environmental Microbiology* **66**, 4514–4517.

Castello, J.D. *et al.* (1999). Detection of tomato mosaic tobamovirus RNA in ancient glacial ice. *Polar Biology* **22**, 207–212.

Cavicchioli, R. (2006). Cold-adapted archaea. *Nature Reviews Microbiology* **4**, 331–343.

Cheng, S.M. and Foght, J.M. (2007) Cultivation-independent and -dependent characterization of bacteria resident beneath John Evans Glacier. *FEMS Microbiology Ecology* **59**, 318–330.

Christner, B.C., Mosley-Thompson, E., Thompson, L.G., and Reeve, J.N. (2001). Isolation of bacteria and 16S rDNAs from Lake Vostok accretion ice. *Environmental Microbiology* **3**, 570–577.

Christner, B.C., Kvitko, B.H., and Reeve, J.N. (2003). Molecular identification of Bacteria and Eukarya inhabiting an Antarctic cryoconite hole. *Extremophiles* **7**, 177–183.

Christner, B.C. *et al.* (2006). Limnological conditions in subglacial Lake Vostok, Antarctica. *Limnology and Oceanography* **51**, 2485–2501.

Cremer, H., Wagner, B., Melles, M., and Hubberten, H.W. (2001). The postglacial environmental development of Raffles Sø, East Greenland: inferences from a 10,000 year diatom record. *Journal of Paleolimnology* **26**, 67–87.

Cremer, H. and Wagner, B. (2003). The diatom flora in the ultra-oligotrophic Lake El'gygytgyn, Chukotka. *Polar Biology* **26**, 105–114.

Cremer, H. and Wagner, B. (2004). Planktonic diatom communities in High Arctic lakes (Store Koldewey, Northeast Greenland). *Canadian Journal of Botany* **82**, 1744–1757.

Cremer, H. *et al.* (2005). Hydrology and diatom phytoplankton of high Arctic lakes and ponds on Store Koldewey, Northeast Greenland. *International Review of Hydrobiology* **90**, 84–99.

Crump, B.C., Kling, G.W., Bahr, M., and Hobbie, J.E. (2003). Bacterioplankton community shifts in an Arctic lake correlate with seasonal changes in organic matter source. *Applied and Environmental Microbiology* **69**, 2253–2268.

De Hoog, G.S. *et al.* (2005). Evolution, taxonomy and ecology of the genus *Thelebolus* in Antarctica. *Studies in Mycology* **51**, 33–76.

Dietrich, D. and Arndt, H. (2004). Benthic heterotrophic flagellates in an Antarctic melt water stream. *Hydrobiologia* **529**, 59–70.

Douglas, M.S.V. and Smol, J.P. (2000). Eutrophication and recovery in the High Arctic: Meretta Lake (Cornwallis Island, Nunavut, Canada) revisited. *Hydrobiologia* **431**, 193–204.

Ellis-Evans, J.C. and Wynn-Williams, D.D. (1985). The interaction of soil and lake microflora at Signy Island. In Siegfried, W.R., Condy, P.R., and Laws, R.M. (eds), *Antarctic Nutrient Cycles and Food Webs*, pp. 662–668. Springer-Verlag, Berlin.

Franzmann, P.D. (1996). Examination of Antarctic prokaryotic diversity through molecular comparisons. *Biodiversity and Conservation* **5**, 1295–1305.

Franzmann, P.D. and Dobson, S.J. (1993). The phylogeny of bacteria from a modern Antarctic refuge. *Antarctic Science* **5**, 267–270.

Galand, P.E., Lovejoy, C., and Vincent, W.F. (2006). Remarkably diverse and contrasting archaeal communities in a large arctic river and the coastal Arctic Ocean. *Aquatic Microbial Ecology* **44**, 115–126.

Galand, P.E., Lovejoy, C., Pouliot, J., Garneau, M.-E., and Vincent, W.F. (2008). Microbial community diversity and heterotrophic production in a coastal arctic ecosystem: a stamukhi lake and its source waters. *Limnology and Oceanography* **53**, 813–823.

Garneau, M.E., Vincent, W.F., Alonso-Sáez, L., Gratton, Y., and Lovejoy, C. (2006). Prokaryotic community structure and heterotrophic production in a river-influenced coastal arctic ecosystem. *Aquatic Microbial Ecology* **42**, 27–40.

Giovannoni, S.J., Britschgi, T.B., Moyer, C.L., and Field, K.G. (1990). Genetic diversity in Sargasso sea bacterioplankton. *Nature* **345**, 60–63.

Glöckner, F.O. *et al.* (2000). Comparative 16S rRNA analysis of lake bacterioplankton reveals globally distributed phylogenetic clusters including an abundant group of actinobacteria. *Applied and Environmental Microbiology* **66**, 5053–5065.

Gordon, D.A., Priscu, J., and Giovannoni, S. (2000). Origin and phylogeny of microbes living in permanent Antarctic lake ice. *Microbial Ecology* **39**, 197–202.

Hahn, M.W. (2003). Isolation of novel ultramicrobacteria classified as Actinobacteria from five freshwater habitats in Europe and Asia. *Applied and Environmental Microbiology* **69**, 1442–1451.

Hooper, D.U. *et al.* (2005). Effects of biodiversity on ecosystem functioning: a consensus of current knowledge. *Ecological Monographs* **75**, 3–35.

Hughes, K.A., McCartney, H.A., Lachlan-Cope, T., and Pearce, D.A. (2004). A preliminary study of airbourne biodiversity over peninsular Antarctica. *Cellular and Molecular Biology* **50**, 537–542.

Izaguirre, I., Mataloni, G., Allende, L., and Vinocur, A. (2001). Summer fluctuations of microbial planktonic communities in a eutrophic lake – Cierva Point, Antarctica. *Journal of Plankton Research* **23**, 1095–1109.

Jung, D.O., Achenbach, L.A., Karr, E.A., Takaichi, S., and Madigan, M.T. (2004). A gas vesiculate planktonic strain of the purple non-sulfur bacterium *Rhodoferax antarcticus* isolated from Lake Fryxell, Dry Valleys, Antarctica. *Archives of Microbiology* **182**, 236–243.

Laybourn-Parry, J. (2002). Survival mechanisms in Antarctic lakes. *Philosphical Transactions of the Royal Society of London Series B Biological Sciences* **357**, 863–869.

Laybourn-Parry, J. and Marshall, W.A. (2003) Photosynthesis, mixotrophy and microbial plankton dynamics in two high Arctic lakes during summer. *Polar Biology* **26**, 517–524.

Laybourn-Parry, J., Henshaw, T., Jones, D.J., and Quayle, W. (2004). Bacterioplankton production in freshwater Antarctic lakes. *Freshwater Biology* **49**, 735–744.

Laybourn-Parry, J., Madan, N.J., Marshall, W.A., Marchant, H.J., and Wright, S.W. (2006). Carbon dynamics in an ultra-oligotrophic epishelf lake (Beaver Lake, Antarctica) in summer. *Freshwater Biology* **51**, 1116–1130.

Lindström, E.S. (1998). Bacterioplankton community composition in a boreal forest lake. *FEMS Microbiology Ecology* **27**, 163–174.

Lindström, E.S. (2000). Bacterioplankton community composition in five lakes differing in trophic status and humic content. *Microbial Ecology* **40**, 104–113.

Lindström, E.S. (2001). Investigating influential factors on bacterioplankton community composition: Results from a field study of five mesoeutrophic lakes. *Microbial Ecology* **42**, 598–605.

Lindström, E.S. and Leskinen, E. (2002). Do neighbouring lakes share common taxa of bacterioplankton? – Comparison of 16S rDNA fingerprints and sequences from three geographic regions. *Microbial Ecology* **44**, 1–9.

Martiny, J.B.H. *et al.* (2006). Microbial biogeography: putting microorganisms on the map. *Nature Reviews Microbiology* **4**, 102–112.

McBee, R.H. and McBee, V.H. (1956) The incidence of thermophilic bacteria in Arctic soils and waters. *Journal of Bacteriology* **71**, 182–185.

McInnes, S.J. (2003). A predatory fungus (Hyphomycetes: *Lecophagus*) attacking Rotifera and Tardigrada in maritime Antarctic lakes. *Polar Biology* **26**, 79–82.

Mikucki, J.A., Foreman, C.M., Sattler, B., Lyons, W.B., and Priscu, J.C. (2005). Geomicrobiology of Blood Falls: an iron-rich saline discharge at the terminus of the Taylor Glacier, Antarctica. *Aquatic Geochemistry* **10**, 199–220.

Moosvi, S.A., McDonald, I.R., Pearce, D.A., Kelly, D.P., and Wood, A.P. (2005). Molecular detection and isolation from Antarctica of methylotrophic bacteria able to grow with methylated sulfur compounds. *Systematic and Applied Microbiology* **28**, 541–554.

Morris, R.M. *et al.* (2002). SAR11 clade dominates ocean surface bacterioplankton communities. *Nature* **420**, 806–810.

Naganuma, T., Hua, P.N., Okamoto, T., Ban, S., Imura, S., and Kanda, H. (2005). Depth distribution of euryhaline halophilic bacteria in Suribati Ike, a meromictic lake in East Antarctica. *Polar Biology* **28**, 964–970.

Pauls, F.P. (1953). Enteric diseases in Alaska. *Arctic* **3**, 205–212.

Pearce, D.A. (2003). Bacterioplankton community structure in a maritime Antarctic oligotrophic lake during a period of holomixis, as determined by denaturing gradient gel electrophoresis (DGGE) and fluorescence in situ hybridisation (FISH). *Microbial Ecology* **46**, 92–105.

Pearce, D.A. (2005). The structure and stability of the bacterioplankton community in Antarctic freshwater lakes, subject to extremely rapid environmental change. *FEMS Microbiology Ecology* **53**, 61–72.

Pearce, D.A. and Butler, H.G. (2002). Short-term stability of the microbial community structure in a maritime Antarctic lake. *Polar Biology* **25**, 479–487.

Pearce, D.A., van der Gast, C., Lawley, B., and Ellis-Evans, J.C. (2003). Bacterioplankton community diversity in a maritime Antarctic lake, as determined by culture dependent and culture independent techniques. *FEMS Microbiology Ecology* **45**, 59–70.

Pearce, D.A. and Wilson, W.H. (2003). Antarctic virus ecology. *Antarctic Science* **15**, 319–331.

Pearce, D.A., van der Gast, C.J., Woodward, K., and Newsham, K.K. (2005). Significant changes in the bacterioplankton community structure of a maritime Antarctic freshwater lake following nutrient enrichment. *Microbiology-SGM* **151**, 3237–3248.

Pearce, D.A., Tranvik, L., Lindström, E., and Cockell, C.S. (2007). First evidence for a bipolar distribution of dominant freshwater lake bacterioplankton. *Antarctic Science* **19**, 245–252.

Petz, W. (2003). Ciliate biodiversity in Antarctic and Arctic freshwater habitats – a bipolar comparison. *European Journal of Protistology* **39**, 491–494.

Petz, W., Valbonesi, A., Schiftner, U., Quesada, A., and Cynan Ellis-Evans, J. (2007). Ciliate biogeography in Antarctic and Arctic freshwater ecosystems: endemism or global distribution of species? *FEMS Microbiology Ecology* **59**, 396–408.

Poglazova, M.N., Mitskevich, I.N., Abyzov, S.S., and Ivanov, M.V. (2001). Microbiological characterization of the accreted ice of subglacial Lake Vostok, Antarctica. *Microbiology* **70**, 723–730.

Price, P.B. (2000). A habitat for psychrophiles in deep Antarctic ice. *Proceedings of the National Academy of SciencesUSA* **97**, 1247–1251.

Priscu, J.C. *et al.* (1999). Geomicrobiology of subglacial ice above Lake Vostok, Antarctica. *Science* **286**, 2141–2144.

Purdy, K.J., Nedwell, D.B., and Embley, T.M. (2003). Analysis of the sulfate-reducing bacterial and methanogenic archaeal populations in contrasting Antarctic sediments. *Applied and Environmental Microbiology* **69**, 3181–3191.

Quayle, W.C., Peck, L.S., Peat, H., EllisEvans, J.C., and Harrington, P.R. (2002). Extreme responses to climate change in Antarctic lakes. *Science* **295**, 645.

Rasch, M. and Caning, K. (2003). *Zackenberg Ecological Research Operations*. 8th Annual Report, 2002, p. 80. Danish Polar Center, Ministry of Science, Technology and Innovation, Copenhagen.

Sawström, C., Graneli, W., Laybourn-Parry, J., and Anesio, A.M. (2007). High viral infection rates in Antarctic and Arctic bacterioplankton. *Environmental Microbiology* **9**, 250–255.

Skidmore, M.L., Foght, J.M., and Sharp, M.J. (2000). Microbial life beneath a high Arctic glacier. *Applied and Environmental Microbiology* **66**, 3214–3220.

Skidmore, M.L., Anderson, S.P., Sharp, M.J., Foght, J.M., and Lanoil, B.D. (2005). Comparison of microbial community compositions of two subglacial environments reveals a possible role for microbes in chemical weathering processes. *Applied and Environmental Microbiology* **71**, 6986–6997.

Spring, S. *et al.* (2003). Characterization of novel psychrophilic clostridia from an Antarctic microbial mat: description of *Clostridium frigoris* sp nov., *Clostridium lacusfryxellense* sp nov., *Clostridium bowmanii* sp nov and *Clostridium psychrophilum* sp nov and reclassification of *Clostridium laramiense* as *Clostridium estertheticum* subsp *laramiense* subsp nov. *International Journal of Systematic and Evolutionary Microbiology* **53**, 1019–1029.

Stackebrandt, E., Brambilla, E., Cousin, S., Dirks, W., and Pukall, R. (2004). Culture-independent analysis of bacterial species from an anaerobic mat from Lake Fryxell, Antarctica: Prokaryotic diversity revisited. *Cellular and Molecular Biology* **50**, 517–524.

Stewart, K.A., Lamoureux, S.F., and Forbes, A.C. (2005). Hydrological controls on the diatom assemblage of a seasonal arctic river: Boothia Peninsula, Nunavut, Canada. *Hydrobiologia* **544**, 259–270.

Taton, A., Grubisic, S., Brambilla, E., De Wit, R., and Wilmotte, A. (2003). Cyanobacterial diversity in natural and artificial microbial mats of Lake Fyyxell (McMurdo Dry Valleys, Antarctica): a morphological and molecular approach. *Applied and Environmental Microbiology* **69**, 5157–5169.

Taton, A. *et al.* (2006a). Biogeographic distribution and ecological ranges of benthic cyanobacteria in East Antarctic lakes. *FEMS Microbiology Ecology* **57**, 272–289.

Taton, A. *et al.* (2006b). Polyphasic study of Antarctic cyanobacterial strains. *Journal of Phycology* **42**, 1257–1270.

Tindall, B.J. (2004). Prokaryotic diversity in the Antarctic: the tip of the iceberg. *Microbial Ecology* **47**, 271–283.

Tong, S., Vors, N., and Patterson, D.J. (1997). Heterotrophic flagellates, centrohelid heliozoa and filose amoebae from marine and freshwater sites in the Antarctic. *Polar Biology* **18**, 91–106.

Tranvik, L.J. and Hansson, L.A. (1997). Predator regulation of aquatic microbial abundance in simple food webs of sub-Antarctic lakes. *Oikos* **79**, 347–356.

Unrein, F., Izaguirre, I., Massana, R., Balague, V., and Gasol, J.M. (2005). Nanoplankton assemblages in maritime Antarctic lakes: characterisation and molecular fingerprinting comparison. *Aquatic Microbial Ecology* **40**, 269–282.

Upton, A.C., Nedwell, D.B., and Wynn-Williams, D.D. (1990). The selection of microbial communities by constant or fluctuating temperatures. *FEMS Microbiology Ecology* **74**, 243–252.

Van Trappen, S. *et al.* (2002). Diversity of 746 heterotrophic bacteria isolated from microbial mats from ten Antarctic lakes. *Systematic and Applied Microbiology* **25**, 603–610.

Vaughan, D.G. *et al.* (2004). Recent rapid regional climate warming on the Antarctic peninsula. *Climatic Change* **60**, 243–274.

Vincent, W.F. (2000). Evolutionary origins of Antarctic microbiota: invasion, selection and endemism. *Antarctic Science* **12**, 374–385.

Waleron, M., Waleron, K., Vincent, W.F., and Wilmotte, A. (2007). Allochthonous inputs of riverine picocyanobacteria to coastal waters in the Arctic Ocean. *FEMS Microbial Ecology* **59**, 356–65.

Wasley, J., Robinson, S.A., Lovelock, C.E., and Popp, M. (2006). Climate change manipulations show Antarctic flora is more strongly affected by elevated nutrients than water. *Global Change Biology* **12**, 1800–1812.

Weckström, J. and Korhola, A. (2001) Patterns in the distribution, composition and diversity of diatom assemblages in relation to ecoclimatic factors in Arctic Lapland. *Journal of Biogeography* **28**, 31–45.

Woese, C.R., Sogin, M.L., Bonen, L., and Stahl, D. (1975). Sequence studies on 16S ribosomal RNA from a blue-green alga. *Journal of Molecular Evolution* **4**, 307–315.

Yannarell, A.C. and Triplett, E.W. (2005) Geographic and environmental sources of variation in lake bacterial community composition. *Applied and Environmental Microbiology* **71**, 227–239.

Zakhia, F., Jungblut, A.-D., Taton, A., Vincent, W.F., and Wilmotte, A. (2007). Cyanobacteria in cold environments. In Margesin, R., Schinner, F., Marx, J.C., and Gerday, C. (eds), *Psychrophiles: From Biodiversity to Biotechnology*, pp. 121–135. Springer-Verlag, Berlin.

Zwart, G. *et al.* (1998). Nearly identical 16S rRNA sequences recovered from lakes in North America and Europe indicate the existence of clades of globally distributed freshwater bacteria. *Systematic and Applied Microbiology* **21**, 546–556.

Zwart, G., Crump, B.C., Agterveld, M., Hagen, F., and Han, S.K. (2002) Typical freshwater bacteria: an analysis of available 16S rRNA gene sequences from plankton of lakes and rivers. *Aquatic Microbial Ecology* **28**, 141–155.

CHAPTER 13

Zooplankton and zoobenthos in high-latitude water bodies

Milla Rautio, Ian A.E. Bayly, John A.E. Gibson, and Marjut Nyman

Outline

Zooplankton and **benthic** invertebrates are important components of polar lakes and often form the highest trophic level in the aquatic food web. Their biodiversity decreases towards high latitudes, although the two poles have different species composition despite environmental similarities in temperature, habitat structure, and light cycle. Different geological history and accessibility largely explain the faunal differences between the poles. The Arctic, with easier colonization from the south, extensive high-latitude glacial refugia and a milder climate, is more species-rich than the Antarctic. Furthermore, some taxonomic groups occurring in the Arctic are entirely missing from the Antarctic. Whereas some species live close to their environmental tolerance in the cold polar regions, others have adapted to life at low temperatures, short growing season, long periods of ice cover, and low food supply. Their abundance can be as high as at lower latitudes, especially in shallow water bodies where microbial mats increase the habitat complexity and food. Lakes, ponds, rivers, and streams are all inhabited by various and often different communities of zooplankton and zoobenthos. Small-bodied zooplankton grazing on small food particles are dominant in lakes, while large, pigmented species are often found in shallow ponds where they can be exposed to high levels of ultraviolet (UV) radiation. In some coastal saline lakes of Antarctica zooplankton can form the highest biomass in the plankton. In the well-oxygenated polar lakes and streams benthic animals can be found at high abundance. The aim of this chapter is to describe the unique cold-water communities of zooplankton and zoobenthos common to both poles. The chapter reviews the most important factors that define the zoogeography and diversity of freshwater invertebrates in the Arctic and Antarctic. It describes the range of aquatic invertebrate communities in different habitats, and introduces in more detail the ecology and life history of some key zooplankton and aquatic insects. The final section considers how climate change, especially elevated temperature and increase in UV radiation, are altering high-latitude aquatic invertebrate communities.

13.1 Introduction

In the Earth's high-latitude regions winter-like conditions last up to 9 months of the year, and these areas are characterized by water bodies that are ice-covered or frozen solid, and that experience limited underwater light intensities, short open-water season, and low food supply. In this harsh and demanding environment aquatic invertebrate communities can be stable and abundant (Hershey *et al.* 1995; O'Brien *et al.* 2004).

An important factor that contributes to the success of polar aquatic invertebrates is the lack of efficient predators. All Antarctic and many Arctic zooplankton and zoobenthic communities live in a fish-free environment in which top-down control is absent or only from other invertebrate species. In such 'two-level' systems zooplankton form the highest trophic level in the water column (Hansson *et al.* 1993) and their abundance is mainly structured by food supply and the ability to survive in cold conditions. There are several successful species that have adapted to a life at low temperatures so well that their abundance and biomass can be as high as the biomass of aquatic invertebrates at lower latitudes. Once an animal can survive in near-0°C water it can potentially live at very high latitudes, as the water temperature in lakes is buffered against the extremes of external temperature.

Persistence of abundant zooplankton communities is further facilitated by well-developed benthic–pelagic coupling. Phytoplankton may be the preferred food source for zooplankton but they have also evolved the ability to use alternative food. Clear water, shallowness, and oligotrophy create good conditions for periphytic communities and they can be responsible for a large proportion of primary productivity in ponds and shallow lakes (Vadeboncoeur *et al.* 2003). Many zooplankton species are capable of feeding directly on these microbial mats and other species can ingest particles from mats that are brought up into suspension (Rautio and Vincent 2006, 2007).

Despite successful adaptations to life at low temperatures, the biodiversity of zooplankton and benthic invertebrates decreases towards high latitudes (Hebert and Hann 1986; Patalas 1990; Gibson and Bayly 2007). Many zooplankton and benthic invertebrates live close to their limit of existence in the cold polar region and limited dispersal capability further keeps the species number low. This has important implications for the resistance of polar ecosystems to climate change.

13.2 The origin of polar fauna

The biogeography of metazoans in the lakes of both the Arctic and Antarctic has been of great interest since early in the twentieth century. Classical taxonomic and distributional studies are now being complemented by modern molecular taxonomic work and paleolimnological research that have provided new perspectives on faunal distributions.

The origins of the fauna of the Arctic and Antarctic lakes can be considered over two time frames: since the development of the current arrangement of the continents, including the separation of Pangaea into Laurasia and Gondwana, and the subsequent break-up of Gondwana, beginning about 170 million years ago; and the reaction of the fauna to glaciation that may have begun as early as 34 million years ago (Ivany *et al.* 2006) and that have had dramatic influence on the landforms of both polar regions. Prior to the onset of glaciation, both poles would probably have harbored a biota that was more diverse than at present. The species and genera present undoubtedly reflected the arrangement of the continents at the time, with Antarctica dominated by Gondwanan species (Antarctica retained direct links to South America until about 32 million years ago; Livermore *et al.* 2007) and the Arctic by Laurasian species. While current-day distributions may reflect the species present during the preglacial period, it may also be that they represent in part or in entirety more recent secondary colonization. For example, tardigrades of Antarctica have a clear Gondwanan origin (McInnes and Pugh 1998), and, as most of the species present are endemic there is an implication of a long association with the continent. The presence of a few cosmopolitan species suggests some later colonization from the north. In contrast, the Arctic fauna appears to have been the result of postglacial recolonization rather than local survival (Pugh and McInnes 1998).

The series of Pliocene/Pleistocene glaciations have clearly had a major influence on the distribution of the lake biota. Many of the lakes now present were under many kilometers of ice at the last glacial maximum (approximately 18 000–20 000 years ago), and have probably undergone a cycle of glaciation and exposure since the start of the Pliocene. Any present-day inhabitants of lakes in previously ice-covered areas would have to have migrated from non-glaciated areas that provided habitats (often termed refugia) in which the animals survived.

For both the Arctic and the Antarctic there were three general types of refugia during glacial periods: relatively low-latitude areas beyond the furthest advance of the ice sheet; high-latitude areas where low precipitation rates or other factors precluded the formation of extensive ice sheets; and marine waters, in particular brackish estuaries and semi-enclosed seas, such as the present-day Gulf of Bothnia.

Glacial ice in the Arctic reached as far south as the Great Lakes in North America, and central France and Germany in Europe. As the ice expanded some lacustrine species were able to migrate to the south. This was in part due to the regular occurrence of large, long-lived proglacial lakes along the southern margins of the ice sheets. Once the ice began to retreat many species were able to migrate north to populate new lakes in recently exposed areas, but also retained their southern populations. A classic example is the copepod *Limnocalanus macrurus*, which occurs from Lake Michigan in the south to the northernmost lakes in far northern Canada (Van Hove *et al.* 2001). Molecular genetic studies indicate that the populations that occur in inland lakes through Ontario, Québec, and in central-western Canada are closely related, and are the result of postglacial northern migration (Dooh *et al.* 2006).

The presence of northerly refugia at first seems counter-intuitive, but major ice-free regions occurred at high latitude throughout the glacial periods. Perhaps the best known has been termed

Beringia, which was located in Alaska and northeastern Siberia and included the land bridge that joined Asia and North America that is now beneath the Bering Strait. Patalas (1990) showed that Canadian zooplankton could be separated into two groups: those such as *L. macrurus* that survived south of the Holocene glaciation and have since radiated north, and a second group, including the cladoceran *Daphnia middendorfiana* and the copepod *Diaptomus pribilofensis*, which has its locus in modern-day Alaska. Species have clearly dispersed from these two areas: biodiversity is greatest in areas that could be colonized from both regions, and is lowest in areas to the northeast well away from both major refugia (Figure 13.1). Greatest penetration of the southern group to the north has occurred along the Mackenzie River drainage basin, while the absence of strong north–south and west–east transport mechanisms appears to have limited the penetration of the Beringean group into new areas. Similar northern refugia are still present today: Peary Land in far north Greenland is largely ice-free, and many lakes are present. If new lakes were to be formed by the melting of the ice sheet they could be populated from these lakes.

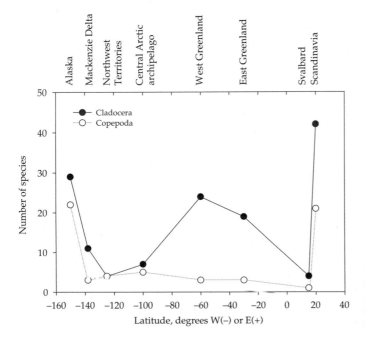

Figure 13.1 Species number of Cladocera and Copepoda along latitude gradient 68–78°N in different arctic regions in North America and northern Europe. Sources: Reed (1962), Rolf and Carter (1972), Hebert and Hann (1986), and Silfverberg (1999).

234 POLAR LAKES AND RIVERS

The final source for the lacustrine animals is marine waters, particularly brackish water such as the Gulf of Bothnia. Molecular genetic studies of species such as *L. macrurus* and *Mysis relicta* (and related species) show that while there are populations that have stemmed from glacial refugia; others, particularly in near-coastal lakes that have formed from the result of isostatic rebound (uplift of the land as a response to the removal of ice), have come from marine populations (Audzijonyte and Väinölä 2006; Dooh *et al.* 2006). Interestingly it does not appear that all these marine populations came from the same genetic stock, but rather that some pre-speciation had occurred in the marine environment prior to isolation in the lakes.

The situation in Antarctica is similar, but with an important difference. In this case there was no land at the margin of the ice during glacial maxima, so that if a species were to populate the continent from temperate refugia it also had to migrate across significant zones of open water. The closest approach of another continent to Antarctica is South America, some 1500 km across the Drake Passage from the Antarctic Peninsula. Other extra-continental refugia may include islands of the Southern Ocean such as Heard Island and Îles Kerguelen. South America (Patagonia) and the Antarctic Peninsula share many species (e.g. the copepod *Boeckella poppei*), and it is easy to conclude that these populations have been sourced from further north. However, at present there is little evidence that this is the case (Gibson and Bayly 2007).

Most east Antarctic tardigrades and nematodes, and many of those on the Antarctic Peninsula, are endemic to the continent, and therefore it becomes more difficult to conclude that there is a strong extra-continental component to the fauna. It is becoming increasingly apparent that many species must have survived glaciation in continental refugia (Gibson and Bayly 2007). A striking example is the copepod *Gladioferens antarcticus*, which is only known from a series of freshwater epishelf lakes in the Bunger Hills (66°10'S, 101°E). The other species of this genus are limited to estuaries of the south coast of Australia and around New Zealand. *G. antarcticus* is the most primitive of its sub-genus, so there is every indication that it has not reinvaded Antarctica recently from an extra-continental source, but

rather has been associated with the Antarctic continent since its separation from Australia (Bayly *et al.* 2003). Survival in continental lakes has been shown directly for the most recent glacial period for two species, the cladoceran *Daphniopsis studeri* (Plate 12) and the rotifer *Notholca* sp., which have been present continually in lakes of the Larsemann Hills for approximately 140 000 years (Cromer *et al.* 2006). It may well be that other species shared with Patagonia have also survived on the continent, and this will be either confirmed or refuted by molecular studies.

The marine habitat has also been a source of lacustrine biota in Antarctica. Modern-day marine biotas occur in saline lakes of the Vestfold Hills, east Antarctica, but these are recent colonizers of the Antarctic lakes (see Section 13.4.2, on saline standing waters, below). Molecular and paleolimnological evidence suggests that a marine copepod closely related to *Paralabidocera antarctica*, one of the species in the modern-day continental saline lakes, had undergone an earlier transition to survival in lakes, eventually occurring in a freshwater lake, Lake Terrasovoje (Bissett *et al.* 2005; Gibson and Bayly 2007).

A further mechanism for transporting animals around the Arctic and Antarctic exists: human involvement. There are many records of deliberate introductions in the Arctic, particularly in Scandinavia. This is not the case, however, in Antarctica, where there is only one known anthropogenic introduction, a chironomid, *Eretmoptera murphyi*.

13.3 Species diversity between poles

Eighteen species of Crustacea, including copepods, cladocerans, ostracods, and an anostracan, are known to inhabit lakes of the Antarctic continent or islands offshore from the Antarctic Peninsula. Of these, 10 are found in maritime Antarctica, with nine on the Antarctic continent proper (Gibson and Bayly 2007). If species apparently recently derived from the marine environment are removed, the numbers drop to eight and five respectively. Only one species, the copepod *B. poppei*, is shared between these regions. New species continue to be discovered in Antarctica as many lake districts have yet to be sampled. New species have been found even in well-studied areas: Roberts *et al.* (2004) recently reported the occurrence of an as-yet-unidentified copepod in Lake Joyce in the McMurdo Dry Valleys, where it had long been thought that no crustaceans occurred. Tadpole shrimps, clam shrimps, and freshwater mysids are entirely missing from the Antarctic.

The geographically less isolated and bigger Arctic has about eight times as many crustacean zooplankton species and an order of magnitude more benthic invertebrate species than the Antarctic (Table 13.1). Generally, the Arctic and Antarctic do not share the same species but some zooplankton genera can be found at both poles, for instance the fairy shrimp *Branchinecta* occurs in the Arctic as *Branchinecta paludosa* and in the Antarctic as *Branchinecta gaini*, and the copepod *Drepanopus* occurs in the Arctic as *Drepanopus bungei* and in the Antarctic as *Drepanopus bispinosus*.

The benthic insect diversity is low for both poles (Table 13.1). Non-biting midges (Chironomidae) are the most numerous benthic invertebrate group and among the best adapted and most successful insect groups in the Arctic. They can form up to 95% of the benthic community species composition (Oliver 1964). Orthocladiinae is the dominant chironomid subfamily in the Arctic zone, decreasing in numbers in warmer regions. The several hundred known Arctic chironomid species contrast with only three (including one introduced) in the maritime Antarctic.

Ostracods, amphipods, mollusks, turbellarians, and oligochaetes contribute to the high-latitude benthic invertebrate community but the number of species present is limited (Table 13.1). For example, there is only one known occurrence of oligochaetes in the Antarctic, on Livingston Island (Toro *et al.* 2007), and only a few species of freshwater bivalves and gastropods occur in the Arctic.

The Arctic climate is more variable than the Antarctic, hence dispersal rates and environmental conditions differ greatly within the Arctic, leading to large regional differences in species composition and number (Figure 13.1). Species richness of cladocerans is poor in Arctic islands and northern Greenland while the region of maximum zooplankton species richness is in Alaska, with 54 reported

Table 13.1 Major freshwater zooplankton and zoobenthos groups and their reported species numbers in the Arctic and Antarctic. Antarctica is here considered as the Antarctic continent and the South Orkney Islands (notably Signy Island), while the Arctic is restricted to the July 10°C isotherm. Arctic excludes Russian Arctic for all other taxa except for nematodes that are reported only from the Russian Arctic and Sub-Arctic. –, No data found.

Taxon	Number of species	
	Antarctica	Arctic
Copepoda	11	48*
Cladocera	4	69
Notostracans	0	2
Anostracans	1	4
Ostracods	2	–
Amphipods	0	–
Mysids	0	4
Tardigrades	10–20	279
Nematodes	<10	179[†]
Rotifers	62[†]	164
Chironomids	2	>94
Gastropods	0	–
Hydracarina	2–5	1
Trichopterans	0	1
Mollusca	0	>6
Oligochaetes	>1	–
Dytiscidae	0	2

* Excluding harpacticoids and parasitic species.
[†] Includes only free-swimming species

Sources: *Antarctic.* Arthropods: Gibson and Bayly (2007); rotifers: Sudzuki (1988); oligochaetes: Toro *et al.* (2007). *Arctic.* Crustaceans: Hebert and Hann (1986), Patalas (1990), Silfverberg (1999), Swadling *et al.* (2001), Audzijonyte and Väinölä (2006); rotifers: Pejler (1995); **nematodes**: Gagarin (2001); **tardigrades**: Pugh and McInnes (1998); **chironomids**, hydracarina, trichopterans, mollusca, and Dytiscidae: Danks (1981).

crustacean zooplankton species. This unusually high diversity derives from the glaciation history of Alaska, as discussed above. The high species number in northern Scandinavia derives from easy dispersal from the south and from a more Sub-Arctic than Arctic-like climate. Calanoid copepods seem to be the most sensitive zooplankton to cold temperatures with only three species, *L. macrurus*, *Drepanopus bungei*, and *Eurytemora* sp. recorded at sites in Canadian Arctic islands with less than 350 degree-days (Hebert and Hann 1986; Van Hove *et al.* 2001). The low number of limnetic zooplankton on Arctic islands derives also from limited dispersal between islands. Central Arctic islands including Devon and Cornwallis islands have the most severe environment for aquatic insect survival and as a consequence the lowest diversity. Abiotic conditions largely define the species distribution while competitive exclusion does not occur among Arctic insects because their densities are usually very low. In most cases parasites and predators are also found in low densities (Strathdee and Bale 1998).

One of the biodiversity hotspots of Antarctica is Byers Peninsula, Livingston Island, which lies offshore from the northern Antarctic Peninsula. However, it still has a relatively low diversity compared to many Arctic systems (Toro *et al.* 2007). The macrozooplankton community in most of the lakes is composed of the copepod *B. poppei* and the fairy shrimp *Br. gaini*, which is the largest freshwater animal in Antarctica, and can play an important role in aquatic food webs. A cladoceran species also occurs in the Byers Peninsula lakes: *Macrothrix ciliata* is a plankto-benthic species associated with aquatic moss.

13.4 Habitats and their key species

13.4.1 Lakes and ponds

The high number of freshwater lakes is a characteristic feature of Arctic and Antarctic landscapes. Within the circum-Arctic treeline region, the most abundant type of water body is a shallow pond often formed in glacial depressions or inside ice-wedge polygons. The ponds freeze solid and lack a biologically active habitat for many months of the year. Most lakes, too, are small and seldom exceed 5 m in depth. Length of the open-water season varies from a few weeks to few months and in both the Antarctic and High Arctic some lakes are frozen all year round. Most Arctic water bodies are dimictic or do not stratify because of their shallowness or perennial low temperatures. The lakes in the Antarctic continent do not usually undergo thermal stratification while some lakes in maritime Antarctica (e.g. Signy Island) may be dimictic. Both polar regions also have permanently stratified meromictic lakes, though they are more common in the Antarctic.

The zooplankton in high-latitude water bodies—rotifers, copepods, cladocerans, fairy shrimps (Anostraca), and mysids—are primarily restricted to lentic environments. Rotifers play important grazing roles in microbial mats with benthic species dominating the communities; for example, several species of *Philodina* occur in the Dry Valley lakes and they have also colonized the open water. Truly planktonic rotifers such as *Keratella*, *Notholca*, and *Filinia* living in the open water are in general very rare, although the genus *Notholca* is well represented in Antarctica.

Early studies on limnology in the Arctic drew attention to the considerable crop of metazooplankton in shallow lakes (e.g. McLaren 1958). More recently the Arctic studies of Rautio and Vincent (2006, 2007) and the Antarctic work of Hansson and Tranvik (2003) have suggested that this biomass is sustained by the well-developed benthic microbial mats, which provide additional food sources to zooplankton and increase the physical complexity in the water body. Ponds with higher benthic primary production subsequently harbor a higher crop of zooplankton than lakes (Figure 13.2). They also have species that are bigger in size. Fairy shrimps (adults can reach >10 mm; Plate 12) and large *Daphnia* (>2 mm; Plate 12) have been recorded in many ponds throughout the Arctic (Stross *et al.* 1980; Rautio 1998) while lakes often have smaller species. Fish predation in the Arctic lakes partly explains the dominance of small and transparent zooplankton but even in lakes without fish or other planktivores small copepods frequently dominate. It is probable that food is the most important single environmental factor explaining the zooplankton size distribution in different polar water bodies. Ponds have higher water-column chlorophyll *a* concentrations (Figure 13.2) and it is likely that many of the potential food sources for zooplankton, such as cyanobacterial trichomes, associated protists, and resuspended benthic mat, are also larger in size than in the lakes. Small food particles can only be ingested by small copepods with correspondingly smaller mouthparts. This may provide the explanation for the marked dwarfing of the copepod *P. antarctica* in such Antarctic saline lakes as Ace Lake (I.A.E. Bayly and J.A.E. Gibson, unpublished results).

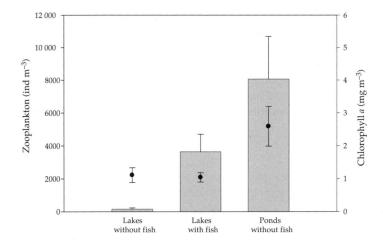

Figure 13.2 Adult crustacean zooplankton abundance (bars) in oligotrophic (chlorophyll *a* <5 μg L⁻¹) high-latitude water bodies above the treeline in northern Canada and Finnish Lapland. The chlorophyll *a* values are given as filled circles. Error bars are SE. Lakes without fish: $n = 7$, mean depth = 7.3 m; lakes with fish: $n = 17$, mean depth = 8.9 m; ponds: $n = 8$, mean depth = 0.7 m. From Rautio and Vincent (2006).

Fairy shrimps (Plate 12) and mysids include the largest pelagic zooplankton species. Fairy shrimps at lower latitudes are primarily restricted to ephemeral environments but the four species present in the Arctic and one in the Antarctic, found in South Orkney Islands, South Shetland Islands and Antarctic Peninsula, are also abundant in small lakes and ponds (Stross *et al.* 1980; Peck 2004). Instead of alternation between wetting and drying the life cycle of fairy shrimps in Arctic or Antarctic waters is regulated by alternation between freezing and thawing. In the Arctic these large animals are strictly missing from lakes with fish. Mysid crustaceans are predominantly marine, but the descendants of Arctic marine *Mysis* are distributed in circumarctic lakes and are included in several 'glacial relict species' of the *M. relicta* group (Audzijonyte and Väinölä 2006).

The benthic invertebrate community is also well developed and abundant in Arctic lakes. In Antarctic fresh waters microbenthos are often the only benthic animals and they occur wherever there are algal mats. The high-energy lake littoral areas in the Arctic provide similar habitats for benthos to those of rivers. The oxygen supply is rich and the detritus accumulation from above is insignificant. Many riverine benthic insects may be found in lake littorals in the area. In the Arctic, insect larvae (especially Chironomidae) constitute most of the macrobenthic fauna, although oligochaete worms, snails, mites, and turbellarians

can also be important. Chironomidae genera *Chironomus, Corynoneura, Cricotopus, Orthocladius, Procladius, Sergentia,* and *Tanytarsini* are often abundant in cold shallow ponds (Walker and Mathewes 1989). In oligotrophic lakes, cold stenotherms, such as *Heterotrissocladius, Paracladius, Parakiefferiella, Protanypus, Pseudodiamesa, Stictochironomus,* and also *Tanytarsini,* may be abundant (Oliver 1964; Hershey 1985).

One of the largest and most abundant benthic invertebrate species occurring in Arctic lakes and ponds is the notostracan *Lepidurus arcticus* (Plate 12). It inhabits numerous localities in Arctic and Sub-Arctic circumpolar areas from approximately 60 to 80°N, and is recorded from all major Arctic regions. The giant amphipod *Gammarus lacustris* with a distribution that spreads from mid- to high latitudes also contributes significantly to the benthic biomass in many Arctic lakes. The high variability in its reproductive traits appears to be a key factor in the maintenance of successful populations in a wide range of aquatic habitats (Wilhelm and Schindler 2000). High-latitude species are generally characterized by biannual or perennial life histories, large body size, delayed maturity, and single or few broods with many, relatively large embryos.

The extreme seasonality in physical features (light, temperature) is reflected in life cycles. In all polar water bodies the perennially low temperatures mean that the species need to be able to complete their life cycle within the period of tolerable physical conditions in the water body. Many high-latitude aquatic invertebrates are therefore monovoltine (Figure 13.3), although some Arctic species exhibit a life-cycle length of up to 3 years (Elgmork and Eie 1989; Wilhelm and Schindler 2000). However, some Antarctic copepods are known to have several generations a year; for example, some *B. poppei* populations are multivoltine. Arctic cladocerans and rotifers alternate between asexual and sexual reproduction and can produce several cohorts during the open-water period. As a result juveniles, adults, and ephippial (resting) eggs can all be present in the water column at the same time. The Antarctic cladoceran *Da. studeri* has adopted varied strategies in lakes of the Vestfold and Larsemann Hills (Gibson and Bayly 2007). While in some lakes parthenogenetic females occur throughout the year in others the overwintering stage is ephippial eggs. Males of this species have been reported, but they may be nonfunctional as is the case in some Arctic species.

13.4.2 Saline standing waters

Lentic limnology in Antarctica has dealt not only with freshwater ponds and lakes, but also standing

waters that are decidedly saline. Although the considerable effort that has been put into the study of saline lakes in east Antarctica has generally been treated as an aspect of limnology, rather than oceanography, these waters are, for the most part, thalassic (marine) rather than athalassic (non-marine), in nature *sensu* Bayly (1967). Saline lakes also occur in the Arctic, particularly at high latitude (Van Hove *et al.* 2006).

When the ice-free 'oases' around the periphery of Antarctica lost their load of continental ice with Holocene amelioration, they rebounded by way of isostatic adjustment (Zwartz *et al.* 1998). As they rebounded, sea water was impounded in valleys and former fiords, and a series of lakes came into existence. The major inorganic ions in these lakes are clearly of marine origin. Where the water budget within the catchment of a lake was positive it became hyposaline with respect to sea water, and where the budget was in deficit it became hypersaline. Such a simple dichotomy makes the mean salinity of sea water, $35 \, g \, L^{-1}$, a key reference point. Bayly (1967) makes the point that it is undesirable to extend a classification based on sea water to athalassic saline waters. However, when dealing, as we are here, with thalassic saline waters it seems natural to do so. Sea water is here considered as having

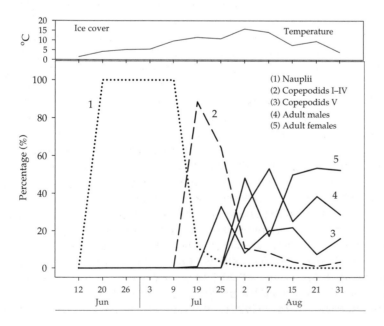

Figure 13.3 Life cycle of a calanoid copepod *Mixodiaptomus laciniatus* in an Arctic pond (69°05′N, 20°87′E) in 1994.

a salinity (S) of 35±2gL^{-1}. It should be noted that the use here of the terms hyposaline (S 3–33gL^{-1}) and hypersaline (S more than 37gL^{-1}) does not correspond with the system of Hammer (1986, p. 15) that has been widely adopted for athalassic saline waters; with the Hammer system hyposaline denotes 3.0–20gL^{-1}, mesosaline 20–50gL^{-1}, and hypersaline more than 50gL^{-1}.

Why is it that most Antarctic saline lakes should be treated as being 'thalassic' (apart from the fact that their major ions are of relict marine origin)? Although they may have permanently lost their connection with the sea, with many there is little evidence that the originally enclosed sea water has, in its entirety, been evaporated to dryness prior to twentieth- and twenty-first-century scientific studies. This lack of desiccation excludes them from Bayly's (1967, 1991) definition of athalassic waters, and means that essentially marine species without proven desiccation-resistant stages in their life cycle may still flourish. In contrast, the Dry Valley saline lakes are much older than the coastal ones, and many were reduced to shallow hypersaline pools about 1000 years ago. This event very probably ensured that the status of most of these lakes is that of true athalassic saline waters.

Microcrustacean species occurring in Antarctic salt lakes include the calanoid copepod *P. antarctica*, the harpacticoid copepods *Idomene scotti* and *Amphiascoides* sp., and the cladoceran *Da. studeri* (Plate 12). *Da. studeri* inhabits only mildly saline lakes (up to about 5gL^{-1}) and also occurs in fresh waters. The copepod *Drepanopus bispinosus* is capable of tolerating hypersaline waters (up to 50gL^{-1}) in lakes, such as Lake Fletcher, Vestfold Hills, with an intermittent connection with the sea.

P. antarctica has been intensively studied. In inshore marine habitats, it is, at least in the earlier stages of its life cycle, sympagic (ice-dwelling). These early stages live in brine channels and small pockets of liquid in the bottom few centimeters of the undersurface of fast ice. The best-studied lacustrine population of *P. antarctica* is that in Ace Lake, Vestfold Hills, where it is non-sympagic. The species was first recorded from the hyposaline portion (6–31gL^{-1}) of this lake by Bayly (1978), who showed that the body size of individuals in this population was much smaller than that of marine individuals.

The vertical distribution of post-naupliar stages swimming freely in the water column during summer was studied by Bayly and Burton (1987). A vertical profile of population density showed two peaks: in the first, located 7–130cm beneath the under surface of the ice, copepodid (C) stages III and IV dominated, but in the second, located just above the anoxylimnion at 10–10.5m depth, stage CV was dominant. Mature males and females were also relatively abundant in the bottom peak. At a depth of 30cm beneath the ice–water interface the total density of all post-naupliar stages combined was close to 20indL^{-1}. Adult females were more abundant relative to adult males and stages CI–V in the water immediately beneath the ice than anywhere else in the water column, suggesting that they may feed in part on epontic algae. Swadling *et al.* (2004) found that the life cycle of the Ace Lake population was less tightly regulated than that of marine populations; developmental synchrony was less marked and earlier naupliar stages were present throughout the year. The seasonal population structure and life cycle is shown in Figure 13.4.

13.4.3 Rivers and streams

There are no permanently flowing streams in Antarctica, and the ephemeral streams that do flow in summer have limited fauna (Young 1989). Protozoans, nematodes, rotifers, and tardigrades may be locally abundant, but there are no insect or crustacean grazers in the Antarctic continent and the total biomass of secondary producers is small. The benthic invertebrate communities of streams in Byers Peninsula, Livingston Island, have a higher number of taxa and functional groups (Toro *et al.* 2007). Three species of chironomids occur along with oligochaetes. Further microinvertebrates, including unidentified species of nematodes, rotifers, and tardigrades, are associated with cyanobacterial mats within the streams.

Arctic streams are mainly first- to fourth-order streams with low habitat diversity. They are not shaded, except occasionally with dwarf willows, hence falling leaves provide a minimum amount of additional food resources to benthic stream invertebrates. Ecotones, according to the river

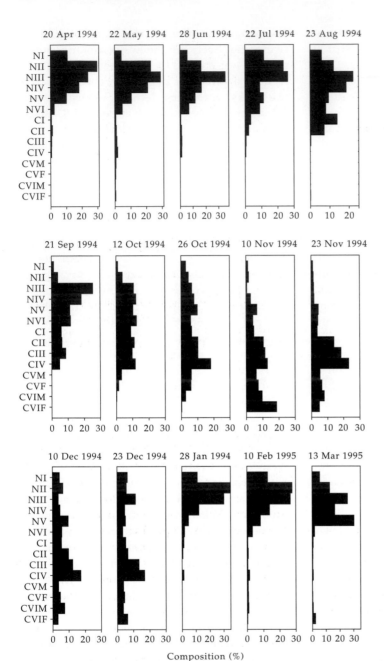

Figure 13.4 *Paralabidocera antarctica*, percentage composition of developmental stages from nauplii (NI–NVI) to copepodites (CI–CVI) in Ace Lake, April 1994–March 1995. Individuals in the last two developmental stages are separated into males and females (CVM, CVF, CVIF, CVIM). From Swadling *et al.* (2004).

continuum concept by Vannote *et al.* (1980), are weakly developed and are of minor functional significance. Fine particulate organic matter is an abundant resource and dominates macroconsumer food webs in a wide range of running-water types (Vannote *et al.* 1980) favoring filter-feeding collectors and collector-gatherer consumers. The communities are subjected to many types of abiotic disturbance. Most streams freeze solid for several months of the year and only insects that have physiological adaptations to survive encased in ice are found there (Hershey *et al.* 1995). In larger

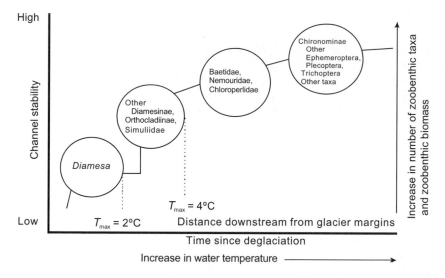

Figure 13.5 Glacial rivers provide generally unfavorable environments for aquatic invertebrates because of their high turbidity, cold water, and unstable channels. In these systems, temperature and channel stability are the principal variables determining temporal and longitudinal patterns of invertebrate colonization. *Diamesa* species are usually the primary invertebrates colonizers of very cold (<2°C) glacial streams in most parts of the world. Schematic presentation of characteristic invertebrate taxa along a glacial river channel of increasing stability and water temperature. Redrawn with permission from Milner and Petts (1994).

streams animals can migrate to deeper water. Most yearly discharge in streams occurs in the spring melt period. This causes major disturbance to the zoobenthos. Fauna in small streams, however, are protected as the stream substrata is still solidly frozen at this time of the year.

The species diversity of insects in the Arctic is lower than in temperate regions and derives from the life-history constraints and habitat characteristics. The species present are largely cold stenotherms and the zoobenthic communities are rhithral; that is, they prefer fast-flowing, well-oxygenated waters with coarse substrata (Figure 13.5). Diptera are the only insects that are well represented in the Arctic, due particularly to the success of Chironomidae and Simuliidae (Oliver 1968). Chironomidae especially are very successful in the Arctic and are nearly as species rich as in temperate streams (Hershey *et al.* 1995). They are able to overwinter in ice, which partly explains their success and high abundance. Sampling over two summer seasons in Alaska yielded 16 species of Simuliidae in an area of 168 km^2 area (Hershey *et al.* 1995). Aquatic Coleoptera, Hemiptera,

Odonata, and Megaloptera are under-represented in the Arctic compared with temperate lotic systems (Oswood 1989). Ephemeroptera tend to be species-poor at high latitudes. For example, there may be a few species in larger, fourth-order streams in Alaska but in first-order streams no mayflies are found (Miller *et al.* 1986; Hershey *et al.* 1995). Plecoptera and Trichoptera are rare or absent in the Arctic and only few species can be found in Sub-Arctic streams. Trichoptera are not species rich but they can be important in the trophic dynamics of Arctic streams due to their large size and aggressive behaviour (Hershey *et al.* 1995). Few species of aquatic beetles have been recorded from small beaded streams and Arctic ponds in Alaska, but they are not found from larger streams.

Low seasonal temperatures and nutrient or food limitation keep the macroinvertebrate production low in Arctic streams; for example, in Alaska the production ranges between 0.5–1 g organic C m^{-2}year^{-1} (Huryn and Wallace 2000). Black flies (Simuliidae) often dominate the secondary production in Alaskan Arctic streams where they can comprise nearly 100% of the insect secondary production of the outlet streams (Hershey et al. 1995). Arctic spring streams, due to their perennially flowing water and high level of substratum stability, support many times higher biomass than the mountain streams (Parker and Huryn 2006). The slowest growth rate and longest-living stream macroinvertebrates are Mollusca, and those with ages over 150 years are known (Ziuganov et al. 2000).

Life cycles longer than 1 year are common for lentic chironomids in the Arctic (Butler 1982; Hershey 1985) but less common among lotic species. In general, four main categories of life cycle have been recognized for northern insects (Ulfstrand 1968): (1) 1-year life cycles, with considerable growth commencing shortly after ovipositioning (e.g. *Ephemerella aurivillii*); intense growth concentrated to a period preceding emergence (e.g. *Ameletus inopinatus* and almost all Simuliidae); or egg and larval stages lasting a short period and a long-lasting pupal stage (e.g. *Glossosoma intermedium*); (2) 2-year cycles, with considerable growth before and during the first winter (e.g. *Arctopsyche ladogensis*); or first winter spent as egg or as quiescent early nymph/larva (e.g. *Rhyacophila nubila*); (3) 3- or 4-year cycle (e.g. *Dinocras cephalotes*); and (4) doubtful cases (e.g. in *Polycentropus flavomaculatus*, which probably has 2-year cycles).

13.5 Implications of climate change

13.5.1 Temperature

Some areas of the Arctic and Antarctic have experienced the greatest regional warming in recent decades (ACIA 2005). Average annual temperatures in areas of both polar regions have risen about 2–3°C since the 1950s and the trend is continuing. The temperature changes have already led to impacts on the environment, for example comparison of satellite images between years 1973 and 1997–1998 have shown that 11% of Siberian lakes have either severely reduced in size or totally disappeared (Smith et al. 2005). Thawing of permafrost and subsequent increase in drainage is driving this process.

Disappearance of a habitat is the ultimate threat to aquatic organisms (Smol and Douglas 2007) but even earlier in the warming process species

respond to temperature change and to its various feedback mechanisms. Most aquatic animals have geographical range limits that are sensitive to temperature. Thermal responses, however, vary markedly among co-occurring species, thus affecting food-web structures and possibly entire ecosystems. Cold-adapted biota will be particularly exposed to strong thermal stress. These species show the greatest risk of local extinction during heating. Two zooplankton species in North America, the copepod *Dia. pribilofensis* and the cladoceran *D. middendorfiana* are restricted in their southern limits by a temperature of around 15°C (Patalas 1990). The calanoid copepod *Mixodiaptomus laciniatus* is also only found in cold lakes and ponds.

Nyman et al. (2005) concluded that factors most affecting chironomids in Finnish Lapland seem to be those that with great probability will also change as climate changes; for example, temperature and accumulation of organic material. Cascading effects of the individual responses (whether direct or indirect) to climate change have an inevitable effect on the composition and structure of communities. Given the thermal constraints that act on Arctic insects it seems likely that for any level of climate warming there will be a disproportionately greater biological response with increasing latitude (Strathdee and Bale 1998).

As the effects of climate change become more pronounced (e.g. shifts in degree-day boundaries or mean temperature isotherms) the more dispersive northern hemisphere species are likely to extend their geographical ranges northward but their southern boundaries may move northwards too because of the expansion of other, warm-adapted aquatic invertebrates that will have a competitive edge. On the other hand, locally adapted Arctic species may be extirpated from some areas as environmental conditions exceed the tolerance of the organisms. Climate warming is expected to alter aquatic insect composition by shifting the locations of thermal optima northward by about 160 km per 1°C increase in surface temperature. As a consequence of new dispersals, the total species number in the Arctic is estimated to increase with warming climate (e.g. Patalas 1990 for zooplankton; Figure 13.6). This greater total species number is gained through a compression and loss of optimal

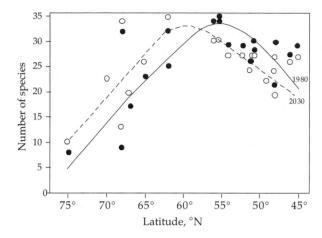

Figure 13.6 Predicted number of crustacean zooplankton species in regions at different latitudes following climate warming (from 3 to 8°C, mean July temperature). Redrawn with permission from Patalas (1990). The greater number of northward dispersing southern species in relation to the number of disappearing northern species causes the potential increase of species diversity in the north.

habitat for native Arctic species and the increase in diversity should be seen as an adverse rather than positive phenomenon.

Temperature effects are likely even more pronounced in the Antarctic Peninsula, which is currently the fastest-warming region in the southern hemisphere. Populations of extant species are likely to expand as a result of climate change. New sites further south will become available for colonization of aquatic invertebrates, as pools further south become unfrozen and algal mats build up. There is strong likelihood that new species will eventually appear on the Antarctic continent as warming continues, and these new species could displace the current endemic groups (Peck 2005). Hahn and Reinhardt (2006) predict by studying the ecology of the chironomid *Parochlus steinenii* on the South Shetland Islands that the species will rapidly colonize habitats that become available in the course of the regional warming of the Antarctic Peninsula region.

In addition to direct effects of warming temperatures on species ecology and distribution, many feedback mechanisms of climate change contribute to the changes in aquatic invertebrate communities. Biological interactions, including trophic structure and food-web composition will very likely be altered. The earlier ice break-up affects supplies of nutrients, sediments, and water quality, but also increases the amount of harmful UV radiation during spring. Increase in surface runoff is expected to

promote higher levels of carbon of allochthonous origin in lakes. The organic supply will alter the structure of benthic habitat and change the light milieu for both benthos and plankton. These multiple stressors will have the additional impact on cold-adapted aquatic biota that they will have to compete with the new invaders from lower latitudes.

13.5.2 UV radiation

Dramatic change in the thickness of the stratospheric ozone layer and corresponding changes in the intensity of solar UV radiation were first observed in Antarctica in the mid-1980s. These were related to the release of industrial chemicals containing chlorine or bromine into the atmosphere. In recent years, rates of ozone loss in the Arctic region have reached values comparable with those recorded over the Antarctic and further increases of 20–90% have been predicted for the period of 2010–2020 (Taalas *et al.* 2000).

Aquatic organisms are protected from UV radiation by colored dissolved organic matter in the water that absorbs this high-energy waveband (Laurion *et al.* 1997), and in the polar regions by the prolonged cover of snow and ice (see Chapter 4). Despite this external physical protection, many aquatic organisms are still exposed to UV radiation at intensities that can be detrimental. Unlike terrestrial organisms, the outer layer of plankton is thin, providing little protection against radiation. Furthermore, Arctic lakes are shallow and clear, with often less than $2\,mg\,L^{-1}$ of UV-attenuating dissolved organic carbon (Rautio and Korhola 2002). In most polar lakes, all functional groups, including the benthos, are exposed to UV radiation.

UV radiation can directly affect the DNA, fecundity, sex ratio, developmental and growth rates, pigmentation, feeding behavior, swimming behavior, and survival of aquatic invertebrates, especially the zooplankton (reviewed by Perin and Lean 2004). Responses to UV radiation differ among taxa and life stages. In general, early developmental stages are most susceptible in all species. Among taxa, cladocerans have shown to have the lowest UV tolerance, with *Daphnia* being one of the most sensitive taxa, whereas adult calanoid and cyclopoid copepods have higher UV tolerances. As a habitat adaptation, species in shallow water bodies are more resistant to UV radiation than deep-lake species and have evolved efficient ways to cope with high UV radiation irradiances.

Some species of zooplankton reduce exposure to UV radiation by vertical migration; that is, organisms avoid the highly irradiated areas by escaping the brightly lit surface zone (Rhode *et al.* 2001). Aquatic organisms can also escape from UV radiation by reducing the effective radiation that penetrates the cell. A variety of UV-screening

compounds have been described; the three major types are carotenoids, mycosporine-like amino acids, and melanin (Figure 13.7, Plate 12). The photoprotective properties of carotenoids are mainly associated with antioxidant mechanisms and inhibition of free radicals as opposed to direct UV screening, whereas mycosporine-like amino acids and melanin, with their absorption maximum in the UV range of the spectrum, function more via direct absorption. Organisms can also repair damage from UV by nucleotide excision repair or by photoreactivation mechanisms such as photoenzymatic repair (Leech and Williamson 2000).

Despite their lines of defense against UV radiation, polar zooplankton remain vulnerable to UV radiation. Furthermore, the defenses may not be efficient enough if environmental changes in high latitudes continue. Most Arctic lakes are shallow, thereby restricting the amplitude of vertical migration. If these lakes become more UV-transparent as a result of reduced precipitation, and hence receive less UV-shielding colored dissolved organic matter from the catchment, vertical migration and other defenses may not provide enough protection from radiation. Similarly, changes in the catchment-derived dissolved organic carbon may occur in the maritime Antarctic where there are reasonable soils with mosses and higher plants. Lakes in the continental Antarctic with their bare rocky surroundings

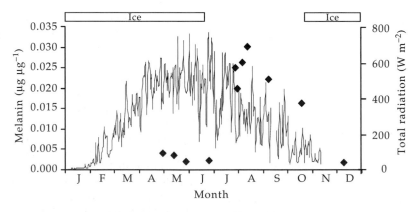

Figure 13.7 Changes over time in the ground-level total radiation and the *Daphnia umbra* melanin concentration in Lake Saanajärvi (69°05'N, 20°87'E; maximum depth=24m, altitude=679m above sea level, dissolved organic carbon=1.6mgL⁻¹), showing the effects of increased underwater light intensity on pigment synthesis. The thickness of the ice exceeds 1m in late spring. From Rautio and Korhola (2002), by permission ogf the American Society of Limnology and Oceanography.

should be less affected by changes in precipitation. Climate warming also leads to a shorter period of ice cover, which has already occurred in the northern hemisphere (Magnuson *et al.* 2000). As a result of earlier thawing, Arctic lakes will be exposed to the most intensive period of UV radiation in mid-June. This may be lethal for those organisms that are not able to adjust their protection against increased UV radiation. Ecosystems in northern Europe are especially exposed to increases in UV radiation. Ozone is being depleted more rapidly over Scandinavia than over most geographical regions at corresponding latitudes. Furthermore, as a result of ocean circulation (Gulf Stream) the climate in this area is warmer than at corresponding latitudes in North America and Asia, resulting in longer open-water period and hence longer exposure to UV radiation.

13.6 Conclusions

Most aquatic animals have geographical range limits that are sensitive to temperature. With decreasing temperatures the species number of aquatic invertebrates decreases towards high latitudes. The abundance can still be high, with both zooplankton and benthic animals benefiting from additional resources of high-biomass microbial mats. Hence, species that are traditionally considered planktonic often exhibit a benthic habitat in polar lakes which makes the separation of the plankton from the benthos difficult. The biggest threat to the future of these animal communities comes from the changes in the temperature and UV radiation. In a warmer climate, cold-adapted species will be restricted to smaller more northern regions in the Arctic. An increase in UV radiation over the poles is a special ecological concern because the biota have adapted to UV conditions that are substantially lower in intensity than those imposed by ozone depletion.

References

ACIA (2005). *Impacts of a Warming Arctic: Arctic Climate Impact Assessment.* Cambridge University Press, Cambridge.

Audzijonyte, A. and Väinölä, R. (2006). Phylogeographic analyses of a circumarctic coastal and a boreal lacustrine mysid crustacean, and evidence of fast postglacial mtDNA rates. *Molecular Ecology* **15**, 3287–3301.

Bayly, I.A.E. (1967). The general biological classification of aquatic environments with special reference to those in Australia. In Weatherley, A.H. (ed.), *Australian Inland Waters and their Fauna*, pp. 78–104. ANU Press, Canberra.

Bayly, I.A.E. (1978). The occurrence of *Paralabidocera antarctica* (I.C. Thompson) (Copepoda: Calanoida: Acartiidae) in an Antarctic saline lake. *Australian Journal of Marine and Freshwater Research* **29**, 817–824.

Bayly, I.A.E. (1991). On the concept and nature of athalassic (non-marine) saline waters. *Salinet* **5**, 76–80.

Bayly, I.A.E. and Burton, H.R. (1987). Vertical distribution of *Paralabidocera antarctica* (Copepoda: Calanoida) in Ace Lake, Antarctica, in Summer. *Australian Journal of Marine and Freshwater Research* **38**, 537–543.

Bayly, I.A.E., Gibson, J.A.E., Wagner, B., and Swadling, K.M. (2003). Taxonomy, ecology and zoogeography of two Antarctic freshwater calanoid copepod species: *Boeckella poppei* and *Gladioferens antarcticus*. *Antarctic Science* **15**, 439–448.

Bissett, A., Gibson, J.A.E., Jarman, S.N., Swadling, K.M., and Cromer, L. (2005). Isolation, amplification and identification of ancient copepod DNA from lake sediments. *Limnology and Oceanography: Methods* **3**, 533–542.

Butler, M.G. (1982). A 7-year life cycle for two *Chironomus* species in arctic Alaskan ponds (Diptera: Chironomidae). *Canadian Journal of Zoology* **60**, 58–70.

Cromer, L., Gibson, J.A.E., Swadling, K.M., and Hodgson, D.A. (2006). Evidence for a lacustrine faunal refuge in the Larsemann Hills, East Antarctica, during the Last Glacial Maximum. *Journal of Biogeography* **33**, 1314–1323.

Danks, H.V. (1981). *Arctic Arthropods. A Review of Systematics and Ecology with Particular Reference to North American Fauna.* Entomological Society of Canada, Ottawa.

Dooh, R.T., Adamowicz, S.J., and Hebert, P.D.N. (2006). Comparative phylogeography of two north American 'glacial relict' crustaceans. *Molecular Ecology* **15**, 4459–4475.

Elgmork, K. and Eie, J.A. (1989). Two- and three-year life cycles in the planktonic copepod *Cyclops scutifer* in two high mountain lakes. *Holoarctic Ecology* **12**, 60–69.

Gagarin, V.G. (2001). Review of free-living nematode fauna from fresh waters of the Russian Arctic and Subarctic. *Biology of Inland Waters* **2**, 32–37.

Gibson, J.A.E. and Bayly, I.A.E. (2007). New insights into the origins of crustaceans of Antarctic lakes. *Antarctic Science* **19**, 157–164.

Hahn, S. and Reinhardt, K. (2006). Habitat preference and reproductive traits in the Antarctic midge *Parochlus steinenii* (Diptera: Chironomidae). *Antarctic Science* **18**, 175–181.

Hammer, U.T. (1986). *Saline Lake Ecosystems of the World.* Junk, Dordrecht.

Hansson, L.-A., Lindell, M., and Tranvik, L. (1993) Biomass distribution among trophic levels in lakes lacking vertebrate predators. *Oikos* **66**, 101–106.

Hansson, L.-A. and Tranvik, L. (2003). Food webs in sub-Antarctic lakes: a stable isotope approach. *Polar Biology* **26**, 783–788.

Hebert, P.D.N. and Hann, B.J. (1986). Patterns in the composition of arctic tundra pond microcrustacean communities. *Canadian Journal of Fisheries and Aquatic Sciences* **43**, 1416–1425.

Hershey, A.E. (1985). Littoral chironomid communities in an arctic Alaskan lake. *Holarctic Ecology* **8**, 39–48.

Hershey, A.E., Merritt, R.W., and Miller, M.C. (1995). Insect diversity, life history, and trophic dynamics in arctic streams, with particular emphasis on black flies (Diptera: Simuliidae). In Chapin, F.S. and Körner, C. (eds), *Arctic and Alpine Biodiversity: Patterns, Causes and Ecosystem Consequences.* Ecological Studies, vol. 113, pp. 283–295. Springer, Berlin.

Huryn, A.D. and Wallace, J.B. (2000). Life history and production of stream insects. *Annual Review of Entomology* **45**, 83–110.

Ivany, L.C., Van Simaeys, S, Domack, E.W., and Samson, S.D. (2006). Evidence for an earliest Oligocene ice sheet on the Antarctic Peninsula. *Geology* 34, 377–380.

Laurion, I., Vincent, W.F., and Lean, D.R.S. (1997). Underwater ultraviolet radiation: Development of spectral models for northern high latitude lakes. *Photochemistry and Photobiology* **65**, 107–114.

Leech, D.M. and Williamson, C.E. (2001). In situ exposure to ultraviolet radiation alters the depth distribution of *Daphnia. Limnology and Oceanography* **46**, 416–420.

Livermore, R., Hillenbrand, C.-D., Meredith, M., and Eagles, G. (2007). Drake Passage and Cenozoic climate: an open and shut case? *Geochemistry Geophysics Geosystems* 8, Q01005, doi:10.1029/2005GC001224.

Magnuson, J.J. *et al.* (2000). Historical trends in lake and river ice cover in the Northern Hemisphere. *Science* **289**, 1743–1746.

McInnes, S.J. and Pugh, P.J.A. (1998). Biogeography of limno-terrestrial Tardigrada, with particular reference to the Antarctic fauna. *Journal of Biogeography* **25**, 31–36.

McLaren, I.A. (1958). Annual report and investigators' summaries. *Journal of Fisheries Research Board of Canada.* **15**, 27–35.

Miller, M.C., Stout, J.R., and Alexander, V. (1986). Effects of a controlled under-ice oil spill on invertebrates on an arctic and subarctic stream. *Environmental Pollution* **42**, 99–132.

Milner, A.M. and Petts, G.E. (1994). Glacial rivers: physical habitat and ecology. *Freshwater Biology* **32**, 295–307.

Nyman, M., Korhola, A., and Brooks, S.J. (2005). The distribution and diversity of Chironomidae (Insecta: Diptera) in western Finnish Lapland, with special emphasis on shallow lakes. *Global Ecology and Biogeography* **14**, 137–153.

O'Brien, W.J. *et al.* (2004) Physical, chemical and biotic effects on arctic zooplankton communities and diversity. *Limnology and Oceanography* **49**, 1250–1261.

Oliver, D.R. (1964). A limnological investigation of a large arctic lake, Nettilling Lake, Baffin Island. *Arctic* **17**, 69–83.

Oliver, D.R. (1968) Adaptations of arctic Chironomidae. *Annales Zoology Fennica* **5**, 111–118.

Oswood, M.W. (1989). Community structure of benthic invertebrates in interior Alaskan (USA) streams and rivers. *Hydrobiologia* **172**, 97–110.

Parker, S.M. and Huryn, A.D. (2006). Food web structure and function in two arctic streams with contrasting disturbance regimes. *Freshwater Biology* **51**, 1249–1263.

Patalas, K. (1990). Diversity of the zooplankton communities in Canadian lakes as a function of climate. *Verhanlungen des Internationalen verein für Limnologie* **24**, 360–368.

Peck, L.S. (2004). Physiological flexibility: the key to success and survival for Antarctic fairy shrimps in highly fluctuating extreme environments. *Freshwater Biology* **49**, 1195–1205.

Peck, L.S. (2005). Prospects for surviving climate change in Antarctic aquatic species. *Frontiers in Zoology* **2**, doi:10.1186/1742-9994-2-9.

Pejler, B. (1995). Relation to habitat in rotifers. *Hydrobiologia* **313**, 267–278.

Perin, S. and Lean, D.R.S. (2004). The effects of ultraviolet-B radiation on freshwater ecosystems of the Arctic: influence from stratospheric ozone depletion and climate change. *Environmental Review* **12**, 1–70.

Pugh, P.J.A. and McInnes, S.J. (1998). The origin of Arctic terrestrial and freshwater tardigrades. *Polar Biology* **19**, 177–182.

Rautio, M. (1998). Community structure of crustacean zooplankton in subarctic ponds—effects of altitude and physical heterogeneity. *Ecography* **21**, 327–335.

Rautio, M. and Korhola, A. (2002). UV-induced pigmentation in subarctic *Daphnia. Limnology and Oceanography* **47**, 295–299.

Rautio, M. and Vincent, W.F. (2006). Benthic and pelagic food resources for zooplankton in shallow high-latitude lakes and ponds. *Freshwater Biology* **51**, 1038–1052.

Rautio, M. and Vincent, W.F. (2007). Isotopic analysis of the sources of organic carbon for zooplankton in shallow subarctic and arctic waters. *Ecography* **30**, 77–87.

Reed, E.B. (1962). Freshwater plankton Crustacea of the Conville river area, northern Alaska. *Arctic* **15**, 27–50.

Rhode, S.C., Pawlowski, M., and Tollrian, R. (2001). The impact of ultraviolet radiation on the vertical distribution of zooplankton on the genus *Daphnia*. *Nature* **412**, 69–72.

Roberts, E.C., Priscu, J.C., Wolf, C., Lyons, W.B., and Laybourn-Parry, J. (2004). The distribution of microplankton in the McMurdo Dry Valley Lakes, Antarctica: response to ecosystem legacy or present-day climatic controls? *Polar Biology* **27**, 238–249.

Rolf, J.C. and Carter, J.C.H. (1972). Life-cycle and seasonal abundance of the copepod *Limnocalanus macrurus* in a high Arctic lake. *Limnology and Oceanography* **17**, 363–370.

Silfverberg, H. (1999). A provisional list of Finnish Crustacea. *Memoranda Societatis pro Fauna et Flora Fennica* **75**, 15–39.

Smith, L.C., Sheng, Y., MacDonald, G.M., and Hinzman, L.D. (2005). Disappearing arctic lakes. *Science* **308**, 1429.

Smol, J.P. and Douglas, M.S.V. (2007). Crossing the final ecological threshold in high Arctic ponds. *Proceedings of the National Academy of Sciences USA* **104**, 12395–12397.

Strathdee, A.T. and Bale, J.S. (1998). Life on the edge: insect ecology in Arctic environments. *Annual Review of Entomology* **43**, 85–106.

Stross, R.G., Miller, M.C., and Daley, R.J. (1980). Zooplankton: communities, life cycles and production. In Hobbie, J.E. (ed.), *Limnology of Tundra Ponds*, pp. 251–296. Hutchinson and Ross, Stroudsburg, PA.

Sudzuki, M. (1988). Comments on the Antarctic Rotifera. *Hydrobiologia* **165**, 89–96.

Swadling, K.M., Gibson, J.A.E., Pienitz, R., and Vincent, W.F. (2001). Biogeography of copepods in lakes of northern Quebec, Canada. *Hydrobiologia* **453/454**, 341–350.

Swadling, K.M., McKinnon, A.D., De'ath, G., and Gibson, J.A.E. (2004). Life cycle plasticity and differential growth and development in marine and lacustrine populations of an Antarctic copepod. *Limnology and Oceanography* **49**, 644–655.

Taalas, P. *et al.* (2000). The impact of greenhouse gases and halogenated species on future solar UV radiation doses. *Geophysical Research Letters* **27**, 1127–1130.

Toro, M. *et al.* (2007). Limnological characteristics of the freshwater ecosystems of Byers Peninsula, Livingston Island, in maritime Antarctica. *Polar Biology* **30**, 635–649.

Ulfstrand, S. (1968). Life cycles of benthic insects in Lapland streams (Ephemeroptera, Plecoptera, Trichoptera, Diptera, Simuliidae). *Oikos* **19**, 167–190.

Vadeboncoeur, Y. *et al.* (2003). From Greenland to green lakes: cultural eutrophication and the loss of benthic pathways in lakes. *Limnology and Oceanography* **48**, 1408–1418.

Van Hove, P., Swadling, K., Gibson, J.A.E., Belzile, C., and Vincent, W.F. (2001). Farthest north lake and fjord populations of calanoid copepods in the Canadian High Arctic. *Polar Biology* **24**, 303–307.

Van Hove, P., Belzile, C., Gibson, J.A.E., and Vincent, W.F. (2006). Coupled landscape-lake evolution in the coastal high Arctic. *Canadian Journal of Earth Sciences* **43**, 533–546.

Vannote, R.L., Minshall, G.V., Cummins, K.W., Sedell, J.R., and Cushing, C.E. (1980). The river continuum concept. *Canadian Journal of Fisheries and Aquatic Sciences* **37**, 130–137.

Walker, I.R. and Mathewes, R.W. (1989). Chironomidae (Diptera) remains in surficial lake sediments from the Canadian Cordillera: analysis of the fauna across an altitudinal gradient. *Journal of Paleolimnology* **2**, 61–80.

Wilhelm, F.M. and Schindler, D.W. (2000). Reproductive strategies of *Gammarus lacustris* (Crustacea: Amphipoda) along elevational gradient. *Functional Ecology* **14**, 413–422.

Young, E.C. (1989). Ecology and conservation of the polar regions. *Ambio* **18**, 23–33.

Ziuganov, V. *et al.* (2000). Life span variation of the freshwater pearl shell: A model species for testing longevity mechanisms in animals. *Ambio* **29**, 102–105.

Zwartz, D., Bird, M., Stone, J., and Lambeck, K. (1998). Holocene sea-level change and ice-sheet history in the Vestfold Hills, East Antarctica. *Earth and Planetary Science Letters* **155**, 131–145.

CHAPTER 14

Fish in high-latitude Arctic lakes

Michael Power, James D. Reist, and J. Brian Dempson

Outline

There is a limited freshwater fish fauna in the High Arctic, with Arctic char (*Salvelinus alpinus*) dominating in most aquatic systems. As a result Arctic char are biologically and culturally important throughout the north polar region. Contrasting theories of observed bimodality in Arctic lacustrine fish population length-frequency distributions suggest bimodality may result from top-down control of fish population dynamics, resulting from the dominance of a few, large piscivorous individuals, or arise as the net consequence of survival and growth trajectories. Although such theories pertain to a number of fish species, they have been most often applied in the context of Arctic char studies. In the High Arctic, Arctic char are the only resident freshwater species. In lake habitats, Arctic char display a complex variety of life-history tactics, varying in growth and feeding patterns to produce ecophenotypes that occupy distinctive niches. Anadromous Arctic char use lake habitats for critical life-history stages, including reproduction, juvenile growth, and over-wintering. Lakes, therefore, provide essential habitat for all Arctic char populations. In addition to Arctic char, lake char (*Salvelinus namaycush*) occupy many Arctic lake environments in the southern parts of the Canadian archipelago and the northern continental margin of North America. Lake char can structure lake fish populations through piscivory and compete with Arctic char, particularly in eastern lake environments. Atlantic salmon (*Salmo salar*) also occur in the Arctic, and are an important food source where they occur. Most other species, with the exception of sticklebacks (*Gasterosteus aculeatus* and *Pungitius pungitius*) occur only as populations at the northern fringes of their distributional range and while their occurrence can complicate the ecology of any given lake, such species are not an integral part of most High Arctic lake environments.

14.1 Introduction

High-latitude lakes present unique challenges to the fish that inhabit them. Resident species must possess the physiological adaptations necessary to deal with low productivity and cold environments that are ice covered for much of the year (see Chapter 1). Within the Antarctic convergence conditions are extreme enough to preclude the occurrence of endemic anadromous and freshwater populations fish altogether, the closest endemic populations of freshwater fish being found in the Auckland, Campbell, and Falkland islands (Hammar 1989).

In the Arctic, the ability of fish to migrate across conveniently defined geographic boundaries and to seasonally use available habitats, however, belies easy definition of what constitutes a true high-latitude species. For example, in their zoogeographic discussion of western North America Lindsey and McPhail (1986) included 69 anadromous and freshwater species in their count of Arctic species. If sub-species occurring only in Great Slave Lake and its tributaries are excluded, the number is reduced to 51 (Power 1997). In the Hudson Bay drainage Scott and Crossman (1973) list 94 species and Power (1997) 101 species. Farther east in the Ungava drainage there are only 18 species, including fourhorn sculpin (*Myoxocephalus quadri-cornis*) originally of marine origin, whereas in the northern part of Labrador Black *et al.* (1986) report only 11 species north of approximately 56° latitude. In the Canadian Arctic archipelago the count is reduced to only eight species of anadromous and freshwater fish.

In any definition of what constitutes a high-latitude species there will be those included that are at the northern limits of their distribution. The number defined as truly high latitude, there-fore, will vary depending on the selection criteria used (Dunbar 1968). The exact number of species is probably not as important as the fact that Arctic

freshwater environments support less than 1% of all anadromous and freshwater fish species (Power 1997) and can be described as being low in both diversity and abundance. Within freshwater envir-onments, Arctic char (*Salvelinus alpinus*) dominate in overall abundance and population number. Thus much of the discussion of high-latitude fish will invariably focus on this species.

Within the limitations of a single chapter it is not possible to exhaustively discuss all species that may be found in Arctic environments. Many of the species that are present north of the 10°C July isotherm (Figure 14.1) occur in low numer-ical abundances in comparison to southern locales

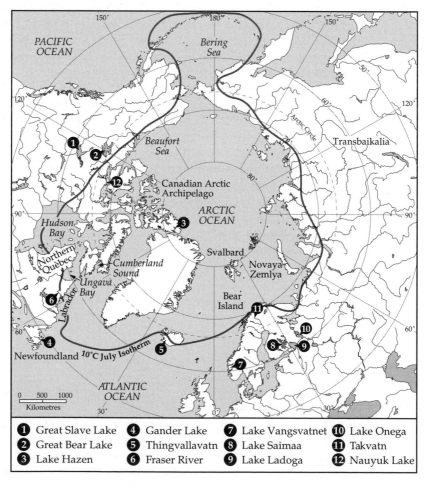

❶ Great Slave Lake	❹ Gander Lake	❼ Lake Vangsvatnet
❷ Great Bear Lake	❺ Thingvallavatn	❽ Lake Saimaa
❸ Lake Hazen	❻ Fraser River	❾ Lake Ladoga

❿ Lake Onega	
⓫ Takvatn	
⓬ Nauyuk Lake	

Figure 14.1 The map delineates the southern boundary of the area occupied by high-latitude fish as discussed in this chapter. The boundary roughly corresponds to the 10° July isotherm, but includes only the northern coastal fringe of continental areas and the Arctic Islands. Major place names mentioned in the chapter are also marked for reference.

more central to the natural distributional range of the species. Many species using high-latitude lakes are also not obligate lacustrine residents. Most use available lentic habitats for feeding or reproductive purposes and many migrate to nearshore marine areas (e.g. anadromous Arctic char) or nearby seasonally available lotic habitats (e.g. Arctic grayling, *Thymallus arcticus*) for summer feeding. Lakes are thus one of many available habitats that fish will use to maximize fitness. Nevertheless, lakes play an important role in the provision of overwintering or juvenile-rearing habitat for most of these species.

It is also important to note that for many species extensive reviews of the ecology and population dynamics exist (e.g. Atlantic salmon, *Salmo salar*; Klemetsen *et al*. 2003), which we will not attempt to synthesize because of space limitations. Rather, in this chapter we intend to provide a summary review of the ecology of the fish species having the greatest significant biological and cultural importance in the High Arctic and along the northern continental margins where indigenous peoples have traditionally gathered for hunting and harvesting purposes. Biological significance is defined to arise from the importance of the species' functional role in lacustrine ecosystems. Cultural significance is tied to the economic (commercial or subsistence) importance of the species. The definition of high latitude adopted here, therefore, precludes discussion of the more complex fish communities found at lower latitudes (e.g. Great Bear Lake, 65°50′N) and concentrates on the distinctively simpler communities characteristic of coastal regions (e.g. Nauyuk Lake, 68°22′N) and the Canadian Arctic archipelago (e.g. Lake Hazen 81°83′N). Accordingly, discussions will focus on species such as Arctic char, that exert top-down control on the community structure of High Arctic lakes and are used as an important food resource and Atlantic salmon that are more important economically than ecologically throughout much of the polar region.

14.2 Fish population structure in Arctic lakes

Arctic lakes have short, simple food chains dominated by a terminal predator, typically Arctic char,

lake char (*Salvelinus namaycush*), or lake whitefish (*Coregonus clupeaformis*). Repeated sampling of Arctic lakes has indicated that in many the dominant fish populations tend to consist of large, mature fish of uniform size, but variable age, that form a stable establishment showing little variability in either mean size or abundance (Johnson 1976). For example, in the continental and Arctic archipelago lakes of North America, lake trout, lake whitefish, and Arctic char can account for more than 95% of the fish biomass (Johnson 1976). Competing views (Johnson 1976; Power 1978) as to the mechanisms responsible for the observed age/length structure of fish populations in Arctic lakes emerged in the late 1970s with the examination of data collected from earlier extensive surveys of isolated lakes.

On the basis of examination of data from 35 lakes in the Northwest Territories, Canada, Johnson (1976) noted the frequency distributions of fish in unexploited Arctic lakes showed similar size and age patterns, with a high degree of clustering around length-frequency modal values that was more pronounced in the length than age distributions. The resulting 'standing waves' were interpreted as characteristic of populations limited by resources. A mortality filter that controlled juvenile recruitment to the established adult population was presumed to operate, whereby mortality among the adults reduced competitive pressures on juveniles (Johnson 1976). The filter imposes top-down stability control on the aquatic ecosystem through control of the fish population by acting to dampen the effect of year-to-year fluctuations in the year-class strength (Johnson 1995). Over successive generations, the recruitment mechanism acts to maintain the population in a steady state or so-called climax condition and will restore the standing wave length and age distributions even after periods of intense fishing (Johnson 2002).

Johnson (1976) proposed that cannibalism may play a part in the suppression of juveniles by adults, but noted the low incidence of cannibalism in lake trout populations, suggesting it was not the main mechanism. Restriction of juveniles to lake peripheries to reduce predation risks was argued as another means by which juveniles were suppressed. The mechanism, however, presupposes that no energetic or thermal advantages are conferred on juveniles through the use of shallow lake margins. In Johnson's view, the salient feature of the Arctic lake was its stability and he viewed isolated Arctic lakes as 'closed' systems, acting within defined boundaries that precluded exchange of materials with the other systems, but allowed exchange of energy and entropy to the extent that the structure and function of these systems was viewed to accord with basic thermodynamic principles (Johnson 2002).

Johnson's view has gained some support in the context of studies with Arctic char populations, where evidence of cannibalistic control of population structure has been found in Arctic Canada (Johnson 1980), Greenland (Sparholt 1985), Bear Island (Klemetsen *et al.* 1985), and Svalbard (Svenning and Borgstrøm 1995). In Svalbard lakes cannibalism is presumed to structure the fish population (Svenning and Borgstrøm 1995) and may be an important strategy for Arctic char despite the substantial cost of parasite load accumulation and eventual mortality (Hammar 2000).

Using supplemental observations on lake trout and whitefish populations in northern Québec, Power (1978) argued that erroneous interpretation of the aging structures routinely used to construct age-frequency distributions contributed to a mistaken characterization of the mechanisms responsible for the characteristic shape of sample length-frequency distributions and the apparent structure and stability of Arctic fish populations. By assuming continuous, asymptotic growth and a survivorship curve dominated by high juvenile mortality that declined rapidly to low adult mortalities, the characteristic Arctic lake fish population length-frequency distributions described by Johnson (1976) were found to result.

The ecological consequence of low adult mortality in the Power model is the provision of a reproductive reservoir likely to yield stability in the face of climatic or other environmental fluctuations, which at intervals may exceed the limits for successful reproduction for the species (Power 1978). The model does not require special invocation of life-history specifics to explain the evolution of bimodality, relying instead on the simple interaction of growth and mortality. Whereas removal of excess juveniles by intra-specific density-dependent predation and possible resource competition may explain observed bimodalities in lake trout and Arctic char populations, lake whitefish populations cannot be stabilized by the same mechanism as a result of the lack of evidence for intra-specific predation (Power 1978). In addition, the Johnson theory presumes that lakes are autonomous systems. For many High Arctic and upland plateau lakes in the Sub-Arctic, the condition may be true. For some coastal plain lakes, particularly those in which anadromous populations occur, the condition does not hold, with anadromy providing a mechanism whereby the effects of external events may be transmitted to an otherwise isolated lake system.

14.3 Adaptations for high-latitude life

Fish living at high latitudes in fresh water have been shaped by the cumulative history of a changing environment, particularly repeated interactions with periods of glaciation. In many areas periodic extinction will have been the norm. In refugia, populations will have survived and been the focal point for range expansion in the ensuing interglacial periods. Occupants of the refugia, therefore, are most likely to have developed the traits necessary for survival through abiotically dominated periods of repeated range expansion and contraction. Many of the refugia will have taken the form of proglacial lakes formed by the impoundment of water ahead of moving ice sheets (Power 2002). Within these lakes, environments capable of supporting fish and other aquatic life will have existed and the cumulative scientific evidence points to many species (e.g. genus *Salvelinus*) having survived and/or diversified in one or more cold periglacial environments. Conditions in proglacial lakes would have been variable, with geologically sudden changes in the directions of outflows, inflows, and lake level having been likely (Power 2002). Furthermore, high turbidity, ice cover, low productivity, and reduced light penetration would also have been the norm. Over time, selection for life in cold, impoverished, highly variable environments with extreme seasonal ice conditions would have occurred. Many of the resulting physiological and behavioral adaptations for life in such an environment are obvious (e.g. cold tolerance, migratory or exploratory tendencies). Many of the adaptations are less obvious (e.g. modified agonistic behavior), but all adaptations have served to promote the continued survival of individuals and facilitated the collective, selective evolution of the genome over the millennia. Table 14.1 lists and describes the life-history traits likely to have been important for survival in proglacial lakes. Most of the traits will also promote individual survival in the high-latitude lakes of today. Although many of the traits will apply to the cold-water fish community as a whole, chars as a group, and Arctic char in particular, exhibit extreme manifestation of the suite of characteristics (Power 2002). Taken collectively, the traits define species suited to the wide variability and harsh biological conditions found in existing high-latitude lakes and species with those traits will have been most able to disperse quickly during glacial retreats to the young environments in which they are found today.

The traits described in Table 14.1 are evident at both the individual and population level. Thus individuals may display variability in year-to-year reproduction and population averages may also vary as a consequence of large numbers of individuals within the population using similar tactics. Responses to given conditions may also determine population characteristics. For example, the ability to deal with low food availability through variable allocations of energy to somatic growth or reproduction has been implicated in the development of stunted populations of Arctic char (Hindar and Jonsson 1982). Traits are also often highly correlated, with changes in one trait holding significant implications for expression of other traits. For example, delayed maturation will tend to increase body size with attendant positive effects on egg size and survivorship (Hindar and Jonsson 1982). Therefore, delayed maturation is often associated with increased body size, with increased longevity acting as a means of offsetting possible consequences for lifetime reproductive fitness.

14.4 Arctic char, *Salvelinus alpinus*

Arctic char form a taxonomically diverse set of salmonid fishes distributed throughout the Holarctic and to latitudes as low as 43°N in North America and 45°N in Europe. The Arctic char is considered a habitat generalist and may be found at appropriate times of the year in a variety of aquatic habitats (lakes, streams, rivers, and the sea). There are fluvial and lacustrine stocks and stocks in which individuals migrate between both habitats. Within a habitat type, more than one life-history form (ecophenotype) can live sympatrically with another. Lake occupancy, however, dominates throughout the Arctic distributional range and north of about 75° latitude Arctic char is the only fish species found in fresh water. Anadromous populations are numerous, particularly in the mid-range of the distribution where access to productive marine coastal margins is relatively easy. Few entirely riverine populations are known or described (Power 2002). A generalized depiction of life-history variation as influenced by environmental and physical factors is given in Figure 14.2.

Within lakes, Arctic char will use all habitat types (e.g. pelagic, littoral, profundal), with usage dependent on age and life stage of the char and co-occurring species in the lake. Arctic char appear pre-adapted to low aggression (Power 2002) and in the face of interspecific competition niche shift is usually observed. In Scandinavia, Arctic char interactively segregate from brown trout (*Salmo trutta*), occupying the littoral zone for feeding in allopatry and shifting to the pelagic or profundal zones in sympatry with brown trout (Klemetsen et al. 2003). Hammar (1998) has noted that mechanisms responsible for dietary segregation in summer in Swedish lakes break down in winter when decreases in water temperature restrict brown trout activity. Reduced brown trout forging activity in winter and early spring permits competitive release, allowing the expansion of the Arctic char dietary niche to include exploitation of littoral benthos typically preyed on by brown trout at other times of the year. Similar separation between Arctic char and sympatric brook char (*Salvelinus fontinalis*) and lake trout populations has been observed in North America where feeding competition and predation risks are reduced by non-overlapping habitat use (Fraser and Power 1989; O'Connell and Dempson 1996). In sympatry with lake trout, Arctic char show faster growth rates, lower survival, and

Table 14.1 Life-history traits promoting survival of fishes in high-latitude lake environments. Traits are grouped by the consequence of their main effect on the basic population processes of reproduction, growth, and survival that determine eventual population dynamics. Traits within each category are not necessarily independent of those in the other categories. For example, variable growth holds direct implications for reproductive plasticity, both of which facilitate maximization of lifetime fitness under varying environmental conditions

Life-history trait	Description of conferred advantage or effect
Reproductive traits	
End of summer spawning	Facilitates maximal energy acquisition for gonadal development, both large egg size and increased number of eggs.
Large egg size	Promotes development of larger progeny, reduced larval mortality and dependence on exogenous food sources during critical larval developmental periods.
Active nest-site selection	Provides protective advantages from biotic (e.g. predation) and abiotic (e.g. freezing) events, thereby increasing egg survival.
Variable reproductive cycles	Allows individuals to tune reproductive events to environmental conditions by postponing reproduction in poor years and accelerating gonadal development in good years. As a population trait, variability allows exploitation of a range of environments along a latitudinal gradient.
Variable age at maturity	Facilitates development of larger body sizes associated with larger egg sizes and improved fecundity.
Iteroparity	Multiple reproductive events facilitate hedging against environmental stochasticity by spreading reproductive effort across variously favorable years. Iteroparity also helps minimizes competition for limited habitat. Overall, iteroparity promotes greater reproductive success and population persistence.
Growth traits	
Variable growth	Occurs both within and between seasons. Within season, variability promotes growth tuned to food availability. Between seasons, variability promotes growth tuned to environmental stochasticity that acts in concert with variable reproductive cycles to facilitate individual survival during adverse environmental periods.
Adaptation to low temperatures	Ability to complete all life processes at low temperatures, with metabolic optima (growth, incubation) typically occurring as among the lowest for freshwater fish.
Diet flexibility	Consumption of all prey types in all habitats, including cannibalistic behavior. With ecophenotypes, individual diet flexibility is reduced, but among group variability is maintained and competition reduced.
Metabolic efficiency/variability	Ability to feed to satiation in summer and rapidly convert calorie intake for either somatic growth or gonadal development and reduce energy conversion requirements in winter via low basal metabolic demands.
Survival	
Habitat generalists	Allows exploitation of all available habitat types at optimal times (e.g. marine summer feeding) and facilitates maximization of growth and reproduction. With ecophenotypes, individual habitat flexibility is reduced, but among-group variability is maintained, survival maximized, and competition minimized.
Facultative life histories	Facilitates optimal use of available habitats in unpredictable and changeable environments either as individuals (e.g. variable migration) or collectively as a population (e.g. development of land-locked populations as a result of isostatic rebound).
Exploratory/migratory tendencies	Allows facultative use of available overwintering or reproductive habitats in the face of environmental stochasticity (e.g. low flows preventing re-entry into freshwater) and opportunistic use of new environments. Established movement between reproductive, feeding and overwintering areas promote familiarity with the physical environment.
Limited agonistic behavior	Low aggression toward conspecifics and others avoids energy use in social conflict, preserving it for somatic and gonadal purposes.

shorter longevities (Fraser and Power 1989). High intraspecific competition may also result in onto-genetic habitat shifts. In Takvatn, northern Norway (Figure 14.3) adult Arctic char occupy the littoral zone while juveniles occupy the profundal zone, shifting to the pelagic zone as they became food limited with increased size (Klemetsen *et al.* 1989).

Stable isotope studies of feeding patterns suggest that resource-use separation within and among size or morphotype groupings has the effect of lowering resource competition, but can impose developmental energetic constraints dictating limits on body size, maturation rate, and fecundity (Guiguer *et al.* 2002; Power *et al.* 2005a).

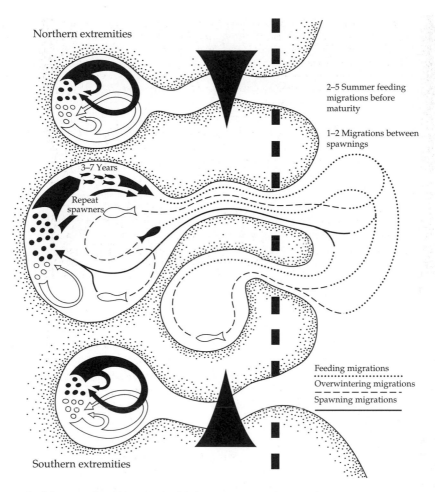

Northern extremities

2–5 Summer feeding
migrations before
maturity

1–2 Migrations between
spawnings

3–7 Years

Repeat
spawners

Feeding migrations
Overwintering migrations
Spawning migrations

Southern extremities

Figure 14.2 A generalized depiction of possible Arctic char life-history strategies. Life-history strategy varies by latitude as a result of physical or ecological barriers that prevent marine feeding migrations. The black triangular symbols at the northern and southern extremities of the distribution represent ecological barriers to migration that may hinder or block char migrations. In the north, river and sea ice, stream-discharge patterns, and low marine temperatures and productivity may all prevent migration or alter the migration/residency trade-off by altering the risks, costs and benefits associated with migration. In the south, high stream and marine temperatures and increased marine competition may similarly alter the costs and benefits of migration and residency. Toward the centre of the distribution, ecological constraints are relaxed, but physical barriers symbolized by the heavy dashed line may prevent migration. Physical barriers take the form of waterfalls too high to navigate, high water velocities associated with high stream gradients, and insufficient water depths at critical in- or out-migration periods. The effectiveness of physical barriers may vary between years as a result of variable weather (e.g. high or low precipitation leading to high or low stream-water levels). Populations established in fresh water that are prevented from migrating may develop ecophenotypes recognizable as distinctive body forms. Ecophenotypes may reproduce selectively (i.e. may also be distinctive genotypes) and have separate reproductive cycles as shown. Ecophenotypes are known to occur at both extremes of the distribution, but may also occur in mid-latitudes. In mid-latitudes, represented here by the middle lakes, a small resident population may co-exist with the dominant **anadromous** form. The latter may use alternative overwintering sites as shown, may spawn in the year of migration, or may spend a year in fresh water before spawning and then spend a second year in fresh water prior to migrating for summer feeding again.

Morphological variation in lacustrine populations of Arctic char is characteristic of the species and has been described extensively, generally from European lakes (Klemetsen *et al.* 2003 and references therein). The term ecophenotypes is used to describe sympatric (i.e. co-occurring) forms of char that may be morphologically, genetically, and/or ecologically distinct. When distinctive forms exist, there is often a small, epibenthic form that feeds on zoobenthos and a large, pelagic form that feeds

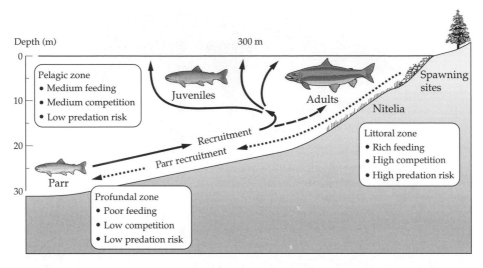

Figure 14.3 A schematic model of the life cycle of lake-resident Arctic char in Takvatn, Norway (adapted from Klemetsen *et al.* 1989). The littoral zone provides rich feeding opportunities (e.g. *Gammarus*, *Eurycercus*, *Lymnaea*, *Planorbis*, insect larvae, surface insects, sticklebacks) and is dominated by large adult fish. After emergence from spawning sites located in the littoral zone, juveniles take up residency in the profundal zone where reduced feeding potential (e.g. *Heterotrissocladius*) is offset by reduced predation risk. Limited feeding opportunities result in stunted body size and unless fish are recruited as small adults to the pelagic zone they will not continue to increase in size. In the pelagic zone, improved feeding opportunities (e.g. *Bosmina*, Copepoda, surface insects) facilitate further increases in size. Once large enough (>25 cm) to avoid predation (e.g. by adult char, brown trout, mergansers, divers), small adults may be recruited to the large adult population in the littoral zone. Horizontal and vertical distances are approximate. A similar pattern of life-stage-specific occupancy of the profundal, pelagic, and littoral zones will occur in many lakes where Arctic char is the only fish species.

on zooplankton (e.g. Lake Vangsvatnet; Hindar and Jonsson 1982). A larger piscivorous form may also exist that predates on the other forms (Jonsson and Jonsson 2001). In Scandinavian lakes form differentiation occurs regularly, with forms being primarily distinguishable by size, body coloration, and habitat association. Small morphological differences may exist, but have not been well documented in most instances. The young of all forms usually start as epibenthic zoobenthivores, becoming pelagic as reductions in predation risk allow. Where piscivory develops, lengths of 20–25 cm must be attained and the proportion of piscivorous fish will increase with increasing body size (Jonsson and Jonsson 2001). Different degrees of reproductive separation exist among sympatric ecophenotypes, varying from a high degree of interbreeding to complete isolation. Genetic studies have indicated that in most cases the forms are repeatedly derived from a single original type in the lake, with form occurrence likely depending on ecological conditions such as the number of available niches (Klemetsen *et al.* 2003).

The morphological diversity observed in Arctic char is thought to be adaptive (Skúlason and Smith 1995). In many cases where distinctive forms exist there is evidence of an ecological basis for its occurrence, with morphologically specialized morphs appearing to feed more effectively than intermediate forms. For example, in Lake Hazen, Ellesmere Island, sympatric large and small forms of Arctic char occur that differ substantially in body form, color, number of pyloric caecae, caudal peduncle length, and pelvic and pectoral fin lengths (Reist *et al.* 1995).

Guiguer *et al.* (2002) demonstrated that small-form, benthic-feeding individuals in Lake Hazen were more $\delta^{13}C$-enriched and $\delta^{15}N$-depleted than large-form, piscivorous conspecifics (Figure 14.4), thereby establishing a trophic basis for the polymorphism. Polymorphisms associated with diet, however, show considerable diversity (Power *et al.* 2005b), and have been associated with differing combinations of morphological, behavioral, and life-history trait differences (Adams *et al.* 2003), and may occur independently of lake morphometry and

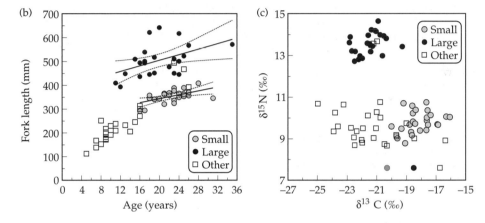

Figure 14.4 (a) Lake Hazen ecophenotypes (large and small forms). Forms differ substantially in body form, color, number of pyloric caecae, caudal peduncle length, and pelvic and pectoral fin lengths. The large form preys more heavily on fish (i.e. is cannibalistic) than the small form, which preys largely on chironomids. (b) Differences in diets are reflected in differences in length at age as shown, where 'other' (squares) represents juveniles and individuals that could not be identified on the basis of morphology. Length-at-age regressions with 95% confidence intervals for small-form (gray circles) and large-form (black circles) Arctic char are also plotted and are significantly different from one another. (c) The nitrogen (δ^{15}N) and carbon (δ^{13}C) stable isotopes of the ecophenotypes with other, small-, and large-form differentiated as above. Higher nitrogen signatures for the large-form char are consistent with the higher occurrence of fish in the diet and indicative of long-term reliance on fish as a food resource. The 3.6–3.4‰ increase in δ^{15}N between large- and small-form, and the large-form and other, respectively, is also indicative of large-form individuals occupying higher trophic positions. On the δ^{13}C axis the small-form is 2–3‰ more enriched than other forms and more reliant on benthic secondary production as indicated by gut-content analysis. Overall, the isotope plots indicate long-term reliance on fish as a food for large Arctic char, with large-form individuals predating on juveniles of all ecophenotypes under 250 mm, but not on small-form Arctic char over 250 mm. A colour version of (a) is shown in Plate 13.

history. For example, documented Canadian Arctic char polymorphisms (e.g. Lake Hazen (Reist *et al.* 1995) and Gander Lake (O'Connell and Dempson 2002)) occur in lakes situated at opposite ends of the latitudinal range that vary markedly in climate, environment, glacial history, and the complexity of resident fish communities. Thus, clearly other factors influence the development of ecophenotypes in char populations.

Geologically young, recently deglaciated habitats (i.e. ≈10 000 years) or species-poor ecosystems

are generally thought to facilitate polymorphic development in Arctic char and a growing number of studies have suggested that relaxation of interspecific competition and open niche availability are essential prerequisites (Skúlason and Smith 1995). Perhaps the greatest ecophenotypic diversity known for Arctic char occurs in Thingvallavatn, Iceland, where four distinct trophic morphs are known to exist (e.g. Skúlason *et al.* 1996 and references therein): small benthic, large benthic, pelagic piscivore, and pelagic planktivorous forms. The forms are quite different with respect to morphology, habitat association, and diet, but are closely related genetically. Thingvallavatn is geologically quite young (i.e. ≈10 000 years) and in an area of tectonic activity (spreading continental plates and high vulcanism), indicating that char ecological complexity can develop rapidly in conditions of high environmental instability. Deep lakes with separate littoral, pelagic, and profundal zones offering unique foraging opportunities have been argued to be characteristic of systems where Arctic char morphotypes develop (Jonsson and Jonsson 2001). Reports from northern Québec (M. Power, unpublished results) and the Transbaikalian region of Russia (S.S. Alexseyev, personal communication), however, belie the generality of this view. In both locales, morphologically distinctive forms of Arctic char occur in shallow lakes (<30 m) with numerous (5–11) co-occurring species. Such observations suggest that mechanisms promoting morphological divergence in Arctic char are not fully understood and generalizations of cause and effect based on limited geographic studies should be avoided.

Where there are no ecophenotypes, variation in size within a single ecophenotype may occur. Many lacustrine Arctic char populations exhibit size polymorphisms in the form of bimodal length-frequency distributions (Johnson 1995), with the frequency of bimodality increasing with latitude (Griffiths 1994). In undisturbed unimodal populations, only a few individuals will grow very large, usually as a result of an early dietary switch to piscivory. Thus in single species lakes, piscivory (cannibalism) is thought to play an important role in population regulation (Svenning and Borgström 1995; Hammar 2000; Johnson 2002), with experimental manipulations in Norway having shown that large, cannibalistic Arctic char are essential for the maintenance of a bimodal population structure (Klemetsen *et al.* 2003). Seasonal prey shortages and slow juvenile growth, when coupled with the fitness effects associated with larger body size, are thought to induce cannibalism (Hammar 2000) and, where it exists, cannibalism among Arctic char has a top-down structuring effect on the whole population. Bimodality may also arise as a result of divergent life-history tactics whereby a portion of a population remains lake-resident while the remainder of the population becomes anadromous.

Morphological and size diversity among Arctic char are important because they provide evolutionary options in the context of the variable and changing environments inhabited by the species. Diversity may also facilitate the structure and functioning of aquatic ecosystems by increasing the number and nature of trophic pathways present (e.g. Folke *et al.* 2004). For Arctic systems that are relatively depauperate at the species level, diversity below the species level may fulfill a stabilizing role in aquatic ecosystems similar to that of community complexity in lower-latitude systems.

Arctic char are known to be a cold-adapted species. Indeed, in a comparative study of freezing resistance in five salmonid species, Arctic char showed the greatest resistance to freezing (Fletcher *et al.* 1988). Arctic char can survive and feed close to 0°C (Brännäs and Wiklund 1992). In several Swedish lakes, benthos has been found in the stomachs of Arctic char actively feeding under the ice during the winter (Hammar 1998). High temperatures, however, can limit both activity and growth as char are among the least resistant of salmonids to high temperatures (Baroudy and Elliott 1994). As early as 1964 the detrimental growth effects of temperatures above 14°C were noted (Swift 1964). Jobling (1983) measured the physiological effects of temperature on specific growth relative to 14°C and found significant positive correlations between growth and temperature up to 14°C, with sharp declines in growth thereafter. Baker (1983) similarly reported an optimal growth temperature of about 13°C for a Labrador and Norwegian strain of Arctic char and noted that there could be intraspecific differences in thermal optima for growth.

More recently, Larsson and Berglund (1998) have extended studies of growth to temperatures above 14°C to estimate an optimum growth temperature of 15°C for age 0+ Arctic char and argued that within the range 9–20°C there are no significant differences in measured growth rates when fish can feed at will. Results provide no indications that southern populations have been selected to tolerate higher temperatures. Recent experimental findings also indicate negligible geographical variation in optimal growth temperatures (Larsson *et al.* 2005). Growth efficiency, however, does decrease with increasing temperature, implying that increased ration is required to sustain growth rates at higher temperatures. Individuals experiencing temperatures in excess of 14–15°C for extended periods of time will experience decreased seasonal growth patterns, with attendant effects on survival and fecundity, unless sufficient increases in ration are available to offset the increased metabolic costs of living at higher temperatures. Warmer temperatures may also limit habitat choice, with Arctic char tending to avoid littoral-zone temperatures in the 16–20°C range (Langeland and L'Abée-Lund 1998). Thus warm epilimnetic water will induce a summer habitat switch by Arctic char, with movement to the hypolimnion in lakes in the southern portion of the distributional range being the rule.

Arctic char are generally a long-lived fish, with northern lake-dwelling populations showing the greatest age range (e.g. to 30+ years; Johnson 1983). Anadromous populations generally have shorter life spans (e.g. 20+ years; Johnson 1989) than lacustrine populations, but are similarly long-lived in comparison to many southern fishes. Among populations age and size at maturation varies greatly (3–10 years), with females tending to mature 1–2 years later than males. Age-at-maturity varies with latitude, increasing toward the north, particularly in North America. Arctic char are iteroparous, potentially reproducing several times during their life. Fecundity is correlated with length, which has been shown to explain much of the observed variation in fecundity for lacustrine and anadromous populations (Power *et al.* 2005a). Fecundity also declines as latitude increases. Lake environments appear to buffer some of the negative effect of latitude, with declines in the fecundity among lacustrine Arctic char being less severe than among anadromous Arctic char. Thus variations in lacustrine fecundity must be influenced by other energy-related considerations as determined by growing season length. Contributory factors may include variations in lake productivity driven by nutrient availability, the presence/absence of predators, the presence/absence of closely related competitive species, or the quantity and quality of available forage species (Power *et al.* 2005a).

Unlike many other salmonids there is no evidence of an effect of egg size on Arctic char fecundity (Power *et al.* 2005a), which may result from the ability of Arctic char to vary reproductive periodicity by 'resting' for a year or more to gather sufficient energy for reproduction. The tactic is more typical of females as a result of the greater energy required to produce ova. Despite their apparent long life spans, limitations imposed by short feeding seasons on acquiring surplus energy for gonadal development reduce the overall lifetime reproductive potential of Arctic char and may have significant implications for the exploitation and sustainability of many populations.

Rather little detailed information is available on spawning behavior and site selection because of the difficulties associated with making direct observations in the Arctic. All spawning occurs in fresh water, either in lakes or in streams and rivers. Arctic char may excavate redds, in which case they would require an unconsolidated gravelly substrate. They may also broadcast spawn their eggs after performing a ritual nest-site cleaning behavior when, as in Thingvallavatn, Iceland, the substrate is unsuitable for redd excavation (Sigurjonsdottir and Gunnarsson 1989). In streams and rivers, groundwater upwellings are probably required to protect the eggs from freezing. For example, Cunjak *et al.* (1986) recorded upwelling ground water in sites used by anadromous Arctic char in rivers in northern Québec. Where lakes are accessible and rivers are unsuitable, anadromous Arctic char will spawn in lakes. Use of lakes may involve a 2-year cycle of migration and maturation marked by entry into fresh water after marine feeding, overwintering in the lake and maturing and spawning in the subsequent summer and autumn such as that first described for the Nauyuk Lake/Willow Lake system in the Canadian Central Arctic (Johnson 1980, 1989). In the Fraser River (Labrador), however, anadromous Arctic char use both lake and stream habitat (Dempson and Green 1985) without first overwintering in fresh water.

Anadromy is observed in most populations with open access to the marine environment and summer marine feeding, usually in near-shore coastal waters (Dempson and Kristofferson 1987), and provides an important opportunity to increase size prior to reproduction. Regardless of life-history type, all Arctic char utilize fresh water during the first few years of life and for overwintering. Groundwater seepages maintain suitable overwintering habitats for some riverine populations of Arctic char, but for most populations lacustrine environments play a critical role in the life cycle by providing the necessary reproductive, juvenile rearing, and overwintering habitats. At the northern fringe of the distribution (north of 75°N) Arctic char populations favor lacustrine residency even when access to the sea exists. Watershed sizes and gradients appear to influence the prevalence of anadromy, with gradients of more than $80\,m\,km^{-1}$ precluding successful upstream migration in Ungava (Power and Barton 1987). Basin sizes of more than $1000\,km^2$ seem to guarantee sufficient river discharge to permit anadromy where gradient or other obstruction barriers (e.g. waterfalls) do not exist (Power and Barton 1987).

Anadromous behavior is facultative and will vary depending on local conditions. Where anadromy does occur, the transition to summer marine feeding begins generally between the ages of 3 and 8 years. Individuals feed annually in the sea for a 6–8-week period and, therefore, spend rather little time actually feeding in marine environments. Nevertheless, marine feeding migrations facilitate rapid growth, with individuals increasing body size by as much as 42% in the period (Johnson 1980). Overall Arctic char populations are suspected of being facultative with respect to anadromy. Factors determining anadromy appear to be primarily environmental (e.g. duration of ice cover versus open water in freshwater systems, productivity in natal fresh waters versus near-shore marine waters, and costs versus benefits of migration). In some Arctic char populations with access to the sea the phenomenon of 'partial migration' exists, with a fraction of individuals remaining resident in fresh water (Nordeng 1983). A recent study from North Norway has provided a proximate explanation for the phenomenon, noting that in comparison with freshwater resident char, anadromous char maintain a high freshwater growth rate in late summer and early autumn prior to migration. In freshwater resident fish, arrested autumn growth, higher lipid levels and low lipid depletion in winter appears to

facilitate maturation the following summer without emigration to marine environments (Rikardsen et al. 2004).

The sources and levels of natural mortality acting on Arctic char populations are generally not well studied (Johnson 1980), with those studies that exist tending to focus on anadromous populations. Sparholt (1985) reported mean annual mortality for ages 3–20 years in the range of 23% on the basis of the study of four interconnected lakes at latitude 62°15'N in Greenland. Results parallel those available for anadromous studies. For example, Moore (1975) estimated average annual mortality of 10–20-year-old char from the Cumberland Sound area of Baffin Island at 16%, with the higher rates (24–30%) occurring in the older age classes. Estimates of mean annual mortality from selected Canadian anadromous and lacustrine populations (M. Power, J.D. Reist, and J.B. Dempson, unpublished results) for 6- to 15-year-old fish (Figure 14.5) yield similar results, with annual mortality rates generally occurring in the 30–45% range, although several

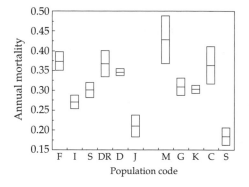

Figure 14.5 Estimated mean annual mortality ±95% confidence interval for selected **anadromous** and lacustrine Arctic char populations in Canada. Estimates were derived from catch curves for fish of approximately equal ages. Anadromous populations analysed are: F, Fraser River, Labrador (56°39'N, ages 7–14 years); I, Ikarut River, Labrador (58°09'N, ages 6–13); S, Sappukait, Nunavik, Québec (59°28'N, ages 6–13); DR, Deception River, Nunavik, Québec (61°46'N, ages 6–13); D, Diana River, Nunavut (62°50'N, ages 7–14); and J, Jayco River, Victoria Island (69°34'N, ages 7–14). Lacustrine populations analysed are: M, Mini-Minto, Québec (57°19'N, ages 7–13); G, Gavia Lake, Booth Penninsula (68°21'N, ages 8–15); K, Keyhole Lake, Victoria Island (69°23'N, ages 8–15); C, Crooked Lake, Prince of Wales Island (72°38'N, ages 8–15); and S, Stanwell Fletcher Lake, Somerset Island (72°45'N, ages 8–15).

populations have rates of less than 25%. Overall, variations in morality rates among anadromous and lacustrine populations appear similar, with variation probably arising from differences in habitat productivity and intraspecific competition.

Old age and post-spawning debilitation were presumed to be the major causes of death. In Nauyuk Lake Johnson (1980) similarly reported post-spawning mortality. Like other salmonids, Arctic char mortality in early life (i.e. egg to age 3+) is relatively high, likely to be associated with the environments occupied and to result from both density-independent (e.g. environmental stochasticity) and density-dependent (e.g. predation) factors. Mortality levels during transition to migratory habits for sea-run forms and throughout the years to first sexual maturity are similarly likely to be fairly high (Power 1978). Mortality for adult fish appears to be low, especially from predation (Power 1978; Johnson 1980) and to be confined largely to post-spawning mortality. However, mortality resulting from subsistence or commercial fisheries can be moderately high and for many populations remains the most significant single source of mortality. Analyses of potential climate change impacts on Arctic char have suggested that climate may increase mortality, particularly in the juvenile age classes (Reist *et al.* 2006a, 2006b).

14.5 Lake char, *Salvelinus namaycush*

The lake char is an important commercial and sportfish across its range and is distributed widely in North America, including the Arctic. Most populations occur on the mainland, but populations of lake char have been reported from the Arctic Islands of Southampton, Victoria, Banks, King William, and Baffin. The major determinant of the existing natural range has been the glacial history of North America, with the extent of the existing distribution lying almost totally within the extent of Pleistocene glaciation (Marten and Olver 1980). The lake char is viewed as an evolutionary stable species specialized for a mode of life within narrow environmental tolerances (Behnke 2002). It is almost an obligatory lacustrine species, but can be river-dwelling for part or all of the year and has been recorded in brackish waters along

the Labrador, Hudson Bay, and Beaufort Sea coasts (Marten and Olver 1980). Limited salinity tolerances have shaped its postglacial dispersion patterns and help to explain why lake char are less widely distributed at high latitudes than Arctic char (Behnke 2002).

A critical problem with trying to understand the ecology and evolutionary history of lake char at high latitudes is that the majority of the scientific literature deals with populations at or near the southern margins of the present distributional range (Power 2002). Like Arctic char, lake char are cold-adapted and in the northern portions of their geographic distribution are not obliged to seek profundal refuge during the summer. In Arctic lakes, lake char typically inhabit simple fish communities, live at low densities, and may be the only fish in the lake (Johnson 1976). In many Arctic lakes lake char co-occur with lake whitefish and together will account for approximately 95% of the total fish population in the lake, with the relative importance of one species falling as other rises (Johnson 1976).

Lake char morphological characteristics have generally been regarded as stable throughout the geographic range, although ecological differentiation and specializations have evolved in some populations in the last 10000 years (Behnke 2002). Among the chars, lake char are the least colorful. Individuals grow slowly and live a long time (30–60 years). Faster growth rates are reported among populations in the vicinity of Great Slave Lake. Lower growth rates predominate in northern Québec, possibly as a result of lower lake productivity. Across the distributional range, a gradual decline in growth with latitude is also evident. Recent studies, have documented the presence of ecophenotypes in some Arctic lakes (e.g. Great Bear Lake; Alfonso 2004) that appears to result from ecological polymorphism (Blackie *et al.* 2003). Currently, it is unknown how widespread this situation may be in Arctic lakes.

Maturity occurs between 8 and 12 years, with spawning occurring at irregular intervals of 2–4 years as conditions permit (Power 2002). Reasons for the interrupted reproductive pattern are not fully understood, but there appears to be a strong relationship between latitude and spawning frequency (Marten and Olver 1980).

Although considerable variability exists in egg counts between individuals of similar size and age from different populations, there is a general trend toward increasing fecundity with both length and age and a strong correlation between weight and fecundity. A large individual (90–100 cm) may produce in excess of 15 000 eggs in a single spawn (Marten and Olver 1980). Spawning is temperature- and light-dependent, with time of spawning varying between populations from late September to October. Wind plays a role in triggering spawning, with spawning events being initiated by pervasive onshore autumn winds. Spawning usually occurs in shallow water over coarse rock or rubble substrates (3–15 cm in diameter) along eroding, wave-washed shoals with sufficient fetch to irrigate eggs. Selected spawning beds must be deep enough to avoid ice-related substrate scour. Eggs are broadcast-spawned over the substrate and sink into interstitial spaces in the substrate where they swell and wedge. The substrate is kept clear of debris by shoreline currents or fetch, with local currents maintaining water quality, supplying oxygen, and removing metabolic wastes during development (Power 2002). There is some evidence to suggest lake char home to spawning areas, but homing is generally considered to be incomplete (Marten and Olver 1980).

The long life-span of high-latitude lake char is associated with relatively low rates of adult mortality in comparison with Arctic char. Low mortality has been argued to be reflective of the trophic status of lake char and their role as terminal predator in lake food webs (Healey 1978). Recruitment to the adult population is typically low due to long lifespan, low adult mortality, and low average fecundity and spawning success (Johnson 1976, Marten and Olver 1980).

14.6 Atlantic salmon, *Salmo salar*

Atlantic salmon is a cold-water species that displays considerable phenotypic plasticity and ecological variability (see review by Klemetsen *et al.* 2003), with the life history of the anadromous form being functionally linked to marine and freshwater environments that include riverine and often lacustrine habitats. Throughout their distribution

salmon occupy a diverse array of physical and biological environments (Elliott *et al.* 1998) and can exist in both anadromous and non-anadromous forms, the latter completing their life cycle entirely within fresh water often by moving between rivers and lakes for spawning and overwintering. Under most circumstances the two forms do not co-exist as a result of interbreeding between forms resulting in the assimilation of the non-anadromous by the anadromous form (Power 1969). Non-anadromous populations are common throughout the entire distributional range of Atlantic salmon in North America, particularly at the northern edge of the distribution in northern Québec and Labrador (Klemetsen *et al.* 2003). In Europe the non-anadromous form is normally associated with larger lakes including Ladoga, Onega, Vänern, and Saima (Klemetsen *et al.* 2003). Populations of small resident freshwater salmon, however, occur in both lacustrine and riverine habitats in Norway, most notably the River Namsen (Berg and Gausen 1988). It is generally believed that non-anadromous or 'land-locked' populations were derived in postglacial times as a result of isolation from anadromous populations owing to isostatic rebound of coastal regions following the last glacial periods (Power 1958).

Non-anadromous forms utilize lakes for growth and maturation purposes in a manner analogous to ocean use by anadromous salmon (Berg 1995), but may be subjected to higher postspawning mortality in marginal lacustrine environments (Power 1969). In North America, non-anadromous salmon typically range in size from 33 to 53 cm and weigh approximately 1.6 kg (Behnke 2002), although smaller sizes have been reported from lacustrine populations in northern Québec (Power 1969). Predation on smelt (*Osmerus eperlanus*) is often associated with larger sizes and nonanadromous salmon in excess of 12 kg have been reported from north European lakes (e.g. Ladoga) where smelt are common.

Along the northern margins of the Atlantic salmon distribution, climate conditions vary from Arctic (north Labrador and northern Québec) to Sub-Arctic (northern Norway, Finland, Russia; Hammar 1989). In all northern regions freshwater areas are ice-covered for long periods. As a consequence, salmon have adapted their existence to extreme environmental conditions where cold conditions and ice play a prominent role in influencing the local environment. In comparison to southern populations, northern populations show increased age and size at smoltification (Power 1981; Klemetsen *et al.* 2003), delayed maturity, and reduced fecundity (Klemetsen *et al.* 2003). Nevertheless, local differences in temperature, length of growing season, and the amounts and types of food available result in variations in observed biological characteristics (e.g. length at age) among both anadromous and nonanadromous northern populations. For example, growth in fresh water is characteristically slower in more northern areas, generally resulting in increased age at smoltification compared with northern European locations (Power 1981; Metcalfe and Thorpe 1990). Indeed, smolt ages of salmon from the Ungava region of northern Québec are among the oldest recorded (Power 1969). It has been suggested that where cold ocean temperatures limit the distribution of salmon, large smolt size is selected owing to greater osmoregulatory ability (Jensen and Johnsen 1986). The possibility of a cold ocean temperature barrier in Ungava Bay has been linked with the phenomenon of some salmon remaining in fresh water and spawning in two successive years (Robitaille *et al.* 1986). Heinimaa and Erkinaro (2004) have also reported that the shorter growing season and long winters in northern areas contribute to a greater decline in lipid reserves and could explain the reduced frequency of mature parr in north European populations. Smolt run timing is also generally believed to be synchronized with ocean temperatures and feeding opportunities such that timing of migration is generally later in more northern latitudes (Hvidsten *et al.* 1998).

The cold environmental conditions that characterize much of the Ungava region in northern Québec have been linked with possible residualization of smolts in some populations (Power 1969), with temperature being the factor most important for controlling the distribution of salmon across the region (Power 1990). Thus parr growth is slow, owing to cold temperatures, but the size of Ungava smolts is quite large. In some instances, the onset of female maturation curtails the normal life cycle such that the marine phase of the life cycle can be eliminated (Power 1969).

Besides the use of lakes by resident nonanadromous salmon, lacustrine habitats are also important rearing areas for juvenile anadromous salmon. This phenomenon was first reported for Newfoundland, but similar occurrences have been reported in Labrador, Iceland, northern Finland, and Norway (Klemetsen *et al.* 2003). Lacustrine-reared parr are often larger in size at the same age than those reared in fluvial areas. Investigations have also shown that the relative amount of lacustrine habitat in a watershed influences migrating smolt size and, for some systems, marine survival is enhanced by greater amounts of lacustrine habitat (Klemetsen *et al.* 2003). Lacustrine rearing enhances freshwater survival during the egg-to-smolt period, with lacustrine-reared parr often having higher protein, lipid, and overall energy levels compared to those reared in fluvial habitats (Dempson *et al.* 2004).

14.7 The coregonines

Coregonine fishes are recognized as a sub-family closely allied to other salmonids. Similar to chars, salmon, and grayling, they are primarily northern fishes found in temperate, Sub-Arctic, and Arctic aquatic systems throughout the circum-Arctic. As for other salmonids, ciscoes and whitefishes are important ecologically, economically, and socio-culturally throughout high latitudes. Similar to the situation described above for chars, coregonines exhibit high diversity ecologically and ecophenotypically both among species and within species resulting in confused taxonomy. Unlike the chars, however, and with a few isolated exceptions in the western Canadian Arctic, almost all arctic coregonines relevant to our discussion are primarily distributed in lakes and rivers extending only to the northern continental mainland coasts (e.g. see Scott and Crossman 1973; Reshetnikov 2003). That is, they are marginally present in the southern portion of the 'high-latitude' area considered herein and do not extend much beyond the southern edges of some of the Canadian Arctic islands; however, they can locally be extremely abundant and important from many perspectives. Exceptions include Arctic cisco (*Coregonus autumnalis*), distributed along the southern coast of Victoria Island, and broad and lake whitefish (*Coregonus nasus* and *Coregonus clupeaformis*), both reported to occur on Victoria Island (Scott and Crossman 1973).

14.8 Other species

Several other fish species are present in freshwater habitats along the continental coastal margin and on the Arctic islands and may be locally abundant and significant for lake ecology and/or humans (see Scott and Crossman 1973; Reshetnikov 2003). These species can be divided into two groups: those that have a more southerly distribution, extending only as far north as the continental coasts of North America and Eurasia (Arctic grayling, *Thymallus arcticus*; European grayling, *Thymallus thymallus*; burbot, *Lota lota*; sculpins, *Cottus* spp.; longnose sucker, *Catostomus catostomus*; rainbow smelt, *Osmerus mordax*; European smelt, *Osmerus eperlanus*; and northern pike, *Esox lucius*), and those that are distributed in the Arctic islands of Canada and elsewhere (threespine sticklebacks, *Gasterosteus aculeatus*; ninespine sticklebacks, *Pungitius pungitius*). All are found in lakes and rivers in this area and will move between lakes and rivers for feeding and/or reproductive purposes. Of all the species, ninespine sticklebacks have the most northerly distribution, reaching to the northern portions of Baffin Island, southern Greenland, and the Eurasian mainland to the north coast and southern areas of Novaya Zemlya.

Sticklebacks, although not important in fisheries, must be noted for their significant ecological roles in northern lakes where they occur. Threespine and ninespine sticklebacks are small (to ≈5 cm), locally abundant northern fishes widely distributed around the circum-Arctic. Sticklebacks exhibit both freshwater and anadromous life histories similar to those described for chars and as euryhaline species have been able to disperse to Canadian Arctic islands. Threespine sticklebacks also exhibit sympatric ecological polymorphisms (benthic and limnetic morphologies and life histories) that parallel those described for Arctic char and are attributable to similar causes in some situations. For example,

polymorphic forms in Icelandic lakes appear to originate in depauperate, geologically young lakes with large unoccupied areas of heterogenous habitat being available (Kristjansson *et al.* 2002a, 2002b). Sticklebacks are largely benthivores and consume large amounts of aquatic invertebrates, with limnetic forms also predating plankton. Sticklebacks are also important prey species for anadromous and freshwater fish including Arctic char and may indirectly affect population structure. For example, Fraser and Power (1989) noted that the interactive feeding segregation of Arctic char and lake trout benefited Arctic char through a reduction in parasite loading as a result of a lower reliance on threespine sticklebacks as a food source.

14.9 Conclusions

Arctic freshwater fishes, in particular Arctic char, have adapted to life in harsh, variable environments with few resources. The dynamics of the ecosystems in which they live, however, are being continually perturbed by the increasing number of human interventions associated with northern development that have resulted in polluted environments, increasing habitat eutrophication, barriers to migration, over-fishing, the introduction of alien species, and—most recently—climate warming (Maitland 1995). Of these climate warming has the greatest potential to alter fish community structure and the population characteristics of Arctic freshwater fish both spatially and temporally. Most other threats will be local in nature, but at a given impact site are equally likely to cause significant changes. The ability of fish to adapt to changing environments will be species-specific and therefore a general predictive scenario is difficult to construct. In cases of rapid warming associated with climate change there are three possible outcomes for existing Arctic fish species: local extirpation due to thermal stress, a northward shift in the geographic range of potential competitors where dispersive pathways and other biotic and abiotic conditions allow, and genetic change within the limits of heredity through rapid natural selection (Reist *et al.* 2006a, 2006b).

Local extirpations are typically difficult to predict without detailed knowledge of critical population parameters (e.g. fecundity, growth, mortality, population age structure, etc.). Dispersal invasions will undoubtedly occur, but will inevitably be constrained by watershed drainage characteristics. In watershed systems draining to the north, increases in temperature will allow potential competitor species to shift their geographic distribution northward. In watershed systems draining to the east or west, increases in temperature may be compensated for by altitudinal shifts in riverine populations where barriers to headwater movement do not exist. Lake populations needing to avoid temperature extremes (e.g. Arctic char) will be confined to the hypolimnion during the warmest months provided that anoxic conditions do not develop. Patterns of seasonal occurrence in shallower littoral zones will almost certainly change, with consequent effects on trophic dynamics. Changes in species dominance will also occur because species are adapted to an ecology with specific spatial, thermal, and temporal characteristics that will inevitably be altered by climate-induced shifts in precipitation and temperature (Reist *et al.* 2006a, 2006b). Throughout much of its northern range, the Arctic char may be increasingly restricted to the occupancy of profundal lake habitats as is typical of its occurrence at southern latitudes. Lake char may also begin to display seasonal patterns of movement to the hypolimnion as seen further south. Atlantic salmon may colonize rivers to the north, as indicated by increasing capture of strays along the southern coasts of Baffin Island. A general increase in the incidence of mortality and decreased growth and productivity from disease and/or parasites is also likely in many high-latitude lake systems (Reist *et al.* 2006a, 2006b).

Unfortunately, our ability to fully understand the likely effects of ecosystem change on high-latitude fish populations is limited by the dearth of long-term studies of existing populations and the lack of readily available spatial comparisons. Upcoming studies in the International Polar Year (2007–2008) promise to correct some of the information deficiencies that exist; nevertheless, continued study of unique high-latitude fish populations will be required if we are to properly document the specializations that have allowed them to flourish at the fringe of the planetary biome.

Acknowledgments

We would like to thank Canadian International Polar Year Program, the Natural Sciences and Engineering Research Council, the Network of Centres of Excellence program ArcticNet, and Fisheries and Oceans Canada for funding support. Significant contributions to the chapter were provided by G. Power (artwork) and C. Sawatzky (research). A critical review by I. Winfield and two anonymous reviewers improved the chapter, as did the editorial guidance by Warwick Vincent.

References

Adams, C.E. *et al.* (2003). Stable isotope analysis reveals ecological segregation in a bimodal size polymorphism in Arctic charr from Loch Tay, Scotland. *Journal of Fish Biology* **62**, 474–481.

Alfonso, N.R. (2004). Evidence for two morphotypes of lake charr, *Salvelinus namaycush*, from Great Bear Lake, Northwest Territories, Canada. *Environmental Biology of Fishes* **71**, 21–32.

Baker, R. (1983). The effects of temperature, ration and size on the growth of Arctic charr (*Salvelinus alpinus* L.). MSc Thesis, University of Manitoba, Winnipeg, Manitoba.

Baroudy, E. and Elliott, J.M. (1994). The critical thermal limits for juvenile Arctic charr, *Salvelinus alpinus*. *Journal of Fish Biology* **45**, 1041–1053.

Behnke, R.J. (2002). *Trout and Salmon of North America*. The Free Press, New York.

Berg, O.K. and Gausen, O.D. (1988). Life history of a riverine, resident Atlantic salmon *Salmo salar* L. *Fauna. Norvegica Series A* **9**, 63–68.

Berg, O.K. (1995). Downstream migration of anadromous Arctic charr (*Salvelinus alpinus* (L.)) in the Vardnes River, northern Norway. *Nordic Journal of Freshwater Research* **71**, 157–162.

Black, G.A., Dempson, J.B., and Bruce, W.J. (1986). Distribution and postglacial dispersal of freshwater fishes of Labrador. *Canadian Journal of Zoology* **64**, 21–31.

Blackie, C.T., Weese, D.J., and Noakes, D.L.G. (2003). Evidence for resource polymorphism in the lake charr (*Salvelinus namaycush*) population of Great Bear Lake, Northwest Territories, Canada. *Ecoscience* **10**, 509–514.

Brännäs, E. and Wiklund, B.-S. (1992). Low temperature growth potential of Arctic charr and rainbow trout. *Nordic Journal of Freshwater Research* **67**, 77–81.

Cunjak, R.A., Power, G., and Barton, D.R. (1986). Reproductive habitat and behaviour of anadromous Arctic charr (*Salvelinus alpinus*) in the Koroc River, Québec. *Naturaliste Canadienne* **113**, 383–387.

Dempson, J.B. and Green, J.M. (1985). Life history of anadromous Arctic charr, *Salvelinus alpinus*, in the Fraser River, northern Labrador. *Canadian Journal of Zoology* **63**, 315–324.

Dempson, J.B. and Kristofferson, A.H. (1987). Spatial and temporal aspects of the ocean migration of anadromous Arctic char. *American Fisheries Society Special Publication* **1**, 340–357.

Dempson, J.B., Schwarz, C.J., Shears, M., and Furey, G. (2004). Comparative proximate body composition of Atlantic salmon with emphasis on parr from fluvial and lacustrine habitats. *Journal of Fish Biology* **64**, 1257–1271.

Dunbar, M.J. (1968). *Ecological Development in Polar Regions*. Prentice-Hall, Englewood Cliffs, NJ.

Elliott, S.R., Coe, T.A., Helfield, J.M., and Naiman, R.J. (1998). Spatial variation in environmental characteristics of Atlantic salmon (*Salmo salar*) rivers. *Canadian Journal of Fisheries and Aquatic Sciences* **55** (suppl. 1), 267–280.

Fletcher, G.L., Kao, M.H., and Dempson, J.B. (1988). Lethal freezing temperatures of Arctic char and other salmonids in the presence of ice. *Aquaculture* **71**, 369–378.

Folke, C. *et al.* (2004). Regime shifts, resilience, and biodiversity in ecosystem management. *Annual Review of Ecology and Systematics* **35**, 557–581.

Fraser, N.C. and Power, G. (1989). Influences of lake trout on lake-resident Arctic char in northern Québec, Canada. *Transactions of the American Fisheries Society* **118**, 36–45.

Griffiths, D. (1994). The size structure of lacustrine Arctic charr (Pisces: Salmonidae) populations. *Biological Journal of the Linnean Society* **51**, 337–357.

Guiguer, K.R.R.A., Reist, J.D., Power, M., and Babaluk, J.A. (2002) Using stable isotopes to confirm the trophic ecology of Arctic charr morphotypes from Lake Hazen, Nunavut, Canada. *Journal of Fish Biology* **60**, 348–362.

Hammar, J. (1989). Freshwater ecosystems of polar regions: vulnerable resources. *Ambio* **18**, 6–22.

Hammar, J. (1998). Interactive asymmetry and seasonal niche shifts in sympatric Arctic char (*Salvelinus alpinus*) and brown trout (*Salmo trutta*): evidence from winter diet and accumulation of radiocesium. *Nordic Journal of Freshwater Research* **74**, 33–64.

Hammar, J. (2000) Cannibals and parasites: conflicting regulators of bimodality in high latitude Arctic char, *Salvelinus alpinus*. *Oikos* **88**, 33–47.

Healey, M.C. (1978). Dynamics of exploited lake trout populations and implications for management. *Journal of Wildlife Management* **42**, 307–328.

Heinimaa, S. and Erkinaro, J. (2004). Characteristics of mature male parr in the northernmost Atlantic salmon populations. *Journal of Fish Biology* **64**, 219–226.

Hindar, K. and Jonsson, B. (1982). Habitat and food segregation of dwarf and normal Arctic charr (*Salvelinus alpinus*) from Vangsvatnet Lake, western Norway. *Canadian Journal of Fisheries and Aquatic Sciences* **39**, 1030–1045.

Hvidsten, N.A., Heggberget, T.G., and Jensen, A.J. (1998). Water temperature at Atlantic salmon smolt entrance. *Nordic Journal of Freshwater Research* **74**, 79–86.

Jensen, A.J. and Johnsen, B.O. (1986). Different adaptation strategies of Atlantic salmon (*Salmo salar*) populations to extreme climates with special reference to some cold Norwegian rivers. *Canadian Journal of Fisheries and Aquatic Sciences* **43**, 980–994.

Jobling, M. (1983). Influence of body weight and temperature on growth rates of Arctic charr, *Salvelinus alpinus* (L.). *Journal of Fish Biology* **22**, 471–475.

Johnson, L. (1976). Ecology of arctic population of lake trout, *Salvelinus namaycush*, lake whitefish, *Coregonus clupeaformis*, Arctic charr, *S. alpinus*, and associated species in unexploited lakes of the Canadian Northwest Territories. *Journal of the Fisheries Research Board of Canada* **33**, 2459–2488.

Johnson, L. (1980). The Arctic charr, *Salvelinus alpinus*. In Balon, E.K. (ed.), *Charrs, Salmonid Fishes of the Genus Salvelinus*, pp. 15–98. Dr. W. Junk Publishers, The Hague.

Johnson, L. (1983). Homeostatic characteristics of single species fish stocks in Arctic lakes. *Canadian Journal of Fisheries and Aquatic Sciences* **40**, 987–1024.

Johnson, L. (1989). The anadromous Arctic charr, *Salvelinus alpinus*, of Nauyuk Lake, N.W.T., Canada. *Physiology and Ecology Japan* **1**, 201–227.

Johnson, L. (1995). Systems for survival: of ecology, morals and Arctic charr (*Salvelinus alpinus*). *Nordic Journal of Freshwater Research* **71**, 9–22.

Johnson, L. (2002). *Imperfect Symmetry: Thermodynamics in Ecology and Evolution*. Torgoch Publishing, Sydney, BC.

Jonsson, B. and Jonsson, N. (2001). Polymorphism and speciation in Arctic charr. *Journal of Fish Biology* **58**, 605–638.

Klemetsen, A., Grotnes, P.E., Holthe, H., and Kristoffersen, K. (1985). Bear Island charr. *Report of the Institute of Freshwater Research, Drottningholm* **62**, 98–119.

Klemetsen, A., Amundsen, P.-A., Muladal, H., Rubach, S., and Solbakken, J.I. (1989). Habitat shifts in a dense, resident Arctic charr *Salvelinus alpinus* population. *Physiology and Ecology Japan* **1**, 187–200.

Klemetsen, A. *et al.* (2003). Atlantic salmon *Salmo salar* L., brown trout *Salmo trutta* L. and Arctic charr *Salvelinus alpinus* (L.): a review of aspects of their life histories. *Ecology of Freshwater Fish* **12**, 1–59.

Kristjansson, B.K., Skúlason, S., and Noakes, D.L.G. (2002a). Morphological segregation of Icelandic three-spine sticklebacks (*Gasterosteus aculeatus* L.). *Biological Journal of the Linnean Society* **76**, 247–257.

Kristjansson, B.K., Skúlason, S., and Noakes, D.L.G. (2002b). Rapid divergence in a recently isolated population of threespine stickleback (*Gasterosteus aculeatus* L.). *Evolutionary Ecology Research* **4**, 1–14.

Langeland, A. and L'Abée-Lund, J.H. (1998). An experimental test of the genetic component of the ontogenetic habitat shift in Arctic charr (*Salvelinus alpinus*). *Ecology of Freshwater Fish* **7**, 200–207.

Larsson, S. and Berglund, I. (1998). Growth and food consumption of 0+ Arctic charr fed on pelleted or natural food at six different temperatures. *Journal of Fish Biology* **52**, 230–242.

Larsson, S. *et al.* (2005). Thermal adaptation of Arctic charr: experimental studies of growth in eleven charr populations from Sweden, Norway and Britain. *Freshwater Biology* **50**, 353–368.

Lindsey, C.C. and McPhail, J.D. (1986). Zoogeography of fishes of the Yukon and Mackenzie basins. In Hocutt, C.H. and Wiley, E.O. (eds), *The Zoogeography of North American Freshwater Fishes*, pp. 639–674. John Wiley and Sons, New York.

Maitland, P.S. (1995). World status and conservation of the Arctic charr *Salvelinus aplinus* (L.). *Nordic Journal of Freshwater Research* **71**, 113–127.

Marten, N.V. and Olver, C.H. (1980). The lake charr, *Salvelinus namaycush*. In Balon, E.K. (ed.), *Charrs, Salmonid Fishes of the Genus Salvelinus*, pp. 207–277. Dr. W. Junk Publishers, The Hague.

Metcalfe, N.B. and Thorpe, J.E. (1990). Determinants of geographical variation in the age of seaward-migrating salmon, *Salmo salar*. *Journal of Animal Ecology* **59**, 135–145.

Moore, J.W. (1975). Reproductive biology of anadromous arctic char, *Salvelinus alpinus* (L.), in the Cumberland Sound area of Baffin Island. *Journal of Fish Biology* **7**, 143–151.

Nordeng, H. (1983). Solution to the "char problem" based on Arctic char (*Salvelinus alpinus*) in Norway. *Canadian Journal of Fisheries and Aquatic Sciences* **40**, 1372–1387.

O'Connell, M.F. and Dempson, J.B. (1996). Spatial and temporal distributions of salmonids in two ponds

in Newfoundland, Canada. *Journal of Fish Biology* **48**, 738–757.

O'Connell, M.F. and Dempson, J.B. (2002). The biology of Arctic charr, *Salvelinus alpinus*, of Gander Lake, a large, deep, oligotrophic lake in Newfoundland, Canada. *Environmental Biology of Fishes* **64**, 115–126.

Power, G. (1958). The evolution of the freshwater races of the Atlantic salmon (*Salmo salar* L.) in eastern North America. *Arctic* **11**, 86–92.

Power, G. (1969). The salmon of Ungava Bay. *Arctic Institute of North America Technical Paper*, no.22. Arctic Institute of North America, Montreal.

Power, G. (1978). Fish population structure in arctic lakes. *Journal of the Fisheries Research Board of Canada* **35**, 53–59.

Power, G. (1981). Stock characteristics and catches of Atlantic salmon (*Salmo salar*) in Québec, and Newfoundland and Labrador in relation to environmental variables. *Canadian Journal of Fisheries and Aquatic Sciences* **38**, 1601–1611.

Power, G. (1990). Salmonid communities in Québec and Labrador: temperature relations and climate change. *Polskie Archiwum Hydrobiologii* **37**, 13–28.

Power, G. (1997). A review of fish ecology in Arctic North America. *American Fisheries Society Symposium* **19**, 13–19.

Power, G. (2002). Charrs, glaciations and seasonal ice. *Environmental Biology of Fishes* **64**, 17–35.

Power, G. and Barton, D.R. (1987). Some effects of physiographic and biotic factors on the distribution of anadromous Arctic char (*Salvelinus alpinus*) in Ungava Bay, Canada. *Arctic* **40**, 198–203.

Power, M., Dempson, J.B., Reist, J.D., Schwarz, C.J., and Power, G. (2005a). Latitudinal variation in fecundity among Arctic char populations in eastern North America. *Journal of Fish Biology* **67**, 255–273.

Power, M., O'Connell, M.F., and Dempson, J.B. (2005b). Ecological segregation within and among Arctic char morphotypes in Gander Lake, Newfoundland. *Environmental Biology of Fishes* **73**, 263–274.

Reist, J.D., Gyselman, E.G., Babaluk, J.A., Johnson, J.D., and Wissink, R. (1995). Evidence for two morphotypes of Arctic char (*Salvelinus alpinus* (L.)) from Lake Hazen, Ellesmere Island, Northwest Territories, Canada. *Nordic Journal of Freshwater Research* **71**, 396–410.

Reist, J.D. *et al.* (2006a). General effects of climate change on Arctic fishes and fish populations. *Ambio* **35**, 370–380.

Reist, J.D. *et al.* (2006b). An overview of effects of climate change on selected Arctic freshwater and anadromous fishes. *Ambio* **35**, 381–387.

Reshetnikov, Yu.S. (2003). *Atlas of Russian Freshwater Fishes, Volume 1.* Academy of Sciences Russia. Nauka, Moscow [in Russian].

Rikardsen, A.H., Thorpe, J.E., and Dempson, J.B. (2004). Modelling the life-history variation of Arctic charr. *Ecology of Freshwater Fish* **13**, 305–311.

Robitaille, J.A., Côté, Y., Shooner, G., and Hayeur, G. (1986). Growth and maturation patterns of Atlantic salmon of Atlantic salmon, *Salmo salar*, in Koksoak River, Ungava, Québec. In Meerburg, D.J. (ed.), *Salmonid Age at Maturity*, pp. 62–69. *Canadian Special Publication of Fisheries and Aquatic Sciences* **89**.

Scott, W.B. and Crossman, E.J. (1973). *Freshwater Fishes of Canada.* Bulletin 184. Fisheries Research Board (Canada), Ottawa.

Sigurjonsdottir, H. and Gunnarsson, K. (1989). Alternative mating tactics of Arctic charr, *Salvelinus alpinus*, in Thingvallavatn, Iceland. *Environmental Biology of Fishes* **26**, 159–176.

Skúlason, S. and Smith, T.B. (1995). Resource polymorphism in vertebrates. *Trends in Ecology and Evolution* **10**, 366–370.

Skúlason, S.S., Snorasson, S.S., Noakes, D.L.G., and Ferguson, M.M. (1996). Genetic basis of life history variation among sympatric morphs of Arctic char, *Salvelinus alpinus. Canadian Journal of Fisheries and Aquatic Sciences* **53**, 1807–1813.

Sparholt H. (1985). The population, survival, growth, reproduction and food of Arctic charr, *Salvelinus alpinus* (L.), in four unexploited lakes in Greenland. *Journal of Fish Biology* **26**, 313–330.

Svenning, M.A. and Borgstrom, R. (1995). Population structure in landlocked Spitsbergen Arctic charr. Sustained by Cannibalism? *Nordic Journal of Freshwater Research* **71**, 424–431.

Swift, D.R. (1964). The effect of temperature and oxygen on the growth of Windermere charr (*Salvelinus alpinus* Wilughbii). *Comparative Biochemistry and Physiology* **12**, 179–183.

CHAPTER 15

Food-web relationships and community structures in high-latitude lakes

Kirsten S. Christoffersen, Erik Jeppesen, Daryl L. Moorhead, and Lars J. Tranvik

Outline

Lakes at high latitudes represent a range of biotic complexity imposed by strong abiotic limitations that are differentially ameliorated by biogeographic factors between the northern and southern latitudes. However, the communities in these lakes are simpler than those typically found in more temperate environments and represent a range of declining biocomplexity as conditions approach the limits to life in these extreme environments. These include changes in the relative importance of keystone predators and higher trophic levels (e.g. fish and birds), adjacent terrestrial communities, colonization between lakes, and benthic–pelagic coupling within lakes in structuring the food webs. Examples and case studies are used to illustrate food-web interactions between microbial and classic food webs, the importance of autochthonous versus allochthonous carbon, and the implications of a changing climate.

15.1 Introduction

15.1.1 Why study high-latitude lake food webs?

High latitudes contain a considerable fraction of the world's lakes. Satellite images show that lakes and ponds are extremely abundant in many parts of the Arctic (Figure 15.1), and the ice-free regions of Antarctica also have numerous lakes and ponds; for example, in the Vestfold Hills, the McMurdo Dry Valleys, and the Byers Peninsula in maritime Antarctica (see Chapter 1 for site descriptions). Hence, there is an obvious need to understand the role of lakes and ponds in a broader ecosystem perspective, within both the current and future climates. In addition, high-latitude lakes offer a range of conditions making them particularly suitable for the investigation of properties that are of general relevance for freshwater food webs.

Continental Antarctic lakes are dominated by microbial-loop organisms; they generally lack higher trophic levels common elsewhere, receive little or no input of allochthonous organic matter from terrestrial ecosystems, and exist under extreme environmental conditions near the limits to life. Lakes in maritime and Sub-Antarctic regions tend to have slightly more complex food webs with more invertebrate taxa and potential allochthonous inputs from terrestrial flora. Within this short continuum, the aquatic food webs in Antarctica may serve as models where fundamental food-web properties may be explored. Studies of these simple ecosystems can provide insights that are difficult or impossible to obtain from larger, more complex systems (Bell and Laybourn-Parry 1999; McKenna *et al.* 2006).

In contrast, Arctic lakes show higher biocomplexity under comparable physical conditions, and they

269

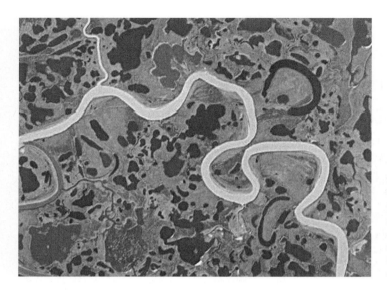

Figure 15.1 Abundant lakes and ponds in the Arctic landscape. This is a satellite image of the Mackenzie River flood-plain showing one of the many small channels of the turbid river, and a small fraction of the 40 000 lakes that are distributed across it (see Chapter 5 and Plate 8). Photograph: NASA GSFC/METI/ERSDAC/JAROS and the US`/Japan ASTER Science Team.

cover a greater range of climatic conditions. Thus, Arctic lakes establish a broader continuum from lakes similar to the microbial-loop communities of Antarctic lakes to the more typical aquatic communities of temperate regions, which include fish, macrophytes, insects, and crustaceans. Although generally simpler than temperate lakes, Arctic lakes more closely resemble lower-latitude lakes than Antarctic lakes, and allow examination of the importance of detrital food webs, impacts of vertebrate predators, and the role of metazoan links between the microbial loop and the overall food web (e.g. Hobbie *et al.* 1999).

15.1.2 Commonalities and differences between Arctic and Antarctic lakes

An extremely cold and often dry climate is the most obvious shared characteristic of polar regions, and it entails a very short growing season, in terms of both temperature and radiant energy regime (Roberts and Laybourn-Parry 1999; Prowse *et al.* 2006). High-latitude lakes and ponds are ice-covered for most or all of the year. Arctic lakes are typically frozen for 8–10 months, and may have a maximum ice thickness of 2.5 m. Antarctic lakes above 70°S often have a roughly 5-m-thick permanent ice cover which only thaws at the margins during brief periods in summer. In both regions, shallow water bodies freeze solid every winter. An

additional layer of snow may accumulate on surface ice, further reducing the light penetration and retarding melt (Howard-Williams *et al.* 1998; see also Chapter 4). Thus, photosynthesis is precluded during the continuous darkness of winter and limited by light attenuation in the overlying ice and snow. In addition, compared to the water inputs in warmer climates, nutrient concentrations in polar lakes are typically low because water inputs largely consist of surface runoff of melt water, and have less contact with soils during transport in the drainage area. For these reasons biomass of the different trophic levels is often relatively low compared with lakes in temperate regions, suggesting that interactions between organisms are less complex (but see Chapter 13).

One of the greatest sources of differences between Arctic and Antarctic lakes may be their evolutionary histories. Antarctica has been geographically isolated from all other continents for millions of years, which limits potential colonization (Battaglia *et al.* 1997; Fogg 1998). The continent offers few refuges for organisms during climatic fluctuations, implying that species present in Antarctic lakes apparently survived strong selection pressures within the region. For example, lakes in the Vestfold Hills formed 10 000 years ago by isostatic uplift after the last major glaciation (Adamson and Pickard 1986) when isolated pockets of sea water evolved into lakes that today range from brackish to hypersaline (Ferris and Burton 1988). The marine communities trapped in these lakes lost many of the original marine protozoa as well as all metazoans, except the copepod *Paralabidocera antarctica* (Laybourn-Parry *et al.* 2002). In contrast, the McMurdo Dry Valleys (78°S, 162°E) in southern Victoria Land have a freshwater history over the past 2 million years, with cycles of greater inundation during glacial maxima and desiccation during interglacial times (Lyons *et al.* 1998). The pelagic food webs in the McMurdo lakes mainly consist of freshwater species of protozoa and a single metazoan, the rotifer *Philodina* sp. (James *et al.* 1998), although Roberts *et al.* (2004) reported collecting unidentified copepod nauplii from one lake. In contrast to the evolutionary isolation of the Antarctic continent, Arctic regions are much more closely linked to ecosystems at lower latitudes, allowing more migration and colonization, although species richness is low in Greenland and other Arctic islands such as Svalbard.

Another major difference between Arctic and Antarctic lakes concerns the relative importance of allochthonous organic matter received from terrestrial ecosystems. In Arctic lakes, the flow of carbon through the detrital food chain may greatly exceed *in situ* primary production, driven by inputs of terrestrial organic materials (Michaelson *et al.* 1998; Sobek *et al.* 2005; Jansson *et al.* 2007) that may be thousands of years old. In contrast, most Antarctic lakes receive little detrital input due to scarce terrestrial vegetation (Bell and Laybourn-Parry 1999; Laybourn-Parry *et al.* 2005). This may be another reason for the large difference in species richness of lakes between the two polar regions. In fact, it seems likely that some aquatic ecosystems in Antarctica export organic matter to surrounding soil communities, reversing the net flow of organic matter between terrestrial and aquatic systems that usually prevails in most temperate as well as in Arctic ecosystems (see below).

15.2 Food webs

15.2.1 Structural features

High-latitude biota are under strong physical controls exerted by extreme variations in radiant energy, temperature, and nutrient availability on an annual basis. Many of the multicellular organisms present in the unfrozen waters of lakes during winter are primarily in a quiescent state, but become active as soon as sufficient light is able to penetrate the ice cover to initiate photosynthesis by phytoplankton. Increases in the populations of phytoplankton species stimulate cascading increases in productivity through zooplankton and other organisms at higher trophic levels. The seasonal peak in lake productivity is normally reached before complete loss of ice cover after which productivity rapidly declines as the limited available nutrients in the water body are depleted.

The carbon and energy transferred through the food webs of Arctic and Antarctic lakes are derived either from *in situ* primary production of phytoplankton in the water column as well as benthic algae, or mobilized by bacteria that utilize organic matter imported from the surrounding watershed. The biomass of benthic algae is often considerable in high-latitude lakes, resulting in extensive benthic algal mats (Wharton *et al.* 1983; Moorhead *et al.* 1999; Vadeboncoeur *et al.* 2006), and benthic algae often dominate total primary production, not least in the shallow lakes and ponds (Vadeboncoeur *et al.* 2003). In turn, these mats are grazed by micro- and macroinvertebrates (Rautio and Vincent 2006) that may be eventually consumed by fish and birds.

The communities of nutrient-poor Arctic lakes include many groups of phytoplankton (diatoms, dinoflagellates, chrysophytes, and cyanobacteria). These algae are grazed by zooplankton (copepods, cladocerans, and rotifers) that in turn are predated by fish or large invertebrates. Algae and other organisms release dissolved organic substances into the water, either as feces and exudates from living organisms, or during senescence. These substances are utilized by bacteria, which in turn become prey for protozoan (flagellates and ciliates) or metazoan bacterivores including cladoceran and rotifer species. The recycling of released organic matter into the food web via bacteria and protozoan bacterivores is commonly referred to as the microbial loop. In addition, bacteria consume organic matter imported from the catchment area (e.g. humic substances leaching from surrounding soils). Hence, bacteria are not only part of an internal loop within lakes, but also constitute a link between terrestrial primary producers and the aquatic food web (e.g. Jansson *et al.* 2007). The heterotrophic microbial metabolism of Arctic lakes is commonly based to a large extent on organic matter imported from the watershed (e.g. Sobek *et al.* 2005). This is certainly the case for tundra lakes surrounded by organic-rich soils, whereas other High Arctic and Antarctic lakes have very barren watersheds providing little or no organic matter via runoff. Hence, the subsidy that lake food webs receive via the import of organic matter may vary greatly among different high-latitude ecosystems. As will be described below, this variability even extends to extreme cases in Antarctica, where the terrestrial environment is subsidized via organic matter produced in lakes.

At high latitudes, the importance of the different links and webs varies considerably among lake ecosystems. A general feature of lakes in Continental Antarctica and on Sub-Antarctic islands is the lack of fish and other freshwater vertebrates. In addition, the invertebrate fauna is less diverse, and insects are very few in comparison with Arctic lakes (see also Chapter 13). Due to the low biodiversity and the lack of higher animals, food-web structure is often less complex than in corresponding High Arctic lakes. The pelagic food webs of high-latitude lakes may be depicted (Figure 15.2) along a gradient of increasing complexity, including (1) food webs in Antarctic lakes where metazoa are represented by only rotifers, (2) food webs where herbivorous copepods constitute the uppermost functional trophic level, as exemplified by Antarctic and Sub-Antarctic lakes, (3) Arctic food webs including *Daphnia*, which graze upon bacteria as well as on phytoplankton, and (4) Arctic food webs where herbivorous zooplankton are under predation pressure from fish and/or invertebrate predators.

15.2.2 Continental Antarctic lakes

In general, the pelagic food webs in lakes at higher latitudes in Antarctica are dominated by microbial-loop organisms, and they usually have few or no metazoans (James *et al.* 1998; Bell and Laybourn-Parry 1999; Roberts and Laybourn-Parry 1999;

Laybourn-Parry 2002). For example, Lake Fryxell, in the McMurdo Dry Valleys, southern Victoria Land, has seven trophic groups: rotifers (*Philodina* sp.), bacterial-feeding ciliates, flagellate-feeding ciliates, heterotrophic nanoflagellates, photosynthetic nanoflagellates, larger phytoplankton, and bacteria (Figure 15.2a). In comparison, Ace Lake in the Vestfold Hills, Princess Elizabeth Land, contains adult and juvenile copepods (*P. antarctica*), ciliates, heterotrophic nanoflagellates, photosynthetic nanoflagellates, *Mesodinium rubrum*, dinoflagellates, and bacteria (Figure 15.2b). A single species of copepod appears to be the top predator in Ace Lake, whereas one group of rotifers is the top predator in Lake Fryxell.

In an effort to determine the relative importance of top-down versus bottom-up controls in these food webs, McKenna *et al.* (2006; unpublished results) modeled the carbon flow through the pelagic food webs of both lakes (see Case studies, Section 15.4). In brief, it appeared likely that these food webs were more strongly influenced by bottom-up than top-down controls. In Lake Fryxell, fluctuations in total community biomass followed the general patterns of seasonal and interannual availability of photosynthetically available irradiance, suggesting a strong energy limitation. Although the presence of adult copepods often had a significant effect on biomass and carbon flows in the Ace Lake food web, detailed analyses showed that both community structure (biomass) and function (productivity) of Ace Lake were more related to nitrates, ammonia, ice thickness, and salinity than to adult copepods. Thus resource availability (radiant energy and nutrients) and conditions (temperature, salinity) appeared to exert stronger control on these food webs than predation.

15.2.3 Maritime Antarctic and Sub-Antarctic lakes

Maritime lakes are common at both Antarctic and Sub-Antarctic latitudes, including the South Orkney, South Shetland, and South Sandwich Islands, as well as the Antarctic Peninsula. Differences in elevation, aspect, exposure, distance from coast, and impacts of sea birds and mammals produce a wide range of oligotrophic

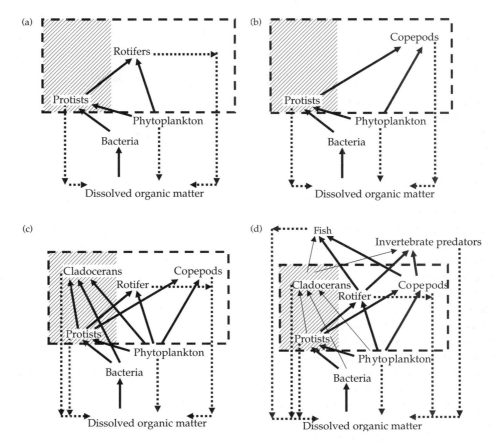

Figure 15.2 Pelagic food webs in high-latitude lakes of differing complexity. Bacterivores are within the hatched areas and herbivores in the rectangles surrounded by broken lines. Solid arrows indicate energy flux by feeding, and dotted arrows indicate release of dissolved organic matter. (a) Antarctic dry valley lakes, with metazoa represented only by rotifers (McKenna *et al.* 2006); (b) Sub-Antarctic lakes with no pelagic rotifers, and herbivorous copepods constituting the highest functionally important trophic level (Tranvik and Hansson 1997); (c) Arctic pond with no fish or invertebrate predators (Christoffersen *et al.* 2008); (d) Arctic lake with fish and/or invertebrate predators, where the biomass of cladocerans is highly suppressed (Christoffersen *et al.* 2008). In addition to internal fluxes, the dissolved organic matter is amended to different extents with substances imported from the catchment area. In all four cases mixotrophic species combine the roles of bacterivorous protists and phytoplankton (the two latter phenomena are not depicted in the illustrations).

to hypertrophic conditions (Rosswall and Heal 1975; Heywood *et al.* 1980; Hansson *et al.* 1996). For example, snow strongly attenuates radiant energy transmission, so that even small differences in amount, distribution, and persistence have large impacts on productivity of both pelagic and benthic communities (Hawes 1985; Moorhead *et al.* 1997). Nutrient regimes are also highly variable, influenced by substrate weathering, atmospheric deposition, and wastes from penguin and seal rookeries. For example, the lakes most remote (a few kilometers) from the sea on South Georgia Island (54°10'S,

36°40'W) are more strongly influenced by meltwater, are not visited by marine mammals, but to a limited extent by marine birds. The lakes closer to the sea are frequently visited by penguins and seals and are additionally influenced by breeding petrels, skuas, and gulls. This results in a productivity gradient showing a negative correlation between total phosphorus content in the water (ranging from 4 to 70 µg L⁻¹) and distance to the sea (ranging from 40 to 3500 m; Hansson *et al.* 1996).

None of these maritime lakes contain fish. The metazoan communities provide interesting contrasts to lakes of the more extreme Antarctic continent and of temperate regions. In general, maritime Antarctic and Sub-Antarctic lakes have higher metazoan diversity than lakes in Antarctic continental locations and metazoan diversity increases with decreasing latitude (Pugh *et al.* 2002). For example, Otero Lake (64°09'S, 60°57'W) on the Antarctic Peninsula has no metazoans (Izaguirre *et al.* 2001) but near Hope Bay (63°24'S, 57°00'W) the omnivorous copepod *Boeckella poppei* and the rotifer *Philodina gregaria* are found in oligotrophic Lake Chico; *B. poppei* is also present in the meso-eutrophic Lake Boeckella; but only bdelloid rotifers occupy the hypertrophic Pingüi Pond (Izaguirre *et al.* 2003). Further north on King George Island (62°14'S, 58°40'W), Pociecha (2007) found the anostracan *Branchinecta gaini* in several lakes, including Lake Wujka (62°09'S, 58°28'W), and Hahn and Reinhardt (2006) found midge larvae *Parochlus steinenii* in lakes with stable water levels. On the Byers Peninsula of Livingston Island (62°36'S, 61°00'W), Toro *et al.* (2007) reported both *B. poppei* and *Branchinecta gaini* in pelagic zones, the nekto-benthic cladoceran *Macrothrix ciliata*, and the chironomids *Belgica antarctica* and *Parochlus steineii*, as well as oligochaete *Lumbricillus* sp. in benthic regions. Still further north, Sombre Lake on Signy Island (60°43'S, 45°38'W) has two copepods, *B. poppei* and the carnivorous *Parabroteus sarsi*, as well as *Br. gaini* (Butler *et al.* 2005). Finally, a recent survey of freshwater habitats on South Georgia Island (54°10'S, 36°40'W) found 29 species of arthropods, including 11 crustaceans and four insects, although not all were pelagic (Dartnall 2005).

Some of the earliest and most intensive studies of Sub-Antarctic lakes focused on the South Orkney Islands in conjunction with the Tundra Biome Program (e.g. Rosswall and Heal 1975; Heywood *et al.* 1980), particularly lakes on Signy Island. This island remains an important site for the study of Sub-Antarctic maritime lakes (e.g. Hawes 1985; Moorhead *et al.* 1997; Butler *et al.* 2005) and serves as a baseline for assessing impacts of climate change on Antarctic ecosystems (Quayle *et al.* 2002). However, a recent suite of studies focusing on 20 lakes surrounding Husvik (54°10'S, 36°41'W) on South Georgia Island provides another excellent example of the low complexity and dynamics of lacustrine food webs on maritime Antarctic and Sub-Antarctic islands.

The South Georgian lakes have thick benthic mats of filamentous algae and cyanobacteria (Dartnall and Heywood 1980; Hansson *et al.* 1996). More than 50 species of rotifers have been identified from the lakes, almost exclusively associated with the benthic habitat. In addition, five species of cladocerans have been found in the benthic environment, with only very occasional recordings from the water column (Figure 15.2c). The lakes harbor few other invertebrates, including nematodes and ostracods, but mollusks are absent. In both the water column and in benthic habitats, copepods are abundant, and a diving beetle, *Lancetes angusticollis*, is common in the benthos of many lakes, but occasionally appears in the water column as well. The invertebrate fauna of the water column is dominated by three species of copepods (Figure 15.2b). The almost 5-mm-long *P. sarsi* is predatory. The other two copepods, *Boeckella michaelseni* and *B. poppei*, are herbivorous and are prey for the predatory copepod (*Parabroteas*) and diving beetle (*Lancetes*). Experimental studies of the impact of these two predators on the abundance of herbivorous copepods reveal that the predatory pressure is weak, with much less than 1% of herbivorous copepods removed per day by the predators (Hansson and Tranvik 1996). Hence, the herbivorous copepods probably constitute the highest functionally important trophic level. Possibly due to long life cycles, the development of *Lancetes* and *Parabroteas* is restricted by the severe climate and short growing season (Nikolai and Droste 1984), which may explain why the predators do not exert stronger control of their prey (Hansson and Tranvik 1996).

In a survey of 19 lakes in South Georgia, Hansson *et al.* (1993) demonstrated that the increase in algal biomass in the water column (measured as chlorophyll *a*) with increasing nutrient (phosphorus) concentration was less than across similar nutrient gradients in temperate lakes. They suggested a strong control of algae by herbivorous zooplankton in Sub-Antarctic and Antarctic lakes, because herbivorous zooplankton constituted the highest functionally important trophic level (see above). In contrast, zooplanktivorous fish or invertebrates typically dominate food webs at lower latitudes and thereby constrain the grazing pressure exerted by herbivorous zooplankton (Hansson *et al.* 1993). The hypothesis that herbivorous copepods occur essentially at the top of the food web is further corroborated by an increasing ratio of herbivorous copepod per carnivorous copepod (*Parabroteas*) with increasing trophic state, but no corresponding increase in algal biomass per herbivorous copepod. Such a pattern is expected when the herbivores are not suppressed by predation and algae are controlled by herbivory. Interestingly, Hansson *et al.* (1993) found indications that the trophic cascade extended down to bacteria because bacterivorous flagellate abundance per unit grazer (copepods) did not increase with increasing trophic state, and flagellated bacterivores thus appeared to be controlled by grazing. In contrast, planktonic bacteria increased along the same gradient, a tentative reason for this being that flagellates were not able to exert control due to predation from copepods.

The trophic cascade from copepods via phago-trophic flagellates to bacteria was tested experimentally in two of the lakes (Tranvik and Hansson 1997). In microcosms incubated *in situ* at different copepod densities, flagellates decreased and bacteria increased over time in the presence of copepods. The small-sized *B. michaelseni* (≈0.7 mm) was able to feed on bacterivorous flagellates, while the larger *B. poppei* (≈2.4 mm) was not able to suppress bacterivorous flagellates, even at high densities, but depended on larger algae as a food source (Hansson and Tranvik 1997). Both species competed for algae roughly 10–30 μm in length, while only *B. poppei* could feed on larger algae as well. A mesocosm experiment, using 0.5-m³ mesocosms in contact with the sediment, and where copepods were either reduced in density by sieving, or re-added to achieve roughly the original abundance, further corroborated the trophic cascade from copepods to bacteria; at reduced copepod abundance flagellates thrived, while bacteria reached low levels, and vice versa.

In conclusion, the cross-system comparison of lakes on South Georgia indicates top-down control of algae and bacterivorous flagellates by copepods functionally occupying the top of the food web. Since these studies have not yet been extended across the full range of Antarctic lakes,

generalizations may be premature. Few attempts have been made to quantify the patterns of energy flow through Antarctic lake food webs, but it is likely that energy flux or at least complexity increases with increasing diversity of the metazoan community, which is consistent with general relationships between productivity and such environmental variables as latitude, local energy regime, and nutrient availability.

15.2.4 Arctic lakes

The food-web structure in Arctic lakes is often more complex than in Antarctica because of a typically higher species diversity of zooplankton (see Chapters 11 and 13), including also key-stone grazers; that is, *Daphnia* species and the presence of fish in many of the lakes (Figure 15.2d). Within the Arctic region, complexity varies between lakes on the American and European continent to lakes on islands, such as Greenland and from north towards south in the Arctic. Complexity generally increases from east to west Greenland (Jeppesen *et al.* 2001; Lauridsen *et al.* 2001) as the latter has closer contact with more species-rich continents.

When fish are present they are key players in structuring the food web. They have a strong impact on the size, abundance, and structure of the zooplankton (O'Brien 1987; O'Brien *et al.* 1992; Hersey *et al.* 1999; Jeppesen *et al.* 2003a, 2003b) and macroinvertebrate communities (Hersey *et al.* 1999). Predation risk for large-bodied zooplankton is high because they grow slowly in these resource-poor and cold environments, and they are therefore exposed to fish predation for a longer period before reproduction than in more nutrient-rich temperate lakes. High water clarity further enhances the foraging success of fish that hunt visually. Moreover, due to high contribution of the benthic community to lake production, benthic–pelagic feeding fish may maintain a potentially high predation capacity on pelagic zooplankton even in periods when zooplankton are scarce (Jeppesen *et al*, 2003a, 2003b). Finally, only large but predation-vulnerable *Daphnia* occur in Greenland and these invertebrates are often more pigmented in Arctic than in temperate lakes, making them more vulnerable to predation by fish (Sægrov *et al.* 1996). In continental

Arctic lakes and in Sub-Arctic lakes, smaller-sized *Daphnia* species like *Daphnia longiremis* in Canada and *Daphnia longispina* in Iceland and northern Norway may co-exist with fish if fish density remains low (Dahl-Hansen 1995).

When fish are present the zooplankton community is typically dominated by small-bodied zooplankton, with potentially smaller cascading effects on phytoplankton and bacterioplankton. Greenland lakes are excellent for studying the cascading effects of fish as a large proportion of the lakes are naturally without fish. Lakes that have not been in contact with the sea since the last glaciation are nearly all without fish, but many lakes that were connected to the sea before the land-lift now host isolated landlocked populations of Arctic char (*Salvelinus arcticus*, one–three morphs; see also Chapter 13) and/or threespined sticklebacks (*Gasterestus aculeatus*). A set of 82 lakes in the High Arctic northeast Greenland and near the Arctic circle in west Greenland recently have been subjected to intensive studies (Christoffersen 2001; Jeppesen *et al.* 2001, 2003a, 2003b; Lauridsen *et al.* 2001) In northeast Greenland, Arctic char is the only fish present, in west Greenland sticklebacks are present in some of the lakes. Neo- and paleo-olimnological analyses show clear differences in the cladoceran community structure among lakes with and without fish, with large pelagic and benthic forms dominating lakes without fish (Lauridsen *et al.* 2001; Jeppesen *et al.* 2003a, 2003b). Body size varies largely among the lakes (Figure 15.3), and the ratio of zooplankton biomass to phytoplankton biomass (expressed as chlorophyll *a*) is much higher in the fishless lakes. Nevertheless, no differences are seen in chlorophyll or the chlorophyll *a*/total phosphorus ratio between lakes with and without fish in northeast Greenland and west Greenland, although canonical correspondence analyses indicate changes in community structure of phytoplankton (K.S. Christoffersen *et al.* unpublished results) along the gradient in fish density. These results suggest a strong cascading effect of fish on zooplankton abundance and that body size is not related to phytoplankton biomass, which is likely determined mainly by nutrient availability. Likewise, no difference was found in the biomass of flagellates, picoplankton, and bacterioplankton.

Biomass but not production of these communities seems determined by resources rather than predation. However, the total ciliate biomasses, its contribution to the microbial carbon pool, and the biomass per cell of ciliates were significantly higher in lakes with fish, most likely as a result of released competition from large-bodied *Daphnia* (E. Jeppesen *et al.*, unpublished results; Figure 15.3).

In Arctic Canadian lakes, which typically are more nutrient-rich than the lakes studied in Greenland, cascading effects of fish also appear to reach down to the phytoplankton level. This is supported by an overall lower chlorophyll *a*/total phosphorus ratio than observed in temperate lakes (Flanagan *et al*, 2003), although other factors may be involved (e.g. low inorganic nitrogen concentrations). However, comparable with studies in Greenland, a set of enclosure experiments in Toolik Lake showed strong effects on zooplankton when adding Arctic grayling (*Thymalu arcticus*), but no effect on phytoplankton biomass (O'Brien *et al.* 1992). Because of the higher variation in species richness of fish on the continents, response patterns for primary consumers show more variation with lake types and fish community structure (Hersey *et al.* 1999; Table 15.1). This may also affect primary consumers and microbial communities differently.

15.2.5 Benthic and pelagic production and the role in the food web

Benthic microbial mats are commonly found in shallow lakes both in Arctic and Antarctic environments. They are the largest source of biomass and productivity in shallow ponds and streams of the McMurdo Sound region of Antarctica, and are also common in the larger lakes (Wharton *et al.* 1983; Vincent *et al.* 1993; Hawes and Howard-Williams 1998; Hawes and Schwarz 2000) and in many Arctic and Sub-Arctic streams, lakes, and ponds (Hecky and Hesslein 1995; Hansson and Tranvik 2003; Vadeboncoeur *et al.* 2003; Wrona *et al.* 2006; Rautio and Vincent 2007). In these systems with low nutrient concentrations in the water, primary production is highly skewed towards the benthic habitat, where nutrients are largely taken up and retained (Vadeboncoeur *et al.* 2003). Accordingly, Bonilla *et al.* (2005) showed that Arctic lake phytoplankton

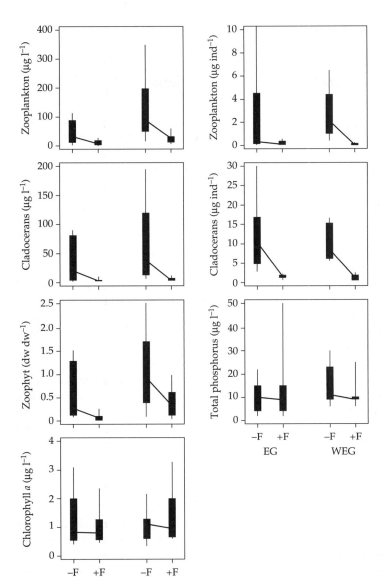

Figure 15.3 Zooplankton, phytoplankton, and phosphorus in Greenland lakes. The boxplots (10, 25, 75, and 90th percentiles) are for zooplankton and cladoceran biomass, the ratio of zooplankton to phytoplankton biomass (Zoophyt), chlorophyll *a*, zooplankton and cladoceran body mass, and total phosphorus. The study was in northeast Greenland (EG) and West Greenland (WEG) in lakes with (+F) and without (−F) fish. The data are from Jeppesen *et al.* (2001, 2003a, 2003b), Lauridsen *et al.* (2001), Jones and Jeppesen (2007), and E. Jeppesen *et al.* (unpublished results).

were strongly nutrient limited, whereas the benthos grew in a nutrient-rich environment, so that light availability may be a limiting factor within the mats but not for overlaying phytoplankton.

A recent modeling analysis compared primary production of pelagic phytoplankton and benthic microbial mats in Lake Hoare, in the McMurdo Dry Valleys, Antarctica (Moorhead *et al.* 2005). In brief, field measurements of incident radiation and optical properties of the overlying ice and water column (Howard-Williams *et al.* 1998) were used to drive photosynthesis and respiration for mats at various depths (Hawes and Schwarz 2000). The predicted maximum daily rates of carbon fixation of 15–20 $\mu g\,C\,cm^{-2}$ at approximately 10 m depth yielded a net annual rate of about 1.2 $mg\,C\,cm^{-2}$. When extrapolated across the entire surface area ($1.94 \times 10^6\,m^2$) and depth (34 m) of Lake Hoare, rates

Table 15.1 Invertebrate communities are determined by the presence of fish and by the distribution of fish species. Exemplified by data from the Toolik Lake Long-Term Ecological Research site (Hersey *et al.* 1999).

Fish community	Invertebrate communities	Ecological implications
Ponds (no fish)	Multiple invertebrate predators	No snails; **benthic** and pelagic invertebrate predators; large zooplankton; moderate chironomid **biomass**
High-gradient lakes (no fish)	Multiple invertebrate predators	Planktonic community resembles that of ponds; **benthic** and pelagic invertebrate predators; large zooplankton; large **Lymnaea**
Grayling-only lakes	Intermediate pelagic predators	Benthic invertebrate predators common; high chironomid **biomass**; small zooplankton; high grayling recruitment
Grayling and sculpin lakes	Intermediate pelagic and benthic predators	Small-bodied zooplankton; low chironomid **biomass**; moderate grayling recruitment; large snails; sculpin abundant on sediments
Lakes with lake trout	Top pelagic and benthic predators	Mixed zooplankton; small snails; high chironomid **biomass**; low grayling recruitment; sculpin distribution restricted shoreward

of production for phytoplankton communities averaged only 15% of calculated rates for mats, further supporting the importance of benthic primary production in high-latitude oligotrophic systems. The role of benthic production typically decreases toward the temperate zone due to higher nutrient input, and, accordingly, it also decreases with eutrophication (Liboriussen and Jeppesen 2003; Vadeboncoeur *et al.* 2003, 2006).

Food-web studies have traditionally been focusing on the pelagic webs, and it has been shown recently that lakes in general exhibit a large share of metabolism in the benthos, as well as strong benthic–pelagic coupling (Vadeboncoeur *et al.* 2002, 2003). This is the case particularly in cold, nutrient-poor environments (Wrona *et al.* 2006; Rautio and Vincent 2006, 2007).

The role of benthos in the food webs is often examined using stomach contents or by undertaking laboratory or field enclosure experiments. Another method is stable isotope analysis, which is particularly useful when food webs are simple. $\delta^{15}N$ can provide information about where a given organism is placed in the food web as the proportion increases roughly 3.4‰ per trophic level, and $\delta^{13}C$ indicates the food sources, as $\delta^{13}C$ typically is more negative organic carbon originating from pelagic primary production, as compared with littoral primary production, and only changes slightly with transfer through the food web (Van der Zanden and Rasmussen 1999). Simple end-member mixing models may be used

to reconstruct food webs from analyses of the nitrogen and carbon isotopes (Van der Zanden and Rasmussen 2001). This it is a powerful tool for polar lakes as shown by Kling and Fry (1992), Jeppesen *et al.* (2002a, 2002b), Sierszen *et al.* (2003), and Rautio and Vincent (2006).

Two lakes from the Toolik Lake region studied by Sierszen *et al.* (2003) illustrate the use of stable isotopes for elucidating food-web structure and the role of the benthic and pelagic systems (Figure 15.4). Trout are top predators in both lakes (highest $\delta^{15}N$), while many organisms that are one trophic level lower have very different $\delta^{13}C$. To help resolve which food the trout depend on, a band encompassing the mean $\pm 1\,SD$ of lake trout $\delta^{13}C$ values, and with a slope of 3.4:1, is shown on each dual isotope plot. Lower trophic levels that make major contributions to assimilated carbon in lake trout would be expected to have carbon isotope ratios lying within or along this band. As might be expected, the bands encompass slimy sculpin and, in the other lake, grayling. Carbon isotope ratios indicate that, for the species and life stages analyzed, assimilated carbon in higher consumers of both lakes passes through benthic invertebrates. Zooplankton appears not to make substantial direct contributions to the carbon in adult fish muscle tissue. Nevertheless, the carbon present in benthos and fish may originally have been fixed by a variety of primary producers. Epiphytic algae, which are directly upon the pathway leading to lake trout and other fish, are apparently an important carbon

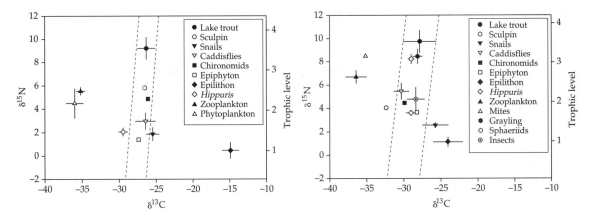

Figure 15.4 The food web of two lakes in the Toolik Lake region. Bivariate isotope ratio plot with means±SD of replicate samples for all taxa analyzed. A sample mean without error bars indicates that only one sample was analyzed (but with multiple subsamples). See text for further explanations. The data are from Sierszen *et al.* (2003).

source. However, macrophyte beds in these lakes are not extensive and the productivity of epiphyton may not be sufficient to fuel the entire food web. Epilithic algae and phytoplankton, which lie on either side of the pathway, may also make substantial contributions to the food web.

Fish may not only affect the pelagic and benthic communities but also interactions within each of these communities, as demonstrated by a paleoecological study (Jeppesen *et al.* 2001) of surface sediment of lakes and ponds in northeast Greenland, some with and some without fish. In lakes with fish present the tadpole shrimp, *Lepidurus arcticus*, is absent and the sediment hosts many remains of the dominating microcrustaceans *Chydorus sphaericus* and *Alona* sp., which take advantage of the lack of competition for food and predation from the omnivorous *Lepidurus*. However, when fish are lacking *Lepidurus* is abundant and there are only few remains of the microcrustaceans. Hence, changes in the abundance of fish will not only directly but also indirectly affect trophic dynamics in the pelagic and the sediment and the interactions among the two environments. Moreover, Rautio and Vincent (2006, 2007) demonstrated by studies of fishless ponds in the Arctic and Sub-Arctic that not only fish and *Lepidurus* but also zooplankton may depend on the benthic production either after resuspension into the water, or for some species by foraging at the sediment surface.

15.3 Climate as a stressor

Climate change can result in major changes in the physical, chemical, and biological characteristics of polar lakes (Doran *et al.* 2002). More maritime conditions and increased precipitation are expected for both the coastal Arctic and the Antarctic regions, which also will influence lakes and ponds. However, interactions among temperature, precipitation, cloudiness, and radiant energy regime are complex when ambient temperatures hover near freezing at higher latitudes and in more continental locations. Clearly a larger amount of snow and the subsequent later ice melt may lead to prolonged ice cover and lower water temperature, which in turn could result in a shorter growing season and lower nutrient input. The lower water temperature itself is hardly a limiting factor as most animals and plants in these ecosystems are adapted to a life at very low temperatures, but a shorter growing season means that less time is available to plants and poikilothermic plants and animals to complete their life cycles. Moreover, zooplankton growth will likely be lower due to the reduced abundance of food, even when the length of the growing season suffices for reproduction.

An increase in precipitation and melting of greater snow accumulations during summer will also result in higher runoff and thus higher nutrient loading of organic and inorganic matter, and possibly also of silt and clay in the lakes receiving glacial water. Higher nutrient loading may lead to increased phytoplankton production, and over a longer time period decomposition of accumulated organic material may lead to oxygen depletion under ice in winter and mortality of Arctic char, especially in shallow lakes where the water volume under ice is modest. However, input of silt in glacial lakes will result in lower light penetration and thus lower primary production. Changes in temperature and precipitation may also lead to the reduction or even loss of permanent ice covers on high-latitude lakes, with impacts that are difficult to extrapolate.

If precipitation also increases in summer, the associated runoff, and possible earlier ice melt, will lead to increased input of humic materials, stimulating growth of bacteria and other microzooplankton, which may support growth of higher trophic levels (Jeppesen *et al.* 2003a). These contributions to organic matter could increase the risk of oxygen depletion in winter when the lakes are covered by ice. Higher humic content, however, will result in improved protection of fish and microzooplankton against ultraviolet (UV) radiation because the attenuation of light in the water column, especially at short wavelengths, will increase due to increased concentrations of absorbing dissolved organic compounds. The result may be a decreased level of pigment production by crustaceans, which perhaps would also reduce predation risk. A reduction of UV radiation in the water column will probably increase hatching success of fish eggs and more fish fry will survive. This would be particularly important in shallow lakes, which because of their low depth are especially vulnerable to effects of UV radiation. If the ice-free season is shortened and there is increased cloudiness, there would be a corresponding reduction in the duration and magnitude of UV radiation.

The overall increase in air temperatures in some polar regions implies that average water temperatures are increasing (e.g. Quayle *et al.* 2002). The thermal conditions in both shallow and deeper lakes undergoing changes have far-reaching implications, such as development of thermal stratification, increased heat capacity of the water mass, thus providing longer ice-free periods, and water-column stabilization during the summer. Also, many biogeochemical and biological processes change with changes in the overall thermal regime. The species composition as well as their interactions may changes as restrictions in tolerance to temperature are undergoing changes.

Large regions of the Arctic tundra are covered by organic soils and peat. With increasing temperatures and permafrost thawing, the previously freeze-preserved organic matter is mobilized (Michaelson *et al.* 1998; Kawahigashi *et al.* 2004) and oxidized by microorganisms, partly under anoxic conditions into methane, or completely oxidized to carbon dioxide. Siberian thaw lakes have been shown to export large amounts of methane to the atmosphere via the degradation of organic carbon that has been stored in the permafrost since the Pleistocene (Walter *et al.* 2006). Warming of the Arctic is expected to enhance these fluxes, which may be a major positive feedback mechanism of global warming. The oxidation of organic matter takes place partly in lakes and ponds, and represents an increasing subsidy to the lake food web via bacterial utilization of allochthonous organic matter. Moreover, the massive production of methane in these lakes probably supports methanotrophic bacteria in the water, which can be prey for bacterivores. A similar food-web connection from methane to metazooplankton via methanotrophic bacteria has previously been described for lakes in the boreal region (Bastviken *et al.* 2003).

In Antarctic lakes, benthic microbial mats are a major source of primary production but are affected by any change in the many physical factors affecting incident radiation, such as snow cover, ice thickness, attenuation in water column, etc. Moorhead *et al.* (2005) evaluated the response of simulated mat production (see below) to changes in incident radiation by varying transmission through the overlying ice as a surrogate for climate change (Doran *et al.* 2002). Current transmission through the permanent ice on Lake Hoare, in the McMurdo Dry Valleys, is less than 5% of incident intensity, but increasing transmittance above 5–7% had little impact because photosynthesis is saturated at this level. A sensitivity analysis of model behavior suggested that uncertainties in mat respiration rates were more important than rates of photosynthesis, implying more sensitivity to changing temperatures than light regimes. However, earlier analyses of Sombre Lake, on maritime Antarctic Signy Island, identified the persistence of ice and snow cover on seasonal ice cover as being the most important determinants of annual mat production, largely due to the high albedo of snow (Moorhead *et al.* 1997). For these reasons, predicting impacts of climate change on Antarctic lakes is very uncertain.

15.4 Case studies

15.4.1 Energy flow in Ace Lake and Lake Fryxell, Antarctica

A simple budget model (K.C. McKenna *et al.*, unpublished results) was used to estimate carbon flow through the pelagic food webs of both Ace Lake and Lake Fryxell (Figure 15.5a) based on field observations collected at several depths from both lakes at approximately 2-week intervals, including

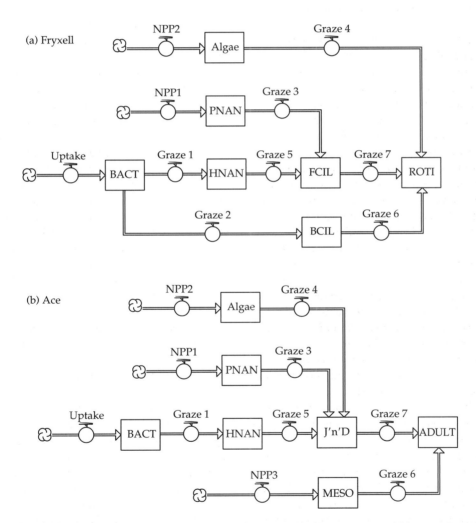

Figure 15.5 Diagrams of carbon flow among trophic groups in the pelagic food webs of (a) Lake Fryxell and (b) Ace Lake. Included are bacteria (BACT), photosynthetic nanoplankton (PNAN), larger-celled algae (Algae), *Mesodinium rubrum* (MESO), heterotrophic nanoplankton (HNAN), bacterial-feeding ciliates (BCIL), flagellate-feeding ciliates (FCIL), rotifers (ROTI), juvenile copepods and dinoflagellates (J'n'D), and adult copepods (ADULT). NPP, net primary production.

two summer seasons (Bell and Laybourn-Parry 1999; Roberts and Laybourn-Parry 1999; Laybourn-Parry et al. 2002, 2005; Madan et al. 2005). Details of the modeling approach are described in McKenna et al. (2006), but in brief, they used an hourly time step to compute daily flows of carbon between trophic groups (Figure 15.5b) assuming steady-state conditions in which losses from compartments were met by inputs:

$$I_i = P_i + R_i + E_i \qquad (15.1)$$

where I is input (uptake), P is predatory loss, R is respiration, and E is egestion for each group. Results showed that relationships between the top predators and other measures of community structure and function differed between the two lakes. For example, the heterotrophic community accounted for only 6% of the total biomass in the food web of Lake Fryxell, with rotifers representing only 14% of this amount. No significant relationship existed between rotifer biomass and any other group. In contrast, heterotrophs accounted for 56% of the total biomass in the pelagic food web in Ace Lake, with adult copepod biomass representing about 20% of this amount and significantly related to biomass of photosynthetic nanoflagellates.

Relationships between carbon uptake rates by the top predator and other trophic groups also were stronger in Ace Lake perhaps because estimates of carbon flux (eqn 15.1) were strongly related to biomass, and the biomass of the top predator in Ace Lake represented a much larger fraction of the total community than in Lake Fryxell. Carbon uptake by rotifers in Lake Fryxell was significantly related to uptake rates for only three other groups: flagellate-feeding ciliates, other heterotrophic ciliates, and photosynthetic nanoflagellates (Figure 15.5a). In contrast, uptake rates for adult copepods in Ace Lake were related to uptake rates of all other groups except heterotrophic ciliates and photosynthetic bacteria (Figure 15.5b). The relative carbon demand of the top predator accounted for less than 2% of total community carbon flow in Lake Fryxell and less than 5% of total community carbon flow in Ace Lake. However, this demand was seldom related to the carbon flow or biomass of any other group in Lake Fryxell. In contrast, demand was significantly related to biomass of both heterotrophic

nanoflagellates and bacteria, and productivity of *M. rubrum*, in Ace Lake. Although these results revealed few indications of top-down control by top predators, significant relationships between the top predators and other measures of community structure and function were more common and stronger in the more productive Ace Lake, which is located almost 10° south of Lake Fryxell. Perhaps this indicates a change in trophic relationships accompanying a gradient in primary production similar to those reported in Sub-Antarctic and Arctic lakes (see above).

15.4.2 Reverse spatial subsidies model for Antarctic landscapes

One of the peculiar characteristics of the Antarctic dry valleys is the balance in exchange of organic material between terrestrial and aquatic habitats. Although many freshwater communities receive relatively large subsidies of allochthonous organic matter from adjacent terrestrial ecosystems, this balance is reversed in the dry valleys. Hopkins et al. (2006) and Moorhead (2007) present similar, conceptual models describing organic-matter production in these freshwater habitats, with subsequent dispersed to adjacent soils. These models are similar to one proposed by Polis et al. (1997) for the dispersal of marine-derived organic materials along desert coastlines, and may be a characteristic of deserts both hot and cold.

Elberling et al. (2006) found that the rate of organic-matter accumulation along the shores of Lake Garwood (latitude 78°01'S; Garwood Valley, McMurdo Dry Valleys) was 30 times greater than the rate of carbon dioxide efflux from these materials, suggesting that turnover largely resulted from physical displacement rather than decay. Moreover, this displaced material could supply the entire pool of organic matter found in the surface 1-cm soil layer of the valley within 1.4–14 years. In comparison, Moorhead et al. (2003) and Moorhead (2007) found organic-matter accumulations along the margins of several ponds and streams in the McMurdo Dry Valleys, partly resulting from wave action and partly exposed by declining water levels. Indeed, soil organic-matter concentrations decline with distance from edges of lakes (Powers *et al.* 1998; Fritsen *et al.* 2000; Elberling *et al.* 2006), ponds (Moorhead *et al.* 2003), and streams (Treonis *et al.* 1999) in the dry valleys.

Both Hopkins *et al.* (2006) and Moorhead *et al.* (2003) suggested that wind movements influenced distributions of organic matter in dry valley soils, and Moorhead (2007) captured more organic matter dispersing from a pond in pitfall traps located in the primary directions of wind movement. An average $0.34 \, \mathrm{g\,C\,m^{-2}\,day^{-1}}$ of organic matter accumulated in pitfall traps, which is about an order of magnitude larger than rates of primary production for benthic microbial mats (Moorhead *et al.* 1997, 2005; Hawes *et al.* 2001). This difference between rates of production and dispersal suggests that primary production within littoral zones of lakes and ponds may concentrate across space and/or time along the margins of water bodies before dispersing to the surrounding landscape (Moorhead *et al.* 2003; Elberling *et al.* 2006; Moorhead 2007).

These spatial subsidies have broad implications to patterns of biodiversity in the dry valleys because soil organic carbon is among the most important determinants of biodiversity in Antarctic soils (Powers *et al.* 1998; Wall and Virginia 1999; Moorhead *et al.* 2003). Anderson and Wait (2001) proposed a subsidized island biogeography hypothesis, in which organic-matter inputs from the surrounding matrix supported a greater diversity per unit productivity on islands than could be explained on the basis of *in situ* primary production. The Antarctic dry valleys simply reverse the usual pattern of a large surrounding aquatic ecosystem subsidizing smaller terrestrial islands, by supplying the surrounding matrix of arid soils with organic subsidies provided by small 'islands of water'. The aquatic subsidy to the terrestrial environment in these Antarctic watersheds is an interesting illustration of lakes being hot spots of carbon cycling in the landscape (Cole *et al.* 2007).

15.4.3 Nine years' monitoring of two small lakes in northeast Greenland

The chemical and biological characteristics of two lakes in northeast Greenland (Zackenberg, 71°N) have been monitored since 1997 (Christoffersen *et al.* 2008). The lakes, Sommerfuglesø and Langemandssø, are shallow (maximum 6 m) and of similar size ($\approx 0.01 \, \mathrm{km^2}$). In both lakes, the 1–2-m-thick ice cover starts melting near-shore in mid-June, but they are usually only ice-free from mid- or late July until the second half of September. One lake (Langemandssø) houses a fish population of dwarf Arctic char that matures sexually at a size of only 11–13 cm. Their density is very low (Jeppesen *et al.* 2001).

The phytoplankton biomass is dominated (often 95% of total biomass in terms of biovolume) by chrysophytes in both lakes. The most important taxa in Sommerfuglesø are chrysophytes, *Stichogloea* spp., and *Dinobryon bavaricum*, while chrysophytes, *Dinobryon bavaricum*, *Dinobryon boreale*, *Kephyrion boreale*, and *Ochromonas* spp. dominate in Langemandssø. In addition, several dinoflagellates (*Gymnodinium* spp. and *Peridinium umbonatum* group) as well as nanoflagellates (Sommerfuglesø) and green algae (Langemandssø) occur. The zooplankton community in Sommerfuglesø consists of the cladoceran *Daphnia pulex*, the copepod *Cyclops abyssorum alpinus*, and the rotifers *Polyarthra dolicopthera* and *Keratella quadrata*. The fact that *D. pulex* dominates the community probably explains the low abundances of copepods and rotifers. *C. abyssorum alpinus* and *P. dolicopthera* dominate the community in the fish-containing Langemandssø.

During the 9-year monitoring period, the date of 50% ice cover in Sommerfuglesø and Langemandssø has varied by almost a month. The earliest was mid-June (2005) and the latest was mid-July (1999). This timing clearly coincides with the average spring snow cover of the entire valley as a correlation exists between snow cover in early summer (as recorded by June 10) and the timing of the ice-melt and thus water temperatures (averaged for the water column and for the summer) (Christoffersen *et al.* 2008). This relationship also applies to the other water bodies in the area. The tight coupling between snow cover, ice-melt, and water temperature also affects the appearance of the planktonic communities. Increased water temperatures correspond to increased chlorophyll *a* and zooplankton abundance.

The concentration of phytoplankton (expressed as chlorophyll) varies from year to year and between the two lakes (Figure 15.6) in accordance with annual changes in the concentrations of nutrients (nitrogen) and water temperature. The low average water temperature and low nutrient concentrations recorded in 1999 led to lower phytoplankton abundance and dominance of dinophytes compared to the warmer season of, for example, 2001 with chrysophyte dominance. In 1999, ice-melt did not occur until late July, and the average water temperature was therefore low. Chrysophytes and dinophytes together contributed 93% of total phytoplankton abundance in Sommerfuglesø, while dinoflagellates constituted 89% of the phytoplankton in the deeper and colder Langemandssø. The average phytoplankton biomass reached its highest level in 2003. From multivariate analyses it seemed that Nostocales (cyanobacteria) and diatoms decreased in importance while dinophytes and chrysophytes increased with rising levels of chlorophyll *a*, temperature, and conductivity in the lakes.

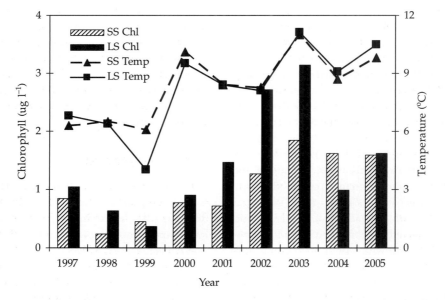

Figure 15.6 The chlorophyll concentrations in two High Arctic lakes in northeast Greenland (SS, Sommerfuglesø; LS, Langemendssø) during 9 successive years. Data from Christoffersen *et al.* (2008). Chl, chlorophyll; Temp, temperature.

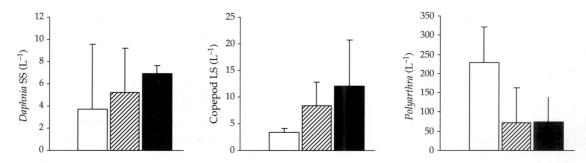

Figure 15.7 Zooplankton community in two lakes in northeast Greenland, with and without fish. The abundances of *Daphnia pulex*, *Cyclops abyssorum alpinus*, and the rotifer *Polyarthra* sp. in cold (white bars; 6–8°C), intermediate (hatched bars; 8–10°C), and warm (black bars; 10–12°C) summer in fishless Lake Sommerfuglesø (SS) and in Langemendssø (LS) hosting a population of landlocked Arctic char. Data from Christoffersen *et al.* (2008).

The abundances of crustaceans increased in years with warm water temperatures, probably explained by improved growth conditions when both food and water temperatures are more favourable. There were significant inter-annual differences between the relative importance of species, copepods being dominant in colder years and daphnids in warmer years (Figure 15.7). Apparently, rotifers react negatively to high temperatures, perhaps because daphnids and copepods increase in importance and become stronger competitors for food in warm years.

The predation pressure on zooplankton in Langemandssø was high despite the low abundances of fish and it follows that phytoplankton are almost completely regulated by nutrients, while grazing is low. However, observed differences are also related to the physical conditions: especially the ice-out date and subsequent rise in water temperature. It is clear that warmer summers favor the reproduction potential of zooplankton and especially *Daphnia* and *Cyclops*.

The combined results from 10 years of survey suggest rather strong differences between the relatively warm (i.e. early ice-out) and cold years with regard to nutrient concentration (especially nitrogen) as well as phytoplankton and zooplankton populations. Warmer years generally stimulated phytoplankton and crustacean production and changed species composition. Higher nutrient concentrations in warm years are probably caused by increased loading of nutrients and humic materials from the catchment as the active layer melts. However, no relationship was found between any of the physicochemical or biological variables and the winter North Atlantic Oscillation index, indicating that local climate variations are more important driving factors than is the North Atlantic Oscillation in this short time series.

15.5 Conclusions

In summary, food webs in high-latitude lakes typically harbor limited species diversity, but still range from very simple with a lack of most metazoans, to communities including a range of invertebrates and vertebrates (fish and birds).

Many of them are subsidized from the surrounding catchment areas via downstream flow of organic matter that provides an additional energy source, in the same manner as lower latitude lakes. However, High Arctic lakes and Antarctic lakes include examples where barren watersheds provide insignificant amounts of organic matter, and in the Antarctic dry valleys there are even cases where lakes subsidize the surrounding terrestrial habitat with organic matter. The polar regions are expected to experience considerable warming in the near future, which will also affect the food webs of lakes. Among these effects we foresee increased allochthonous organic matter concentrations in the water, resulting in an increased terrestrial energy subsidy, but also enhanced risk of winter fish kills due to higher oxygen demand, and decreased penetration of UV in the water column, which decreases UV stress on zooplankton and other organisms.

Acknowledgments

The work was supported by the Global Climate Change Programme (no. 9700195), the Commission for Scientific Research in Greenland, The North Atlantic Research Programme 1998–2000, the Arctic Programme, 1998–2002, the Nordic Council of Ministers, the Research in the North by the National Research Council, and the EU project Eurolimpacs. We thank M. Rautio and J. Laybourn-Parry for helpful comments on the manuscript.

References

Adamson, D.A. and Pickard, J. (1986). Cainozoic history of the Vestfold Hills. In Pickard, J. (ed.), *Antarctic Oasis*, pp. 63–97. Academic Press, Sydney.

Anderson, W.B. and Wait, D.A. (2001). Subsidized island biogeography hypothesis: another new twist on an old theory. *Ecology Letters* **4**, 289–291.

Bastviken, D., Ejlertsson, J., Sundh, I., and Tranvik, L.J. (2003). Methane oxidation in lakes: Implications for methane emission and energy transfer to pelagic food webs. *Ecology* **84**, 969–981.

Battaglia, B., Valencia, K., and Walton, D.W.H. (1997). *Antarctic Communities: Species, Structure and Survival.* Cambridge University Press, Cambridge.

Bell, E.M. and Laybourn-Parry, J. (1999). Annual plankton dynamics in an Antarctic saline lake. *Freshwater Biology* **41**, 507–519.

Bonilla, S., Villeneuve, V., and Vincent, W.F. (2005). Benthic and planktonic algal communities in a high arctic lake: pigment structure and contrasting responses to nutrient enrichment. *Journal of Phycology* **41**, 1120–1130.

Butler, H., Atkinson, A., and Gordon, M. (2005). Omnivory and predation impact of the calanoid copepod *Boeckella poppei* in a maritime Antarctic lake. *Polar Biology* **28**, 815–821.

Christoffersen, K. (2001). Predation on *Daphnia pulex* by *Lepidurus arcticus*. *Hydrobiologia* **442**, 223–229.

Christoffersen, K.S., Amsinck, S.L., Landkildehus, F., Lauridsen, T.L., and Jeppesen, E. (2008). Lake flora and fauna in relation to ice-melt, water temperature and chemistry. In Meltofte, H., Christensen T.R, Elberling, B., Forchhammer, M. C., and Rasch, M. (eds), *High-Arctic Ecosystem Dynamic in a Changing Climate. Ten years of monitoring and research at Zackenberg Research Station, Northeast Greenland*, pp. 371–390. Advances in Ecological Research, vol. 40, Academic Press, San Diego, CA.

Cole, J.J. *et al.* (2007). Plumbing the global carbon cycle: integrating inland waters into the terrestrial carbon budget. *Ecosystems* **10**, 171–184.

Dahl-Hansen, G.A.P. (1995). Long-term changes in crustacean zooplankton – the effect of a mass removal of Arctic charr, *Salvelinus alpinus* (L.), from an oligotrophic lake. *Journal of Plankton Research* **17**, 1819–1833.

Dartnall, H.J.G. (2005). Freshwater invertebrates of subantarctic South Georgia. *Journal of Natural History* **39**, 3321–3342.

Dartnall, H.J.G. and Heywood, R.B. (1980). The freshwater fauna of South Georgia. *British Antarctic Survey Bulletin* **50**, 115–118.

Doran, P.T. *et al.* (2002). Recent climate cooling and ecosystem response in the McMurdo Dry Valleys, Antarctica. *Nature* **415**, 517–520.

Elberling, B. *et al.* (2006). Distribution and dynamics of soil organic matter in an Antarctic dry valley. *Soil, Biology and Biochemistry* **38**, 3095–3106.

Ferris, J.M., Burton, H.J., Johnstone, G.W., and Bayley, I.A.E. (eds) (1988). *Biology of the Vestfold Hills, Antarctica*. Kluwer Academic Publishers, Dordrecht.

Flanagan, K.M., McCauley, E., Wrona, F., and Prowse, T. (2003). Climate change: the potential for latitudinal effects on algal biomass in aquatic ecosystems. *Canadian Journal of Fisheries and Aquatic Sciences* **60**, 635–639.

Fogg, G.E. (1998). *The Biology of Polar Habitats*. Oxford University Press, Oxford.

Fritsen, C.H., Grue, A.M., and Priscu, J.C. (2000). Distribution of organic carbon and nitrogen in surface soils in the McMurdo Dry Valleys, Antarctica. *Polar Biology* **23**, 121–128.

Hahn, S. and Reinhardt, K. (2006). Habitat preference and reproductive traits in the Antarctic midge *Parochlus steinenii* (Diptera: Chironomidae). *Antarctic Science* **18**, 175–181.

Hansson, L.-A., Lindell, M., and Tranvik, L. (1993). Biomass distribution among trophic levels in lakes lacking vertebrate predators. *Oikos* **66**, 101–106.

Hansson, L.-A. and Tranvik, L. (1996). Quantification of invertebrate predation and herbivory in food webs of low complexity. *Oecologia* **108**, 542–551.

Hansson, L.-A. and Tranvik, L. (1997). Algal species composition and phosphorus recycling at contrasting grazing pressure: an experimental study in sub-Antarctic lakes with two trophic levels. *Freshwater Biology* **37**, 45–53.

Hansson, L.-A. and Tranvik, L. (2003). Food webs in sub-Arctic lakes: a stable isotope approach. quantification of invertebrate predation and herbivory in food webs of low complexity. *Polar Biology* **26**, 783–788.

Hansson, L.-A., Dartnall, H.J.G., Ellis-Evans, J.C., MacAlister, H., and Tranvik, L.J. (1996). Temporal and spatial fluctuations in physical, chemical and biological variables in the sub-Antarctic lakes of South Georgia. *Ecography* **19**, 393–403.

Hawes, I. (1985). Light climate and phytoplankton photosynthesis in maritime Antarctic lakes. *Hydrobiologia* **123**, 69–79.

Hawes, I. and Howard-Williams, C. (1998). Primary production processes in streams of the McMurdo Dry Valleys, Antarctica. In Priscu, J.C. (ed.), *Ecosysten Dynamics in a Polar Desert, the McMurdo Dry Valleys, Antarctic*, pp. 129–140. Antarctica Research Series vol. 72, American Geophysical Union, Washington D.C.

Hawes, I. and Schwarz, A.-M. (2000). Absorption and utilization of irradiance by cyanobacterial mats in two ice-covered Antarctic lakes with contrasting light climates. *Journal of Phycology* **37**, 5–15.

Hecky, R.E. and Hesslein, R.H. (1995). Contribution of benthic algae to lake food webs as revealed by stable isotopes. *Journal of North American Benthological Society* **14**, 631–653.

Hersey, A.E. *et al.* (1999). A geomorphic-trophic model for landscape control of arctic lake food webs. *Bioscience* **49**, 887–897.

Heywood, R.B., Dartnall, H.J.G., and Priddle, J. (1980). Characteristics and classification of the lakes of Signy Island, South Orkney Islands, Antarctica. *Freshwater Biology* **10**, 47–59.

Hobbie, J.E., Bahr, M., and Rublee, P.A. (1999). Controls on microbial food webs in oligotrophic arctic lakes. *Archiv für Hydrobiologie Special Issues in Advanced Limnology* **54**, 61–76.

Hopkins, D.W. *et al.* (2006). Controls on the distribution of productivity and organic resources in Antarctic Dry Valley soils. *Proceedings of the Royal Society London Series B Biological Sciences* **273**, 2687–2695.

Howard-Williams, C., Schwarz, A.-M., and Hawes, I. (1998). Optical properties of the McMurdo Dry Valley Lakes, In Priscu, J.C. (ed.), *Ecosysten Dynamics in a Polar Desert, the McMurdo Dry Valleys, Antarctica*, pp. 189–203. Antarctic Research Series vol. 72, American Geophysical Union, Washington D.C.

Izaguirre, I., Mataloni, G., Allende, L., and Vinocur, A. (2001). Summer fluctuations of microbial planktonic communities in a eutrophic lake – Cierva Point, Antarctica. *Journal of Plankton Research* **23**, 1095–1109.

Izaguirre, I., Allende, L., and Marinone, M.C. (2003). Comparative study of the planktonic communities of three lakes of contrasting trophic status at Hope Bay (Antarctic Peninsula). *Journal of Plankton Research* **25**, 1079–1097.

James, M.R., Hall, J.A., and Laybourn-Parry, J. (1998). Protozooplankton of the Dry Valley Lakes of Southern Victoria Land. In Priscu, J.C. (ed.), *Ecosystem Dynamics in a Polar Desert, The McMurdo Dry Valleys, Antarctica*, pp. 225–268. Antarctic Research Series vol. 72. American Geophysical Union, Washington DC.

Jansson, M., Perssson, L., De Roos, A.M., Jones, R.I., and Tranvik, L.J. (2007). Terrestrial carbon and intraspecific size-variation shape lake ecosystems. *Trends in Ecology and Evolution* **22**, 316–322.

Jeppesen, E., Christoffersen, K., Landkildehus, F., Lauridsen, T.L., and Amsinck, S. (2001). Fish and crustaceans in northeast Greenland lakes with special emphasis on interactions between Arctic charr (*Salvelinus alpinus*), *Lepidurus arcticus* and benthic chydorids. *Hydrobiologia* **44**, 329–337.

Jeppesen, E. *et al.* (2002a). Food web interactions in five Faroese lakes using stable isotopes. *Annales Societatis Scientiarum Færoensis Supplementum* **36**, 14–125.

Jeppesen, E., Christoffersen, K., Malmquist, H., Faafeng, B., and Hansson, L.-A. (2002b). Ecology of five Faroese Lakes: summary and synthesis. *Annales Societatis Scientiarum Færoensis Supplementum* **36**, 126–139.

Jeppesen, E. *et al.* (2003a). The impact of nutrient state and lake depth on top-down control in the pelagic zone of lakes: a study of 466 lakes from the temperate zone to the Arctic. *Ecosystems* **6**, 313–325.

Jeppesen, E. *et al.* (2003b). Sub-fossils in the surface sediment as proxies for community structure and dynamics of zooplankton in lakes: a study of 150 lakes

from Denmark, New Zealand, the Faroe Islands and Greenland. *Hydrobiologia* **491**, 321–330.

Jones, J.I. and Jeppesen, E. (2007): Body size and trophic cascades in lakes. In Hildrew, A.G., Rafaelli, D.G., and Edmonds-Brown, R. (eds), *Body Size and the Structure and Function of Aquatic Ecosystsems*, pp. 118–139. Cambridge University Press, Cambridge.

Kawahigashi, M., Kaiser, K., Kalbitz, K., Rodionov, A., and Guggenberger, G. (2004). Dissolved organic matter in small streams along a gradient from discontinuous to continuous permafrost. *Global Change Biology* **10**, 1576–1586.

Kling, G.W. and Fry, B. (1992). Stable isotopes and planktonic trophic structure in Arctic lakes. *Ecology* **73**, 561–566.

Lauridsen, T.L., Jeppesen, E., Landkildehus, F., Christoffersen, K., and Søndergaard, M. (2001). Horizontal distribution of cladocerans in arctic Greenland lakes. *Hydrobiologia* **442**, 107–116.

Laybourn-Parry, J. (2002). Survival mechanisms in Antarctic lakes. *Philosophical Transactions of the Royal Society of London Series B Biological Sciences* **357**, 863–869.

Laybourn-Parry, J., Quayle, W., and Henshaw, T. (2002). The biology and evolution of Antarctic saline lakes in relation to salinity and trophy. *Polar Biology* **25**, 542–552.

Laybourn-Parry, J., Marshall, W., and Marchant, H. (2005). Flagellate nutritional versatility as a key to survival in two contrasting Antarctic saline lakes. *Freshwater Biology* **50**, 830–838.

Liboriussen, L.T. and Jeppesen, E. (2003). Temporal dynamics in epipelic, pelagic and epiphytic algal production in a clear and a turbid shallow lake. *Freshwater Biology* **48**, 418–431.

Lyons, W.B., Tyler, S.W., Wharton, Jr, R.A., McKnight, D.M., and Vaughn, B.H. (1998). A late Holocene desiccation of Lake Hoare and Lake Fryxell, McMurdo Dry Valleys, Antarctica. *Antarctic Science* **10**, 247–265.

Madan, N.J., Marshall, W.A., and Laybourn-Parry, J. (2005). Virus and microbial loop dynamics over an annual cycle in three contrasting Antarctic lakes. *Freshwater Biology* **50**, 1291–1300.

McKenna, K.C., Moorhead, D.L., Roberts, E.C., and Laybourn-Parry, J. (2006). Simulated patterns of carbon flow in the pelagic food web of Lake Fryxell, Antarctica: little evidence of top-down control. *Ecological Modelling* **192**, 457–472.

Michaelson, G.J., Ping, C.L., Kling, G.W., and Hobbie, J.E. (1998). The character and bioactivity of dissolved organic matter at thaw and in the spring runoff waters of the arctic tundra north slope, Alaska. *Journal of Geophysical Research – Atmospheres* **D22**, 28939–28946.

Moorhead, D.L. (2007). Mesoscale dynamics of ephemeral wetlands in the Antarctic Dry Valleys: Implications to production and distribution of organic matter. *Ecosystems* **10**, 86–94.

Moorhead, D.L., Wolf, C.F., and Wharton, Jr, R.A. (1997). Impact of light regimes on productivity patterns of benthic microbial mats in an Antarctic lake: a modeling study. *Limnology and Oceanography* **42**, 1561–1569.

Moorhead, D.L. *et al.* (1999). Ecological legacies: impacts on ecosystems of the McMurdo Dry Valleys. *BioScience* **49**, 1009–1019.

Moorhead DL., Barrett, J.E., Virginia, R.A., Wall, D.H., and Porazinska, D. (2003). Organic matter and soil biota of upland wetlands in Taylor Valley, Antarctica. *Polar Biology* **26**, 567–576.

Moorhead, D., Schmeling, J., and Hawes, I. (2005). Contributions of benthic microbial mats to net primary production in Lake Hoare, Antarctica. *Antarctic Science* **17**, 33–45.

Nikolai, V. and Droste, M. (1984). The ecology of *Lancetes claussi* (Müller) (*Coleoptera, Dytiscidae*), the subantarctic water beetle of South Georgia. *Polar Biology* **3**, 39–44.

O'Brien, W.J. (1987). Planktivory by freshwater fish: thrust and parry in the pelagic. In Kerfoot, W.C. and Sih, A. (eds), *Predation: Direct and Indirect Impacts on Aquatic Communities*, pp. 3–16. The University Press of New England, Lebanon, NH.

O'Brien, W.J. *et al.* (1992). Control mechanisms of arctic lake ecosystems: a limnocorral experiment. *Hydrobiologia* **240**, 143–188.

Pociecha, A. (2007). Effect of temperature on the respiration of an Antarctic freshwater anostracan, *Branchinecta gaini* Daday 1910, in field experiments. *Polar Biology* **30**, 731–734.

Polis, G.A., Anderson, W.B., and Holt, R.D. (1997). Toward an integration of landscape and food web ecology: The dynamics of spatially subsidized food webs. *Annual Review of Ecology and Systematics* **28**, 289–316.

Powers, L.E., Ho, M., Freckman, D.W., and Virginia, R.A. (1998). Distribution, community structure and microhabitats of soil biota along an elevational gradient in Taylor Valley, Antarctica. *Arctic and Alpine Research* **30**, 133–141.

Prowse, T.D. *et al.* (2006). Climate change effects on hydroecology of Arctic freshwater ecosystems. *Ambio* **35**, 347–358.

Pugh, P.J.A., Dartnall, H.J.G., and McInnes, S.J. (2002). The non-marine Crustacea of Antarctica and the islands of the southern ocean: biodiversity and biogeography. *Journal of Natural History* **36**, 1047–1103.

Quayle, W.C., Peck, L.S., Peat, H., Ellis-Evans, J.C., and Harrigan, P.R. (2002). Extreme responses to climate change in Antarctic lakes. *Science* **295**, 645.

Rautio, M. and Vincent, W.F. (2006). Benthic and pelagic food resources for zooplankton in shallow high-latitude lakes and ponds. *Freshwater Biology* **51**, 1038–1052.

Rautio, M. and Vincent, W.F. (2007). Isotopic analysis of the sources of organic carbon of zooplankton in shallow subarctic and arctic waters. *Ecography* **30**, 77–87.

Roberts, E.C. and Laybourn-Parry, J. (1999). Mixotrophic cryptophytes and their predators in the Dry Valley lakes of Antarctica. *Freshwater Biology* **41**, 737–746.

Roberts, E.C., Priscu, J.C., Wolf, C., Lyons, W.B., and Laybourn-Parry, J. (2004). The distribution of microplankton in the McMurdo Dry Valley lakes, Antarctica: response to ecosystem legacy or present-day climatic controls? *Polar Biology* **27**, 238–249.

Rosswall, T. and Heal, O. (1975). Structure and function of tundra ecosystems. *Ecological Bulletin (Stockholm)* **20**, 425–448.

Sægrov, H., Hobæk, A., and Lábe-Lund, H.H. (1996). Vulnerability of melanic *Daphnia* to brown trout predation. *Journal of Plankton Research* **18**, 2113–2128.

Sierszen, M.E., McDonald, M.E., and Jensen, D.A. (2003). Benthos as the basis for arctic lake food webs. *Aquatic Ecology* **37**, 437–445.

Sobek, S., Cole, J.J., and Tranvik, L.J. (2005). Temperature independence of carbon dioxide suprsaturation in global lakes. *Global Biogeochemical Cycles* **19**, GB2003, doi:10.1029/2004GB002264.

Toro, M. *et al.* (2007). Limnological characteristics of the freshwater ecosystems of Byers Peninsula, Livingston Island, in maritime Antarctica. *Polar Biology* **30**, 635–649.

Tranvik, L.J. and Hansson, L.-A. (1997). Predator regulation of aquatic microbial abundance in simple food webs of sub-Antarctic lakes. *Oikos* **79**, 347–356.

Treonis, A.M., Wall, D.H., and Virginia, R.A. (1999). Invertebrate biodiversity in Antarctic Dry Valley soils and sediments. *Ecosystems* **2**, 482–492.

Vadeboncoeur, Y., Jeppesen, E., Van der Zanden, M.J., Schierup, H.-H., Christoffersen, K., and Lodge, D. (2003). From Greenland to green lakes: Cultural eutrophication and the loss of benthic pathways. *Limnology and Oceanography* **48**, 1408–1418.

Vadeboncoeur, Y., Kalff, J., Christoffersen, K., and Jeppesen, E. (2006). Substratum as a driver of variation in periphyton chlorophyll in lakes. *The Journal of the North American Benthological Society* **25**, 379–392.

Vadeboncoeur, Y., Van der Zanden, M.J., and Lodge, D.M. (2002). Putting the lake back together: reintegrating benthic pathways into lake food web models. *Bioscience* **52**, 44–54.

Van der Zanden, M.J, and Rasmussen, J.B. (1999). Primary consumer $\delta^{13}C$ and $\delta^{15}N$ and the trophic position of aquatic consumers. *Ecology* **80**, 1395–1403.

Van der Zanden, M.J. and Rasmussen, J.B. (2001). Variation in $\delta^{15}N$ and $\delta^{13}C$ trophic fractionation: implications for aquatic food web studies. *Limnology and Oceanography* **46**, 2061–2066.

Vincent, W.F., Downes, M.T., Castenholz, R.W., and Howard-Williams, C. (1993). Community structure and pigment organisation of cyanobacteria-dominated microbial mats in Antarctica. *European Journal of Phycology* **28**, 213–221.

Wall, D.H. and Virginia, R.A. (1999). Controls on soil biodiversity: insights from extreme environments. *Applied Soil Ecology* **13**, 137–150.

Walter, K.M., Zimov, S.A., Chanton, J.P., Verbyla, D., and Chapin, III, F.S. (2006). Methane bubbling from Siberian thaw lakes as a positive feedback to climate warming. *Nature* **443**, 71–75.

Wharton, Jr, R.A., Parker, B.C., and Simmons, Jr, G.M. (1983). Distribution, species composition and morphology of algal mats in Antarctic dry valley lakes. *Phycologia* **22**, 355–365.

Wrona, F.J., Prowse, T.D., and Reist, J.D. (eds) (2006). Climate change impacts on Arctic freshwater ecosystems and fisheries. *Ambio* **35**, 325–415.

CHAPTER 16

Direct human impacts on high-latitude lakes and rivers

Martin J. Riddle and Derek C.G. Muir

Outline

In this chapter we review and compare possible causes of environmental impacts to lake and river systems in the Arctic and Antarctica, with emphasis on the direct effects of local human activities. Potential impacts fall into three broad categories: physical processes; chemical contamination; and the introduction of exotic species, including microbial contaminants, and in this chapter we focus on physical and chemical impacts. By way of examples, we illustrate similarities and differences between the Arctic and Antarctic and identify characteristics of high-latitude lake and river systems that may influence their susceptibility to anthropogenic disturbance, including potential cumulative or synergistic effects of local disturbance during a time of rapid global change.

16.1 Introduction

The polar regions are among the most remote and the least populated parts of the planet; however, they are not immune from the impacts of human activity occurring both within their own regions and elsewhere on the planet. Until relatively recently human activities in Antarctica have been limited to occasional visits; as a consequence, the potential for direct environmental impacts has also been limited. In the last few decades growing interest in the Antarctic for both science and tourism has significantly increased the numbers of visitors and the infrastructure required to support them.

Thousands of years of traditional activities of indigenous people in the Arctic have given way to increasing industrial and military activities over the past century. The abundant resources of the Arctic and their relative proximity to populated areas of Europe and North America have resulted in industrial activity such as mining in the Arctic, beginning in the 1920s in Russia, even earlier in

Svalbard and northern Scandinavia, and by the mid-twentieth century in North America. Military activity during World War II and the Cold War was another major factor, bringing airports and military bases to locations that previously would have been technically impractical or uneconomic. Thus for different reasons human activity has increased quickly in both polar regions. Irrespective of the driver for growth, increased activity brings with it growing risks of environmental disturbance and impacts.

Just as distance has been no barrier to human activity at high latitudes, neither can it prevent the by-products of the modern world reaching these regions. Despite the great distances that separate the polar regions from the major population and industrial centres, global circulation processes link the Arctic and Antarctic with the rest of the world and deliver to them many of the atmospheric changes and pollutants that are the consequences of modern life.

Table 16.1 Processes that have contaminated or could contaminate high-latitude lake and river systems and their effects.

	Impacting process	Effect on water bodies
Local human activity		
Physical processes	Catchment disturbance leading to erosion and sedimentation	Increased turbidity and reduced light penetration
	Turbulence and mixing created by divers or instrument deployment	Destruction of water-column stratification
	Use for cooling waters	Change in the thermal balance
	Damming, diversion of rivers, and removal of lakes	Change in hydrodynamics, loss of fish habitat
Chemical contamination	Fuel spill	Smothering (high concentrations)
		Direct toxicity (moderate concentrations)
		Provision of microbial substrate (low concentrations)
	Industrial waste disposal sites, mining effluent, oil and gas drilling wastes, nuclear reprocessing effluent	Direct toxicity from range of metals and metalloids, radionuclides, pharmaceuticals and personal care products, and hydrocarbons
	Municipal sewage effluent disposal	C, N, P enrichment, eutrophication
		High biological oxygen demand, shift towards anaerobic conditions
	Airborne emissions from smelters, energy generation and other combustion sources	Acidification of regional lakes from SO_2 and NO_x deposition, contamination from metal deposition, e.g. mercury
	Radiation releases from industry, nuclear weapons testing, from nuclear processing and storage sites, from underground non-military uses (mining, river-diversion projects)	Exposure of humans and wildlife in arctic and Sub-Arctic Russia; localized contamination of freshwaters near sources in Russia
	Petroleum hydrocarbon discharges to land and fresh water from pipeline breaks and other accidents	Direct toxicity to algae, macrophytes, and wildlife, fouling of river habitat
Biological invasions	Sewage effluent disposal	Competition, predation, and grazing on native communities
	Translocation on equipment	Competition, predation, and grazing on native communities
	Habitat modification and climate change	Arrival of pathogens and competing species
Global processes		
Physical processes	Climate change (local warming or cooling)	Changes to ice cover and subsequent changes to light penetration and wind-driven mixing
		Changes to precipitation regimes and consequent changes to hydrodynamics
		Changes to wind regimes and consequent changes to wind-driven mixing and aeolian sediment flux
	Ozone depletion	Increased exposure to ultraviolet radiation
Chemical contamination	Global transport of volatile organic chemicals	Main concern is exposure of wildlife and humans to persistent and bioaccumulative chemicals that biomagnify in the food chain; direct toxicity is generally not an issue
	Arctic haze and acid rain	pH change
	Atmospheric transport and deposition of radionuclides	Emissions from the Chernobyl nuclear accident; emissions deposited in Scandinavia and Russian surface waters
Biological introductions	Climate warming and amelioration	Competition, predation, and grazing on native communities

Three broad categories of environmental impacts may be defined (Table 16.1):

- physical perturbations, such as the addition of inert sediment particles, turbulence or mixing, modification of incident light, addition of heat;
- chemical contamination caused by the introduction of pollutants; for example, via oil spills or waste disposal, as well as by atmospheric deposition; for example, acidifying substances, radionuclides, industrial chemicals, and pesticides;
- the addition of biological elements, such as the invasion of non-native species or the introduction of genetic material.

Although these categories are convenient, we recognize that in many situations impacts are complex and may be caused by combinations of physical, chemical, and biological processes. For example, release of sewage into a lake may change thermal stratification and reduce light penetration by increasing suspended sediment loads (physical), while increasing naturally occurring nutrients to levels well above normal (chemical), and introducing non-native microorganisms (biological). Beyond documenting the temporal and spatial extent of contamination, the biological or ecological impacts of most chemical and biological pollutants have generally not been investigated in the polar freshwater environments.

16.2 Physical impacts

Lake and river systems can be affected by physical processes occurring both in their catchments and in the water bodies themselves. Catchment processes include ground disturbance leading to compaction, channeling, erosion and increased sedimentation, and changes to topography, perhaps through the establishment of buildings or the creation of snow banks during road maintenance leading to downwind changes in snow accumulation and subsequent changes to water budgets. Water-column stratification is a common feature of many high-latitude lakes and, for those lakes in which it occurs, it is perhaps the defining feature controlling the characteristics of all other ecosystem processes. Stratification is vulnerable to physical disturbance, such as mixing caused by removal

of water in large quantities for operational reasons, or by rising bubbles generated by the use of open-circuit breathing apparatus during diving operations. However, the one study to have examined the effects of diving in a closed-basin, ice-covered lake was unable to detect any changes in a range of physicochemical properties or in microbial distributions attributable to diving (Kepner et al. 2000).

High-latitude soils are particularly prone to structural damage from even slight physical disturbance because they are typically low in organic material, lack cohesion, and are either loose or only weakly consolidated (Campbell and Claridge 2004). Natural deposits in the High Antarctic are predominantly coarse sand and gravel with little or no silt or clay (Burgess et al. 1992) and once damaged the freeze–thaw process may take many years to recreate thermokarst soil structure. The presence of a permafrost layer below the surface provides some protection against erosion but also provides a surface on which free water can flow, leading to channelization (Burgess et al. 1992). If roads or tracks are created in steeply sloping catchments the lack of cohesion of surface material and the presence of permafrost prevents the construction of side cuts and encourages routes perpendicular to the contours, which further enhances erosion (Burgess et al. 1992).

In the Larsemann Hills (East Antarctica) increased turbidity of some lakes has been attributed to damage to soil structure caused by vehicles, and even by foot traffic, causing changing drainage patterns and increasing sediment runoff in catchments (Ellis-Evans et al. 1997; Kaup and Burgess 2003). Increased turbidity in a previously transparent oligotrophic lake (No Worries Lake) caused a reversal of the inverse temperature stratification seen in the other natural lakes in the area into direct stratification with higher water temperatures in the upper layers (Kaup and Burgess 2003). Before the Zhongshan Station was established in its catchment in 1989, the No Worries Lake in the Larsemann Hills was a slightly brackish, clear, shallow lake very similar to others in the area (Ellis-Evans et al. 1997) with a gravel catchment and virtually no terrestrial vegetation. By 1993 the lake was surrounded by buildings and vehicles, and had planktonic microbial communities markedly different from other lakes in the Larsemann Hills. These changes were attributed to human activities in the catchment, altering meltwater and sediment-input patterns and to recycling of lake water to facilitate cooling of the station generators resulting in bottom-water warming, which appeared to promote heterotrophic microbial activity even under winter ice cover (Ellis-Evans *et al.* 1997).

Most of the large Arctic rivers begin their flow south of the Arctic, including the major rivers of Siberia (Ob, Yenisey, and Lena) and the Mackenzie River in Canada while numerous smaller rivers have headwaters within the Arctic. Flows are largely influenced by precipitation and spring time ice/snow melt because permafrost under-lying most of Arctic river drainages limits water-storage capacity (Woo *et al.* 1992). Although most north-flowing rivers have not been dammed or modified, major exceptions are the La Grande hydroelectric project on the eastern side of James Bay in Québec, the Churchill River diversion and hydroelectric plants on the Nelson River in Manitoba on the western side of Hudson Bay, and the Kransnoyarskoye Reservoir on the Yenisey River in Siberia. The Mackenzie River drainage in western Canada has seen extensive development in its southern reaches (which include northern Alberta and northeastern British Columbia due to the Athabasca River and Peace River drainages). The Athabasca River flows through the Athabasca oil sands, large deposits of oil-rich bitumen located in northeastern Alberta, which are the subject of massive surface mining and *in situ* steam injection of deeper bitumen deposits. However, waste water emissions to the river from bitumen upgrader facilities is minimized by large reservoirs and there is little evidence yet of downstream contamination (Evans *et al.* 2003).

Physical disturbances to lakes and smaller rivers in the Arctic are too numerous to list all locations where they occur. Causes include road building, oil and natural gas pipeline construction, and surface and subsurface mining, oil and gas drilling. Associated with these activities have occasionally been drainage of small lakes and construction of bridges, small dams, and weirs. Oil-exploration wastes are usually discharged to small reservoirs (sumps). The efficiency of these sumps varies widely

depending on climatic conditions and permafrost stability (Robertson *et al.* 1998).

Most of the road building and physical disturbances related to coal- and metal-mining activity, military activities, oil and gas exploration, and hydroelectric development have been in regions linked geographically to more southerly Sub-Arctic and temperate regions (e.g. Kola Peninsula, the Pechora River basin, and southern Taimyr regions in Russia, Mackenzie Delta (oil and gas) and Lac La Gras (diamonds) in Canada, and Prudhoe Bay (oil and gas) in Alaska). This kind of development is more limited in the High Arctic islands. Nevertheless there has been mining activity (mainly for coal) in Svalbard since the early twentieth century by Russian, North American, and Norwegian companies. Base-metal mines have also have operated or continue to operate in Canada (northern Baffin Island, Little Cornwallis Island) and in west Greenland.

16.3 Chemical impacts

Chemical contamination of freshwater systems in Antarctica is limited to a few locations very close to station infrastructure (a few hundred meters). These water bodies are effectively part of station facilities, being used for water supply, cooling, or receipt of brine from desalination plants. In general, water bodies more than 1000 m from the station infra-structure are at uncontaminated background levels. Commonly, if pollution is present in Antarctica, several contaminants are involved; for this reason we discuss Antarctic contamination principally by location rather than by type. The story of con-tamination in Arctic lakes and rivers is much more complex (more extensive and more severe) than in the Antarctic and to simplify the Arctic story we discuss it by categories of pollutants.

16.3.1 Chemical contamination in Antarctica

Studies of chemical pollution in Antarctic fresh-water systems provide a mixed story of contamina-tion. Some studies have indicated that trace metals in water bodies close to Antarctic stations are at naturally occurring levels and are not related to human activities, for example, Jubany Station, King

George Island (Vodopivez *et al.* 2001). Other studies indicate some contamination of water bodies from human activities, for example, in the Larsemann Hills polyaromatic hydrocarbons (PAHs) were elevated in some water samples from tarns with station facilities in their catchments (Goldsworthy *et al.* 2003) and trace metal concentrations were slightly higher in lakes close to station facilities (Gasparon and Burgess 2000). Biological activity and freeze–thaw cycles were found to be important for regulation of trace metal concentrations in the Larsemann Hills, rather than the amount and supply of clastic sediment (Gasparon and Matschullat 2006).

Oil-derived hydrocarbons are the most common fuel in most remote areas and there is a risk of oil spill wherever they are used. At Australia's Casey Station a visible sheen appears each summer on the surface of a shallow tarn (area ≈1000 m²) down slope from a storage facility that leaked about 5000 L of Special Antarctic Blend diesel fuel in 1999. In the mid-1990s at Davis Station, also operated by Australia, oil leaked from fuel drums into the tarn used for station water supply in concentrations high enough to damage filters in the reverse-osmosis water-treatment plant. The same tarn has been subject to gradually rising salinity because of returned saline waste water from the reverse osmosis plant. On the Fildes Peninsula, King George Island, concentrations of heavy metals and organic matter in water close to stations were reported to be higher than in reference locations (Zhao *et al.* 1998).

Early recognition of a potential problem and intervention (Hayward *et al.* 1994) appears to have averted significant contamination of Lake Bonney (Parker *et al.* 1978) and Lake Vanda (Webster *et al.* 2003), both in the McMurdo Dry Valleys, where rising lake levels threatened to engulf sites previously occupied by field stations. The accommodation hut at Lake Bonney was removed and the chemistry laboratory relocated to higher ground in October 1977 after 15 years of occupation. A subsequent assessment of the environmental impacts at the Lake Bonney site indicated that there had been both accidental spills and leaks, and wilful violations of environmental constraints, such as disposal of spent lubricants, human waste, and kitchen wastes to soils and streams (Parker 1978). New Zealand's

Lake Vanda station was removed in 1994, leaving residual chemical contamination including discrete fuel spills, locally elevated heavy metals in soils, and elevated phosphate concentrations in supra-permafrost fluids in a gully formerly used for domestic washing water disposal (Webster *et al.* 2003). Overall, it was concluded that although some contaminants, particularly zinc, could be selectively leached from flooded soil during rising lake levels, the small area of contaminated soils exposed and low level of contamination was not likely to adversely affect shallow-lake water quality.

16.3.2 Acidifying substances and nutrients

Arctic haze (Kerr 1979) and acid rain (Koerner and Fisher 1982) are both caused primarily by industrial and urban emissions of sulphur dioxide and the nitrogen oxides (NO$_x$) forming sulphuric acid and nitric acid in the atmosphere which reach the ground as precipitation (Iversen and Joranger 1985). Acid rain is a major environmental concern in the European Arctic but as yet has not been recorded in Antarctica (Legrand and Mayewski 1997). The difference between the two polar regions is explained by the concentration of industry in the north and Antarctica's physical separation from the industrial world by the Southern Ocean.

Skjelkvåle *et al.* (2006) and Kämäri *et al.* (1998) have reviewed the spatial extent, concentrations and sensitivity of Arctic lakes to acidifying pollutants (mainly sulphate and nitrate). In the 1960–1970s direct damage to forests and fish was reported in the industrial areas of Nikel, Zapolyarnyy, and Monchegorsk in the Kola Penninsula and lake acidification related to acid deposition from mid-latitude European sources (e.g. the UK and Germany) became evident throughout Scandinavia at about the same time. Long-term monitoring indicates that lakes in the Kola Peninsula region are recovering from acidification and that this is due to reduced sulphur deposition originating both from central Europe and from smelters in the Kola region. Fortunately, much of the Arctic—for example Canadian and Alaska tundra, the Pechora basin and the Ob River basin in Siberia—has relatively high buffering capacity due to high base saturation in soils. As a consequence there have been fewer impacts on lakes near Norilsk in the southern Taimyr peninsula region, site of the largest base metal smelter, and the largest source of sulphur dioxide in the Arctic (Blais *et al.* 1999).

The water chemistry of Arctic lakes is not well documented except in Finland, Norway, and Sweden where national lake surveys have been conducted since 1986, with historical data going back to the mid-1960s. A significant proportion of the lakes in the Kola Pennisula in Russia, as well as in Iceland and Svalbard, have also been surveyed. There are relatively few published data on water chemistry of lakes in Siberia. Duff *et al.* (1999) surveyed 81 Arctic lakes from the Taymyr Peninsula, the Lena River, and the Pechora River basin. Lake pH values were neutral to slightly alkaline in the Taymyr Peninsula and the Lena River areas (pH 6.9–9.2) but some lakes in the Pechora basin were acidified (lowest pH = 4.2).

In North America, a large number of lakes and ponds have been surveyed (Skjelkvåle *et al.* 2006) but the ability to make regional generalizations from available data is much more limited due to the vast area and hundreds of thousands of lakes and ponds. For example, in Canada's Northwest Territories 70 lakes spanning the treeline were categorized as slightly acidic to slightly alkaline (pH range = 5.5–8.6) and nutrient-poor (total phosphorus ranged from <0.002 to 0.022 mg L^{-1}), and a strong trend toward high ion concentrations in densely forested areas was observed relative to tundra regions (Rühland and Smol 1998) More recently the same research group has published studies for small lakes on Bathurst Island (Lim *et al.* 2001), Victoria Island and Alex Heiberg Island (Michelutti *et al.* 2002a, 2002b), and Ellef Ringnes Island (Antoniades *et al.* 2003).

There are only a few reports of high nutrient loading to lakes in the North American Arctic. A well-documented example is the eutrophication of Meretta Lake, a small (30-ha) lake at Resolute Bay (Nunavut), which received runoff from sewage-treatment ponds from a nearby airport facility for nearly 50 years. Total phosphorus concentrations in the lake exceeded 70 µg L^{-1} in the early 1970s (Schindler *et al.* 1974) but by the 1990s had dropped to 5–15 µg L^{-1}, reflecting reduced inputs (Douglas and Smol 2000). The eutrophication of the lake

resulted in marked shifts in diatom species assemblages coincident with sewage inputs beginning in the late 1940s (Michelutti *et al.* 2002c).

Some high-latitude lakes and ponds are naturally eutrophic as a result of marine animals; for example, lakes and ponds in penguin rookeries in Antarctica (Vincent and Vincent 1982) and near seabird colonies in the Arctic (Evenset *et al.* 2004; Blais *et al.* 2005). In the latter studies, the extent of seabird influence, using stable nitrogen isotope ratios ($\delta^{15}N$) as a proxy, was strongly correlated with polychlorinated biphenyl (PCB) concentrations in sediments.

16.3.3 Heavy metals

Ford *et al.* (2005) and Dietz *et al.* (1998) have summarized the information available on heavy metals in Arctic lake and river sediments with an emphasis on the toxic metals mercury, cadmium, and lead. As with lake waters, the most detailed surveys of Arctic lake sediment enrichment factors (the ratio of concentrations or fluxes in recent sediment horizons compared to pre-industrial horizons) for heavy metals are from Scandinavia; however, a circumpolar picture can be developed due to relatively good coverage throughout the Arctic. Contamination of Arctic lakes by heavy metals is closely associated with deposition of sulphate and nitrogen oxides. Very elevated concentrations of copper and nickel were reported in lakes and rivers in the Murmansk region of the Kola Peninsula with values of both metals generally exceeding 10 µg L^{-1} within a 30-km radius of the mining and smelting complexes in the region (Dietz *et al.* 1998; Skjelkvåle *et al.* 2001). The contamination of two large lakes in the region, Lake Kuetsjärvi in the Pasvik river near the border with Norway, and Lake Imandra in the Niva river basin has been well documented (Lukin *et al.* 2003; Moiseenko *et al.* 2006). In Lake Imandra, nickel and copper concentrations in water have declined during the 1990s due to reduced emissions but remain elevated and as of 2003 still exceeded background concentrations in other Kola and Scandinavian lakes by five to 10 times (Moiseenko *et al.* 2006). The most recent Scandinavian survey of metals in surface waters (1995) concluded that bedrock and surficial geology were important factors for controlling the concentrations of nickel and copper while long-range transported air pollution was the major important factor for distribution of lead, cadmium, zinc, and cobalt (Skjelkvåle *et al.* 2001). It also found that heavy metal pollution in Scandinavian Arctic lakes was a minor ecological problem on a regional scale except for the Kola region lakes (Skjelkvåle *et al.* 2001). By comparison to Scandinavia/Kola, where about 3000 lakes were included in the 1995 survey (Skjelkvåle *et al.* 2001), relatively little information is available on metals in lakes in North America, Greenland, and the central and eastern Russian Arctic (Skjelkvåle *et al.* 2006).

Concentrations of heavy metals in major Arctic rivers tend to be high by comparison with lake waters. Concentrations of copper in the Ob, Yenisey, and Lena rivers, the major Russian rivers draining into the Arctic Ocean, averaged 4–8 $\mu g\,L^{-1}$, slightly higher than in the Mackenzie River and other smaller rivers in Canada (3.1 and 1 $\mu g\,L^{-1}$, respectively), and much higher than smaller rivers in northern Norway (0.6 $\mu g\,L^{-1}$) (Alexeeva *et al.* 2001). Dissolved and suspended concentrations of metals vary seasonally and are highest in the spring freshet (Hölemann *et al.* 2005). Kimstach and Dutchak (2004) found that mercury, lead, and cadmium fluxes near the outflows of the Yenisey and Pechora rivers were greatly influenced by the high flows in the spring flood period. They concluded that local contamination originating from the Norilsk industrial area explained most of the observed fluxes in the Yenisey.

Nickel, copper, and mercury concentrations in lake sediments in the Kola Peninsula area of northwestern Russia exceed background concentrations by 10–380 times (Lukin *et al.* 2003). However, metal deposition in lake sediments declines rapidly with distance from these sources, reflecting the association of the metals with the coarse aerosol fraction which declines exponentially from the smelters (Shevchenko *et al.* 2003).

The extent of deposition of anthropogenic lead (i.e. from smelting and use as a gasoline additive) has been studied in remote lake sediments from Greenland and Canada using the isotope ratio $^{206}Pb:^{207}Pb$. Outridge *et al.* (2002) found major latitudinal differences in inputs of recent anthropogenic and natural lead in Canada with only limited anthropogenic lead in High Arctic sediments. In southwestern Greenland, lead concentrations in recent sediments were about 2.5-fold higher than background and, based on lead isotope ratios indicated the source of lead was mainly from western Europe (Bindler *et al.* 2001a). In contrast, a lake sediment core from the eastern Taymir Peninsula in central Siberia showed no recent lead enrichment (Allen-Gil *et al.* 2003), suggesting that emissions from the Norilsk smelting complex 300 km to the southwest was not a significant regional source of lead.

Mercury is the most studied heavy metal in Arctic lake sediments. Indigenous peoples in the Arctic have elevated blood and hair mercury particularly in Greenland and Nunavut. Although the source for humans is thought to be mainly due to a marine diet, freshwater fish could be a contributor in some regions (Hansen *et al.* 2003). Results from dated sediment cores are useful for examining current and historical inputs to freshwater environments because they permit the history of deposition to be inferred and fluxes to be estimated (Muir and Rose 2004). In general the enrichment factors for mercury in dated sediment cores from the North American Arctic are lower than those of southwest Greenland and in the European Arctic (Landers *et al.* 1998). Within Canada, mercury enrichment factors are weakly negatively correlated with latitude; that is, they are higher in Sub-Arctic Québec than in lakes in the Arctic archipelago or Alaska (Muir *et al.* 2007). There is a lot of variation between lakes which may be due to differences in catchment to lake area ratios, sedimentation rates, and extent of erosional inputs from the watershed (Fitzgerald *et al.* 2005). Mercury deposition has not declined in the 1990s in the Canadian Arctic, Alaska, or Greenland, unlike what has been observed in Sweden (Bindler *et al.* 2001b) and in some lakes near sources in mid-latitude North America (Lockhart *et al.* 1998; Engstrom and Swain 1997) as well as in mid-latitude glaciers (Schuster *et al.* 2002).

There has been speculation that annual mercury-depletion events (MDEs) may enhance inputs of Hg^{2+} to lake surfaces and catchments. MDEs occur both in the Arctic and Antarctic at polar sunrise as a result of interaction of sunlight and bromine oxide (Schroeder *et al.* 1998). However, St. Louis *et al.* (2005) concluded that while MDEs temporarily elevated concentrations of Hg(II) in the upper layers of snowpacks much of this deposited Hg(II) was rapidly photoreduced to Hg(0), which then diffused back into the atmosphere. Thus the overall significance of MDEs in terms of mercury trends in Arctic lakes is still a matter of scientific debate.

Another factor that might contribute to the observed lack of decline of mercury in Arctic sediments is the warming trend which has become apparent especially in the North American Arctic (Hassol 2004; Smol *et al.* 2005). Longer open-water seasons and higher temperatures may lead to increased phytoplankton productivity thereby increasing the rate of mercury scavenging via sedimenting particles from the water column into sediments (as illustrated in Plate 13). Outridge *et al.* (2005, 2007) have proposed that a significant fraction of recent mercury deposition in High Arctic lakes can be attributed to this mechanism.

16.3.4 Radionuclides

The Antarctic Treaty (1959) prohibits military activities or the testing of nuclear weapons south of 60°S. Since then the use of radionuclides in the region have been limited to one nuclear power plant operated at McMurdo Station in the Ross Sea during 1962–1972 and conservative tracers used for research. The Antarctic is subject to fallout of radioactive substances liberated to the atmosphere from nuclear testing and accidents elsewhere in the world and as a consequence anthropogenic radionuclides such as ^{90}Sr, ^{137}Cs, ^{238}Pu, and ^{239,240}Pu are present in low but measurable quantities in Antarctic lake systems (Hedgpeth 1972; Triulzi *et al.* 2001).

Contamination by long-lived, bioaccumulative, gamma-ray-emitting radionuclides ^{137}Cs and ^{90}Sr has been a major concern throughout the Arctic since the 1960s when the first measurements were made. Contamination by these and other fission products as well as products from nuclear reactors is a particularly important issue in the Russian Arctic due to past nuclear weapons testing, non-military nuclear explosions, and waste reprocessing and storage. Two nuclear waste reprocessing plants at the Krasnoyarsk Mining and Chemical Industrial Complex near Zheleznogorsk on Yenisey River and the Siberian Chemical Combine (SCC) on the Romashka River near Tomsk in Siberia are sources of radionuclides to the Yenisey and Ob rivers, respectively (Kryshev and Riazantsev 2000). Although most of the waste is injected into deep wells or buried on site, historically there have been substantial discharges of contaminated water from both complexes which has led to significant accumulation of radionuclides in bottom sediments and biota. In the case of SCC, most of the radionuclides discharged are removed from the water as it travels first to the Romashka River and Tom Rivers and finally along the Ob River system, and they are not detectable in the lower reaches of the Ob or in the Ob estuary (Strand *et al.* 2004). Global atmospheric fallout is the predominant source of the long-lived radionuclides ^{137}Cs and ^{239,240}Pu in bottom sediments of the Ob delta and estuary (Panteleyev *et al.* 1995).

Soils in the Russian Arctic have also been contaminated by above-ground nuclear explosions carried out at Novaya Zemlya test site and also by underground nuclear explosions for non-military purposes carried out by the former Soviet Union. A total of 130 tests were conducted on Novaya Zemlya, 88 in the atmosphere, three underwater, and 39 underground. Research has shown that these explosions led to considerable contamination of localized areas. The USA also conducted underground nuclear explosions on Amchitka Island between 1965 and 1971. However, radionuclides in Amchitka freshwater and terrestrial environments are essentially of fallout origin (Dasher *et al.* 2002).

The Chernobyl nuclear reactor accident in the Ukraine in 1986 also increased ^{137}Cs in soils and surface waters in Scandinavia immediately after the incident but this isotope and ^{90}Sr (which had ten times lower releases at Chernobyl) are now both less than $5\,Bq\,m^{-3}$ in two Finnish Arctic rivers (Saxen 2003). In the northern parts of Scandinavia and Finland and northwestern Russia ground deposition after the Chernobyl accident was approximately $1–2\,kBq\,m^{-2}$, which was similar to that due to global fallout from nuclear weapons. Lower radionuclide deposition was recorded east of the White Sea and in North America (Strand *et al.* 1998).

16.3.5 Petroleum hydrocarbons

Significant hydrocarbon exploration and development, to date mostly land-based, has taken place in the Arctic particularly in Alaska, Canada, and Russia. The main aquatic environmental issues are physical disturbances associated with the exploration and drilling. However, pipeline ruptures are also a concern. The largest Arctic crude oil spill on land, and one of the world's largest oil spills, occurred due to the leakage of a pipeline in the Usinsk region of the Komi Republic in Russia. Released oil flowed into the Kolva and Usa rivers in the Pechora River basin. West Siberia is the most highly developed and oldest oil and gas region, producing about 78% of all Russian oil. In general, all Russian Arctic rivers in western Siberia (lower Ob River and Pechora River basins) are impacted by petroleum hydrocarbon pollution due to oil and gas exploration and production (Robertson *et al.* 1998; Iwaco Consultants 2001). Kimstach and Dutchak (2004) also found four-ring PAHs (fluoranthene and pyrene) in the lower reaches of the Pechora River and concluded that it was affected by local sources of PAHs.

Oil spills have also occurred at Prudhoe Bay, Alaska, and in the Mackenzie River basin in Canada but on a much smaller scale and with limited environmental impact. There are anecdotal reports of contamination of the Mackenzie River from oil production at Norman Wells, an oil field development on an island in the river, with evidence from fish tainting (Lockhart and Danell 1992). Dissolved alkanes and PAHs in the Mackenzie River occur at sub-nanogram- to low nanogram-per-liter concentrations and are considered to be mainly of natural origin (Nagy *et al.* 1987; Yunker *et al.* 1994). High PAH values have also been found in suspended particulates from the Mackenzie River and are thought to be mainly of petrogenic origin (Yunker and Macdonald 1995).

16.3.6 Combustion-related hydrocarbons and particles

While contamination by petroleum hydrocarbons has been relatively well documented, particularly in northern Russia, there is far less information on combustion-related products such as PAHs and spherical carbonaceous particles (SCPs). Sediment cores from lakes in the Pechora Basin showed high SCPs (up to 5000 counts g^{-1} dry weight), indicative of industrial and power hydrocarbon combustion sources, but only near industrial centers (Solovieva *et al.* 2002). Rose (1995) produced a SCP profile for a sediment core taken from Stepanovichjärvi, a lake within 20 km of Nikel on the Kola Peninsula. This core had a peak SCP concentration of 12 000 counts g^{-1}, much higher than cores from the Pechora Basin or from remote lakes in Scandinavia or Svalbard (Muir and Rose 2004).

Fernández *et al.* (1999, 2000) found that PAHs in cores from Lake Arresjøen on northwestern Svalbard (79°40'N) were primarily pyrolytic in origin (i.e. resulting from combustion of coal and hydrocarbon fuel). Rose *et al.* (2004) found highest fluxes of PAHs on Svalbard in Lake Tenndammen, a lake within 20 km of the coal-mining towns of Barentsburg and Longyearbyen on Svalbard. Total PAH fluxes in southern Yukon lakes ranged from 9.1 to 174 µg m^{-2} year^{-1} whereas two remote lakes in the Mackenzie River basin had much higher PAH fluxes (68 and 140 µg m^{-2} year^{-1}) (Lockhart 1997; Lockhart *et al.* 1997). In general, higher fluxes of substituted PAHs have been found in the Mackenzie River valley and Yukon lakes in Canada compared with the European Arctic lakes (Muir and Rose 2004). This may be related to natural sources of PAHs rather than to combustion sources.

16.3.7 Persistent organic pollutants

Elevated concentrations of PCBs and hexachlorobenzene, as well as chlorinated pesticides such as DDT and chlordane, in indigenous peoples and in top predators in the Arctic were first recognized in the late 1980s and were the stimulus for the Stockholm Convention on Persistent Organic Pollutants (POPs) (UNEP 2001). The main concern with POPs is that they can undergo long-range atmospheric transport from urban and agricultural areas in the mid-latitudes and, following deposition, can enter aquatic and terrestrial food webs where they biomagnify (de Wit *et al.* 2004). POPs undergo global distillation and cold condensation, which are processes whereby volatile and persistent chemicals, such as POPs, evaporate in the warmer latitudes and condense out in cooler places including at high altitude and high latitudes (Goldberg 1975; Wania and Mackay 1993). Traces of these wholly anthropogenic chemicals have been found, however, in the environment and biota of both polar regions (Martin 2002; Priddle 2002). However, with a few exceptions, concentrations of POPs are very low in Arctic and Antarctic freshwater environments (sub-nanogram per liter in water, low nanogram per gram in fish), because of low water solubility, and most exposure of humans comes from consumption of marine mammals (Hansen *et al.* 2003). POPs such as PCBs, DDT, and chlorinated dioxin/furans are adsorbed onto particulate organic carbon and tend to settle in profundal sediments.

There are relatively few studies of POPs in Arctic river water. There are major challenges of measuring very low concentrations of hydrophobic contaminants in these large river systems and recent measurements have yielded far lower concentrations than previously reported (Carroll *et al.* 2008). Kimstach and Dutchak (2004) determined total DDT as well as other organochlorine pesticides and PCBs in pooled samples representative of cross-sectional sampling of the northern reaches of the Yenisey and Pechora rivers in 2001–2002. They estimated annual fluxes of 3760 kg year^{-1} for PCBs and 1200 kg year^{-1} for total DDT using long-term average flows for the Yenisey. However, more recent measurements near the mouth of the Yenisey (Carroll *et al.* 2008) estimated an annual flux of only 37 kg year^{-1} of PCBs and 8 kg year^{-1} for total DDT, reflecting much lower concentrations in water. These differences may be due to particle settling, which is an important removal process occurring near river mouths as the rivers meet the Arctic Ocean. Consistent with this is the observation by Sericano *et al.* (2001) of PCBs and DDT in surficial bottom sediments near the mouths of the Ob and Yenisey at low nanogram-per-gram-of-dry-weight concentrations.

Much less information is available on concentrations of POPs in Scandinavian and North American Arctic rivers. Measurements in the early 1990s suggested very low (sub-nanogram per liter) concentrations (de March *et al.* 1998). Hexachlorocyclohexanes

(HCHs) remain the predominant organochlorine contaminants in Arctic lake waters (de Wit *et al.* 2004). Law *et al.* (2001) surveyed 13 small temperate and nine Sub-Arctic and Arctic lakes in Canada and found that α-HCH concentrations in the Arctic and Sub-Arctic lakes were significantly greater (0.6–1.7 ng L^{-1}) than those of the small temperate lakes (0.09–0.43 ng L^{-1}). The ubiquitous presence of α-HCH, which is also the major organochlorine pesticide in Arctic sea water, probably reflects a global distillation effect; that is, colder water and air temperatures combined with greater degradation rates in temperate lakes (Law *et al.* 2001).

Although lake sediments are the environmental compartment of choice for examining spatial and temporal trends of POPs in Arctic freshwater environments, there are relatively few studies compared with heavy metals (Muir and Rose 2004). Circumpolar geographic coverage is very incomplete, with almost no results from Scandinavia, Russia, or Greenland. In the European and Russian Arctic the emphasis has been on analysis of surface sediments while in the North American Arctic, Alaska, and western Greenland full core profiles have been analysed to assess fluxes and temporal trends (de March *et al.* 1998; Muir and Rose 2004; de Wit *et al.* 2004).

Concentrations of POPs such as PCBs and DDT are typically very low (low nanogram per gram of dry weight) in most Arctic lake sediments and large samples and 'clean' techniques are needed for reliable results. In terms of geographical breadth, the largest survey was conducted in the mid-1990s by Skotvold and Savinov (2003) who determined total PCBs in surface sediments (0–2 cm depth) collected from remote lakes in northern Norway, the Svalbard Archipelago, Bjørnøya (Bear Island), and tundra lakes along the southeastern Barents Sea and the east Siberian coast of Russia. PCB concentrations were generally very low, 0.2–1.4 ng g dry weight^{-1}, except for Lake Ellasjøen on Bjørnøya which had much higher concentrations (35 ng g dry weight^{-1}). Subsequent studies showed that this lake was impacted by guano from seabirds using the lake as a resting area, creating an additional source of POPs (Evenset *et al.* 2004).

Other studies have used north–south transects of isolated lakes to infer latitudinal gradients for

PCBs, DDT, and brominated diphenyl ethers. Breivik *et al.* (2006) analyzed decabromodiphenyl ether (BDE-209) along an eastern North American transect and found sedimentation fluxes declined exponentially with latitude. The empirical half-distance for BDE-209 derived from surface flux data were approximately half that of the PCBs which was consistent with the transport of BDE-209 mainly on particles in the atmosphere.

Whereas most Arctic lake sediment studies have focused on long-range atmospheric transport and deposition, a small number have investigated local/regional transport as a source. Rose *et al.* (2004) examined surface and pre-industrial concentrations of PCBs in five cores from Svalbard and concluded that there were higher concentrations near Longyearbyen and Barentsburg, which are major coal-mining towns and the site of power-generation facilities. PCB use at military sites has been shown to have contaminated lakes and terrestrial environments in the vicinity (de Wit *et al.* 2004). Total PCB concentrations in sediment cores from lakes near a former radar facility at Saglek Bay on the northern Labrador coast in Canada were found to decline with increasing distance from the site (Betts-Piper 2001; Paterson *et al.* 2003). The PCB congener profile in both the near-site and remote lakes had a clear Aroclor 1260 profile similar to the product used at the radar facility. Similar contamination problems have been identified in Alaska (Hurwich and Chary 2000; United States Environmental Protection Agency 2002).

16.4 Conclusions

As this chapter demonstrates, polar regions and their freshwater systems are impacted by locally, regionally, and hemisphere-specific global-scale activities despite their separation from the world's population centers, with the extent of impacts broadly reflecting the different levels of human activity in the Arctic and Antarctic. High-latitude regions have the potential to be disproportionately affected by many of the global changes brought by the industrial world. Indeed, several large-scale impacts in fresh waters and terrestrial environments were first observed, or are most evident, in polar regions, such as changes in lake and pond productivity (Smol *et al.* 2005; Antoniades *et al.* 2007), long-range atmospheric transport of pollutants by global distillation and on particles, deposition of acidifying substances, and contamination by radionuclides.

The potential for global processes, such as climate change and increased ultraviolet radiation, to exacerbate the environmental impacts of local human activity in the Arctic is now well recognized and has been the subject of several extensive reviews (Macdonald *et al.* 2003; Schindler and Smol 2006; Wrona *et al.* 2006). However, there is still considerable uncertainty around any predictions of the consequences of these interactions at the ecosystem level and it is likely that some outcomes will be counter-intuitive (Schindler 1997).

Contaminant pathways, both to and within high-latitude regions, can be influenced by physical and biological processes that are sensitive to climate and also by climate-driven changes in human activity (see Plate 13 and Macdonald *et al.* 2003). Physical processes such as changes to winds, surface air temperature, precipitation rates or timing, evaporation, runoff, and ice cover all have the potential to influence uptake, transport, and deposition of contaminants. Similarly, climate-sensitive biological processes such as shifts in the relative importance of different sources of primary productivity, change in food-web structure affecting the degree of biomagnification, changes in the feeding ecology of top-level consumers, and changes to migration ranges all have the potential to influence biologically mediated transport, biomagnification, and biodegradation of chemical contaminants (Blais *et al.* 2007). Climate change may extend the spatial extent of some activities, both within the region or elsewhere, with implications for transport or use of persistent chemicals used in manufacturing, agriculture and silviculture as well as for greater oil/gas and mining development.

Acknowledgments

The authors thank ArcticNet, the Northern Contaminants Program, Environment Canada (Northern Ecosystem Initiative), and the Australian Antarctic Division for financial support which enabled them to conduct research in the polar regions.

References

Alexeeva, L.B. *et al.* (2001). Organochlorine pesticides and trace metal monitoring of Russian rivers flowing to the Arctic Ocean, 1990–1996. *Marine Pollution Bulletin* **43**, 71–85.

Allen-Gil, S.M. *et al.* (2003). Heavy metal contamination in the Taimyr Peninsula, Siberian Arctic. *Science of the Total Environment* **301**, 119–138.

Antoniades, D., Douglas, M.S.V., and Smol, J.P. (2003). The physical and chemical limnology of 24 ponds and one lake from Isachsen, Ellef Ringnes Island, Canadian High Arctic. *International Review of Hydrobiology* **88**, 519–538.

Antoniades, D. *et al.* (2007). Abrupt environmental change in Canada's northernmost lake inferred from diatom and fossil pigment stratigraphy. *Geophysical Research Letters* **34**, L18708, doi:10.1029/2007GL030947.

Betts-Piper, A.A. (2001). *Chrysophyte Stomatosyst-based Paleolimnological Investigations of Environmental Changes in Arctic and Alpine Environments*. MSc Thesis, Department of Biology, Queen's University, Kingston, ON.

Bindler, R., Olofsson, C., Renberg, I., and Frech, W. (2001a). Temporal trends in mercury accumulation in lake sediments in Sweden. *Water, Air, and Soil Pollution Focus* **1**, 343–355.

Bindler, R. *et al.* (2001b). Pb isotope ratios of lake sediment in West Greenland: inferences on pollution sources. *Atmospheric Environment* **35**, 4675–4685.

Blais, J.M., Duff, K., Laing, T.E., and Smol, J.P. (1999). Regional contamination in lakes from the Noril'sk region in Siberia, Russia. *Water, Air and Soil Pollution* **110**, 389–405.

Blais, J.M. *et al.* (2005). Arctic seabirds transport marine-derived contaminants. *Science* **309**, 445.

Blais, J.M. *et al.* (2007). Biologically mediated transport of contaminants to aquatic systems. *Environmental Science & Technology* **41**, 1075–1084

Breivik, K. *et al.* (2006). Empirical and modeling evidence of the long-range atmospheric transport of decabromodiphenyl ether. *Environmental Science & Technology* **40**, 4612–4618.

Burgess, J.S., Spate, A.P., and Norman, F.I. (1992). Environmental impacts of station development in the Larsemann Hills, Princess Elizabeth Land, Antarctica. *Journal of Environmental Management* **26**, 287–299.

Campbell, I.B. and Claridge, G.G.C. (2004). Soil properties and relationships in Cryosols of the region of the Transantarctic Mountains in Antarctica. In Kimble, J.M. (ed.), *Cryosols; Permafrost-affected Soils*, pp. 713–726, Springer-Verlag, Berlin.

Carroll, J.L. *et al.* (2008). PCBs, PBDEs and pesticides released to the Arctic Ocean by the Russian Rivers Ob and Yenisei. *Environmental Science & Technology* **42**, 69–74.

Dasher, D. *et al.* (2002). An assessment of the reported leakage of anthropogenic radionuclides from the underground nuclear test sites at Amchitka Island, Alaska, USA to the surface environment. *Journal of Environmental Radioactivity* **60**, 165–187.

de March, B.G.E. *et al.* (1998). Persistent organic pollutants. In Kämäri, J. and Joki-Heiskala, P. (eds), *AMAP Assessment Report, Arctic Pollution Issues*, pp. 183–372. Arctic Monitoring and Assessment Programme, Oslo.

de Wit, C.A. *et al.* (2004). *AMAP Assessment 2002, Persistent Organic Pollutants in the Arctic*. Arctic Monitoring and Assessment Programme, Oslo.

Dietz, R., Pacyna, J., and Thomas, D.J. (1998). Heavy metals. In Kämäri, J. and Joki-Heiskala, P. (eds), *AMAP Assessment Report, Arctic Pollution Issues*, pp. 373–524. Arctic Monitoring and Assessment Programme. Oslo.

Douglas, M.S.V. and Smol, J.P. (2000). Eutrophication and recovery in the High Arctic: Meretta lake (Cornwallis Island, Nunavut, Canada) revisited. *Hydrobiologia* **431**, 193–204.

Duff, K.E., Laing, T.E., Smol, J.P., and Lean, D.R.S. (1998). Limnological characteristics of lakes located across arctic treeline in northern Russia. *Hydrobiologia* **391**, 205–222.

Ellis-Evans, J.C., Laybourn-Parry, J., Bayliss, P.R., and Perriss, S.T. (1997). Human impact on an oligotrophic lake in the Larsemann Hills. In Battaglia, B., Valencia, J., and Walton, D.W.H. (eds), *Antarctic Communities, Species, Structure and Survival*, pp. 396–404. Cambridge University Press, Cambridge.

Engstrom, D.R. and Swain, E.B. (1997). Recent declines in atmospheric mercury deposition in the upper Midwest. *Environmental Science and Technology* **31**, 960–967.

Evans, M.S. *et al.* (2003). PAH sediment studies in Lake Athabasca and the Athabasca River ecosystem related to the Fort McMurray oil sands operations, sources and trends. In Brebbia, C.A. (ed.), *Oil and Hydrocarbon Spills, I.I.I., Modelling, Analysis and Control*, pp. 365–374. WIT Press, Southampton.

Evenset, A. *et al.* (2004). A comparison of organic contaminants in two high Arctic lake ecosystems, Bjørnøya (Bear Island), Norway. *Science of the Total Environment* **318**, 125–141.

Fernández, P., Vilanova, R.M., and Grimalt, J.O. (1999). Sediment fluxes of polycyclic aromatic hydrocarbons in European high altitude mountain lakes. *Environmental Science & Technology* **33**, 3716–3722.

Fernández, P., Vilanova, R.M., Martinez, C., Appleby, P., and Grimalt, J.O. (2000). The historical record of atmospheric pyrolytic pollution over Europe registered in the sedimentary PAH from remote mountain lakes. *Environmental Science & Technology* **34**, 1906–1913.

Fitzgerald, W.F. *et al.* (2005). Modern and historic atmospheric mercury fluxes in northern Alaska: global sources and Arctic depletion. *Environmental Science & Technology* **39**, 557–568.

Ford, J., Borg, H., Dam, M., and Riget, F. (2005). Spatial patterns. In Marcy, S. and Ford, J. (eds), *AMAP Assessment 2002, Heavy Metals in the Arctic*, pp. 42–83. Arctic Monitoring and Assessment Programme, Oslo.

Gasparon, M. and Burgess, J.S. (2000). Human impacts in Antarctica, trace-element geochemistry of freshwater lakes in the Larsemann Hills, East Antarctica. *Environmental Geology* **39**, 963–976.

Gasparon, M. and Matschullat, J. (2006). Geogenic sources and sinks of trace metals in the Larsemann Hills, East Antarctica, Natural processes and human impact. *Applied Geochemistry* **21**, 318–334.

Goldberg, E.D. (1975). Synthetic organohalides in the sea. *Proceedings of the Royal Society of London, Series, B.,* **189**, 277–289.

Goldsworthy, P.M., Canning, E.A., and Riddle, M.J. (2003). Soil and water contamination in the Larsemann Hills, East Antarctica. *Polar Record* **39**, 319–337.

Hansen, J.C. *et al.* (2003) *AMAP Assessment 2002, Human Health in the Arctic*. Arctic Monitoring and Assessment Programme, Oslo.

Hassol, S.J. (ed.) (2004). *Impacts of a Warming Arctic, Arctic Climate Impacts Assessment Secretariat*. Cambridge University Press, Fairbanks, AK.

Hayward, J., Macfarlane, M.J., Keys, J.R., and Campbell, I.B. (1994). *Decommissioning Vanda Station, Wright Valley, Antarctica*. Initial Environment Evaluation, March 1994. New Zealand Antarctic Programme, Wellington.

Hedgpeth, J.W. (1972). Radioactive contamination in the Antarctic. In Parker, B.C. (ed.), *Conservation Problems in Antarctica*, pp. 97–109. Allen Press, Lawrence, NJ.

Hölemann, J.A., Schirmacher, M., and Prange, A. (2005). Seasonal variability of trace metals in the Lena River and the southeastern Laptev Sea, Impact of the spring freshet. *Global and Planetary Change* **48**, 112–125.

Hurwich, E.M. and Chary, L.K. (2000). *Persistent Organic Pollutants in Alaska, What Does Science Tell Us?* Circumpolar Conservation Union, Washington DC.

Iversen, T. and Joranger, E. (1985) Arctic air pollution and large scale atmospheric flows. *Atmospheric Environment* **19**, 2099–2108.

Iwaco Consultants (2001). *West Siberia Oil Industry Environmental and Social Profile*. Final report. Iwaco Consultants, Rotterdam.

Kämäri, J. *et al.* (1998). Acidifying pollutants, Arctic haze, and acidification in the Arctic. In Kämäri, J. and Joki-Heiskala, P. (eds), *AMAP Assessment Report, Arctic Pollution Issues*, pp. 621–659. Arctic Monitoring and Assessment Programme, Oslo.

Kaup, E. and Burgess, J.S. (2003). Natural and human impacted stratification in the lakes of the Larsemann Hills, Antarctica. In Huiskes, A.H.L. *et al.* (eds), *Antarctic Biology in a Global Context*, pp. 313–318. Backhuys Publishers, Leiden.

Kepner, Jr, R. *et al.* (2000). Effects of research diving on a stratified Antarctic lake. *Water Research* **34**, 71–84.

Kerr, R.A. (1979). Global pollution: is the Arctic haze actually industrial smog? *Science* **205**, 290–293.

Kimstach, V. and Dutchak, S. (2004). *Persistant Toxic Substances (PTS) Sources and Pathways*. Persistent Toxic Substances, Food Security and Indigenous Peoples of the Russian North. Final report, pp. 33–80. Arctic Monitoring and Assessment Programme, Oslo.

Koerner, R.M. and Fisher, D. (1982). Acid snow in the Canadian high Arctic. *Nature* **295**, 137–140.

Kryshev, I.I. and Riazantsev, E.P. (2000). *Ecological Safety of Nuclear-Energy Complex in Russia*. IzdAT, Moscow.

Landers, D.H. *et al.* (1998). Using lake sediment mercury flux ratios to evaluate the regional and continental dimensions of mercury deposition in arctic and boreal ecosystems. *Atmospheric Environment* **32**, 919–928.

Law, S.A., Diamond, M.L., Helm, P.A., Jantunen, L.M., and Alaee M. (2001). Factors affecting the occurrence and enantiomeric degradation of hexachlorocyclohexane isomers in northern and temperate aquatic systems. *Environmental Toxicology and Chemistry* **20**, 2690–2698.

Legrand, M. and Mayewski (1997). Glaciochemistry of polar ice cores, a review. *Reviews of Geophysics* **35**, 219–244.

Lim, D.S.S., Douglas, M.S.V., Smol, J.P., and Lean, D.R.S. (2001). Physical and chemical limnological characteristics of 38 lakes and ponds on Bathurst Island, Nunavut, Canadian High Arctic. *International Review of Hydrobiology* **86**, 1–22.

Lockhart, W.L. (1997). Depositional trends - lake and marine sediments In Jensen, J. (ed.), *Synopsis of Research Conducted Under the 1995/96 and 1996/97 Northern Contaminants Program*, no. 74, pp. 71–87. Indian and Northern Affairs Canada, Ottawa.

Lockhart, W.L. and Danell, R.W. (1992). Field and experimental tainting of arctic freshwater fish by crude and refined petroleum products. *Proceedings of the 15th*

Arctic and Marine Oil Spill Program Technical Seminar, pp.763–771. Environment Canada: Ottawa.

Lockhart, W.L., Stern, G., and Muir, D.C.G. (1997). Food chain accumulation, biochemical effects and sediment contamination in Lake Laberge and other Yukon Lakes. In Jensen, J. (ed.), *Synopsis of Research Conducted Under the 1995–97 Northern Contaminants Program, Environmental Studies*, no. 74, pp. 191–206. Indian and Northern Affairs Canada, Ottawa.

Lockhart, W.L. *et al.* (1998). Fluxes of mercury to lake sediments in central and northern Canada inferred from dated sediment cores. *Biogeochemistry* **40**, 163–173.

Lukin, A. *et al.* (2003). Assessment of copper-nickel industry impact on a subarctic lake ecosystem. *The Science of the Total Environment* **306**, 73–83.

Macdonald, R.W., Harner, T., Fyfe, J., Loeng, H., and Weingartner, T. (2003). *AMAP Assessment 2002, The Influence of Global Change on Contaminant Pathways to, within, and from the Arctic*. Arctic Monitoring and Assessment Programme, Oslo.

Martin, H. (2002). *Regionally Based Assessment of Persistent Toxic Substances*. Arctic Regional Report. UNEP Chemicals, Geneva.

Michelutti, N., Douglas, M.S.V., Lean, D.R.S., and Smol, J.P. (2002a). Physical and chemical limnology of 34 ultra-oligotrophic lakes and ponds near Wynniatt Bay, Victoria Island, Arctic Canada. *Hydrobiologia* **482**, 1–13.

Michelutti, N., Douglas, M.S.V., Muir, D.C.G., Wang, X.W., and Smol, J.P. (2002b). Limnological characteristics of 38 lakes and ponds on Axel Heiberg Island, High Arctic Canada. *International Review of Hydrobiology* **87**, 385–399.

Michelutti, N., Douglas, M.S.V., and Smol, J.P. (2002c). Tracking recent recovery from eutrophication in a high arctic lake (Meretta Lake, Cornwallis Island, Nunavut, Canada) using fossil diatom assemblages. *Journal of Paleolimnology* 28, 377–381.

Moiseenko, T.I. *et al.* (2006). Ecosystem and human health assessment to define environmental management strategies, the case of long-term human impacts on an Arctic lake. *Science of the Total Environment* **369**, 1–20.

Muir, D.C.G. and Rose, N.L. (2004). Lake sediments as records of Arctic and Antarctic pollution by heavy metals, persistent organics, and anthropogenic particles. In Pienitz, R., Douglas, M.S.V., and Smol, J.P. (eds), *Long-term Environmental Change in Arctic and Antarctic Lakes*, pp. 209–239. Developments in Paleoenvironmental Research, vol. 8. Kluwer Academic Publishers, Dordrecht.

Muir, D.C.G. *et al.* (2007). *Spatial Trends and Historical Inputs of Mercury, Lead and Arsenic in Eastern and Northern*

Canada Inferred from Lake Sediment Cores. International Congress of Theoretical and Applied Limnology. Montreal [abstract].

Nagy, E. Carey, J.H., Hart, J.H., and Angley, E.D. (1987). *Hydrocarbons in the Mackenzie River*. NWRI Report Series. National Water Research Institute, Burlington.

Outridge, P.M., Hermanson, M.H., and Lockhart, W.L. (2002). Regional variations in atmospheric deposition and sources of anthropogenic lead in lake sediments across the Canadian Arctic. *Geochimica Cosmochimica Acta* **66**, 3521–3531.

Outridge, P.M. *et al.* (2005). Trace metal profiles in the varved sediment of an Arctic lake. *Geochimica Cosmochimica Acta* **69**, 4881–4894.

Outridge, P.M., Sanei, H., Stern, G.A., Hamilton, P.B., and Goodarzi, F. (2007). Evidence for control of mercury accumulation rates in Canadian High Arctic lake sediments by variations of aquatic primary productivity. *Environmental Science & Technology* **41**, 5259–5265.

Panteleyev, G.P., Livingston, H.D., Sayles, F.L., and Medkova, ON (1995). Deposition of plutonium isotopes and Cs-137 in sediments of the Ob delta from the beginning of the nuclear age. In Strand, P. and Cooke, A. (eds), *Proceedings of the Second International Conference on Environmental Radioactivity in the Arctic*, pp. 57–64. Norwegian Radiation Protection Authority, Østerås.

Parker, B.C. (1978) Potential impact on Lake Bonney of activities associated with modelling freshwater Antarctic ecosystems. In Parker, B.C. and Holliman, M.C. (eds), *Environmental Impact in Antarctica*, pp. 255–278, Virginia Polytechnic Institute and State University, Blacksburg.

Parker, B.C., Howard, R.V., and Allnutt, F.C.T. (1978). Summary of environmental assessment and impact assessment of the DVDP. In Parker, B.C. and Holliman, M.C. (eds), *Environmental Impact in Antarctica*, pp. 211–251. Virginia Polytechnic Institute and State University, Blacksburg.

Paterson, A.M., Betts-Piper, A.A., Smol, J.P., and Zeeb, B.A. (2003). Diatom and chrysophyte algal response to long-term PCB contamination from a point-source in northern Labrador, Canada. *Water Air and Soil Pollution* **145**, 377–393.

Priddle, J. (2002). *Regionally Based Assessment of Persistent Toxic Substances*. Antarctica Regional Report. UNEP Chemicals, Geneva.

Robertson, A. *et al.* (1998). Petroleum hydrocarbons. In Kämäri, J. and Joki-Heiskala, P. (eds), *AMAP Assessment Report, Arctic Pollution Issues*, pp. 661–716. Arctic Monitoring and Assessment Programme, Oslo.

Rose, N.L., Harlock, S., Appleby, P.G., and Battarbee, R.W. (1995). Dating of recent lake sediments in the United

Kingdom and Ireland using spheroidal carbonaceous particles (SCP) concentration profiles. *Holocene* **5**, 328–335.

Rose, N.L., Rose, C.L., Boyle, J.F., and Appleby, P.G. (2004). Lake-sediment evidence for local and remote sources of atmospherically deposited pollutants on Svalbard. *Journal of Paleolimnology* **31**, 499–513.

Rühland, K. and Smol, J.P. (1998). Limnological characteristics of 70 lakes spanning Arctic treeline from coronation gulf to Great Slave Lake in the Central Northwest Territories, Canada. *International Review of Hydrobiology* **83**, 183–203.

Saxen R. (2003). Long-term behaviour of 90Sr and 137Cs in surface waters in Finland. In Paile, V. (ed.), *Radiological Protection in the 2000s – Theory and Practice*, pp. 479–483. 25–29 August 2002. Nordic Society for Radiation Protection, Turku.

Schindler, D.W. (1997). Widespread effects of climatic warming on freshwater ecosystems in North America. *Hydrological Processes* **11**, 1043–1067.

Schindler, D.W. and Smol, J.P. (2006). Cumulative effects of climate warming and other human activities on freshwaters of Arctic and subarctic North America. *Ambio* **35**, 160–168.

Schindler, D.W. et al. (1974). Eutrophication in the high arctic – Meretta Lake, Cornwallis Island (75°N lat.). *Journal of the Fisheries Research Board of Canada* **31**, 647–662.

Schroeder, W.H. et al. (1998). Arctic springtime depletion of mercury. *Nature* **394**, 331–332.

Schuster, P.F. et al. (2002). Atmospheric mercury deposition during the last 270 years, a glacial ice core record of natural and anthropogenic sources. *Environmental Science & Technology* **36**, 2303–2310.

Sericano, J.L., Brooks, J.M., Champ, M.A., Kennicutt, II, M.C., and Makeyev, V.V. (2001). Trace contaminant concentrations in the Kara Sea and its adjacent rivers, Russia. *Marine Pollution Bulletin* **42**, 1017–1030.

Shevchenko, V., Lisitzin, A., Vinogradova, A., and Stein, R. (2003). Heavy metals in aerosols over the seas of the Russian Arctic. *The Science of the Total Environment* **306**, 11–25.

Skjelkvåle, B.L. et al. (2001). Heavy metal surveys in Nordic lakes, Concentrations, geographic patterns and relation to critical limits. *Ambio* **30**, 2–10.

Skjelkvåle, B.L. et al. (2006). Effects on freshwater ecosystems. In Forsius, M. (ed.), *AMAP Assessment 2006, Acidifying Pollutants, Arctic Haze, and Acidification in the Arctic*, pp. 64–90. Arctic Monitoring and Assessment Programme, Oslo.

Skotvold, T. and Savinov, V. (2003). Regional distribution of PCBs and presence of technical PCB mixtures in sediments from Norwegian and Russian Arctic Lakes. *The Science of the Total Environment* **306**, 85–97.

Smol, J.P. et al. (2005). Climate-driven regime shifts in the biological communities of arctic lakes. *Proceedings of the National Academy of Sciences USA* **102**, 4397–4402.

Solovieva, N., Jones, V.J., Appleby, P.G., and Kondratenok, B.M. (2002). Extent, environmental impact and long-term trends in atmospheric contamination in the Usa basin of East-European Russian Arctic. *Water, Air, and Soil Pollution* **139**, 237–260.

St Louis, V.L. et al. (2005). Some sources and sinks of monomethyl and inorganic mercury on Ellesmere island in the Canadian high arctic. *Environmental Science & Technology* **39**, 2686–2701.

Strand, P. et al. (1998). Radioactivity. In Kämäri, J. and Joki-Heiskala, P. (eds), *AMAP Assessment Report, Arctic Pollution Issues*, pp. 525–620. Arctic Monitoring and Assessment Programme, Oslo.

Strand, P. et al. (2004). Radioactive Contamination and Vulnerability of Arctic Ecosystems. In Strand, P. and Tsaturov, Y. (eds), *AMAP Assessment 2002, Radioactivity in the Arctic*, pp. 19–39. Arctic Monitoring and Assessment Programme, Oslo.

Triulzi, C., Nonnis Marzano, F., Casoli, A., Mori, A., and Vaghi, M. (2001). Radioactive and stable isotopes in abiotic and biotic components of Antarctic ecosystems surrounding the Italian base. In Cescon, P. (ed.), *Environmental Chemistry in Antarctica*, pp. 55–60. Gordon and Breach Science Publishers, Amsterdam.

UNEP (2001). *Final Act of the Plenipotentiaries on the Stockholm Convention on Persistent Organic Pollutants*. Chemicals 445. United Nations Environment Program, Geneva.

United States Environmental Protection Agency (2002). *National Priorities List Sites in Alaska*. www.epa.gov/superfund/sites/npl/ak.htm

Vincent, W.F. and Vincent, C.L. (1982). Nutritional state of the plankton in Antarctic coastal lakes and the inshore Ross Sea. *Polar Biology* **1**, 159–165.

Vodopivez, C., Smichowski, P., and Marcovecchio, J. (2001). Trace metals monitoring as a tool for characterisation of Antarctic ecosystems and environmental management. The Argentine programme at Jubany Station. In Caroli, S., Cescon, P., and Walton, D.W.H. (eds), *Environmental Contamination in Antarctica, A Challenge to Analytical Chemistry*, pp. 155–180. Elsevier Science, Oxford.

Wania, F. and Mackay, D. (1993). Global fractionation and cold condensation of low volatility organochlorine compounds in polar regions. *Ambio* **22**, 10–18.

Webster, J., Webster, K., Nelson, P., and Waterhouse, E. (2003). The behaviour of residual contaminants at a former station site, Antarctica. *Environmental Pollution* **123**, 163–179.

Woo, M.-K., Lewkowicz, A.G., and Rouse, W.R. (1992). Response of the Canadian permafrost environment to climatic change. *Physical Geography* **13**, 287–317.

Wrona, F.J. *et al.* (2006). Effects of ultraviolet radiation and contaminant-related stressors on arctic freshwater ecosystems. *Ambio* **35**, 388–401.

Yunker, M.B., Macdonald, R.W., and Whitehouse, B.G. (1994). Phase associations and lipid distribution in the seasonally ice-covered Arctic estuary of the Mackenzie Shelf. *Organic Geochemistry* **22**, 651–669.

Yunker, M.B. and Macdonald, R.W. (1995). Composition and origins of polycyclic aromatic hydrocarbons in the Mackenzie River and on the Beaufort Sea Shelf. *Arctic* **48**, 118–129.

Zhao, Y., Li, T.J., and Zhao, J.L. (1998). Human impacts on the environment of Fildes Peninsula of King George Island, Antarctica. *Jidi Yanjiu* **10**, 262–271.

CHAPTER 17

Future directions in polar limnology

Johanna Laybourn-Parry and Warwick F. Vincent

Outline

Polar limnology is in a phase of rapid knowledge acquisition, and in this concluding chapter we identify some of the emerging concepts and technologies that will drive future advances. These include the increased awareness of climate and related impacts in the polar regions, and the importance of high-latitude lakes and rivers as sentinels of global change; the increased availability of wireless network technology to obtain data-sets of high temporal resolution from these remote sites; the emergence of new sensor technologies and underwater platforms, including robotic systems that may be used in the future to explore subglacial lakes and other polar waters; and the development of surface imagery approaches ranging from local-scale observations by unmanned aerial vehicles to circumpolar-scale measurements by satellite, for example by synthetic aperture radar. We briefly summarize some of the new opportunities provided by environmental genomics, including application towards bioprospecting for novel extremophiles and biomolecules of pharmaceutical and industrial interest. Finally, we address the value of Arctic and Antarctic aquatic ecosystems as sites for developing models of broad application.

17.1 Introduction

The knowledge base for polar lakes and rivers has grown exponentially over the last two decades, and it continues to expand rapidly. In part this is because of the inherent attractiveness of high-latitude aquatic environments and their biota for scientific inquiry, and the value of these extreme ecosystems as ecological models. It is also because of the increasing interest in the polar regions by governments, international bodies, and the public at large. After a long period of debate, there is now a broad consensus that climate change is occurring and that it has begun to have severe impacts on the cryosphere, with likely feedback effects on the global environment (Meehl et al. 2007). Climate has changed substantially in the geological past, but human activities are now exacerbating a natural process. The consequences for freshwater supply, food production, disease propagation, and

sea-level rise, which will result in hazardous events such as storm surges and flooding, are challenging the way that the ordinary global citizen lives his or her life. Large areas in the polar regions are experiencing more rapid warming than elsewhere (see Chapter 1 and Plate 16) and given that vast quantities of water are locked up in the polar ice sheets, the melt of this ice into the oceans will have major consequences on the ocean current systems and climate. Arctic and Antarctic lakes are proving to be sensitive barometers of climate change (e.g. Quayle et al. 2002; Antoniades et al. 2007), and Arctic rivers are also showing the onset of climate impacts (e.g. Peterson et al. 2006; Lesack and Marsh 2007). Some of these changes may have strong, positive feedback effects on the global climate, for example the accelerated release of methane from arctic thermokarst lakes (Walter et al. 2006). Long-term databases now exist for several ecological research sites, providing the opportunity to undertake detailed analysis, modeling, and pre-diction. This represents an example of the value of long-term monitoring programmes that at their inception were not intended to address the issue of climate change, but rather to understand physical processes, biogeochemistry, biodiversity, and eco-system structure and function. Future research in the polar regions will have a strong emphasis on the impacts of climate warming, with applications to all latitudes.

Much of the polar latitudes are inaccessible and subject to weather conditions that are chal-lenging to the acquisition of detailed time-series data. The problem of infrequent sampling renders interpretation difficult, analogous to trying to decipher a complex plot from reading a book in which only one page in every 10 or 20 is avail-able. Inevitably there is a great deal of educated surmise. Remote sensing offers a solution to some of these problems. It can range from remote sam-pling devices that take and preserve small sam-ples over a winter, as was done in Lake Fryxell (McMurdo Dry Valleys, Antarctica) by McKnight *et al.* (2000), enabling a picture of how patterns of phytoplankton abundance changed seasonally, to complex satellite imagery that provides large-scale information on phenomena such as the extent of lake ice cover across Arctic regions (e.g. Latifovic and Pouliot 2007). Remote sensing is being applied much more widely in polar science as the demand grows for more detailed data-sets that allow large-scale, accurate modeling. A variety of other emerging technologies such as robotic systems and genomic tools also offer enormous potential for rapid advances in polar **limnology**.

17.2 Wireless networks

Wireless networks is the term applied to systems that collect and transmit data from remote envir-onments back to the laboratory (Porter *et al.* 2005). One example of this from Antarctica is a platform of instruments deployed on the frozen surface of Crooked Lake in the Vestfold Hills (Palethorpe *et al.* 2004). Crooked Lake is a large freshwater body of water (9 km², 160 m deep) in the Vestfold Hills that loses most or all of its ice cover for a few weeks each year. The platform (Figure 17.1) contained

Figure 17.1 Wireless remote-sensing platform on Crooked Lake, Antarctica. Note the cables running out from the platform that support light and temperature sensors suspended in the water column (Palethorpe *et al.* 2004).

a small meteorological station measuring wind speed and air temperature, photosynthetically active radiation (PAR) sensors measuring surface PAR and levels in the water column from sensors suspended at 3, 5, 15, and 20 m, UV-B radiation sensors at the surface and at 3 m in the water col-umn, water-column temperature at 3 and 5 m and ice thickness by means of an upward-directed SONAR. A small video camera allowed the nature of the under-ice surface to be assessed. The system was powered by a battery charged by solar panels and a wind generator. Data were logged every 5 min and periodically uploaded via Iridium satel-lite phone to Davis Station. Figure 17.2 shows some typical daily data sets for PAR and Figure 17.3 ice thickness during the period up to ice breakout in late summer.

Not only does this type of remote sensing pro-vide extensive data-sets that fill an otherwise temporal void, it also allows much more meaning-ful data acquisition from *in situ* experiments. For example, when measuring photosynthesis at vari-ous depths in the water column one can acquire real-time PAR data every minute for the course of the incubations, thereby facilitating much more accurate photosynthetic efficiency calculations than is usually the case based on a few readings of ambient water-column PAR.

(a)

PAR in Crooked Lake, 19/01/03

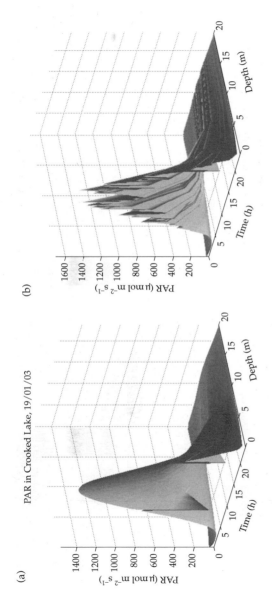

(b)

Figure 17.2 (a) Total photosynthetically active radiation (PAR) over a 24-h period on a sunny day. Note the smoothness of the curve. (b) Total PAR over a 24-h period on a cloudy day. Note the irregularity of the curve resulting from clouds obscuring sunlight. From Palethorpe *et al.* (2004), by permission of the American Society of Limnology and Oceanography.

Ice thickness from sonar measurement

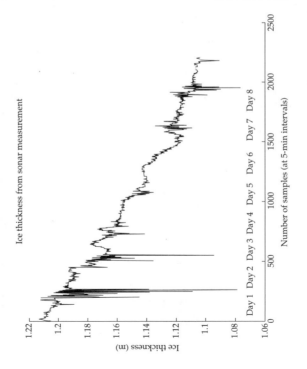

Figure 17.3 Decrease in ice thickness derived from SONAR measurements in Crooked Lake over an 8-day period in late summer as the ice was breaking out. From Palethorpe *et al.* (2004), by permission of the American Society of Limnology and Oceanography.

measurements such as temperature and conductivity, which provide important insights into lake hydrodynamics and the physicochemical constraints on biota in high-latitude waters (see Chapter 4). Temperature microstructure (at millimeter intervals) profiling systems are increasingly used in lakes and oceans to capture short-term mixing

17.3 Underwater sensors and imaging systems

Wireless networks, as well as other underwater deployments, are only as good as the sensors and cameras making the *in situ* observations. This technology began with relatively straightforward dynamics, and spatially detailed fine-structure profiling (at centimeter intervals) will continue to yield new information about the physical structure and dynamics of high-latitude aquatic ecosystems. New optical sensors are allowing high-frequency measurements of nitrate, colored dissolved organic matter (CDOM), and other chemical constituents, as well as rapid synoptic sampling for these variables at many sites within a lake or river. Microelectrode techniques are providing ways to measure biological processes *in situ*; for example, measurement of microbial mat photosynthesis in an ice-covered lake by oxygen microelectrodes deployed by scientific divers (Vopel and Hawes 2006). A variety fluorescence probes are becoming available to detect not only microbial biomass, but also composition and activity (e.g. Bay *et al.* 2005). Underwater digital imaging systems are also developing rapidly, for example for deployment on remotely operated vehicles (ROVs).

A driving force for advances in underwater sensor and imaging technologies is the rapid development and application of robotic systems in oceanography, especially autonomous underwater vehicles (AUVs). These include propeller-driven systems, as well as gliders that move via lifting surfaces (wings) and changes in buoyancy, with artificial intelligence control systems. AUVs are now commercially available from an increasing number of companies, and are proving invaluable for measurements of underwater weather and physical-biotic interactions, for high-intensity measurements required for model applications, and for the exploration of otherwise inaccessible sites (Bellingham and Rajan 2007). These robotic systems have been applied to studies of large (Kumagai *et al.* 2002) and small (Laval *et al.* 2000) lakes and are of interest for the exploration of Antarctic subglacial aquatic environments (National Research Council 2007).

17.4 Surface imagery

Unmanned aerial vehicles (UAVs) were developed during World War II for spying purposes but have subsequently been used in a range of civil applications. Low-flying UAVs have an application in surveying agricultural facilities such as vineyards and have recently been used in surveying cryoconites

on glaciers (Hodson *et al.* 2007). In this investigation, aimed at quantifying respiration carbon flux of cryoconites across the Midre Lovenbreen glacier in Svalbard, a HighSpy helicopter UAV with a gyroscopically mounted camera was used to photograph the glacier surface. Repeated transects across the glacier between 150 and 375 m altitude allowed a detailed analysis of the number and aerial coverage of cryoconites. In conjunction with ground truthing, the contribution of cryoconite communities to carbon flux across the glacier could be accurately computed. UAVs offer a valuable opportunity to developing a much more detailed approach to quantifying the role of glaciers in biogeochemical cycling. The technology exists, using artificial intelligence, to enable a UAV to 'talk' to a remotely operated all-terrain vehicle that could be equipped with sampling devices. This would enable a greater degree of precision in sampling than could be achieved by a field scientist. It is now possible for these devices to prove invaluable in unravelling the contributions of these summer functioning ecosystems to fluxes of carbon and nitrogen not only in the polar regions, but also at high altitudes at lower latitudes where glacial retreat is occurring in response to climate change.

On a wider scale satellite imagery is of great value as high-latitude lakes and rivers are distributed across vast remote regions. While ground truthing is still essential, remotely sensed imagery can help select the appropriate sites for field measurements, as well as allow such measurements to be scaled up to regional or even circumpolar dimensions. There is an increasing realization in limnology that small-sized lakes and ponds may play a significant role in global biogeochemistry because of their intense microbial activities as well as numerical abundance (Downing *et al.* 2006). Landsat imagery applied to the Arctic tundra has begun to provide inventories for these abundant, active ecosystems, for example tundra thaw lakes (Frohn *et al.* 2005) and to check the stability of flood-plain lakes (Emmerton *et al.* 2007). Similarly in Antarctica satellite imagery has allowed the detection and inventory of subglacial

widely accepted that glaciers represent another type of aquatic ecosystem in the cryosphere during summer melt phases (Hodson *et al.* 2008). The application of remote sensing technology will